## Moments of Inertia of Common Geometric Shapes

### Rectangle

$$\bar{I}_{x'} = \frac{1}{12}bh^3$$

$$\bar{I}_{y'} = \frac{1}{12}b^3h$$

$$I_x = \frac{1}{3}bh^3$$

$$I_y = \frac{1}{3}b^3h$$

$$J_C = \frac{1}{12}bh(b^2 + h^2)$$

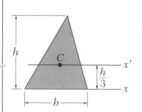

### Triangle

$$\bar{I}_{x'} = \frac{1}{36}bh^3$$

$$I_x = \frac{1}{12}bh^3$$

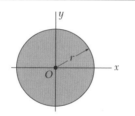

### Circle

$$\bar{I}_x = \bar{I}_y = \frac{1}{4}\pi r^4$$

$$J_O = \frac{1}{2}\pi r^4$$

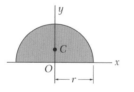

### Semicircle

$$I_x = I_y = \frac{1}{8}\pi r^4$$

$$J_O = \frac{1}{4}\pi r^4$$

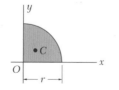

### Quarter circle

$$I_x = I_y = \frac{1}{16}\pi r^4$$

$$J_O = \frac{1}{8}\pi r^4$$

### Ellipse

$$\bar{I}_x = \frac{1}{4}\pi ab^3$$

$$\bar{I}_y = \frac{1}{4}\pi a^3b$$

$$J_O = \frac{1}{4}\pi ab(a^2 + b^2)$$

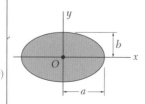

## Mass Moments of Inertia of Common Geometric Shapes

### Slender rod

$$I_y = I_z = \frac{1}{12}mL^2$$

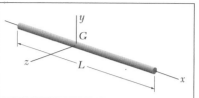

### Thin rectangular plate

$$I_x = \frac{1}{12}m(b^2 + c^2)$$

$$I_y = \frac{1}{12}mc^2$$

$$I_z = \frac{1}{12}mb^2$$

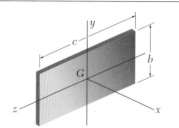

### Rectangular prism

$$I_x = \frac{1}{12}m(b^2 + c^2)$$

$$I_y = \frac{1}{12}m(c^2 + a^2)$$

$$I_z = \frac{1}{12}m(a^2 + b^2)$$

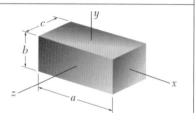

### Thin disk

$$I_x = \frac{1}{2}mr^2$$

$$I_y = I_z = \frac{1}{4}mr^2$$

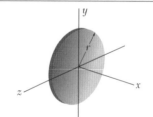

### Circular cylinder

$$I_x = \frac{1}{2}ma^2$$

$$I_y = I_z = \frac{1}{12}m(3a^2 + L^2)$$

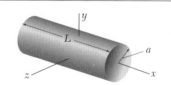

### Circular cone

$$I_x = \frac{3}{10}ma^2$$

$$I_y = I_z = \frac{3}{5}m\left(\frac{1}{4}a^2 + h^2\right)$$

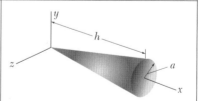

### Sphere

$$I_x = I_y = I_z = \frac{2}{5}ma^2$$

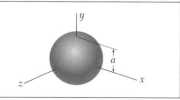

## McGraw-Hill
## Higher Education

MECHANICS FOR ENGINEERS, STATICS, FIFTH EDITION

Published by McGraw-Hill, a business unit of The McGraw-Hill Companies, Inc., 1221 Avenue of the
Americas, New York, NY 10020. Copyright © 2008 by The McGraw-Hill Companies, Inc. Previous editions ©
1976, 1987. All rights reserved. No part of this publication may be reproduced or distributed in any form or by
any means, or stored in a database or retrieval system, without the prior written consent of The McGraw-Hill
Companies, Inc., including, but not limited to, in any network or other electronic storage or transmission, or
broadcast for distance learning.

Some ancillaries, including electronic and print components, may not be available to customers outside the
United States.

This book is printed on acid-free paper.

2 3 4 5 6 7 8 9 0 CCW/CCW 0 9 8

ISBN 978–0–07–246478–8
MHID 0–07–246478–X

Global Publisher: *Raghothaman Srinivasan*
Executive Editor: *Michael Hackett*
Director of Development: *Kristine Tibbetts*
Developmental Editor: *Lora Kalb*
Executive Marketing Manager: *Michael Weitz*
Senior Project Manager: *Kay J. Brimeyer*
Senior Production Supervisor: *Sherry L. Kane*
Associate Media Producer: *Christina Nelson*
Senior Designer: *David W. Hash*
Cover Designer: *John Joran*
Compositor: *Aptara*
Typeface: 10.5/12 *New Caledonia*
Printer: *Courier Westford*

**Library of Congress Cataloging-in-Publication Data**

Beer, Ferdinand Pierre, 1915-
  Mechanics for engineers : statics / Ferdinand P. Beer, E. Russell Johnston, Ralph E. Flori. -- 5th ed.
     p. cm.
  Includes index.
  ISBN  978–0–07–246478–8 --- ISBN  0–07–246478–X (hard copy : alk. paper)
     1. Mechanics, Applied.  I. Johnston, E. Russell (Elwood Russell), 1925-  II. Flori, Ralph E.  III. Title.
TA350.B355 2008
620.1'03--dc22
                                    2007023355

www.mhhe.com

**Fifth Edition**

# MECHANICS FOR ENGINEERS

# Statics

**FERDINAND P. BEER**

Late of Lehigh University

**E. RUSSELL JOHNSTON, JR.**

University of Connecticut

**RALPH E. FLORI, JR.**

University of Missouri—Rolla

 **McGraw-Hill Higher Education**

Boston    Burr Ridge, IL    Dubuque, IA    New York    San Francisco    St. Louis
Bangkok    Bogotá    Caracas    Kuala Lumpur    Lisbon    London    Madrid    Mexico City
Milan    Montreal    New Delhi    Santiago    Seoul    Singapore    Sydney    Taipei    Toronto

# CONTENTS

# CHAPTER 5 Distributed Forces: Centroids and Centers of Gravity 158

# CHAPTER 6 Analysis of Structures 210

# CHAPTER 7 Forces in Beams and Cables 264

CHAPTER 8 Friction 305

CHAPTER 9 Distributed Forces: Moments of Inertia 351

## CHAPTER 10  Method of Virtual Work  406

The main objective of a first course in mechanics should be to develop in the engineering student the ability to analyze any problem in a simple and logical manner and to apply to its solution a few, well-understood, basic principles. It is hoped that this text, designed for the first course in statics offered in the sophomore year, and the volume that follows, *Mechanics for Engineers: Dynamics*, will help the instructor achieve this goal.[†]

In this fifth edition, the vectorial character of mechanics has again been emphasized. The concept of vectors and the laws governing the addition and the resolution of vectors into components are discussed with care at the beginning of Chap. 2, and, throughout the text, forces and other vector quantities have been clearly distinguished from scalar quantities through the use of boldface type. Products of vectors, however, are not used in this text.[‡]

One of the characteristics of the approach used in these volumes is that the mechanics of *particles* has been clearly separated from the mechanics of *rigid bodies*. This approach makes it possible to consider simple practical applications at an early stage and to postpone the introduction of more difficult concepts. In this volume, for example, the statics of particles is treated first (Chap. 2), and the principle of equilibrium is immediately applied to practical situations involving only concurrent forces. The statics of rigid bodies is considered in Chaps. 3 and 4, where the principle of transmissibility and the associated concept of moment of a force are introduced. In the volume on dynamics, the same division is observed. The basic concepts of force, mass, and acceleration, of work and energy, and of impulse and momentum are introduced and first applied to problems involving only particles. Thus the students may familiarize themselves with the three basic methods used in dynamics and learn their respective advantages before facing the difficulties associated with the motion of rigid bodies.

Since this text is designed for a first course in statics, new concepts have been presented in simple terms and every step explained in detail. On the other hand, by discussing the broader aspects of the problems considered, a definite maturity of approach has been achieved. For example, the concepts of partial constraints and of static indeterminacy are introduced early in the text and used throughout.

[†]Both texts are also available in a single volume, *Mechanics for Engineers: Statics and Dynamics*.

[‡]In a parallel text, *Vector Mechanics for Engineers: Statics*, vector and scalar products are introduced at an early stage and used to solve three-dimensional problems.

The fact that mechanics is essentially a *deductive* science based on a few fundamental principles has been stressed. Derivations have been presented in their logical sequence and with all the rigor warranted at this level. However, the learning process being largely *inductive,* simple applications have been considered first. Thus the statics of particles precedes the statics of rigid bodies; coplanar forces are introduced before forces in space; and problems involving internal forces are postponed until Chap. 6.

Free-body diagrams are introduced early, and their importance is emphasized throughout the text. Force is distinguished from other elements of the free-body diagrams by using a lighter shade of blue. This makes it easier for the students to identify the forces acting on a given particle or rigid body and to follow the discussion of sample problems and other examples given in the text. Free-body diagrams are used not only to solve equilibrium problems but also to express the equivalence of two systems of forces or, more generally, of two systems of vectors. This approach is particularly useful as a preparation for the study of the dynamics of rigid bodies. As will be shown in the volume on dynamics, by placing the emphasis on "free-body-diagram equations" rather than on the standard algebraic equations of motion, a more intuitive and more complete understanding of the fundamental principles of dynamics may be achieved.

Because engineers use both the international system of units (SI metric units) and U.S. customary units, both are used in this text. Approximately half the sample problems and 60 percent of the problems to be assigned have been stated in SI units, while the remainder retain U.S. customary units. Since the SI system of units is an absolute system based on the units of time, length, and mass, whereas the U.S. customary system is a gravitational system based on the units of time, length, and force, different approaches are required for the solution of many problems. For example, when SI units are used, a body is generally specified by its mass expressed in kilograms; in most problems of statics it will be necessary to determine the weight of the body in newtons, and an additional calculation will be required for this purpose. On the other hand, when U.S. customary units are used, a body is specified by its weight in pounds and, in dynamics problems, an additional calculation will be required to determine its mass in slugs (or $lb \cdot Sec^2/ft$). The authors, therefore, believe that problem assignments should include both systems of units. The actual distribution of assigned problems between the two systems of units, however, has been left to the instructor, and a sufficient number of problems of each type have been provided so that four complete lists of assignments may be selected with the proportion of problems stated in SI units set anywhere between 50 and 75 percent. If so desired, two complete lists of assignments may also be selected from problems stated in SI units only and two others from problems stated in U.S. customary units.

A large number of optional sections have been included. These sections are indicated by asterisks and may thus easily be distinguished from those which form the core of the basic statics course. They may be omitted without impairing understanding of the rest of the text. Topics covered in these additional sections are fluid statics, shear and bending-moment diagrams for beams, equilibrium of cables, products of inertia and Mohr's circle, and the method of virtual work. The sections on beams are especially useful when the course in statics is immediately followed by a course in mechanics of materials.

The material presented in the text and most of the problems require no previous mathematical knowledge beyond algebra, trigonometry, and elementary calculus. A greater emphasis has been placed on the correct understanding of the basic mathematical concepts involved than on the nimble manipulation of mathematical formulas. For example, determining centroids of composite areas precedes the calculation of centroids by integration, thus making it possible to establish the concept of moment of area firmly before introducing the use of integration.

Each chapter begins with an introductory section setting the purpose and goals of the chapter and describing in simple terms the material to be covered and its application to the solution of engineering problems. The body of the text has been divided into units, each consisting of one or several theory sections, one or several sample problems, and a large number of problems to be assigned. Each unit corresponds to a well-defined topic and generally may be covered in one lesson. In a number of cases, however, the instructor will find it desirable to devote more than one lesson to a given topic. Each chapter ends with a review section summarizing the material covered in that chapter. Marginal notes have been added to help students organize their review work, and cross-references have been included to help them find the portions of material requiring their special attention.

The sample problems have been set up in much the same form that students will use in solving the assigned problems. They thus serve the double purpose of amplifying the text and demonstrating the type of neat and orderly work that students should cultivate in their own solutions. Most of the problems to be assigned are of a practical nature and should appeal to engineering students. They are primarily designed, however, to illustrate the material presented in the text and to help students understand the basic principles of mechanics. The problems have been grouped according to the portions of material they illustrate and have been arranged in order of increasing difficulty. Problems requiring special attention have been indicated by asterisks. Answers to all even-numbered problems are given at the end of the book.

In this new edition of *Mechanics for Engineers,* a group of four problems designed to be solved with a computer has been added to the review problems at the end of each chapter. In *Statics,* these problems may involve the analysis of a structure for various configurations or loadings of the structure, or the determination of the equilibrium positions of a given mechanism. In *Dynamics,* they may involve the determination of the motion of a particle under various initial conditions, the kinematic or kinetic analysis of mechanisms in successive positions, or the numerical integration of various equations of motion. Developing the algorithm required to solve a given mechanics problem will help students gain a better understanding of the mechanics principles involved and will provide them with an opportunity to apply their computer skills to the solution of a meaningful engineering problem.

The authors wish to acknowledge gratefully the many helpful comments and suggestions offered by the users of the previous editions of *Mechanics for Engineers* and of *Vector Mechanics for Engineers*.

**Ferdinand P. Beer**
**E. Russell Johnston, Jr.**
**Ralph E. Flori, Jr.**

# LIST OF SYMBOLS

| | | | | |
|---|---|---|---|---|
| $a$ | Constant; radius; distance | | Q | Force; vector |
| **A, B, C, . . .** | Reactions at supports and connections | | r | Position vector |
| $A, B, C, \ldots$ | Points | | $r$ | Radius; distance; polar coordinate |
| $A$ | Area | | R | Resultant force; resultant vector; reaction |
| $b$ | Width; distance | | R | Radius of earth |
| $c$ | Constant | | s | Position vector |
| $C$ | Centroid | | $s$ | Length of arc; length of cable |
| $d$ | Distance | | S | Force; Vector |
| $e$ | Base of natural logarithms | | $t$ | Thickness |
| F | Force; friction force | | T | Force |
| $g$ | Acceleration of gravity | | $T$ | Tension |
| $G$ | Center of gravity; constant of gravitation | | $U$ | Work |
| $h$ | Height; sag of cable | | V | Shearing force |
| $I, I_x, \ldots$ | Moment of inertia | | $V$ | Volume; potential energy; shear |
| $\bar{I}$ | Centroidal moment of inertia | | $w$ | Load per unit length |
| $I_{xy}, \ldots$ | Product of inertia | | W, $W$ | Weight; load |
| $J$ | Polar moment of inertia | | $x, y, z$ | Rectangular coordinates; distances |
| $k$ | Spring constant | | $\bar{x}, \bar{y}, \bar{z}$ | Rectangular coordinates of centroid or |
| $k_x, k_y, k_o$ | Radius of gyration | | | center of gravity |
| $\bar{k}$ | Centroidal radius of gyration | | $\alpha, \beta, \gamma$ | Angles |
| $l$ | Length | | $\gamma$ | Specific weight |
| $L$ | Length; span | | $\delta$ | Elongation |
| $m$ | Mass | | $\delta s$ | Virtual Displacement |
| M | Couple | | $\delta U$ | Virtual work |
| $M$ | Moment; mass of earth | | $\eta$ | Efficiency |
| $M_o$ | Moment about point $O$ | | $\theta$ | Angular coordinate; angle; polar coordinate |
| N | Normal Component of reaction | | $\mu$ | Coefficient of friction |
| O | Origin of coordinates | | $\rho$ | Density |
| $p$ | Pressure | | $\phi$ | Angle of friction; angle |
| P | Force; vector | | | |

# CHAPTER 1

# Introduction

## 1.1. WHAT IS MECHANICS?

Mechanics may be defined as that science which describes and predicts the conditions of rest or motion of bodies under the action of forces. It is divided into three parts: mechanics of *rigid bodies*, mechanics of *deformable bodies*, and mechanics of *fluids*.

The mechanics of rigid bodies is subdivided into *statics* and *dynamics*, the former dealing with bodies at rest, the latter with bodies in motion. In this part of the study of mechanics, bodies are assumed to be perfectly rigid. Actual structures and machines, however, are never absolutely rigid and deform under the loads to which they are subjected. But these deformations are usually small and do not appreciably affect the conditions of equilibrium or motion of the structure under consideration. They are important, though, as far as the resistance of the structure to failure is concerned and are studied in mechanics of materials, which is a part of the mechanics of deformable bodies. The third division of mechanics, the mechanics of fluids, is subdivided into the study of *incompressible fluids* and of *compressible fluids*. An important subdivision of the study of incompressible fluids is *hydraulics*, which deals with problems involving liquids.

Mechanics is a physical science, since it deals with the study of physical phenomena. However, some associate mechanics with mathematics, while many consider it as an engineering subject. Both these views are justified in part. Mechanics is the foundation of most engineering sciences and is an indispensable prerequisite to their study. However, it does not have the *empiricism* found in some engineering sciences, i.e., it does not rely on experience or observation alone; by its rigor and the emphasis it places on deductive reasoning it resembles mathematics. But, again, it is not an *abstract* or even a *pure* science; mechanics is an *applied* science. The purpose of mechanics is to explain and predict physical phenomena and thus to lay the foundations for engineering applications.

1

## 1.2. FUNDAMENTAL CONCEPTS AND PRINCIPLES

Although the study of mechanics goes back to the time of Aristotle (384–322 B.C.) and Archimedes (287–212 B.C.), one has to wait until Newton (1642–1727) to find a satisfactory formulation of its fundamental principles. These principles were later expressed in a modified form by d'Alembert, Lagrange, and Hamilton. Their validity remained unchallenged, however, until Einstein formulated his *theory of relativity* (1905). While its limitations have now been recognized, *newtonian mechanics* still remains the basis of today's engineering sciences.

The basic concepts used in mechanics are *space, time, mass,* and *force.* These concepts cannot be truly defined; they should be accepted on the basis of our intuition and experience and used as a mental frame of reference for our study of mechanics.

The concept of *space* is associated with the notion of the position of a point *P.* The position of *P* may be defined by three lengths measured from a certain reference point, or *origin,* in three given directions. These lengths are known as the *coordinates* of *P.*

To define an event, it is not sufficient to indicate its position in space. The *time* of the event should also be given.

The concept of *mass* is used to characterize and compare bodies on the basis of certain fundamental mechanical experiments. Two bodies of the same mass, for example, will be attracted by the earth in the same manner; they will also offer the same resistance to a change in translational motion.

A *force* represents the action of one body on another. It may be exerted by actual contact or at a distance, as in the case of gravitational forces and magnetic forces. A force is characterized by its *point of application,* its *magnitude,* and its *direction;* a force is represented by a *vector* (Sec. 2.3).

In newtonian mechanics, space, time, and mass are absolute concepts, independent of each other. (This is not true in *relativistic mechanics,* where the time of an event depends upon its position, and where the mass of a body varies with its velocity.) On the other hand, the concept of force is not independent of the other three. Indeed, one of the fundamental principles of newtonian mechanics listed below indicates that the resultant force acting on a body is related to the mass of the body and to the manner in which its velocity varies with time.

We shall study the conditions of rest or motion of particles and rigid bodies in terms of the four basic concepts we have introduced. By *particle* we mean a very small amount of matter which may be assumed to occupy a single point in space. A *rigid body* is a combination of a large number of particles occupying fixed positions with respect to each other. The study of the mechanics of particles is obviously a prerequisite to that of rigid bodies. Besides, the results obtained for a particle may be used directly in a large number of problems dealing with the conditions of rest or motion of actual bodies.

The study of elementary mechanics rests on six fundamental principles based on experimental evidence.

**The Parallelogram Law for the Addition of Forces.**    This states that two forces acting on a particle may be replaced by a single force, called their *resultant*, obtained by drawing the diagonal of the parallelogram which has sides equal to the given forces (Sec. 2.2).

**The Principle of Transmissibility.**    This states that the conditions of equilibrium or of motion of a rigid body will remain unchanged if a force acting at a given point of the rigid body is replaced by a force of the same magnitude and same direction, but acting at a different point, provided that the two forces have the same line of action (Sec. 3.3). In other words, sliding a force along its line of action to a new point on a rigid body does not change the body's motion or equilibrium state.

**Newton's Three Fundamental Laws.**    Formulated by Sir Isaac Newton in the latter part of the seventeenth century, these laws may be stated as follows:

FIRST LAW.    If the resultant force acting on a particle is zero, the particle will remain at rest (if originally at rest) or will move with constant speed in a straight line (if originally in motion) (Sec. 2.10).

SECOND LAW.    If the resultant force acting on a particle is not zero, the particle will have an acceleration proportional to the magnitude of the resultant and in the direction of this resultant force.

As we shall see in Sec. 12.2, this law may be stated as

$$\mathbf{F} = m\mathbf{a} \tag{1.1}$$

where $\mathbf{F}$, $m$, and $\mathbf{a}$ represent, respectively, the resultant force acting on the particle, the mass of the particle, and the acceleration of the particle, expressed in a consistent system of units.

THIRD LAW.    The forces of action and reaction between bodies in contact have the same magnitude, same line of action, and opposite sense (Sec. 6.1).

**Newton's Law of Gravitation.**    This states that two particles of mass $M$ and $m$ are mutually attracted with equal and opposite forces $\mathbf{F}$ and $\mathbf{F'}$ (Fig. 1.1) of magnitude $F$ given by the formula

$$F = G\frac{Mm}{r^2} \tag{1.2}$$

where    $r$ = distance between the two particles
        $G$ = universal constant called the *constant of gravitation*

Newton's law of gravitation introduces the idea of an action exerted at a distance and extends the range of application of Newton's third law: the action $\mathbf{F}$ and the reaction $\mathbf{F'}$ in Fig. 1.1 are equal and opposite, and they have the same line of action.

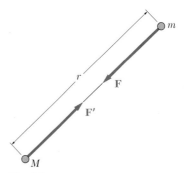

**Fig. 1.1**

A particular case of great importance is that of the attraction of the earth on a particle located on its surface. The force **F** exerted by the earth on the particle is then defined as the *weight* **W** of the particle. Taking $M$ equal to the mass of the earth, $m$ equal to the mass of the particle, and $r$ equal to the radius $R$ of the earth, and introducing the constant

$$g = \frac{GM}{R^2} \tag{1.3}$$

the magnitude $W$ of the weight of a particle of mass $m$ may be expressed as[†]

$$W = mg \tag{1.4}$$

The value of $R$ in formula (1.3) depends upon the elevation of the point considered; it also depends upon its latitude, since the earth is not truly spherical. The value of $g$ therefore varies with the position of the point considered. As long as the point actually remains on the surface of the earth, it is sufficiently accurate in most engineering computations to assume that $g$ equals $9.81 \text{ m/s}^2$ or $32.2 \text{ ft/s}^2$.

The principles we have just listed will be introduced in the course of our study of mechanics as they are needed. The study of the statics of particles, carried out in Chap. 2, will be based on the parallelogram law of addition and on Newton's first law alone. The principle of transmissibility will be introduced in Chap. 3 as we begin the study of the statics of rigid bodies, and Newton's third law in Chap. 6 as we analyze the forces exerted on each other by the various members forming a structure. It should be noted that the four aforementioned principles underlie the entire study of the statics of particles, rigid bodies, and systems of rigid bodies. In the study of dynamics, Newton's second law and Newton's law of gravitation will be introduced. It will then be shown that Newton's first law is a particular case of Newton's second law (Sec. 12.2). It can also be shown that the principle of transmissibility could be derived from the other principles and thus eliminated.[‡] However, Newton's first and third laws, the parallelogram law of addition, and the principle of transmissibility will provide us with the necessary and sufficient foundation for the study of the entire field of statics.

As noted earlier, the six fundamental principles listed above are based on experimental evidence. Except for Newton's first law and the principle of transmissibility, they are independent principles which cannot be derived mathematically from each other or from any other elementary physical principle. On these principles rests most of the intricate structure of newtonian mechanics. For more than two centuries a tremendous number of problems dealing with the conditions of rest and motion of rigid bodies, deformable bodies, and fluids have been solved by applying these

[†] A more accurate definition of the weight **W** should take into account the rotation of the earth.

[‡] Cf. F. P. Beer and E. R. Johnston, "Vector Mechanics for Engineers," 4th ed., sec. 16.5, McGraw-Hill Book Company, 1984.

fundamental principles. Many of the solutions obtained could be checked experimentally, thus providing a further verification of the principles from which they were derived. It is only recently that Newton's mechanics was found at fault, in the study of the motion of atoms and in the study of the motion of certain planets, where it must be supplemented by the theory of relativity. But on the human or engineering scale, where velocities are small compared with the velocity of light, Newton's mechanics has yet to be disproved.

## 1.3. SYSTEMS OF UNITS

With the four fundamental concepts introduced in the preceding section are associated the so-called *kinetic units*, i.e., the units of *length, time, mass,* and *force*. These units cannot be chosen independently if Eq. (1.1) is to be satisfied. Three of the units may be defined arbitrarily; they are then referred to as *base units*. The fourth unit, however, must be chosen in accordance with Eq. (1.1) and is referred to as a *derived unit*. Kinetic units selected in that way are said to form a *consistent system of units*.

**International System of Units (SI Units[†]).**  In this system, the base units are the units of length, mass, and time, and are called, respectively, the *meter* (m), the *kilogram* (kg), and the *second* (s). All three are arbitrarily defined. The second, which is supposed to represent the 1/86 400 part of the mean solar day, is actually defined as the duration of 9 192 631 770 cycles of the radiation associated with a specified transition of the cesium atom. The meter, originally intended to represent one-ten-millionth of the distance from the equator to the pole, is now defined as 1 650 763.73 wavelengths of the orange-red line of krypton 86. The kilogram, which is approximately equal to the mass of 0.001 $m^3$ of water, is actually defined as the mass of a platinum standard kept at the International Bureau of Weights and Measures at Sèvres, near Paris, France. The unit of force is a derived unit. It is called the *newton* (N) and is defined as the force which gives an acceleration of 1 $m/s^2$ to a mass of 1 kg (Fig. 1.2). From Eq. (1.1) we write

$$1 \text{ N} = (1 \text{ kg})(1 \text{ m/s}^2) = 1 \text{ kg} \cdot \text{m/s}^2 \tag{1.5}$$

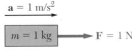

**Fig. 1.2**

The SI units are said to form an *absolute* system of units. This means that the three base units chosen are independent of the location where measurements are made. The meter, the kilogram, and the second may be used anywhere on the earth; they may even be used on another planet. They will always have the same significance.

The *weight* of a body, or *force of gravity* exerted on that body, should, like any other force, be expressed in newtons. From Eq. (1.4) it follows that the weight of a body of mass 1 kg (Fig. 1.3) is

$$W = mg$$
$$= (1 \text{ kg})(9.81 \text{ m/s}^2)$$
$$= 9.81 \text{ N}$$

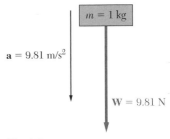

**Fig. 1.3**

[†] SI stands for *Système International d'Unités* (French).

**TABLE 1.1  SI Prefixes**

| Multiplication Factor | Prefix† | Symbol |
|---|:---:|:---:|
| $1\ 000\ 000\ 000\ 000 = 10^{12}$ | tera | T |
| $1\ 000\ 000\ 000 = 10^{9}$ | giga | G |
| $1\ 000\ 000 = 10^{6}$ | mega | M |
| $1\ 000 = 10^{3}$ | kilo | k |
| $100 = 10^{2}$ | hecto‡ | h |
| $10 = 10^{1}$ | deka‡ | da |
| $0.1 = 10^{-1}$ | deci‡ | d |
| $0.01 = 10^{-2}$ | centi‡ | c |
| $0.001 = 10^{-3}$ | milli | m |
| $0.000\ 001 = 10^{-6}$ | micro | $\mu$ |
| $0.000\ 000\ 001 = 10^{-9}$ | nano | n |
| $0.000\ 000\ 000\ 001 = 10^{-12}$ | pico | p |
| $0.000\ 000\ 000\ 000\ 001 = 10^{-15}$ | femto | f |
| $0.000\ 000\ 000\ 000\ 000\ 001 = 10^{-18}$ | atto | a |

† The first syllable of every prefix is accented so that the prefix will retain its identity. Thus, the preferred pronunciation of kilometer places the accent on the first syllable, not the second.

‡ The use of these prefixes should be avoided, except for the measurement of areas and volumes and for the nontechnical use of centimeter, as for body and clothing measurements.

Multiples and submultiples of the fundamental SI units may be obtained through the use of the prefixes defined in Table 1.1. The multiples and submultiples of the units of length, mass, and force most frequently used in engineering are, respectively, the *kilometer* (km) and the *millimeter* (mm); the *megagram*† (Mg) and the *gram* (g); and the *kilonewton* (kN). According to Table 1.1, we have

$$1\ \text{km} = 1000\ \text{m} \qquad 1\ \text{mm} = 0.001\ \text{m}$$
$$1\ \text{Mg} = 1000\ \text{kg} \qquad 1\ \text{g} = 0.001\ \text{kg}$$
$$1\ \text{kN} = 1000\ \text{N}$$

The conversion of these units into meters, kilograms, and newtons, respectively, can be effected by simply moving the decimal point three places to the right or to the left. For example, to convert 3.82 km into meters, one moves the decimal point three places to the right:

$$3.82\ \text{km} = 3820\ \text{m}$$

Similarly, 47.2 mm is converted into meters by moving the decimal point three places to the left:

$$47.2\ \text{mm} = 0.0472\ \text{m}$$

Using the scientific notation, one may also write

$$3.82\ \text{km} = 3.82 \times 10^{3}\ \text{m}$$
$$47.2\ \text{mm} = 47.2 \times 10^{-3}\ \text{m}$$

† Also known as a *metric ton*.

The multiples of the unit of time are the *minute* (min) and the *hour* (h). Since

$$1 \text{ min} = 60 \text{ s} \qquad \text{and} \qquad 1 \text{ h} = 60 \text{ min} = 3600 \text{ s}$$

these multiples cannot be converted as readily as the others.

By using the appropriate multiple or submultiple of a given unit, one may avoid writing very large or very small numbers. For example, one usually writes 427.2 km rather than 427 200 m, and 2.16 mm rather than 0.002 16 m.[†]

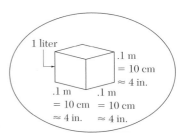

**Units of Area and Volume.** The unit of area is the *square meter* ($m^2$), which represents the area of a square of side 1 m; the unit of volume is the *cubic meter* ($m^3$), equal to the volume of a cube of side 1 m. In order to avoid exceedingly small or large numerical values in the computation of areas and volumes, one uses systems of subunits obtained by respectively squaring and cubing not only the millimeter but also two intermediate submultiples of the meter, namely, the *decimeter* (dm) and the *centimeter* (cm). Since, by definition,

$$1 \text{ dm} = 0.1 \text{ m} = 10^{-1} \text{ m}$$
$$1 \text{ cm} = 0.01 \text{ m} = 10^{-2} \text{ m}$$
$$1 \text{ mm} = 0.001 \text{ m} = 10^{-3} \text{ m}$$

the submultiples of the unit of area are

$$1 \text{ dm}^2 = (1 \text{ dm})^2 = (10^{-1} \text{ m})^2 = 10^{-2} \text{ m}^2$$
$$1 \text{ cm}^2 = (1 \text{ cm})^2 = (10^{-2} \text{ m})^2 = 10^{-4} \text{ m}^2$$
$$1 \text{ mm}^2 = (1 \text{ mm})^2 = (10^{-3} \text{ m})^2 = 10^{-6} \text{ m}^2$$

and the submultiples of the unit of volume are

$$1 \text{ dm}^3 = (1 \text{ dm})^3 = (10^{-1} \text{ m})^3 = 10^{-3} \text{ m}^3$$
$$1 \text{ cm}^3 = (1 \text{ cm})^3 = (10^{-2} \text{ m})^3 = 10^{-6} \text{ m}^3$$
$$1 \text{ mm}^3 = (1 \text{ mm})^3 = (10^{-3} \text{ m})^3 = 10^{-9} \text{ m}^3$$

It should be noted that when the volume of a liquid is being measured, the cubic decimeter ($dm^3$) is usually referred to as a *liter* (L).

Other derived SI units used to measure the moment of a force, the work of a force, etc., are shown in Table 1.2. While these units will be introduced in later chapters as they are needed, we should note an important rule at this time: When a derived unit is obtained by dividing a base unit by another base unit, a prefix may be used in the numerator of the derived unit but not in its denominator. For example, the constant $k$ of a spring which stretches 20 mm under a load of 100 N will be expressed as

$$k = \frac{100 \text{ N}}{20 \text{ mm}} = \frac{100 \text{ N}}{0.020 \text{ m}} = 5000 \text{ N/m} \qquad \text{or} \qquad k = 5 \text{ kN/m}$$

but never as $k = 5$ N/mm.

[†] It should be noted that when more than four digits are used on either side of the decimal point to express a quantity in SI units—as in 427 200 m or 0.002 16 m—spaces, never commas, should be used to separate the digits into groups of three. This is to avoid confusion with the comma which is used in many countries in place of a decimal point.

**TABLE 1.2  Principal SI Units Used in Mechanics**

| Quantity | Unit | Symbol | Formula |
|---|---|---|---|
| Acceleration | Meter per second squared | . . . | $m/s^2$ |
| Angle | Radian | rad | † |
| Angular acceleration | Radian per second squared | . . . | $rad/s^2$ |
| Angular velocity | Radian per second | . . . | $rad/s$ |
| Area | Square meter | . . . | $m^2$ |
| Density | Kilogram per cubic meter | . . . | $kg/m^3$ |
| Energy | Joule | J | $N \cdot m$ |
| Force | Newton | N | $kg \cdot m/s^2$ |
| Frequency | Hertz | Hz | $s^{-1}$ |
| Impulse | Newton-second | . . . | $kg \cdot m/s$ |
| Length | Meter | m | ‡ |
| Mass | Kilogram | kg | ‡ |
| Moment of a force | Newton-meter | . . . | $N \cdot m$ |
| Power | Watt | W | $J/s$ |
| Pressure | Pascal | Pa | $N/m^2$ |
| Stress | Pascal | Pa | $N/m^2$ |
| Time | Second | s | ‡ |
| Velocity | Meter per second | . . . | $m/s$ |
| Volume, solids | Cubic meter | . . . | $m^3$ |
|    Liquids | Liter | L | $10^{-3}\ m^3$ |
| Work | Joule | J | $N \cdot m$ |

† Supplementary unit (1 revolution $= 2\pi$ rad $= 360°$).

‡ Base unit.

**U.S. Customary Units.**  Most practicing American engineers still commonly use a system in which the base units are the units of length, force, and time. These units are, respectively, the *foot* (ft), the *pound* (lb), and the *second* (s). The second is the same as the corresponding SI unit. The foot is defined as 0.3048 m. The pound is defined as the *weight* of a platinum standard, called the *standard pound* and kept at the National Bureau of Standards in Washington, the mass of which is 0.453 592 43 kg. Since the weight of a body depends upon the gravitational attraction of the earth, which varies with location, it is specified that the standard pound should be placed at sea level and at the latitude of 45° to properly define a force of 1 lb. Clearly the U.S. customary units do not form an absolute system of units. Because of their dependence upon the gravitational attraction of the earth, they form a *gravitational* system of units.

While the standard pound also serves as the unit of mass in commercial transactions in the United States, it cannot be so used in engineering computations since such a unit would not be consistent with the base units defined in the preceding paragraph. Indeed, when acted upon by a force of 1 lb, that is, when subjected to the force of gravity, the standard pound receives the acceleration of gravity, $g = 32.2$ ft/s² (Fig. 1.4), not the unit acceleration required by Eq. (1.1). The unit of mass consistent with the foot, the pound, and the second is the mass which receives an acceleration of 1 ft/s² when a force of 1 lb is applied to it (Fig. 1.5). This unit,

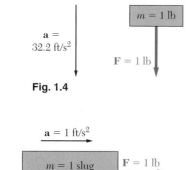

$\mathbf{a} = 32.2$ ft/s²

$m = 1$ lb

$\mathbf{F} = 1$ lb

**Fig. 1.4**

$\mathbf{a} = 1$ ft/s²

$m = 1$ slug
$(= 1$ lb $\cdot$ s²/ft$)$

$\mathbf{F} = 1$ lb

**Fig. 1.5**

sometimes called a *slug,* can be derived from the equation $F = ma$ after substituting 1 lb and 1 ft/s² for $F$ and $a$, respectively. We write

$$F = ma \qquad 1\text{ lb} = (1\text{ slug})(1\text{ ft/s}^2)$$

and obtain

$$1\text{ slug} = \frac{1\text{ lb}}{1\text{ ft/s}^2} = 1\text{ lb}\cdot\text{s}^2/\text{ft} \qquad\qquad (1.6)$$

Comparing Figs. 1.4 and 1.5, we conclude that the slug is a mass 32.2 times larger than the mass of the standard pound.

The fact that in the U.S. customary system of units bodies are characterized by their weight in pounds rather than by their mass in slugs will be a convenience in the study of statics, where we shall constantly deal with weights and other forces and only seldom with masses. However, in the study of dynamics, where forces, masses, and accelerations are involved, we shall have to express the mass $m$ in slugs of a body of which the weight $W$ has been given in pounds. Recalling Eq. (1.4), we shall write

$$m = \frac{W}{g} \qquad\qquad (1.7)$$

where $g$ is the acceleration of gravity ($g = 32.2$ ft/s²).

Other U.S. customary units frequently encountered in engineering problems are the *mile* (mi), equal to 5280 ft; the *inch* (in.), equal to $\frac{1}{12}$ ft; and the *kilopound* (kip), equal to a force of 1000 lb. The *ton* is often used to represent a mass of 2000 lb but, like the pound, must be converted into slugs in engineering computations.

The conversion into feet, pounds, and seconds of quantities expressed in other U.S. customary units is generally more involved and requires greater attention than the corresponding operation in SI units. If, for example, the magnitude of a velocity is given as $v = 30$ mi/h, we shall proceed as follows to convert it to ft/s. First we write

$$v = 30\,\frac{\text{mi}}{\text{h}}$$

Since we want to get rid of the unit miles and introduce instead the unit feet, we should multiply the right-hand member of the equation by an expression containing miles in the denominator and feet in the numerator. But, since we do not want to change the value of the right-hand member, the expression used should have a value equal to unity. The quotient (5280 ft)/(1 mi) is such an expression. Operating in a similar way to transform the unit hour into seconds, we write

$$v = \left(30\,\frac{\text{mi}}{\text{h}}\right)\left(\frac{5280\text{ ft}}{1\text{ mi}}\right)\left(\frac{1\text{ h}}{3600\text{ s}}\right)$$

Carrying out the numerical computations and canceling out units which appear both in the numerator and the denominator, we obtain

$$v = 44\,\frac{\text{ft}}{\text{s}} = 44\text{ ft/s}$$

## 1.4. CONVERSION FROM ONE SYSTEM OF UNITS TO ANOTHER

There are many instances when an engineer wishes to convert into SI units a numerical result obtained in U.S. customary units, or vice versa. Because the unit of time is the same in both systems, only two kinetic base units need be converted. Thus, since all other kinetic units can be derived from these base units, only two conversion factors need be remembered.

**Units of Length.** By definition the U.S. customary unit of length is

$$1 \text{ ft} = 0.3048 \text{ m} \tag{1.8}$$

It follows that

$$1 \text{ mi} = 5280 \text{ ft} = 5280(0.3048 \text{ m}) = 1609 \text{ m}$$

or

$$1 \text{ mi} = 1.609 \text{ km} \tag{1.9}$$

Also

$$1 \text{ in.} = \tfrac{1}{12} \text{ ft} = \tfrac{1}{12}(0.3048 \text{ m}) = 0.0254 \text{ m}$$

or

$$1 \text{ in.} = 25.4 \text{ mm} \tag{1.10}$$

**Units of Force.** Recalling that the U.S. customary unit of force (pound) is defined as the weight of the standard pound (of mass 0.4536 kg) at sea level and at the latitude of 45° (where $g = 9.807 \text{ m/s}^2$), and using Eq. (1.4), we write

$$W = mg$$
$$1 \text{ lb} = (0.4536 \text{ kg})(9.807 \text{ m/s}^2) = 4.448 \text{ kg} \cdot \text{m/s}^2$$

or, recalling (1.5),

$$1 \text{ lb} = 4.448 \text{ N} \tag{1.11}$$

**Units of Mass.** The U.S. customary unit of mass (slug) is a derived unit. Thus, using (1.6), (1.8), and (1.11), we write

$$1 \text{ slug} = 1 \text{ lb} \cdot \text{s}^2/\text{ft} = \frac{1 \text{ lb}}{1 \text{ ft/s}^2} = \frac{4.448 \text{ N}}{0.3048 \text{ m/s}^2} = 14.59 \text{ N} \cdot \text{s}^2/\text{m}$$

and, recalling (1.5),

$$1 \text{ slug} = 1 \text{ lb} \cdot \text{s}^2/\text{ft} = 14.59 \text{ kg} \tag{1.12}$$

Although it cannot be used as a consistent unit of mass, we recall that the mass of the standard pound is, by definition,

$$1 \text{ pound mass} = 0.4536 \text{ kg} \tag{1.13}$$

This constant may be used to determine the *mass* in SI units (kilograms) of a body which has been characterized by its *weight* in U.S. customary units (pounds).

To convert into SI units a derived U.S. customary unit, one simply multiplies or divides the appropriate conversion factors. For example, to convert into SI units the moment of a force which was found to be $M = 47$ lb·in., we use formulas (1.10) and (1.11) and write

$$M = 47 \text{ lb·in.} = 47(4.448 \text{ N})(25.4 \text{ mm})$$
$$= 5310 \text{ N·mm} = 5.31 \text{ N·m}$$

The conversion factors given in this section may also be used to convert into U.S. customary units a numerical result obtained in SI units. For example, if the moment of a force was found to be $M = 40$ N·m, we write, following the procedure used in the last paragraph of Sec. 1.3,

$$M = 40 \text{ N·m} = (40 \text{ N·m})\left(\frac{1 \text{ lb}}{4.448 \text{ N}}\right)\left(\frac{1 \text{ ft}}{0.3048 \text{ m}}\right)$$

Carrying out the numerical computations and canceling out units which appear both in the numerator and the denominator, we obtain

$$M = 29.5 \text{ lb·ft}$$

The U.S. customary units most frequently used in mechanics are listed in Table 1.3 with their SI equivalents.

## 1.5. METHOD OF PROBLEM SOLUTION

The student should approach a problem in mechanics as he would approach an actual engineering situation. By drawing on his own experience and on his intuition, he will find it easier to understand and formulate the problem. Once the problem has been clearly stated, however, there is no place in its solution for the student's particular fancy. *The solution must be based on the six fundamental principles stated in Sec. 1.2 or on theorems derived from them.* Every step taken must be justified on that basis. Strict rules must be followed, which lead to the solution in an almost automatic fashion, leaving no room for the student's intuition or "feeling." After an answer has been obtained, it should be checked. Here again, the student may call upon his common sense and personal experience. If not completely satisfied with the result obtained, he should carefully check his formulation of the problem, the validity of the methods used for its solution, and the accuracy of his computations.

The *statement* of a problem should be clear and precise. It should contain the given data and indicate what information is required. A neat drawing showing all quantities involved should be included. Separate diagrams should be drawn for all bodies involved, indicating clearly the forces acting on each body. These diagrams are known as *free-body diagrams* and are described in detail in Secs. 2.11 and 3.14.

**TABLE 1.3  U.S. Customary Units and Their SI Equivalents**

| Quantity | U.S. Customary Unit | SI Equivalent |
|---|---|---|
| Acceleration | $ft/s^2$ | $0.3048 \ m/s^2$ |
| | $in./s^2$ | $0.0254 \ m/s^2$ |
| Area | $ft^2$ | $0.0929 \ m^2$ |
| | $in^2$ | $645.2 \ mm^2$ |
| Energy | ft·lb | 1.356 J |
| Force | kip | 4.448 kN |
| | lb | 4.448 N |
| | oz | 0.2780 N |
| Impulse | lb·s | 4.448 N·s |
| Length | ft | 0.3048 m |
| | in. | 25.40 mm |
| | mi | 1.609 km |
| Mass | oz mass | 28.35 g |
| | lb mass | 0.4536 kg |
| | slug | 14.59 kg |
| | ton | 907.2 kg |
| Moment of a force | lb·ft | 1.356 N·m |
| | lb·in. | 0.1130 N·m |
| Moment of inertia | | |
|   Of an area | $in^4$ | $0.4162 \times 10^6 \ mm^4$ |
|   Of a mass | $lb·ft·s^2$ | $1.356 \ kg·m^2$ |
| Momentum | lb·s | 4.448 kg·m/s |
| Power | ft·lb/s | 1.356 W |
| | hp | 745.7 W |
| Pressure or stress | $lb/ft^2$ | 47.88 Pa |
| | $lb/in^2$ (psi) | 6.895 kPa |
| Velocity | ft/s | 0.3048 m/s |
| | in./s | 0.0254 m/s |
| | mi/h (mph) | 0.4470 m/s |
| | mi/h (mph) | 1.609 km/h |
| Volume | $ft^3$ | $0.02832 \ m^3$ |
| | $in^3$ | $16.39 \ cm^3$ |
|   Liquids | gal | 3.785 L |
| | qt | 0.9464 L |
| Work | ft·lb | 1.356 J |

    The *fundamental principles* of mechanics listed in Sec. 1.2 *will be used to write equations* expressing the conditions of rest or motion of the bodies considered. Each equation should be clearly related to one of the free-body diagrams. The student will then proceed to solve the problem, observing strictly the usual rules of algebra and recording neatly the various steps taken.

    After the answer has been obtained, it should be *carefully checked*. Mistakes in *reasoning* may often be detected by checking the units. For example, to determine the moment of a force of 50 N about a point 0.60 m from its line of action, we would have written (Sec. 3.12)

$$M = Fd = (50 \text{ N})(0.60 \text{ m}) = 30 \text{ N} \cdot \text{m}$$

The unit $\text{N} \cdot \text{m}$ obtained by multiplying newtons by meters is the correct unit for the moment of a force; if another unit had been obtained, we would have known that some mistake had been made.

Errors in *computation* will usually be found by substituting the numerical values obtained into an equation which has not yet been used and verifying that the equation is satisfied. The importance of correct computations in engineering cannot be overemphasized.

## 1.6. NUMERICAL ACCURACY

The accuracy of the solution of a problem depends upon two items: (1) the accuracy of the given data; (2) the accuracy of the computations performed.

The solution cannot be more accurate than the less accurate of these two items. For example, if the loading of a bridge is known to be 75,000 lb with a possible error of 100 lb either way, the relative error which measures the degree of accuracy of the data is

$$\frac{100 \text{ lb}}{75,000 \text{ lb}} = 0.0013 = 0.13 \text{ percent}$$

In computing the reaction at one of the bridge supports, it would then be meaningless to record it as 14,322 lb. The accuracy of the solution cannot be greater than 0.13 percent, no matter how accurate the computations are, and the possible error in the answer may be as large as $(0.13/100)(14,332 \text{ lb}) \approx 20$ lb. The answer should be properly recorded as $14,320 \pm 20$ lb.

In engineering problems, the data are seldom known with an accuracy greater than 0.2 percent. It is therefore seldom justified to write the answers to such problems with an accuracy greater than 0.2 percent. A practical rule is to use 4 figures to record numbers beginning with a "1" and 3 figures in all other cases. Unless otherwise indicated, the data given in a problem should be assumed known with a comparable degree of accuracy. A force of 40 lb, for example, should be read 40.0 lb, and a force of 15 lb should be read 15.00 lb.

The use of calculators at times tempts students to record many digits. However, the student should not record more significant figures than can be justified, merely because they are easily obtained. As noted above, an accuracy greater than 0.2 percent is seldom necessary or meaningful in the solution of practical engineering problems.

# CHAPTER 2

# Statics of Particles

## 2.1. INTRODUCTION

In this chapter we shall study the effect of forces acting on particles. First we shall learn how to replace two or more forces acting on a given particle by a single force having the same effect as the original forces. This single equivalent force is the *resultant* of the original forces acting on the particle. Later we shall derive the relations which exist among the various forces acting on a particle in a state of *equilibrium* and use these relations to determine some of the forces acting on the particle.

The use of the word "particle" does not imply that we shall restrict our study to that of miniscule objects. What it means is that the size and shape of the bodies under consideration will not significantly affect the solution of the problems treated in this chapter, and that all the forces acting on a given body will be assumed applied at the same point. Since such an assumption is verified in many practical applications, we shall be able to solve a number of engineering problems in this chapter.

The first part of the chapter is devoted to the study of forces contained in a single plane, and the second part to the analysis of forces in three-dimensional space.

## FORCES IN A PLANE

## 2.2. FORCE ON A PARTICLE. RESULTANT OF TWO FORCES

A force represents the action of one body on another and is generally characterized by its *point of application,* its *magnitude,* and its *direction.* Forces acting on a given particle, however, have the same point of application. Each force considered in this chapter will thus be completely defined by its magnitude and direction.

The magnitude of a force is characterized by a certain number of units. As indicated in Chap. 1, the SI units used by engineers to measure the magnitude of a force are the newton (N) and its multiple the kilonewton (kN), equal to 1000 N, while the U.S. customary units used for the same purpose are the pound (lb) and its multiple the kilopound (kip, or k), equal to 1000 lb. The direction of a force is defined by the *line of action* and the *sense* of the force. The line of action is the infinite straight line along which the force acts; it is characterized by the angle it forms with some fixed axis (Fig. 2.1). The force itself is represented by a segment of that line; through the use of an appropriate scale, the length of this segment may be chosen to represent the magnitude of the force. Finally, the sense of the force should be indicated by an arrowhead. It is important in defining a force to indicate its sense. Two forces, such as those shown in Fig. 2.1*a* and *b*, having the same magnitude and the same line of action but different sense, will have directly opposite effects on a particle.

Experimental evidence shows that two forces **P** and **Q** acting on a particle *A* (Fig. 2.2*a*) may be replaced by a single force **R** which has the same effect on the particle (Fig. 2.2*c*). This force is called the *resultant* of the forces **P** and **Q** and may be obtained, as shown in Fig. 2.2*b*, by constructing a parallelogram, using **P** and **Q** as two sides of the parallelogram. *The diagonal that passes through A represents the resultant.* This is known as the *parallelogram law* for the addition of two forces. This law is based on experimental evidence; it cannot be proved or derived mathematically.

## 2.3. VECTORS

Forces do not obey the rules of addition defined in ordinary arithmetic or algebra. For example, two forces acting at a right angle to each other, one of 4 lb and the other of 3 lb, add up to a force of 5 lb, *not* to a force of 7 lb. Forces are not the only expressions which follow the parallelogram law of addition. As we shall see later, *displacements, velocities, accelerations, momenta* are other examples of physical quantities possessing magnitude and direction and which are added according to the parallelogram law. All these quantities may be represented mathematically by *vectors*, while those physical quantities which do not have direction, such as *volume, mass,* or *energy,* are represented by ordinary numbers or *scalars*.

Vectors are defined as *mathematical expressions possessing magnitude and direction, which add according to the parallelogram law.* Vectors are represented by arrows in the illustrations and will be distinguished from scalar quantities in this text through the use of boldface type (**P**). In longhand writing, a vector may be characterized by drawing a short arrow above the letter used to represent it ($\vec{P}$). The magnitude of a vector defines the length of the arrow used to represent the vector. In this text, italic type will be used to denote the magnitude of a vector. Thus, the magnitude of the vector **P** will be referred to as *P*.

A vector used to represent a force acting on a given particle has a well-defined point of application, namely, the particle itself. Such a vector is said to be a *fixed*, or *bound*, vector and cannot be moved without modifying the conditions of the problem. Other physical quantities, however, such as couples (see Chap. 4), are represented by vectors which may be freely moved in space; these vectors are called *free* vectors. Still other

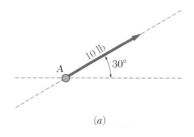

(a)

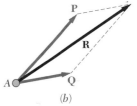

(b)

**Fig. 2.1**

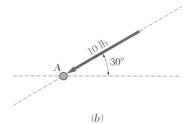

(a)

Draw construction lines parallel to the two vectors to be added, forming a parallelogram.

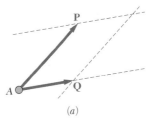

(b)

This parallelogram defines the length and direction of the resultant, **R**.

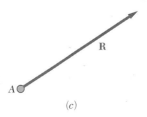

(c)

**Fig. 2.2**

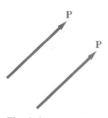

**Fig. 2.4**

**Fig. 2.5**

physical quantities, such as forces acting on a rigid body (see Chap. 3), are represented by vectors which may be moved, or slid, along their line of action; they are known as *sliding* vectors.[†]

Two vectors which have the same magnitude and the same direction are said to be *equal*, whether or not they also have the same point of application (Fig. 2.4); equal vectors may be denoted by the same letter. Two vectors which have the same magnitude, parallel lines of action, and opposite sense are said to be *equal and opposite* (Fig. 2.5).

## 2.4. ADDITION OF VECTORS

We saw in the preceding section that, by definition, vectors add according to the parallelogram law. Thus the sum of two vectors **P** and **Q** is obtained by attaching the two vectors to the same point $A$ and constructing a parallelogram, using **P** and **Q** as two sides of the parallelogram (Fig. 2.6). The diagonal that passes through $A$ represents the sum of the vectors **P** and **Q**, and this sum is denoted by **P** + **Q**. The fact that the sign + is used to denote both vector and scalar addition should not cause any confusion if vector and scalar quantities are always carefully distinguished. Thus, we should note that the magnitude of the vector **P** + **Q** is *not*, in general, equal to the sum $P + Q$ of the magnitudes of the vectors **P** and **Q**.

Since the parallelogram constructed on the vectors **P** and **Q** does not depend upon the order in which **P** and **Q** are selected, we conclude that the addition of two vectors is *commutative*, and we write

$$\mathbf{P} + \mathbf{Q} = \mathbf{Q} + \mathbf{P} \tag{2.1}$$

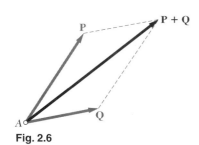

**Fig. 2.6**

[†] Some expressions have magnitude and direction, but do not add according to the parallelogram law. While these expressions may be represented by arrows, they *cannot* be considered as vectors.

A group of such expressions are the finite rotations of a rigid body. Place a closed book on a table in front of you, so that it lies in the usual fashion, with its front cover up and its binding to the left. Now rotate it through 180° about an axis parallel to the binding (Fig. 2.3a); this rotation may be represented by an arrow of length equal to 180 units and oriented as shown. Picking up the book as it lies in its new position, rotate it now through 180° about a horizontal

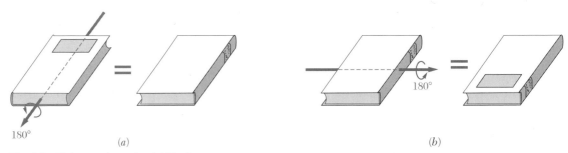

(a)                    (b)

**Fig. 2.3** Finite rotations of a rigid body.

From the parallelogram law, we can derive an alternate method for determining the sum of two vectors. This method, known as the *triangle rule*, is derived as follows: Consider Fig. 2.6, where the sum of the vectors **P** and **Q** has been determined by the parallelogram law. Since the side of the parallelogram opposite **Q** is equal to **Q** in magnitude and direction, we could draw only half of the parallelogram (Fig. 2.7a). The sum of the two vectors may thus be found by *arranging **P** and **Q** in tip-to-tail fashion and then connecting the tail of **P** with the tip of **Q***. In Fig. 2.7b, the other half of the parallelogram is considered, and the same result is obtained. This confirms the fact that vector addition is commutative.

The *subtraction* of a vector is defined as the addition of the corresponding negative vector. Thus, the vector **P** − **Q** representing the difference between the vectors **P** and **Q** is obtained by adding to **P** the vector **Q**′ equal and opposite to **Q** (Fig. 2.8). Here again we should observe that, while the same sign is used to denote both vector and scalar subtraction, confusion will be avoided if care is taken to distinguish between vector and scalar quantities.

We shall now consider the *sum of three or more vectors*. The sum of three vectors **P**, **Q**, and **S** will, *by definition*, be obtained by first adding the vectors **P** and **Q**, and then adding the vector **S** to the vector **P** + **Q**. We thus write

$$\mathbf{P} + \mathbf{Q} + \mathbf{S} = (\mathbf{P} + \mathbf{Q}) + \mathbf{S} \tag{2.2}$$

Similarly, the sum of four vectors will be obtained by adding the fourth vector to the sum of the first three. It follows that the sum of any number of vectors may be obtained by applying repeatedly the parallelogram law to successive pairs of vectors until all the given vectors are replaced by a single vector.

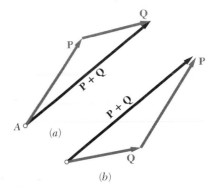

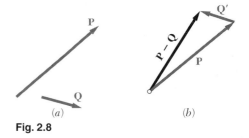

**Fig. 2.7**

**Fig. 2.8**

---

axis perpendicular to the binding (Fig. 2.3b); this second rotation may be represented by an arrow 180 units long and oriented as shown. But the book could have been placed in this final position through a single 180° rotation about a vertical axis (Fig. 2.3c). We conclude that the sum of the two 180° rotations represented by arrows directed respectively along the *z* and *x* axes is a 180° rotation represented by an arrow directed along the *y* axis (Fig. 2.3d). Clearly, the finite rotations of a rigid body *do not* obey the parallelogram law of addition; therefore they *cannot* be represented by vectors.

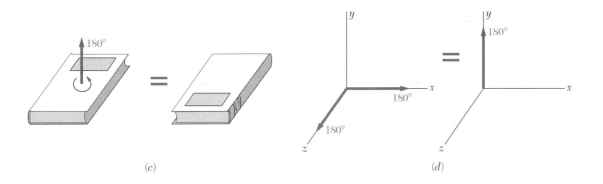

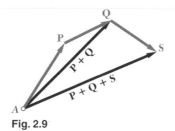

**Fig. 2.9**

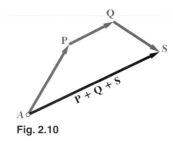

**Fig. 2.10**

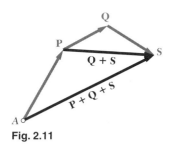

**Fig. 2.11**

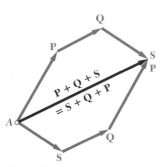

**Fig. 2.12**

If the given vectors are *coplanar*, i.e., if they are contained in the same plane, their sum may be easily obtained graphically. In that case, the repeated application of the triangle rule will be preferred to the application of the parallelogram law. In Fig. 2.9 the sum of three vectors **P**, **Q**, and **S** was obtained in that manner. The triangle rule was first applied to obtain the sum **P** + **Q** of the vectors **P** and **Q**; it was applied again to obtain the sum of the vectors **P** + **Q** and **S**. The determination of the vector **P** + **Q**, however, could have been omitted and the sum of the three vectors could have been obtained directly, as shown in Fig. 2.10, by *arranging the given vectors in tip-to-tail fashion and connecting the tail of the first vector with the tip of the last one*. This is known as the *polygon rule* for the addition of vectors.

We observe that the result obtained would have been unchanged if, as shown in Fig. 2.11, the vectors **Q** and **S** had been replaced by their sum **Q** + **S**. We may thus write

$$\mathbf{P} + \mathbf{Q} + \mathbf{S} = (\mathbf{P} + \mathbf{Q}) + \mathbf{S} = \mathbf{P} + (\mathbf{Q} + \mathbf{S}) \qquad (2.3)$$

which expresses the fact that vector addition is *associative*. Recalling that vector addition has also been shown, in the case of two vectors, to be commutative, we write

$$\mathbf{P} + \mathbf{Q} + \mathbf{S} = (\mathbf{P} + \mathbf{Q}) + \mathbf{S} = \mathbf{S} + (\mathbf{P} + \mathbf{Q})$$
$$= \mathbf{S} + (\mathbf{Q} + \mathbf{P}) = \mathbf{S} + \mathbf{Q} + \mathbf{P} \qquad (2.4)$$

This expression, as well as others which may be obtained in the same way, shows that the order in which several vectors are added together is immaterial (Fig. 2.12).

## 2.5. RESULTANT OF SEVERAL CONCURRENT FORCES

Consider a particle *A* acted upon by several coplanar forces, i.e., by several forces contained in the same plane (Fig. 2.13*a*). Since the forces considered here all pass through *A*, they are also said to be *concurrent*. The vectors representing the forces acting on *A* may be added by the polygon rule (Fig. 2.13*b*). Since the use of the polygon rule is equivalent to the repeated application of the parallelogram law, the vector **R** thus obtained represents the resultant of the given concurrent forces, i.e., the single force

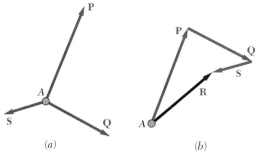

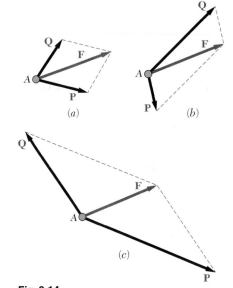

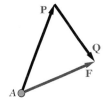

(a)

Fig. 2.13 A concurrent force system: Vectors **P**, **Q**, and **S** act at the same point, *A*.

which has the same effect on the particle *A* as the given forces. As indicated above, the order in which the vectors **P**, **Q**, and **S** representing the given forces are added together is immaterial.

## 2.6. RESOLUTION OF A FORCE INTO COMPONENTS

We have seen that two or more forces acting on a particle may be replaced by a single force which has the same effect on the particle. Conversely, a single force **F** acting on a particle may be replaced by two or more forces which, together, have the same effect on the particle. These forces are called the *components* of the original force **F**, and the process of substituting them for **F** is called *resolving the force* **F** *into components*.

Clearly, for each force **F** there exist an infinite number of possible sets of components. Sets of *two components* **P** *and* **Q** are the most important as far as practical applications are concerned. But, even then, the number of ways in which a given force **F** may be resolved into two components is unlimited (Fig. 2.14). Two cases are of particular interest:

1. *One of the Two Components,* **P**, *Is Known.* The second component, **Q**, is obtained by applying the triangle rule and joining the tip of **P** to the tip of **F** (Fig. 2.15); the magnitude and direction of **Q** are determined graphically or by trigonometry. Once **Q** has been determined, both components **P** and **Q** should be applied at *A*.

2. *The Line of Action of Each Component Is Known.* The magnitude and sense of the components are obtained by applying the parallelogram law and drawing lines, through the tip of **F**, parallel to the given lines of action (Fig. 2.16). This process leads to two well-defined components, **P** and **Q**, which may be determined graphically, or trigonometrically, by applying the law of sines.

Many other cases may be encountered; for example, the direction of one component may be known while the magnitude of the other component is to be as small as possible (see Sample Prob. 2.2). In all cases the appropriate triangle or parallelogram is drawn, which satisfies the given conditions.

Fig. 2.14

Fig. 2.15

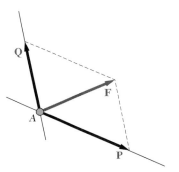

Fig. 2.16

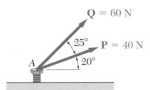

# SAMPLE PROBLEM 2.1

The two forces **P** and **Q** act on a bolt *A*. Determine their resultant.

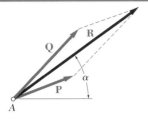

**Graphical Solution.** A parallelogram with sides equal to **P** and **Q** is drawn to scale. The magnitude and direction of the resultant are measured and found to be

$$R = 98 \text{ N} \qquad \alpha = 35° \qquad \mathbf{R} = 98 \text{ N} \measuredangle 35° \ ◀$$

The triangle rule may also be used. Forces **P** and **Q** are drawn in tip-to-tail fashion. Again the magnitude and direction of the resultant are measured.

$$R = 98 \text{ N} \qquad \alpha = 35° \qquad \mathbf{R} = 98 \text{ N} \measuredangle 35° \ ◀$$

**Trigonometric Solution.** The triangle rule is again used; two sides and the included angle are known. We apply the law of cosines.

$$R^2 = P^2 + Q^2 - 2PQ \cos B$$
$$R^2 = (40 \text{ N})^2 + (60 \text{ N})^2 - 2(40 \text{ N})(60 \text{ N}) \cos 155°$$
$$R = 97.73 \text{ N}$$

Now, applying the law of sines, we write

$$\frac{\sin A}{Q} = \frac{\sin B}{R} \qquad \frac{\sin A}{60 \text{ N}} = \frac{\sin 155°}{97.73 \text{ N}} \tag{1}$$

Solving Eq. (1) for sin *A*, we have

$$\sin A = \frac{(60 \text{ N}) \sin 155°}{97.73 \text{ N}}$$

*Using a calculator,* we first compute the quotient, then its arc sine, and obtain

$$A = 15.04° \qquad \alpha = 20° + A = 35.04°$$

We use 3 significant figures to record the answer (cf. Sec. 1.6):

$$\mathbf{R} = 97.7 \text{ N} \measuredangle 35.0° \ ◀$$

**Alternate Trigonometric Solution.** We construct the right triangle *BCD* and compute

$$CD = (60 \text{ N}) \sin 25° = 25.36 \text{ N}$$
$$BD = (60 \text{ N}) \cos 25° = 54.38 \text{ N}$$

Then, using triangle *ACD*, we obtain

$$\tan A = \frac{25.36 \text{ N}}{94.38 \text{ N}} \qquad A = 15.04°$$

$$R = \frac{25.36}{\sin A} \qquad R = 97.73 \text{ N}$$

Again,

$$\alpha = 20° + A = 35.04° \qquad \mathbf{R} = 97.7 \text{ N} \measuredangle 35.0° \ ◀$$

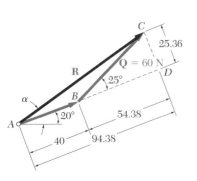

# SAMPLE PROBLEM 2.2

A barge is pulled by two tugboats. If the resultant of the forces exerted by the tugboats is a 5000-lb force directed along the axis of the barge, determine (a) the tension in each of the ropes, knowing that $\alpha = 45°$, (b) the value of $\alpha$ such that the tension in rope 2 is minimum.

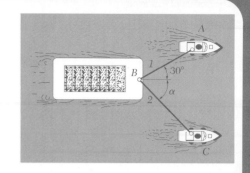

**a. Tension for $\alpha = 45°$. Graphical Solution.** The parallelogram law is used; the diagonal (resultant) is known to be equal to 5000 lb and to be directed to the right. The sides are drawn parallel to the ropes. If the drawing is done to scale, we measure

$$T_1 = 3700 \text{ lb} \qquad T_2 = 2600 \text{ lb} \blacktriangleleft$$

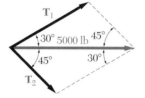

**Trigonometric Solution.** The triangle rule may be used. We note that the triangle shown represents half of the parallelogram shown above. Using the law of sines, we write

$$\frac{T_1}{\sin 45°} = \frac{T_2}{\sin 30°} = \frac{5000 \text{ lb}}{\sin 105°}$$

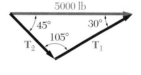

*With a calculator,* we first compute and store the value of the last quotient. Multiplying this value successively by sin 45° and sin 30°, we obtain

$$T_1 = 3660 \text{ lb} \qquad T_2 = 2590 \text{ lb} \blacktriangleleft$$

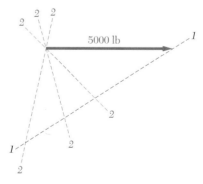

**b. Value of $\alpha$ for Minimum $T_2$.** To determine the value of $\alpha$ such that the tension in rope 2 is minimum, the triangle rule is again used. In the sketch shown, line *1-1* is the known direction of $\mathbf{T}_1$. Several possible directions of $\mathbf{T}_2$ are shown by the lines *2-2*. We note that the minimum value of $T_2$ occurs when $\mathbf{T}_1$ and $\mathbf{T}_2$ are perpendicular. The minimum value of $T_2$ is

$$T_2 = (5000 \text{ lb}) \sin 30° = 2500 \text{ lb}$$

Corresponding values of $T_1$ and $\alpha$ are

$$T_1 = (5000 \text{ lb}) \cos 30° = 4330 \text{ lb}$$
$$\alpha = 90° - 30° \qquad\qquad \alpha = 60° \blacktriangleleft$$

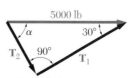

21

# PROBLEMS†

**2.1 and 2.2**   Determine graphically the magnitude and direction of the resultant of the two forces shown, using (*a*) the parallelogram law, (*b*) the triangle rule.

Fig. P2.1                                      Fig. P2.2

**2.3**   Two structural members *B* and *C* are riveted to bracket *A*. Knowing that both members are in compression and that the force is 1200 lb in member *B* and 1600 lb in member *C*, determine graphically the magnitude and direction of the resultant force exerted on the bracket.

**2.4**   Two structural members *B* and *C* are riveted to bracket *A*. Knowing that both members are in compression and that the force is 8 kN in member *B* and 12 kN in member *C*, determine graphically the magnitude and direction of the resultant force exerted on the bracket.

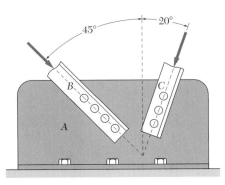

**Fig. P2.3 and P2.4**

**2.5**   The force **F** of magnitude 500 lb is to be resolved into two components along lines *a-a* and *b-b*. Determine by trigonometry the angle $\alpha$, knowing that the component of **F** along line *a-a* is to be 400 lb.

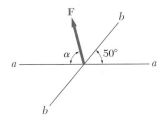

**Fig. P2.5 and P2.6**

**2.6**   The force **F** of magnitude 400 N is to be resolved into two components along lines *a-a* and *b-b*. Determine by trigonometry the angle $\alpha$, knowing that the component of **F** along line *b-b* is to be 150 N.

† Answers to all even-numbered problems are given at the end of the book.

**2.7**  A stake is pulled out of the ground by means of two ropes as shown. (*a*) Knowing that $\alpha = 30°$, determine by trigonometry the magnitude of the force **P** so that the resultant force exerted on the stake is vertical. (*b*) What is the corresponding magnitude of the resultant?

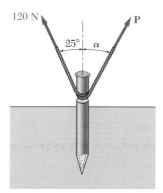

**Fig. P2.7 and P2.12**

**2.8**  A disabled automobile is pulled by means of two ropes as shown. The tension in *AB* is 400 lb and the angle $\alpha$ is 20°. Knowing that the resultant of the two forces applied at *A* is directed along the axis of the automobile, determine by trigonometry (*a*) the tension in rope *AC*, (*b*) the magnitude of the resultant of the two forces applied at *A*.

**2.9**  Solve Prob. 2.8, assuming that the tension in rope *AB* is 2.4 kN and that $\alpha = 25°$.

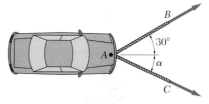

**Fig. P2.8 and P2.11**

**2.10**  Solve Prob. 2.7, assuming that $\alpha = 40°$.

**2.11**  A disabled automobile is pulled by means of two ropes as shown. Knowing that the tension in rope *AB* is 600 lb, determine by trigonometry the tension in rope *AC* and the value of $\alpha$ so that the resultant force exerted at *A* is an 800-lb force directed along the axis of the automobile.

**2.12**  A stake is pulled out of the ground by means of two ropes as shown. Knowing that the tension in one rope is 120 N, determine by trigonometry the magnitude and direction of the force **P** so that the resultant is a vertical force of 160 N.

**2.13**  Solve Prob. 2.1 by trigonometry.

**2.14**  Determine by trigonometry the magnitude and direction of the resultant of the two forces shown.

**2.15**  If the resultant of the two forces exerted on the stake of Prob. 2.7 is to be vertical, find (*a*) the value of $\alpha$ for which the magnitude of **P** is minimum, (*b*) the corresponding magnitude of **P**.

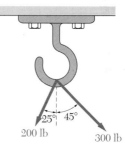

**Fig. P2.14**

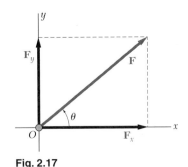

**Fig. 2.17**

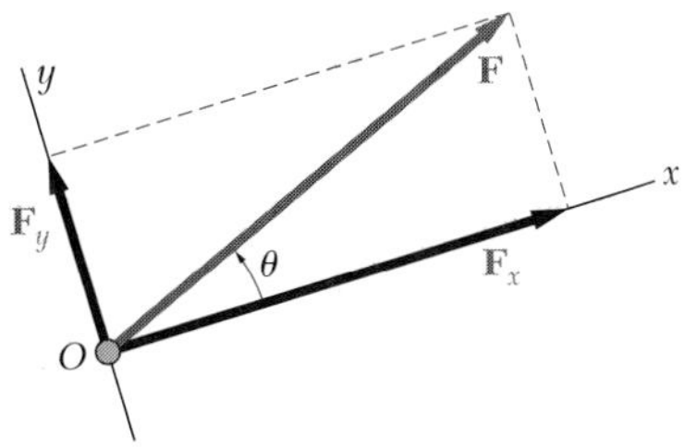

**Fig. 2.18**

## 2.7. RECTANGULAR COMPONENTS OF A FORCE[†]

In many problems it will be found desirable to resolve a force into two components which are perpendicular to each other. In Fig. 2.17, the force **F** has been resolved into a component $\mathbf{F}_x$ along the $x$ axis and a component $\mathbf{F}_y$ along the $y$ axis. The parallelogram drawn to obtain the two components is a *rectangle*, and $\mathbf{F}_x$ and $\mathbf{F}_y$ are called *rectangular components*.

The $x$ and $y$ axes are usually chosen horizontal and vertical, respectively, as in Fig. 2.17; they may, however, be chosen in any two perpendicular directions, as shown in Fig. 2.18. In determining the rectangular components of a force, the student should think of the construction lines shown in Figs. 2.17 and 2.18 as being *parallel* to the $x$ and $y$ axes, rather than *perpendicular* to these axes. This practice will help avoid mistakes in determining *oblique* components as in Sec. 2.6.

Denoting by $F$ the magnitude of the given force **F**, by $\theta$ the angle between **F** and the $x$ axis, and by $F_x$ and $F_y$, respectively, the magnitude of the components $\mathbf{F}_x$ and $\mathbf{F}_y$, we write

$$F_x = F \cos \theta \qquad F_y = F \sin \theta \qquad (2.5)$$

The magnitudes $F_x$ and $F_y$ of the components of the force **F** are called the *scalar components* of **F**, while the actual component forces $\mathbf{F}_x$ and $\mathbf{F}_y$ should be referred to as the *vector components* of **F**. However, when there exists no possibility of confusion, the vector as well as the scalar components of **F** may be referred to simply as the *components* of **F**.

We note that, if the angle $\theta$ is measured counterclockwise from the positive $x$ axis, it will take values ranging from 0 to 360°. It follows from the relations (2.5) that the scalar components $F_x$ and $F_y$ may have negative as well as positive values. More precisely, $F_x$ will be positive when **F** falls in the first or fourth quadrants, and it will be negative when **F** falls in the second or third quadrants. Similarly, $F_y$ will be positive when **F** falls in the first or second quadrants and negative when **F** falls in the third or fourth quadrants. We conclude that *the scalar component $F_x$ is positive when the vector component $\mathbf{F}_x$ has the same sense as the x axis and negative when $\mathbf{F}_x$ has the opposite sense.* A similar conclusion may be drawn regarding the sign of the scalar component $F_y$.

It appears from the above that the rectangular vector components of a force acting on a particle are completely defined by the corresponding scalar components. The vector component $\mathbf{F}_x$, for example, has by definition the $x$ axis for its line of action, and both its sense and its magnitude are defined by the scalar component $F_x$.

[†] The properties established in Secs. 2.7 and 2.8 may be readily extended to the rectangular components of any vector quantity.

***Example 1.***   A force of 800 N is exerted on a bolt $A$ as shown in Fig. 2.19. Determine the horizontal and vertical components of the force.

In order to obtain the correct sign for the scalar components $F_x$ and $F_y$, the value $\theta = 180° - 35° = 145°$ should be substituted for $\theta$ in the relations (2.5). However, it will be found more practical to determine by inspection the signs of $F_x$ and of $F_y$ (Fig. 2.20) and to use the trigonometric functions of the angle $\alpha = 35°$. We write therefore

$$F_x = -F \cos \alpha = -(800 \text{ N}) \cos 35° = -655 \text{ N}$$
$$F_y = +F \sin \alpha = +(800 \text{ N}) \sin 35° = +459 \text{ N}$$

The vector components of **F** are thus

$$\mathbf{F}_x = 655 \text{ N} \leftarrow \qquad \mathbf{F}_y = 459 \text{ N} \uparrow$$

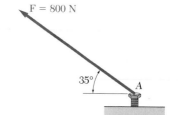

**Fig. 2.19**

***Example 2.***   A man pulls with a force of 300 N on a rope attached to a building, as shown in Fig. 2.21. What are the horizontal and vertical components of the force exerted by the rope at point $A$?

It is seen from Fig. 2.22 that

$$F_x = +(300 \text{ N}) \cos \alpha \qquad F_y = -(300 \text{ N}) \sin \alpha$$

Observing that $AB = 10$ m, we find from Fig. 2.21

$$\cos \alpha = \frac{8 \text{ m}}{AB} = \frac{8 \text{ m}}{10 \text{ m}} = \frac{4}{5} \qquad \sin \alpha = \frac{6 \text{ m}}{AB} = \frac{6 \text{ m}}{10 \text{ m}} = \frac{3}{5}$$

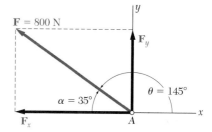

**Fig. 2.20**

We thus obtain

$$F_x = +(300 \text{ N})\tfrac{4}{5} = +240 \text{ N} \qquad F_y = -(300 \text{ N})\tfrac{3}{5} = -180 \text{ N}$$

and the vector components of **F** are

$$\mathbf{F}_x = 240 \text{ N} \rightarrow \qquad \mathbf{F}_y = 180 \text{ N} \downarrow$$

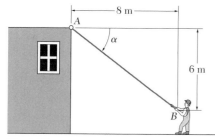

**Fig. 2.21**

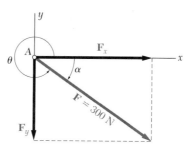

**Fig. 2.22**

When a force $\mathbf{F}$ is defined by its rectangular components $\mathbf{F}_x$ and $\mathbf{F}_y$ (Fig. 2.23), the angle $\theta$ defining its direction can be obtained by writing

$$\tan \theta = \frac{F_y}{F_x} \qquad (2.6)$$

The magnitude $F$ of the force may be obtained by applying the Pythagorean theorem and writing

$$F = \sqrt{F_x^2 + F_y^2} \qquad (2.7)$$

or by solving one of the equations (2.5) for $F$.

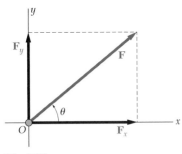

**Fig. 2.23**

***Example 3.*** The horizontal and vertical components of a force $\mathbf{F}$ acting on a bolt $A$ are, respectively, $F_x = +700$ lb and $F_y = +1500$ lb (Fig. 2.24). Determine the magnitude of the force and the angle $\theta$ it forms with the horizontal.

First we draw a diagram showing the two rectangular components of the force and the angle $\theta$ (Fig. 2.24). From Eq. (2.6), we write

$$\tan \theta = \frac{F_y}{F_x} = \frac{1500 \text{ lb}}{700 \text{ lb}}$$

*Using a calculator,*[†] we enter 1500 lb and divide by 700 lb; computing the arc tangent of the quotient, we obtain $\theta = 65.0°$. Solving the second of Eqs. (2.5) for $F$, we have

$$F = \frac{F_y}{\sin \theta} = \frac{1500 \text{ lb}}{\sin 65.0°} = 1655 \text{ lb}$$

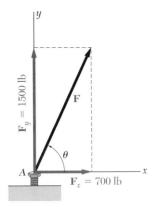

**Fig. 2.24**

The last calculation is facilitated if the value of $F_y$ is stored when originally entered; it may then be recalled to be divided by $\sin \theta$.

## 2.8. ADDITION OF FORCES BY SUMMING x AND y COMPONENTS

It was seen in Sec. 2.2 that forces should be added according to the parallelogram law. From this law, two other methods, more readily applicable to the *graphical* solution of problems, were derived in Secs. 2.4 and 2.5: the triangle rule for the addition of two forces and the polygon rule for the addition of three or more forces. It was also seen that the force triangle used to define the resultant of two forces could be used to obtain a *trigonometric* solution.

When three or more forces are to be added, no practical trigonometric solution may be obtained from the force polygon which defines the

---

[†] It is assumed that the calculator used has keys for the computation of trigonometric and inverse trigonometric functions. Some calculators also have keys for the direct conversion of rectangular coordinates into polar coordinates, and vice versa. Such calculators eliminate the need for the computation of trigonometric functions in Examples 1, 2, and 3 and in problems of the same type.

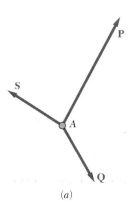

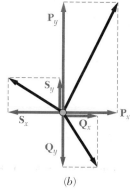

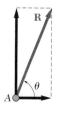

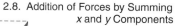

(a)                    (b)                    (c)                    (d)

**Fig. 2.25**

resultant of the forces. In this case, an *analytic* solution of the problem may
be obtained by resolving each force into two rectangular components. Consider, for instance, three forces **P**, **Q**, and **S** acting on a particle $A$ (Fig.
2.25a). In Fig. 2.25b, each of these forces has been replaced by its horizontal and vertical components; thus the forces of Fig. 2.25b produce the same
effect on $A$ as the forces of Fig. 2.25a. Now the horizontal components
may be added into a single force $\mathbf{R}_x$ and the vertical components into a
single force $\mathbf{R}_y$ (Fig. 2.25c). This should be done by applying the parallelogram law; but since, in each case, the forces considered lie in the same line,
the actual computation reduces to the algebraic addition of the scalar components. We write

$$R_x = P_x + Q_x + S_x \qquad R_y = P_y + Q_y + S_y$$

or, for short,

$$R_x = \Sigma F_x \qquad R_y = \Sigma F_y \tag{2.8}$$

The forces $\mathbf{R}_x$ and $\mathbf{R}_y$ may then be added vectorially into the resultant **R** of
the given system (Fig. 2.25d) by the method of Sec. 2.7.

From Eq. (2.8), we conclude that *the scalar components $R_x$ and $R_y$ of
the resultant* **R** *of several forces acting on a particle are obtained by adding
algebraically the corresponding scalar components of the given forces.*[†] The
procedure we have described will be carried out most efficiently if the computations are arranged in a table. While it is the only practical analytic method for adding three or more forces, it is also often preferred to
the trigonometric solution in the case of the addition of two forces.

[†] Clearly, this result also applies to the addition of other vector quantities, such as velocities, accelerations, or momenta.

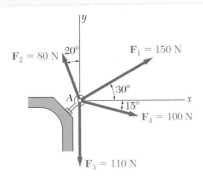

# SAMPLE PROBLEM 2.3

Four forces act on bolt $A$ as shown. Determine the resultant of the forces on the bolt.

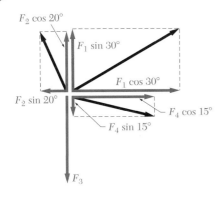

**Solution.** The $x$ and $y$ components of each force are determined by trigonometry as shown and are entered in the table below. According to the convention adopted in Sec. 2.7, the scalar number representing a force component is positive if the force component has the same sense as the corresponding coordinate axis. Thus, $x$ components acting to the right and $y$ components acting upward are represented by positive numbers.

| Force | Magnitude, N | x Component, N | y Component, N |
|:-----:|:------------:|:--------------:|:--------------:|
| $\mathbf{F}_1$ | 150 | +129.9 | +75.0 |
| $\mathbf{F}_2$ | 80 | −27.4 | +75.2 |
| $\mathbf{F}_3$ | 110 | 0 | −110.0 |
| $\mathbf{F}_4$ | 100 | +96.6 | −25.9 |
| | | $R_x = +199.1$ | $R_y = +14.3$ |

Thus, the components of the resultant are

$$\mathbf{R}_x = 199.1 \text{ N} \rightarrow \qquad \mathbf{R}_y = 14.3 \text{ N} \uparrow$$

The magnitude and direction of the resultant may now be determined. From the triangle shown, we have

$$\tan \alpha = \frac{R_y}{R_x} = \frac{14.3 \text{ N}}{199.1 \text{ N}} \qquad \alpha = 4.1°$$

$$R = \frac{14.3 \text{ N}}{\sin \alpha} = 199.6 \text{ N} \qquad \mathbf{R} = 199.6 \text{ N} \angle 4.1° \blacktriangleleft$$

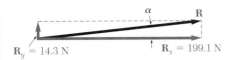

*With a calculator,* the last computation may be facilitated if the value of $R_y$ is stored when originally entered; it may then be recalled to be divided by $\sin \alpha$. (Also see the footnote on p. 26.)

# PROBLEMS

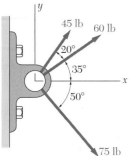

**Fig. P2.16**

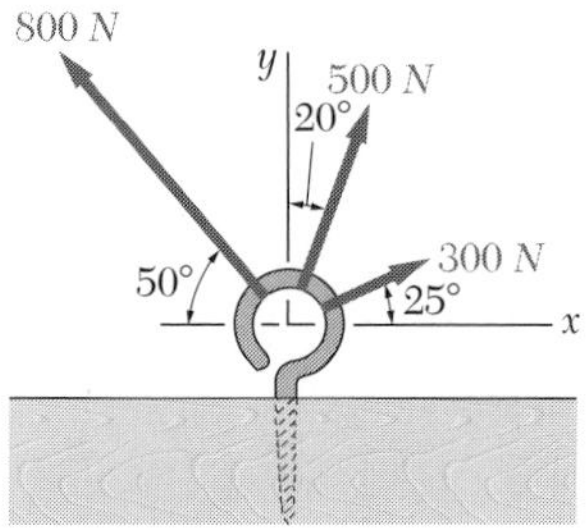

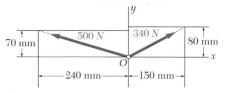

**Fig. P2.17**

**Fig. P2.18**

**2.16 through 2.19** Determine the x and y components of each of the forces shown.

**2.20** Member CB of the vise shown exerts on block B a force **P** directed along line CB. Knowing that **P** must have a 200-lb horizontal component, determine (a) the magnitude of the force **P**, (b) its vertical component.

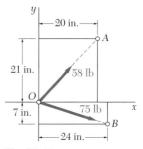

**Fig. P2.19**

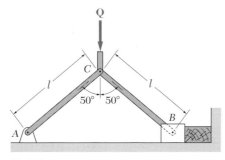

**Fig. P2.20**

**2.21** The hydraulic cylinder GE exerts on member DF a force **P** directed along line GE. Knowing that **P** must have a 600-N component perpendicular to member DF, determine (a) the magnitude of the force **P**, (b) its component parallel to DF.

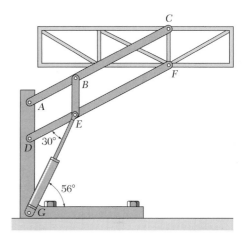

**Fig. P2.21**

**2.22** The tension in the telephone-pole guy wire is 370 lb. Determine the horizontal and vertical components of the force exerted on the anchor at $C$.

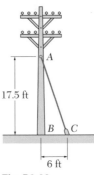

17.5 ft

$B$  $C$

6 ft

**Fig. P2.22**

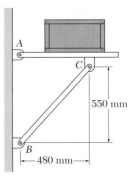

$A$

$C$

550 mm

$B$

480 mm

**Fig. P2.23**

**2.23** The compression member $BC$ exerts on the pin at $C$ a 365-N force directed along line $BC$. Determine the horizontal and vertical components of that force.

**2.24** Using $x$ and $y$ components, solve Prob. 2.2.

**2.25** Using $x$ and $y$ components, solve Prob. 2.1.

**2.26** Find the resultant of the three forces of Prob. 2.17.

**2.27** Find the resultant of the two forces of Prob. 2.18.

**2.28** Determine the resultant of the two forces of Prob. 2.18.

**2.29** Determine the resultant of the two forces of Prob. 2.19.

**2.30** Two cables which have known tensions are attached to the top of pylon $AB$. A third cable $AC$ is used as a guy wire. Determine the tension in $AC$, knowing that the resultant of the forces exerted at $A$ by the three cables must be vertical.

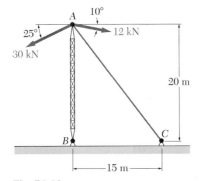

$A$  10°

25°  12 kN

30 kN

20 m

$B$

$C$

15 m

**Fig. P2.30**

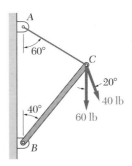

$A$

60°

$C$

20°

40 lb

40°

60 lb

$B$

**Fig. P2.31**

**2.31** Two loads are applied as shown to the end $C$ of boom $BC$. Determine the tension in cable $AC$, knowing that the resultant of the three forces exerted at $C$ must be directed along $BC$.

**2.32**   A hoist trolley is subjected to the three forces shown. Determine (*a*) the value of the angle $\alpha$ for which the resultant of the three forces is vertical, (*b*) the corresponding magnitude of the resultant.

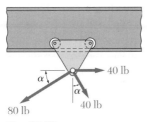

**Fig. P2.32**

**2.33**   A collar which may slide on a vertical rod is subjected to the three forces shown. Determine (*a*) the value of the angle $\alpha$ for which the resultant of the three forces is horizontal, (*b*) the corresponding magnitude of the resultant.

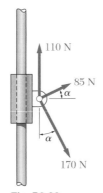

**Fig. P2.33**

## 2.9. EQUILIBRIUM OF A PARTICLE

In the preceding sections, we discussed the methods for determining the resultant of several forces acting on a particle. Although it has not occurred in any of the problems considered so far, it is quite possible for the resultant to be zero. In such a case, the net effect of the given forces is zero, and the particle is said to be in equilibrium. We thus have the following definition: *When the resultant of all the forces acting on a particle is zero, the particle is in equilibrium.*

A particle which is acted upon by two forces will be in equilibrium if the two forces have the same magnitude, same line of action, and opposite sense. The resultant of the two forces is then zero. Such a case is shown in Fig. 2.26.

**Fig. 2.26**

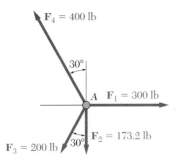

**Fig. 2.27**

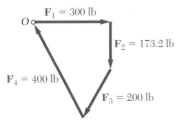

**Fig. 2.28**

Another case of equilibrium of a particle is represented in Fig. 2.27, where four forces are shown acting on A. In Fig. 2.28, the resultant of the given forces is determined by the polygon rule. Starting from point O with $\mathbf{F}_1$ and arranging the forces in tip-to-tail fashion, we find that the tip of $\mathbf{F}_4$ coincides with the starting point O. Thus the resultant **R** of the given system of forces is zero, and the particle is in equilibrium.

The closed polygon drawn in Fig. 2.28 provides a *graphical* expression of the equilibrium of A. To express *algebraically* that a particle is in equilibrium, we shall write that the two rectangular components $R_x$ and $R_y$ of the resultant are zero. Since $R_x$ is equal to the sum $\Sigma F_x$ of the x components of the given forces and $R_y$ is equal to $\Sigma F_y$, we have the following necessary and sufficient conditions for the equilibrium of a particle:

$$\Sigma F_x = 0 \qquad \Sigma F_y = 0 \qquad (2.9)$$

Returning to the particle shown in Fig. 2.27 we check that the equilibrium conditions are satisfied. We write

$$\Sigma F_x = 300 \text{ lb} - (200 \text{ lb}) \sin 30° - (400 \text{ lb}) \sin 30°$$
$$= 300 \text{ lb} - 100 \text{ lb} - 200 \text{ lb} = 0$$
$$\Sigma F_y = -173.2 \text{ lb} - (200 \text{ lb}) \cos 30° + (400 \text{ lb}) \cos 30°$$
$$= -173.2 \text{ lb} - 173.2 \text{ lb} + 346.4 \text{ lb} = 0$$

## 2.10. NEWTON'S FIRST LAW OF MOTION

In the latter part of the seventeenth century, Sir Isaac Newton formulated three fundamental laws upon which the science of mechanics is based. The first of these laws can be stated as follows:

*If the resultant force acting on a particle is zero, the particle will remain at rest (if originally at rest) or will move with constant speed in a straight line (if originally in motion).*

From this law and from the definition of equilibrium given in Sec. 2.9, it is seen that a particle in equilibrium either is at rest or is moving in a straight line with constant speed. In the following section, various problems concerning the equilibrium of a particle will be considered.

## 2.11. PROBLEMS INVOLVING THE EQUILIBRIUM OF A PARTICLE. FREE-BODY DIAGRAM

In practice, a problem in engineering mechanics is derived from an actual physical situation. A sketch showing the physical conditions of the problem is known as a *space diagram*.

The methods of analysis discussed in the preceding sections apply to a system of forces acting on a particle. A large number of problems involving actual structures, however, may be reduced to problems concerning the equilibrium of a particle. This is done by choosing a significant particle and drawing a separate diagram showing this particle and all the forces acting on it. Such a diagram is called a *free-body diagram*.

As an example, consider the 75-kg crate shown in the space diagram of Fig. 2.29a. This crate was lying between two buildings, and it is now being lifted onto a truck, which will remove it. The crate is supported by a vertical cable, which is joined at A to two ropes which pass over pulleys attached to the buildings at B and C. It is desired to determine the tension in each of the ropes AB and AC.

In order to solve this problem, a free-body diagram must be drawn, showing a particle in equilibrium. Since we are interested in the rope tensions, the free-body diagram should include at least one of these tensions and, if possible, both tensions. Point A is seen to be a good free body for this problem. The free-body diagram of point A is shown in Fig. 2.29b. It shows point A and the forces exerted on A by the vertical cable and the two ropes. The force exerted by the cable is directed downward and is equal to the weight W of the crate. Recalling Eq. (1.4), we write

$$W = mg = (75 \text{ kg})(9.81 \text{ m/s}^2) = 736 \text{ N}$$

and indicate this value in the free-body diagram. The forces exerted by the two ropes are not known. Since they are respectively equal in magnitude to the tension in rope AB and rope AC, we denote them by $\mathbf{T}_{AB}$ and $\mathbf{T}_{AC}$ and draw them away from A in the directions shown in the space diagram. No other detail is included in the free-body diagram.

Since point A is in equilibrium, the three forces acting on it must form a closed triangle when drawn in tip-to-tail fashion. This *force triangle* has been drawn in Fig. 2.29c. The values $T_{AB}$ and $T_{AC}$ of the tension in the ropes may be found graphically if the triangle is drawn to scale, or they may be found by trigonometry. If the latter method of solution is chosen, we use the law of sines and write

$$\frac{T_{AB}}{\sin 60°} = \frac{T_{AC}}{\sin 40°} = \frac{736 \text{ N}}{\sin 80°}$$

$$T_{AB} = 647 \text{ N} \qquad T_{AC} = 480 \text{ N}$$

When a particle is in *equilibrium under three forces,* the problem may always be solved by drawing a force triangle. When a particle is in *equilibrium under more than three forces,* the problem may be solved graphically by drawing a force polygon. If an analytic solution is desired, the *equations of equilibrium* given in Sec. 2.9 should be solved:

$$\Sigma F_x = 0 \qquad \Sigma F_y = 0 \qquad (2.9)$$

These equations may be solved for no more than *two unknowns;* similarly, the force triangle used in the case of equilibrium under three forces may be solved for two unknowns.

The more common types of problems are those where the two unknowns represent (1) the two components (or the magnitude and direction) of a single force, (2) the magnitude of two forces each of known direction. Problems involving the determination of the maximum or minimum value of the magnitude of a force are also encountered (see Probs. 2.44 and 2.45).

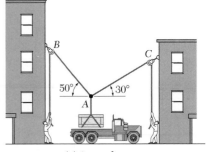

(a) Space diagram

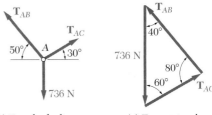

(b) Free-body diagram        (c) Force triangle

**Fig. 2.29**

## SAMPLE PROBLEM 2.4

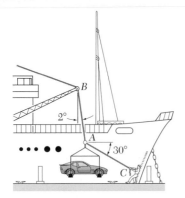

In a ship-unloading operation, a 3500-lb automobile is supported by a cable. A rope is tied to the cable at $A$ and pulled in order to center the automobile over its intended position. The angle between the cable and the vertical is 2°, while the angle between the rope and the horizontal is 30°. What is the tension in the rope?

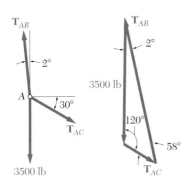

**Solution.** Point $A$ is chosen as a free body, and the complete free-body diagram is drawn. $T_{AB}$ is the tension in the cable $AB$, and $T_{AC}$ is the tension in the rope.

**Equilibrium Condition.** Since only three forces act on the free body, we draw a force triangle to express that it is in equilibrium. Using the law of sines, we write

$$\frac{T_{AB}}{\sin 120°} = \frac{T_{AC}}{\sin 2°} = \frac{3500 \text{ lb}}{\sin 58°}$$

*With a calculator,* we first compute and store the value of the last quotient. Multiplying this value successively by sin 120° and sin 2°, we obtain

$$T_{AB} = 3570 \text{ lb} \qquad\qquad T_{AC} = 144 \text{ lb} \blacktriangleleft$$

## SAMPLE PROBLEM 2.5

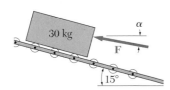

Determine the magnitude and direction of the smallest force $\mathbf{F}$ which will maintain the package shown in equilibrium. Note that the force exerted by the rollers on the package is perpendicular to the incline.

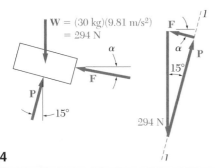

**Solution.** We choose the package as a free body, assuming that it may be treated as a particle. We draw the corresponding free-body diagram.

**Equilibrium Condition.** Since only three forces act on the free body, we draw a force triangle to express that it is in equilibrium. Line *1-1* represents the known direction of $\mathbf{P}$. In order to obtain the minimum value of the force $\mathbf{F}$, we choose the direction of $\mathbf{F}$ perpendicular to that of $\mathbf{P}$. From the geometry of the triangle obtained, we find

$$F = (294 \text{ N}) \sin 15° = 76.1 \text{ N} \qquad \alpha = 15°$$
$$\mathbf{F} = 76.1 \text{ N} \searrow 15° \blacktriangleleft$$

34

# SAMPLE PROBLEM 2.6

As part of the design of a new sailboat, it is desired to determine the drag force which may be expected at a given speed. To do so, a model of the proposed hull is placed in a test channel and three cables are used to keep its bow on the centerline of the channel. Dynamometer readings indicate that for a given speed, the tension is 40 lb in cable $AB$ and 60 lb in cable $AE$. Determine the drag force exerted on the hull and the tension in cable $AC$.

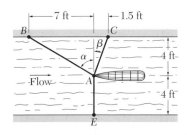

**Solution.**  First, the angles $\alpha$ and $\beta$ defining the direction of cables $AB$ and $AC$ are determined. We write

$$\tan \alpha = \frac{7 \text{ ft}}{4 \text{ ft}} = 1.75 \qquad \tan \beta = \frac{1.5 \text{ ft}}{4 \text{ ft}} = 0.375$$
$$\alpha = 60.26° \qquad\qquad \beta = 20.56°$$

Choosing the hull as a free body, we draw the free-body diagram shown. It includes the forces exerted by the three cables on the hull, as well as the drag force $\mathbf{F}_D$ exerted by the flow. Since more than three forces are involved, we shall resolve the forces into $x$ and $y$ components. Components directed to the right or upward are assigned a positive sign. We write

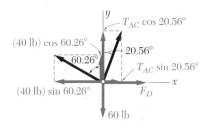

$$(T_{AB})_x = -(40 \text{ lb}) \sin 60.26° \qquad (T_{AB})_y = +(40 \text{ lb}) \cos 60.26°$$
$$= -34.73 \text{ lb} \qquad\qquad\qquad = +19.84 \text{ lb}$$

$$(T_{AC})_x = +T_{AC} \sin 20.56° \qquad (T_{AC})_y = +T_{AC} \cos 20.56°$$
$$= +0.3512 T_{AC} \qquad\qquad\quad = +0.9363 T_{AC}$$

$$(T_{AE})_x = 0 \qquad\qquad\qquad\quad (T_{AE})_y = -60 \text{ lb}$$

$$(F_D)_x = +F_D \qquad\qquad\qquad (F_D)_y = 0$$

**Equilibrium Condition.**  Since the hull is in equilibrium, the resultant of the forces must be zero. Hence

$$\Sigma F_x = 0 \qquad \Sigma F_y = 0$$

Substituting the values obtained for the components, we write

$$\Sigma F_x = 0: \qquad\qquad -34.73 \text{ lb} + 0.3512 T_{AC} + F_D = 0 \qquad (1)$$

$$\Sigma F_y = 0: \qquad\qquad 19.84 \text{ lb} + 0.9363 T_{AC} - 60 \text{ lb} = 0 \qquad (2)$$

From Eq. (2) we find $\qquad\qquad\qquad\qquad\qquad T_{AC} = +42.9 \text{ lb}$ ◀

and, substituting this value into Eq. (1), $\qquad\qquad F_D = +19.66 \text{ lb}$ ◀

In drawing the free-body diagram, we assumed a sense for each unknown force. A positive sign in the answer indicates that the assumed sense is correct. The complete force polygon may be drawn to check the results.

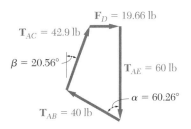

# PROBLEMS

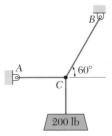

**Fig. P2.34**

**2.34 through 2.36** Two cables are tied together at *C* and loaded as shown. Determine the tension in *AC* and *BC*.

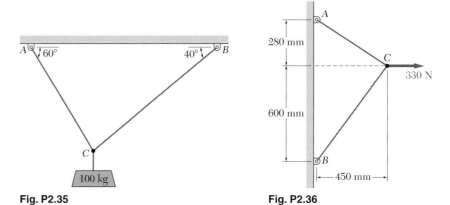

**Fig. P2.35**

**Fig. P2.36**

**2.37** Two weights are suspended from the three-cable system shown. Determine the tensions in cables *AB*, *BC*, and *CD*, and the angle, *α*, as required for equilibrium.

**2.38** A tent pole *BC* sits *on* the ground (not buried, free to pivot) at *C*, held vertical by a guy line *AB*. The rain fly of the tent exerts a 120-N force, as shown at *B*. Determine the tension in cable *AB* and the force developed along the pole *BC*.

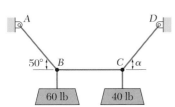

**Fig. P2.37**

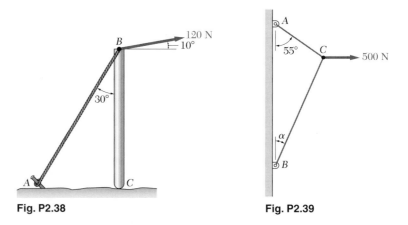

**Fig. P2.38**

**Fig. P2.39**

**2.39** Two cables are tied together at *C* and loaded as shown. Knowing that *α* = 25°, determine the tension in *AC* and *BC*.

**2.40** Two forces **P** and **Q** of magnitude $P = 600$ lb and $Q = 800$ lb are applied as shown to an aircraft connection. Knowing that the connection is in equilibrium, determine the tension in rods $A$ and $B$.

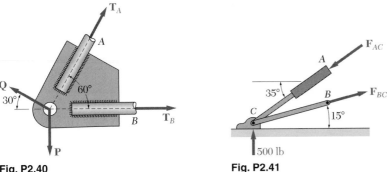

**Fig. P2.40**

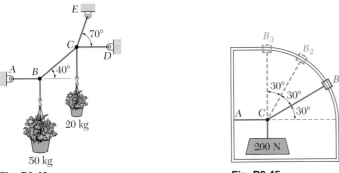

**Fig. P2.41**

**2.41** An outrigger used to stabilize an electric utility's service truck is shown. It consists of a solid member $BC$, a hydraulic cylinder $AC$, and a foot designed to sit securely without slipping on the ground. If the ground exerts a force of 500 lb upward on the pad at $C$, determine the forces in members $AC$ and $BC$. (*Note.* The foot shown here is idealized, with the members intersecting at a single point $C$ to ensure that the force system is concurrent. This is not always the case on outriggers.)

**2.42** Two cables are tied together at $A$ and loaded as shown. Knowing that $P = 640$ N, determine the tension in each cable.

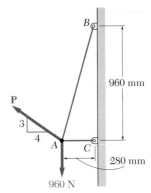

**Fig. P2.42**

**2.43** Two masses, 50 kg and 20 kg, are suspended from the cable system shown. Determine the tensions in cables $AB$, $BC$, $CD$, and $CE$.

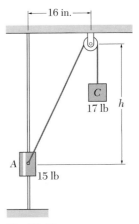

**Fig. P2.46**

**Fig. P2.43**

**Fig. P2.45**

**2.44** For the cables of Prob. 2.39, find the value of $\alpha$ for which the tension is as small as possible (*a*) in cable $BC$, (*b*) in both cables simultaneously. In each case determine the tension in both cables.

**2.45** Two cables, $AC$ and $BC$, suspend a 200-N weight. Cable $BC$'s position $B$ can be varied, as shown. (*a*) Determine the tensions in cables $BC$ and $AC$ at all three positions. (*b*) Why does $T_{AC}$ vary? (*c*) Is there a position of the cables that seems to be a "better" design?

**2.46** The 15-lb collar $A$ may slide on a frictionless vertical rod and is connected as shown to a 17-lb counterweight $C$. Determine the value of $h$ for which the system is in equilibrium.

**2.47** A movable bin and its contents weigh 960 lb. Determine the shortest chain sling *ACB* which may be used to lift the loaded bin if the tension in the chain is not to exceed 730 lb.

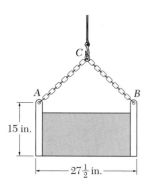

**Fig. P2.47**

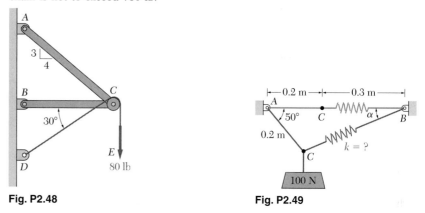

**Fig. P2.48**

**Fig. P2.49**

**2.48** Two slender links, *AC* and *BC*, extend from a wall and support a small pulley at *C*. A cable, tied at *D*, passes over the small pulley at *C* and experiences an 80-lb force at its free end *E*. If the pulley is frictionless and of negligible size, determine the forces developed in members *AC* and *BC*.

**2.49** An inextensible (cannot stretch) cable *AC* and spring *CB* are originally horizontal. A 100-N weight is attached to point *C*, and the system is slowly lowered until it reaches static equilibrium at the new position. If the unstretched length of the spring is 0.3 m, determine the spring constant, *k*.

**2.50** The force **P** is applied to a small wheel which rolls on the cable *ACB*. Knowing that the tension in both parts of the cable is 750 N, determine the magnitude and direction of **P**.

**2.51** A 600-lb crate is to be supported by the rope-and-pulley arrangement shown. Determine the magnitude and direction of the force **F** which should be exerted on the free end of the rope.

*2.52** The collar *A* may slide freely on the horizontal frictionless rod. The spring attached to the collar has a constant of 10 lb/in. and is undeformed when the collar is directly below support *B*. Determine the magnitude of the force **P** required to maintain equilibrium when (*a*) *c* = 9 in., (*b*) *c* = 16 in.

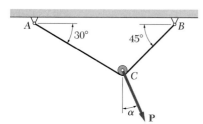

**Fig. P2.50**

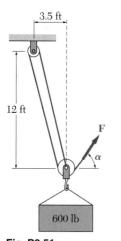

**Fig. P2.51**

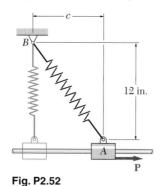

**Fig. P2.52**

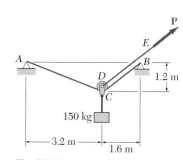

**Fig. P2.53**

*2.53   A 150-kg block is attached to a small pulley which may roll on the cable *ACB*. The pulley and load are held in the position shown by a second cable *DE* which is parallel to the portion *CB* of the main cable. Determine (*a*) the tension in cable *ACB*, (*b*) the tension in cable *DE*. Neglect the radius of the pulleys and the weight of the cables.

## FORCES IN SPACE

### 2.12. RECTANGULAR COMPONENTS OF A FORCE IN SPACE

The problems considered in the first part of this chapter involved only two dimensions; they could be formulated and solved in a single plane. In this section and in the remaining sections of the chapter, we shall discuss problems involving the three dimensions of space.

Consider a force **F** acting at the origin *O* of the system of rectangular coordinates *x*, *y*, *z*. To define the direction of **F**, we may draw the vertical plane *OBAC* containing **F** and shown in Fig. 2.30*a*. This plane passes through the vertical *y* axis; its orientation is defined by the angle $\phi$ it forms with the *xy* plane, while the direction of **F** within the plane is defined by the angle $\theta_y$ that **F** forms with the *y* axis. The force **F** may be resolved into a vertical component $\mathbf{F}_y$ and a horizontal component $\mathbf{F}_h$; this operation, shown in Fig. 2.30*b*, is carried out in plane *OBAC* according to the rules developed in the first part of the chapter. The corresponding scalar components are

$$F_y = F \cos \theta_y \qquad F_h = F \sin \theta_y \tag{2.10}$$

But $\mathbf{F}_h$ may be resolved into two rectangular components $\mathbf{F}_x$ and $\mathbf{F}_z$ along the *x* and *z* axes, respectively. This operation, shown in Fig. 2.30*c*, is carried out in the *xz* plane. We obtain the following expressions for the corresponding scalar components:

$$\begin{aligned} F_x &= F_h \cos \phi = F \sin \theta_y \cos \phi \\ F_z &= F_h \sin \phi = F \sin \theta_y \sin \phi \end{aligned} \tag{2.11}$$

The given force **F** has thus been resolved into three rectangular vector components $\mathbf{F}_x$, $\mathbf{F}_y$, $\mathbf{F}_z$, directed along the three coordinate axes.

Applying the Pythagorean theorem to the triangles *OAB* and *OCD* of Fig. 2.30, we write

$$\begin{aligned} F^2 &= (OA)^2 = (OB)^2 + (BA)^2 = F_y^2 + F_h^2 \\ F_h^2 &= (OC)^2 = (OD)^2 + (DC)^2 = F_x^2 + F_z^2 \end{aligned}$$

Eliminating $F_h^2$ from these two equations and solving for *F*, we obtain the following relation between the magnitude of **F** and its rectangular scalar components:

$$F = \sqrt{F_x^2 + F_y^2 + F_z^2} \tag{2.12}$$

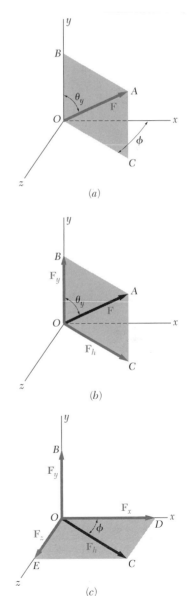

(*a*)

(*b*)

(*c*)

**Fig. 2.30**

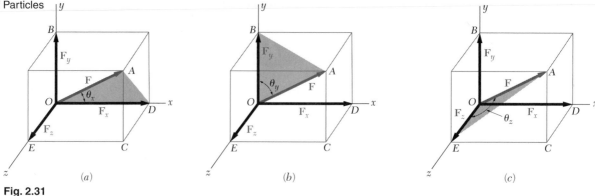

**Fig. 2.31**

The relationship existing between the force **F** and its three components $F_x$, $F_y$, $F_z$ is more easily visualized if a "box" having $F_x$, $F_y$, $F_z$ for edges is drawn as shown in Fig. 2.31. The force **F** is then represented by the diagonal $OA$ of this box. Figure 2.31$b$ shows the right triangle $OAB$ used to derive the first of the formulas (2.10): $F_y = F \cos \theta_y$. In Fig. 2.31$a$ and $c$, two other right triangles have also been drawn: $OAD$ and $OAE$. These triangles are seen to occupy in the box positions comparable with that of triangle $OAB$. Denoting by $\theta_x$ and $\theta_z$, respectively, the angles that **F** forms with the $x$ and $z$ axes, we may derive two formulas similar to $F_y = F \cos \theta_y$. We thus write

$$F_x = F \cos \theta_x \qquad F_y = F \cos \theta_y \qquad F_z = F \cos \theta_z \qquad (2.13)$$

The three angles $\theta_x$, $\theta_y$, $\theta_z$ define the direction of the force **F**; they are more commonly used for this purpose than the angles $\theta_y$ and $\phi$ introduced at the beginning of this section. The cosines of $\theta_x$, $\theta_y$, $\theta_z$ are known as the direction cosines of the force **F**.

**Example 1.** A force of 500 N forms angles of 60, 45, and 120°, respectively, with the $x$, $y$, and $z$ axes. Find the components $F_x$, $F_y$, and $F_z$ of the force.

Substituting $F = 500$ N, $\theta_x = 60°$, $\theta_y = 45°$, $\theta_z = 120°$ into formulas (2.13), we write

$$F_x = (500 \text{ N}) \cos 60° = +250 \text{ N}$$
$$F_y = (500 \text{ N}) \cos 45° = +354 \text{ N}$$
$$F_z = (500 \text{ N}) \cos 120° = -250 \text{ N}$$

As in the case of two-dimensional problems, the plus sign indicates that the component has the same sense as the corresponding axis, and the minus sign that it has the opposite sense.

The angle a force **F** forms with an axis should be measured from the positive side of the axis and will always be comprised between 0 and 180°. An angle $\theta_x$ smaller than 90° (acute) indicates that **F** (assumed attached to

$O$) is on the same side of the $yz$ plane as the positive $x$ axis; $\cos \theta_x$ and $F_x$ will then be positive. An angle $\theta_x$ larger than 90° (obtuse) would indicate that **F** is on the other side of the $yz$ plane; $\cos \theta_x$ and $F_x$ would then be negative. In Example 1 the angles $\theta_x$ and $\theta_y$ are acute, while $\theta_z$ is obtuse: consequently, $F_x$ and $F_y$ are positive, while $F_z$ is negative.

It should also be observed that the values of the three angles $\theta_x$, $\theta_y$, $\theta_z$ are not independent. Substituting for $F_x$, $F_y$, $F_z$ from (2.13) into (2.12), we find that the following identity must be satisfied:

$$\cos^2 \theta_x + \cos^2 \theta_y + \cos^2 \theta_z = 1 \tag{2.14}$$

In Example 1, for instance, once the values $\theta_x = 60°$ and $\theta_y = 45°$ have been selected, the value of $\theta_z$ *must* be equal to 60° or 120° in order to satisfy identity (2.14).

When the components $F_x$, $F_y$, $F_z$ of a force **F** are given, the magnitude $F$ of the force is obtained from (2.12).[†] The relations (2.13) may then be solved for the direction cosines,

$$\cos \theta_x = \frac{F_x}{F} \qquad \cos \theta_y = \frac{F_y}{F} \qquad \cos \theta_z = \frac{F_z}{F} \tag{2.15}$$

and the angles $\theta_x$, $\theta_y$, $\theta_z$ characterizing the direction of **F** may be found.

**Example 2.**   A force **F** has the components $F_x = 20$ lb, $F_y = -30$ lb, $F_z = 60$ lb. Determine its magnitude $F$ and the angles $\theta_x$, $\theta_y$, $\theta_z$ it forms with the axes of coordinates.

From formula (2.12) we obtain[†]

$$F = \sqrt{F_x^2 + F_y^2 + F_z^2}$$
$$= \sqrt{(20 \text{ lb})^2 + (-30 \text{ lb})^2 + (60 \text{ lb})^2}$$
$$= \sqrt{4900} \text{ lb} = 70 \text{ lb}$$

Substituting the values of the components and magnitude of **F** into Eqs. (2.15), we write

$$\cos \theta_x = \frac{F_x}{F} = \frac{20 \text{ lb}}{70 \text{ lb}} \qquad \cos \theta_y = \frac{F_y}{F} = \frac{-30 \text{ lb}}{70 \text{ lb}} \qquad \cos \theta_z = \frac{F_z}{F} = \frac{60 \text{ lb}}{70 \text{ lb}}$$

Calculating successively each quotient and its arc cosine, we obtain

$$\theta_x = 73.4° \qquad \theta_y = 115.4° \qquad \theta_z = 31.0°$$

The computations indicated may easily be carried out with a calculator.

---

[†] With a calculator programmed to convert rectangular coordinates into polar coordinates, the following procedure will be found more expeditious for computing $F$: First determine $F_h$ from its two rectangular components $F_x$ and $F_z$ (Fig. 2.30c), then determine $F$ from its two rectangular components $F_h$ and $F_y$ (Fig. 2.30b). The actual order in which the three components $F_x$, $F_y$, $F_z$ are entered is immaterial.

## 2.13. FORCE DEFINED BY ITS MAGNITUDE AND TWO POINTS ON ITS LINE OF ACTION

Another method of defining the direction of a force **F** acting at the origin *O* of the system of coordinates is to specify a second point *N* through which the line of action of **F** passes (Fig. 2.32). Through *N* we may draw lines parallel to the axes of coordinates and thus define a new box. The sides of this box, denoted, respectively, by $d_x$, $d_y$, $d_z$, measure the components of the vector joining *O* to *N* in directions parallel to the coordinate axes. This vector is denoted by $\overrightarrow{ON}$. Since, in the case considered, *O* is the origin of the system of coordinates, the components $d_x$, $d_y$, $d_z$ are also the coordi-

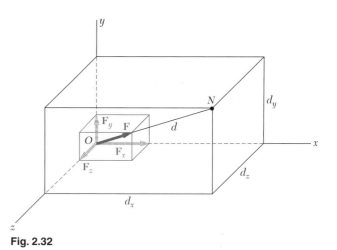

**Fig. 2.32**

nates of point *N*. Observing that the angles that $\overrightarrow{ON}$ forms with the axes of coordinates are $\theta_x$, $\theta_y$, $\theta_z$, and denoting by *d* the distance from *O* to *N*, we proceed as in Sec. 2.12, and write

$$d_x = d \cos \theta_x \qquad d_y = d \cos \theta_y \qquad d_z = d \cos \theta_z \qquad (2.16)$$

$$d = \sqrt{d_x^2 + d_y^2 + d_z^2} \qquad (2.17)$$

Dividing member by member the relations (2.13) and (2.16), we obtain

$$\frac{F_x}{d_x} = \frac{F_y}{d_y} = \frac{F_z}{d_z} = \frac{F}{d} \qquad (2.18)$$

Formula (2.18) states that the components and magnitude of the force **F** are, respectively, proportional to the components and magnitude of the vector $\overrightarrow{ON}$. This result could have been accepted intuitively by observing that the two boxes shown in Fig. 2.32 have the same shape.

The relations (2.18) considerably simplify the determination of the components of a force **F** of given magnitude $F$ and acting at $O$ when the coordinates $d_x$, $d_y$, $d_z$ of a point $N$ of its line of action are also given. From (2.17) we may determine the distance $d$ from $O$ to $N$. Carrying $F$, $d_x$, $d_y$, $d_z$, and $d$ into (2.18), we obtain the components $F_x$, $F_y$, $F_z$ by simple proportions. This computation is easily performed with a calculator, as shown in Sample Prob. 2.7.

The angles $\theta_x$, $\theta_y$, $\theta_z$ that **F** forms with the coordinate axes may then be obtained from Eqs. (2.15). They may also be determined directly from the components and magnitude of the vector $\overrightarrow{ON}$ by solving Eqs. (2.16) for the direction cosines of line $ON$:

$$\cos \theta_x = \frac{d_x}{d} \qquad \cos \theta_y = \frac{d_y}{d} \qquad \cos \theta_z = \frac{d_z}{d} \qquad (2.19)$$

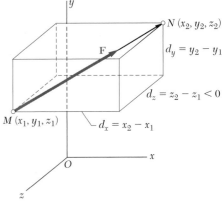

Fig. 2.33

When the line of action of the force **F** is defined by two points $M$ and $N$, neither of which is at the origin (Fig. 2.33), the components $d_x$, $d_y$, $d_z$ are equal to the difference between the coordinates of $M$ and $N$. These components should have the same sign as the corresponding force components. The sign of $d_x$, therefore, will be positive if the $x$ coordinate increases when the line $MN$ is described in the sense defined by **F** (from $M$ to $N$ in the case represented in Fig. 2.33); it will be negative if the $x$ coordinate decreases. The signs of $d_y$ and $d_x$ are determined in a similar way.

## 2.14. ADDITION OF CONCURRENT FORCES IN SPACE

We shall determine the resultant **R** of two or more forces in space by summing their rectangular components. Graphical or trigonometric methods are generally not practical in the case of forces in space.

The method followed here is similar to that used in Sec. 2.8 with coplanar forces. First we resolve each of the given forces into rectangular components. Adding algebraically all $x$ components, we obtain the component $R_x$ of the resultant; proceeding in the same manner with the $y$ and $z$ components, we write

$$R_x = \Sigma F_x \qquad R_y = \Sigma F_y \qquad R_z = \Sigma F_z \qquad (2.20)$$

The magnitude of the resultant and the angles $\theta_x$, $\theta_y$, $\theta_z$ it forms with the axes of coordinates are obtained by the method of Sec. 2.12. We write

$$R = \sqrt{R_x^2 + R_y^2 + R_z^2} \qquad (2.21)$$

$$\cos \theta_x = \frac{R_x}{R} \qquad \cos \theta_y = \frac{R_y}{R} \qquad \cos \theta_z = \frac{R_z}{R} \qquad (2.22)$$

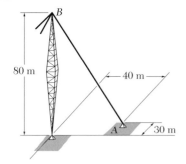

# SAMPLE PROBLEM 2.7

A tower guy wire is anchored by means of a bolt at $A$. The tension in the wire is 2500 N. Determine (a) the components $F_x$, $F_y$, $F_z$ of the force acting on the bolt, (b) the angles $\theta_x$, $\theta_y$, $\theta_z$ defining the direction of the force.

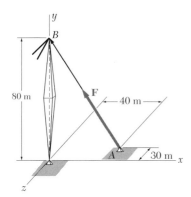

**a. Components of the Force.**   The line of action of the force acting on the bolt passes through $A$ and $B$, and the force is directed from $A$ to $B$. The components of the vector $\overrightarrow{AB}$, which has the same direction as the force, are

$$d_x = -40 \text{ m} \qquad d_y = +80 \text{ m} \qquad d_z = +30 \text{ m}$$

The total distance from $A$ to $B$ is

$$AB = d = \sqrt{d_x^2 + d_y^2 + d_z^2} = 94.3 \text{ m}$$

Since the components of the force **F** are proportional to the components of the vector $\overrightarrow{AB}$, we write

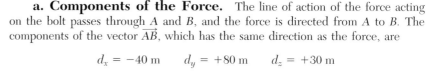

$$\frac{F_x}{-40 \text{ m}} = \frac{F_y}{+80 \text{ m}} = \frac{F_z}{+30 \text{ m}} = \frac{2500 \text{ N}}{94.3 \text{ m}}$$

*With a calculator* we first compute and store the value of the last quotient. Multiplying this value successively by $-40$, $+80$, and $+30$, we obtain

$$F_x = -1060 \text{ N} \qquad F_y = +2120 \text{ N} \qquad F_z = +795 \text{ N} \quad \blacktriangleleft$$

**b. Direction of the Force.**   Using Eqs. (2.15), we write

$$\cos \theta_x = \frac{F_x}{F} = \frac{-1060 \text{ N}}{2500 \text{ N}} \qquad \cos \theta_y = \frac{F_y}{F} = \frac{+2120 \text{ N}}{2500 \text{ N}}$$

$$\cos \theta_z = \frac{F_z}{F} = \frac{+795 \text{ N}}{2500 \text{ N}}$$

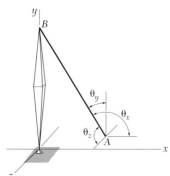

Calculating successively each quotient and its arc cosine, we obtain

$$\theta_x = 115.1° \qquad \theta_y = 32.0° \qquad \theta_z = 71.5° \quad \blacktriangleleft$$

(*Note.* This result could have been obtained by using the components and magnitude of the vector $\overrightarrow{AB}$ rather than those of the force **F**.)

# SAMPLE PROBLEM 2.8

A precast-concrete wall section is temporarily held by the cables shown. Knowing that the tension is 840 lb in cable $AB$ and 1200 lb in cable $AC$, determine the magnitude and direction of the resultant of the forces exerted by cables $AB$ and $AC$ on stake $A$.

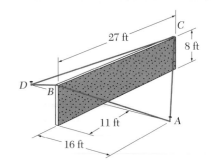

**Solution.** The force exerted by each cable on stake $A$ will be resolved into $x$, $y$, and $z$ components. We first determine the components and magnitude of the vectors $\overrightarrow{AB}$ and $\overrightarrow{AC}$, measuring them from $A$ toward the wall section.

$$\overrightarrow{AB}: \quad d_x = -16 \text{ ft} \quad d_y = +8 \text{ ft} \quad d_z = +11 \text{ ft} \quad d = 21 \text{ ft}$$
$$\overrightarrow{AC}: \quad d_x = -16 \text{ ft} \quad d_y = +8 \text{ ft} \quad d_z = -16 \text{ ft} \quad d = 24 \text{ ft}$$

In order to determine the force components, it is necessary to express that force and distance components are proportional. This is most efficiently done in tabular form. The information entered in the table includes the components and the magnitude of the distances as well as the magnitude of the given forces. The force components are then found by proportions from Eqs. (2.18) and added to obtain the components $R_x$, $R_y$, and $R_z$ of the resultant.

| Force | Distance Components, ft | | | $d$, ft | Force Components, lb | | | $F$, lb |
|---|---|---|---|---|---|---|---|---|
| | $d_x$ | $d_y$ | $d_z$ | | $F_x$ | $F_y$ | $F_z$ | |
| Cable $AB$ | $-16$ | $+8$ | $+11$ | 21 | $-640$ | $+320$ | $+440$ | 840 |
| Cable $AC$ | $-16$ | $+8$ | $-16$ | 24 | $-800$ | $+400$ | $-800$ | 1200 |
| | | | | | $R_x =$ $-1440$ | $R_y =$ $+720$ | $R_z =$ $-360$ | |

The magnitude and direction of the resultant are now determined:

$$R = \sqrt{R_x^2 + R_y^2 + R_z^2} = \sqrt{(-1440)^2 + (720)^2 + (-360)^2}$$
$$R = 1650 \text{ lb} \quad \blacktriangleleft$$

From Eqs. (2.22) we obtain

$$\cos\theta_x = \frac{R_x}{R} = \frac{-1440 \text{ lb}}{1650 \text{ lb}} \qquad \cos\theta_y = \frac{R_y}{R} = \frac{+720 \text{ lb}}{1650 \text{ lb}}$$

$$\cos\theta_z = \frac{R_z}{R} = \frac{-360 \text{ lb}}{1650 \text{ lb}}$$

Calculating successively each quotient and its arc cosine, we have

$$\theta_x = 150.8° \qquad \theta_y = 64.1° \qquad \theta_z = 102.6° \quad \blacktriangleleft$$

**45**

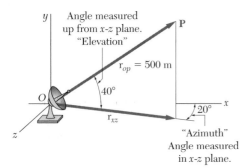

Fig. P2.54

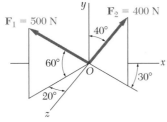

Fig. P2.55

# PROBLEMS

**2.54**   One way to represent vectors in three dimensions is by what is essentially spherical coordinates. In this type of problem, the length of the vector is given plus two angles. One angle is from the vector either to the vertical axis or to the plane (in this case 40° to the plane). The other angle (in this case 20°) is measured in the horizontal plane from (usually) the $x$ axis to the projection ($\mathbf{r}_{xy}$ or $\mathbf{F}_{proj}$) of the vector in the $x$-$z$ plane. In the satellite industry, and in surveying, an angle measured in a horizontal plane is an "azimuth," and an angle measured upward from the horizontal plane is an "elevation." For the radar dish shown (tracking a low-flying small plane 500 m away), determine the $x$, $y$, and $z$ components of the 500-m distance vector. (*Hint.* First find the projection $xz$ of the vector into the $x$-$z$ plane, and then project this projection onto the $x$ and $z$ axes using the 20° angle.)

**2.55**   Shown are two forces, $\mathbf{F}_1$ and $\mathbf{F}_2$, expressed in (essentially) spherical coordinates. Determine (*a*) the $x$, $y$, and $z$ components of each force, (*b*) the force resultant, in Cartesian form, of the two forces.

**2.56**   Shown is the same 49-lb vector in three different coordinate systems. In (*a*), the vector is defined as acting along a line whose distance components are given. Write this vector in the three coordinate systems: (*a*) Cartesian form (from the line coordinate information), (*b*) as a force magnitude plus $\theta$ and $\phi$ angles; (*c*) as a force magnitude plus three coordinate direction angles $\theta_x$, $\theta_y$, and $\theta_z$.

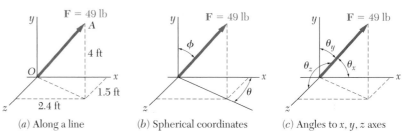

(*a*) Along a line          (*b*) Spherical coordinates          (*c*) Angles to $x, y, z$ axes

Fig. P2.56

**2.57**   Shown is the same 90-N vector in three different coordinate systems. In the center picture, the vector is defined in (essentially) spherical coordinates. Write this vector in the two other coordinate systems: (*a*) Cartesian form ($x$, $y$, and $z$ components), (*b*) as a force magnitude plus three coordinate direction angles $\theta_x$, $\theta_y$, and $\theta_z$.

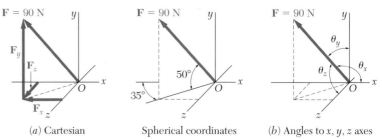

(*a*) Cartesian          Spherical coordinates          (*b*) Angles to $x, y, z$ axes

Fig. P2.57

**2.58**  Shown is the same 120-N vector in three different coordinate systems. In (c), the vector is defined by two coordinate direction angles, $\theta_x$ and $\theta_z$. First, determine $\theta_y$. Then write this vector in the two other coordinate systems: (a) Cartesian form (x, y, and z components), (b) as a force magnitude plus an azimuth $\theta$ and an angle $\phi$ from the y axis. (*Note.* The three coordinate direction angles $\theta_x$, $\theta_y$, and $\theta_z$ are related by the equation ($\cos^2 \theta_x + \cos^2 \theta_y + \cos^2 \theta_z = 1$). If two angles are known, with this equation you can calculate the third. Just remember to check, based on the picture, whether the third angle (in this case $\theta_y$) is less than or greater than 90°.)

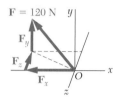

(a) Cartesian

(b) Spherical coordinates

(c) Angles to x, y, z axes

**Fig. P2.58**

**2.59**  Cable AC is 70 ft long and the tension in that cable is 5250 lb. Determine (a) the x, y, and z components of the force exerted by the cable on the anchor C, (b) the angles $\theta_x$, $\theta_y$, and $\theta_z$ defining the direction of the force exerted at C.

**2.60**  The components of a force are known to be $F_x = +650$ N, $F_y = -320$ N, and $F_z = +760$ N. Determine the force magnitude and the angles $\theta_x$, $\theta_y$, and $\theta_z$ for the force.

**2.61**  The components of a force are known to be $F_x = +580$ lb, $F_y = +690$ lb, and $F_z = -300$ lb. Determine the force magnitude and the angles $\theta_x$, $\theta_y$, and $\theta_z$ for the force.

**2.62**  A force acts at the origin in a direction defined by the angles $\theta_x = 75°$ and $\theta_z = 130°$. Knowing that the y component of the force is +300 lb, determine (a) the other components and the magnitude of the force, (b) the value of $\theta_y$.

**2.63**  A force acts at the origin in a direction defined by the angles $\theta_y = 55°$ and $\theta_z = 45°$. Knowing that the x component of the force is −500 N, determine (a) the other components and the magnitude of the force, (b) the value of $\theta_x$.

**2.64**  Shown are two vectors, $\mathbf{F}_1$ and $\mathbf{F}_2$. Determine the resultant vector in Cartesian form.

**2.65**  Cable AB is 80 ft long and the tension in that cable is 4000 lb. Determine (a) the x, y, and z components of the force exerted by the cable on the anchor B, (b) the angles $\theta_x$, $\theta_y$, and $\theta_z$ defining the direction of the force exerted at B.

**2.66**  Knowing that the tension in cable AB is 285 lb, determine the components of the force exerted on the plate at B.

**2.67**  Knowing that the tension in cable AC is 426 lb, determine the components of the force exerted on the plate at C.

**2.68**  Knowing that the tension is 285 lb in cable AB and 426 lb in cable AC, determine the magnitude and direction of the resultant of the forces exerted at A by the two cables.

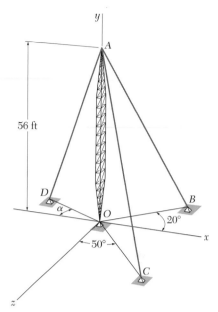

**Fig. P2.59 and P2.65**

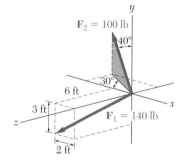

**Fig. P2.64**

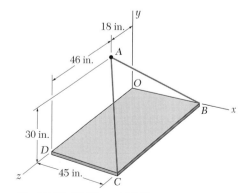

**Fig. P2.66, P2.67, and P2.68**

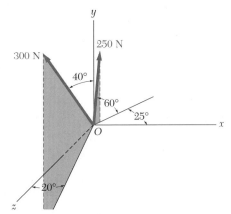

**Fig. P2.69 and P2.70**

**2.69**   The 300 N vector is given in spherical coordinates. (*a*) Write this vector in Cartesian vector form, (*b*) determine the angles $\theta_x$, $\theta_y$, and $\theta_z$ defining the direction of the force.

**2.70**   Determine the resultant of the two forces shown.

**2.71**   Shown are two vectors, $\mathbf{F}_1$ and $\mathbf{F}_2$. Determine the resultant vector in Cartesian form. (*Remember:* The three coordinate direction angles $\theta_x$, $\theta_y$, and $\theta_z$ are related by the equation ($\cos^2 \theta_x + \cos^2 \theta_y + \cos^2 \theta_z = 1$). If two angles are known, with this equation you can calculate the third. Just remember to check, based on the picture, whether the third angle [in this case $\theta_x$] is less than or greater than 90°.)

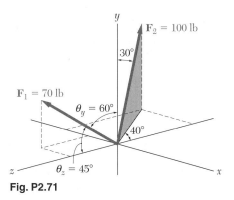

**Fig. P2.71**

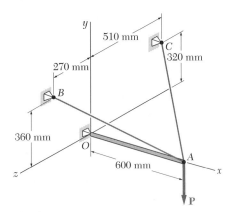

**Fig. P2.72**

**2.72**   The boom *OA* carries a load **P** and is supported by two cables as shown. Knowing that the tension in cable *AB* is 850 N and that the resultant of the load **P** and of the forces exerted at *A* by the two cables must be directed along *OA*, determine the tension in cable *AC*.

**2.73**   Shown are two vectors, $\mathbf{F}_1$ and $\mathbf{F}_2$, each expressed as forces along lines whose line coordinate distances are given. Determine the resultant vector in Cartesian form.

## 2.15. EQUILIBRIUM OF A PARTICLE IN SPACE

According to the definition given in Sec. 2.9, a particle *A* is in equilibrium if the resultant of all the forces acting on *A* is zero. The components $R_x$, $R_y$, $R_z$ of the resultant are given by the relations (2.20); expressing that the components of the resultant are zero, we write

$$\Sigma F_x = 0 \qquad \Sigma F_y = 0 \qquad \Sigma F_z = 0 \tag{2.23}$$

Equations (2.23) represent the necessary and sufficient conditions for the equilibrium of a particle in space. They may be used to solve problems dealing with the equilibrium of a particle involving no more than three unknowns.

To solve such problems, we first should draw a free-body diagram showing the particle in equilibrium and *all* the forces acting on it. We may then write the equations of equilibrium (2.23) and solve them for three unknowns. In the more common types of problems, these unknowns will represent (1) the three components of a single force or (2) the magnitude of three forces each of known direction.

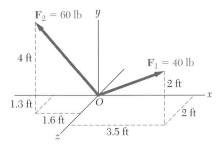

**Fig. P2.73**

# SAMPLE PROBLEM 2.9

A 200-kg cylinder is hung by means of two cables $AB$ and $AC$, which are attached to the top of a vertical wall. A horizontal force **P** perpendicular to the wall holds the cylinder in the position shown. Determine the magnitude of **P** and the tension in each cable.

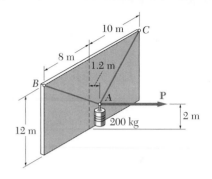

**Solution.** Point $A$ is chosen as a free body; point $A$ is subjected to four forces as shown. The magnitudes of three of the forces, **P**, $\mathbf{T}_{AB}$, and $\mathbf{T}_{AC}$, are unknown; the magnitude of the weight **W** is found to be

$$W = mg = (200 \text{ kg})(9.81 \text{ m/s}^2) = 1962 \text{ N}$$

The $x$, $y$, and $z$ components of each of the unknown forces must be expressed in terms of the unknown magnitudes $P$, $T_{AB}$, and $T_{AC}$. In the case of $\mathbf{T}_{AB}$ and $\mathbf{T}_{AC}$ it is necessary to consider first the corresponding distance components.

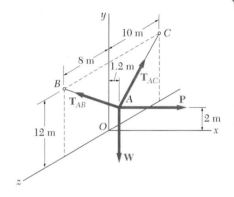

Cable $AB$ ($A$ to $B$):
$$d_x = -1.2 \text{ m} \qquad d_y = 12 - 2 = 10 \text{ m} \qquad d_z = 8 \text{ m} \qquad d = 12.86 \text{ m}$$

Cable $AC$ ($A$ to $C$):
$$d_x = -1.2 \text{ m} \qquad d_y = 12 - 2 = 10 \text{ m} \qquad d_z = -10 \text{ m} \qquad d = 14.19 \text{ m}$$

The force components are now obtained by proportions and the results are shown in tabular form.

| Force | Distance Components, m | | | $d,$ m | Force Components, N | | | Force, N |
|---|---|---|---|---|---|---|---|---|
| | $d_x$ | $d_y$ | $d_z$ | | $F_x$ | $F_y$ | $F_z$ | |
| $\mathbf{T}_{AB}$ | $-1.2$ | $+10$ | $+8$ | 12.86 | $-0.0933T_{AB}$ | $+0.778T_{AB}$ | $+0.622T_{AB}$ | $T_{AB}$ |
| $\mathbf{T}_{AC}$ | $-1.2$ | $+10$ | $-10$ | 14.19 | $-0.0846T_{AC}$ | $+0.705T_{AC}$ | $+0.705T_{AC}$ | $T_{AC}$ |
| **P** | | | | | $+P$ | 0 | 0 | $P$ |
| **W** | | | | | 0 | $-1962$ | 0 | $-1962$ |

***Equilibrium Equations.*** The equilibrium equations may be written directly by adding successively the force-components columns of the table.

$$\Sigma F_x = 0: \qquad -0.0933T_{AB} - 0.0846T_{AC} + P = 0$$
$$\Sigma F_y = 0: \qquad +0.778T_{AB} + 0.705T_{AC} - 1962 \text{ N} = 0$$
$$\Sigma F_z = 0: \qquad +0.622T_{AB} - 0.705T_{AC} = 0$$

Solving these equations we obtain

$$P = 235 \text{ N} \qquad T_{AB} = 1401 \text{ N} \qquad T_{AC} = 1236 \text{ N} \quad \blacktriangleleft$$

# PROBLEMS

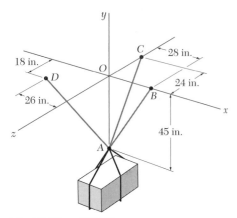

Fig. P2.74 and P2.75

**2.74** A crate is supported by three cables as shown. Determine the weight W of the crate, knowing that the tension in cable *AD* is 924 lb.

**2.75** A crate is supported by three cables as shown. Determine the weight W of the crate, knowing that the tension in cable *AB* is 1378 lb.

**2.76** Shown is a 100-N flower pot suspended from three cables. Determine the tensions in cables *AB*, *BC*, and *BD*.

**2.77** Shown is a 40-lb traffic light suspended from three cables. Determine the tensions in cables *AB*, *BC*, and *BD*.

**2.78** Shown is a 20-lb bucket suspended from three cables. Determine the tensions in cables *AB*, *BC* and *BD*.

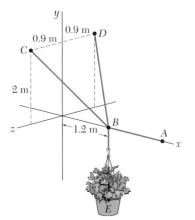

Fig. P2.76

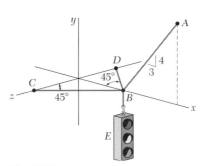

Fig. P2.77

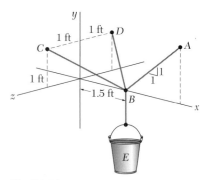

Fig. P2.78

**2.79** A force **F** is applied at point *B* in the cable system shown. If the cable tension in *BD* is known to be 79.2 N, Determine the tensions in cables *AB* and *BC*.

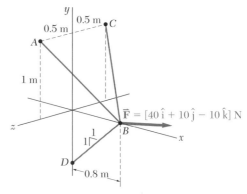

Fig. P2.79

**2.80** Shown is a cable system, suspended from a construction crane, lifting a 20-kN concrete pre-form into place. Determine the tensions in cables *AB*, *AC*, and *AD*. Assume the cables are massless; thus, *T* = 20 kN.

**2.81** Shown is a cable system supporting a 30-lb potted plant. Determine the tensions in cables *AB*, *BC*, and *BD*.

**2.82** A 12-lb circular plate of 7-in. radius is supported as shown by three wires, each of 25-in. length. Determine the tension in each wire, knowing that $\alpha = 30°$.

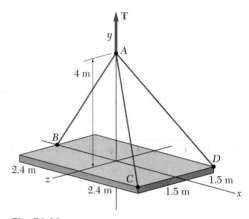

**Fig. P2.80**

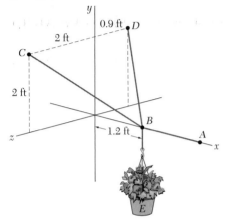

**Fig. P2.81**

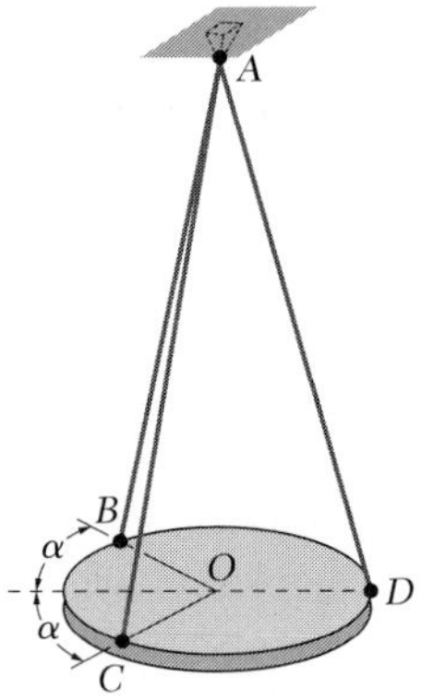

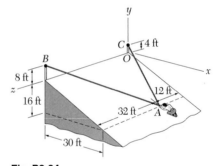

**Fig. P2.82**

**2.83** Solve Prob. 2.82, assuming that $\alpha = 45°$.

**2.84** In trying to move across a slippery icy surface, a 180-lb man uses two ropes *AB* and *AC*. Knowing that the force exerted on the man by the icy surface is perpendicular to that surface, determine the tension in each rope.

**2.85** Shown is a cable system supporting a 400-N load of lumber. Determine the tensions in cables *AB*, *BC*, and *BD*.

**Fig. P2.84**

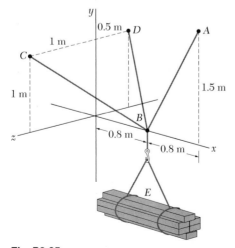

**Fig. P2.85**

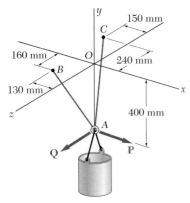

Fig. P2.86 and P2.88

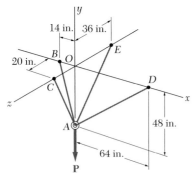

**Fig. P2.90**

**2.86** A container of weight $W = 400$ N is supported by cables $AB$ and $AC$, which are tied to ring $A$. Knowing that $Q = 0$, determine $(a)$ the magnitude of the force $\mathbf{P}$ which must be applied to the ring to maintain the container in the position shown, $(b)$ the corresponding values of the tension in cables $AB$ and $AC$.

**2.87** Solve Prob. 2.86, knowing that $Q = 80$ N.

**2.88** A container is supported by a single cable which passes through a frictionless ring $A$ and is attached to fixed points $B$ and $C$. Two forces $\mathbf{P}$ and $\mathbf{Q}$ are applied to the ring to maintain the container in the position shown. Knowing that the weight of the container is $W = 376$ N, determine the magnitudes of $\mathbf{P}$ and $\mathbf{Q}$. (*Hint.* The tension must be the same in portions $AB$ and $AC$ of the cable.)

**2.89** Determine the weight $W$ of the container of Prob. 2.88, knowing that $P = 164$ N.

**2.90** Cable $BAC$ passes through a frictionless ring $A$ and is attached to fixed supports at $B$ and $C$, while cables $AD$ and $AE$ are both tied to the ring and are attached, respectively, to supports at $D$ and $E$. Knowing that a 150-lb vertical load $\mathbf{P}$ is applied to ring $A$, determine the tension in each of the three cables.

**2.91** Knowing that the tension in cable $AE$ of Prob. 2.90 is 50 lb, determine $(a)$ the magnitude of the load $\mathbf{P}$, $(b)$ the tension in cables $BAC$ and $AD$.

**2.92** A 10-kg circular plate of 250-mm radius is supported as shown by three wires of equal length $L$. Knowing that $\alpha = 30°$, determine the smallest permissible value of the length $L$ if the tension is not to exceed 50 N in any of the wires.

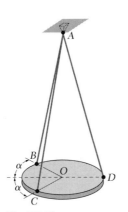

Fig. P2.92

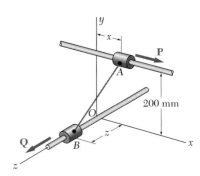

Fig. P2.93 and P2.94

**2.93** Collars $A$ and $B$ are connected by a 250-mm wire and may slide freely on frictionless rods. If a force $\mathbf{Q}$ of magnitude 100 N is applied to collar $B$ as shown, determine $(a)$ the tension in the wire when $x = 90$ mm, $(b)$ the corresponding magnitude of the force $\mathbf{P}$ required to maintain the equilibrium of the system.

**\*2.94** Collars $A$ and $B$ are connected by a 250-mm wire and may slide freely on frictionless rods. Determine the distances $x$ and $z$ for which the equilibrium of the system is maintained when $P = 200$ N and $Q = 100$ N.

# REVIEW AND SUMMARY
# FOR CHAPTER 2

In this chapter we have studied the effect of forces on particles, i.e., on bodies of such shape and size that all forces acting on them may be assumed applied at the same point.

Forces are *vector quantities;* they are characterized by a *point of application,* a *magnitude,* and a *direction,* and add according to the *parallelogram law* (Fig. 2.34). The magnitude and direction of the resultant **R** of two forces **P** and **Q** may be determined either graphically or by trigonometry, using successively the law of cosines and the law of sines [Sample Prob. 2.1].

**Resultant of two forces**

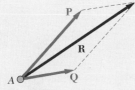

**Fig. 2.34**

Any given force acting on a particle may be resolved into two or more *components,* i.e., it may be replaced by two or more forces which have the same effect on the particle. A force **F** may be resolved into two components **P** and **Q** by drawing a parallelogram which has **F** for its diagonal; the components **P** and **Q** are then represented by the two adjacent sides of the parallelogram (Fig. 2.35) and may be determined either graphically or by trigonometry [Sec. 2.6].

**Components of a force**

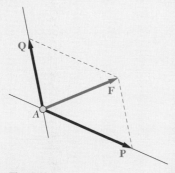

**Fig. 2.35**

A force **F** is said to have been resolved into two *rectangular components* if its components $\mathbf{F}_x$ and $\mathbf{F}_y$ are perpendicular to each other and directed along the $x$ and $y$ axes, respectively (Fig. 2.36). Denoting by $\theta$ the angle between **F** and the $x$ axis, we write

**Rectangular components**

$$F_x = F \cos \theta \qquad F_y = F \sin \theta \qquad (2.5)$$

where the scalar components $F_x$ and $F_y$ may be positive or negative [Sec. 2.7].

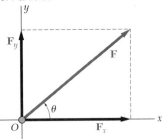

**Fig. 2.36**

Resultant of several coplanar forces

Forces in space

When the rectangular components $F_x$ and $F_y$ of a force **F** are given, the angle $\theta$ defining the direction of the force may be obtained by writing

$$\tan \theta = \frac{F_y}{F_x} \tag{2.6}$$

The magnitude $F$ of the force may then be obtained by solving one of the equations (2.5) for $F$ or by applying the Pythagorean theorem and writing

$$F = \sqrt{F_x^2 + F_y^2} \tag{2.7}$$

When *three or more coplanar forces* act on a particle, the rectangular components of their resultant **R** may be obtained by adding algebraically the corresponding components of the given forces [Sec. 2.8]. We have

$$R_x = \Sigma F_x \qquad R_y = \Sigma F_y \tag{2.8}$$

The magnitude and direction of **R** may then be determined from relations similar to Eqs. (2.6) and (2.7) [Sample Prob. 2.3].

A force **F** in *three-dimensional space* may be resolved into rectangular components $F_x$, $F_y$, and $F_z$ [Sec. 2.12]. Denoting by $\theta_x$, $\theta_y$, and $\theta_z$, respectively, the angles that **F** forms with the $x$, $y$, and $z$ axes (Fig. 2.37), we have

$$F_x = F \cos \theta_x \qquad F_y = F \cos \theta_y \qquad F_z = F \cos \theta_z \tag{2.13}$$

The cosines of $\theta_x$, $\theta_y$, $\theta_z$ are known as the *direction cosines* of the force **F** and must satisfy the relation

$$\cos^2 \theta_x + \cos^2 \theta_y + \cos^2 \theta_z = 1 \tag{2.14}$$

When the rectangular components $F_x$, $F_y$, $F_z$ of a force **F** are given, the magnitude $F$ of the force is found by writing

$$F = \sqrt{F_x^2 + F_y^2 + F_z^2} \tag{2.12}$$

and the direction cosines of **F** are obtained from Eqs. (2.13). We have

$$\cos \theta_x = \frac{F_x}{F} \qquad \cos \theta_y = \frac{F_y}{F} \qquad \cos \theta_z = \frac{F_z}{F} \tag{2.15}$$

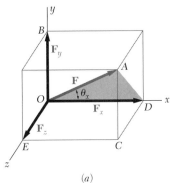

(a)

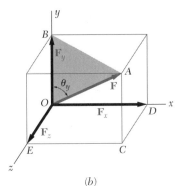

(b)

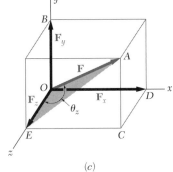

(c)

**Fig. 2.37**

When a force **F** is defined in three-dimensional space by its magnitude $F$ and two points $M$ and $N$ on its line of action [Sec. 2.13], its components $F_x$, $F_y$, $F_z$ may be obtained from the relations

$$\frac{F_x}{d_x} = \frac{F_y}{d_y} = \frac{F_z}{d_z} = \frac{F}{d} \tag{2.18}$$

where $d$, $d_x$, $d_y$, and $d_z$ represent, respectively, the magnitude and the components of the vector $\overrightarrow{MN}$ joining points $M$ and $N$ (Fig. 2.38) [Sample Prob. 2.7].

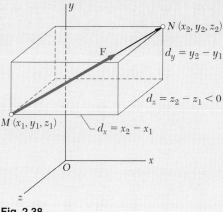

**Fig. 2.38**

When *two or more forces* act on a particle in *three-dimensional space*, the rectangular components of their resultant **R** may be obtained by adding algebraically the corresponding components of the given forces [Sec. 2.14]. We have

$$R_x = \Sigma F_x \qquad R_y = \Sigma F_y \qquad R_z = \Sigma F_z \tag{2.20}$$

The magnitude and direction of **R** may then be determined from relations similar to Eqs. (2.12) and (2.15) [Sample Prob. 2.8].

A particle is said to be in *equilibrium* when the resultant of all the forces acting on it is zero [Sec. 2.9]. The particle will then remain at rest (if originally at rest) or move with constant speed in a straight line (if originally in motion) [Sec. 2.10].

To solve a problem involving a *particle in equilibrium,* one first should draw a *free-body diagram* of the particle showing all the forces acting on it [Sec. 2.11]. If *only three coplanar forces* act on the particle, a *force triangle* may be drawn to express that the particle is in equilibrium. This triangle may be solved graphically or by trigonometry for no more than two unknowns [Sample Prob. 2.4]. If *more than three coplanar forces* are involved, the equations of equilibrium

$$\Sigma F_x = 0 \qquad \Sigma F_y = 0 \tag{2.9}$$

should be used and solved for no more than two unknowns [Sample Prob. 2.6].

Resultant of forces in space

Equilibrium of a particle
Free-body diagram

Equilibrium in space

When a particle is in *equilibrium in three-dimensional space* [Sec. 2.15], the three equations of equilibrium

$$\Sigma F_x = 0 \qquad \Sigma F_y = 0 \qquad \Sigma F_z = 0 \qquad (2.23)$$

should be used and solved for no more than three unknowns [Sample Prob. 2.9].

# REVIEW PROBLEMS

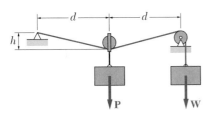

**Fig. P2.95**

**2.95** (*a*) Express the weight $W$ required to maintain equilibrium in terms of $P$, $d$, and $h$. (*b*) For $W = 500$ N, $P = 150$ N, and $d = 400$ mm, determine the value of $h$ consistent with equilibrium.

**2.96** A chain loop of length 1.25 m is placed around a $250 \times 250$ mm piece of timber. Knowing that the mass of the timber is 175 kg, determine the tension in the chain for each of the arrangements shown.

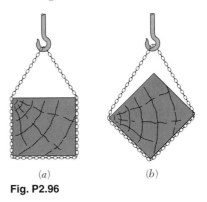

(*a*)          (*b*)

**Fig. P2.96**

**2.97** Knowing that the magnitude of the force **P** is 100 lb, determine the resultant of the three forces applied at $A$.

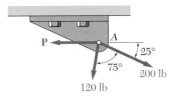

**Fig. P2.97 and P2.98**

**2.98** Determine the range of values of $P$ for which the resultant of the three forces applied at $A$ does not exceed 225 lb.

**2.99**  A cylinder of weight $W = 650$ N is supported by two cables $AC$ and $BC$, which are attached to the top of vertical posts. A horizontal force $\mathbf{P}$, perpendicular to the plane containing the posts, holds the cylinder in the position shown. Determine (*a*) the magnitude of $\mathbf{P}$, (*b*) the tension in each cable.

**2.100**  In Prob. 2.99, determine the angles $\theta_x$, $\theta_y$, and $\theta_z$ for the force exerted at $B$ by cable $BC$.

**2.101**  Knowing that $P = 400$ N, determine the tension in cables $AC$ and $BC$.

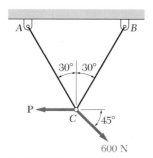

Fig. P2.101 and P2.102

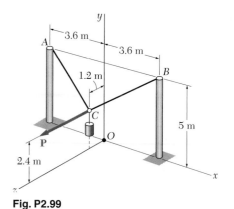

**Fig. P2.99**

**2.102**  Determine the range of values of $\mathbf{P}$ for which both cables remain taut.

**2.103**  The directions of the 75-lb forces may vary, but the angle between the forces is always 50°. Determine the value of $\alpha$ for which the resultant of the forces acting at $A$ is directed horizontally to the left.

**2.104**  Determine the resultant of the three forces acting at $A$ when (*a*) $\alpha = 0$, (*b*) $\alpha = 25°$.

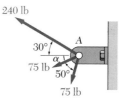

**Fig. P2.103 and P2.104**

**2.105**  An 8-kN force acts at the origin in a direction defined by the angles $\theta_y = 35°$ and $\theta_z = 65°$. It is also known that the $x$ component of the force is positive. Determine the value of $\theta_x$ and the components of the force.

**2.106**  Three wires are connected at point $D$, which is located 9 in. below the T-shaped pipe support $ABC$. Determine the tension in each wire when a 60-lb cylinder is suspended from point $D$ as shown.

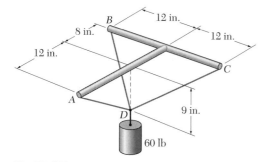

**Fig. P2.106**

# COMPUTER PROBLEMS

**2.C1** Write a computer program which can be used to determine the magnitude and direction of the resultant of $n$ forces which act at a fixed point $A$. Use this program to solve (a) Sample Prob. 2.3, (b) Prob. 2.27.

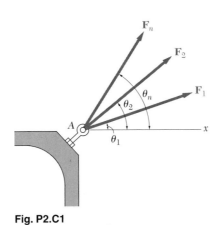

**Fig. P2.C1**

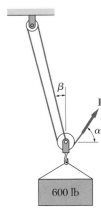

**Fig. P2.C2**

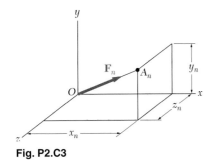

**Fig. P2.C3**

**2.C2** A 600-lb crate is supported by the rope-and-pulley arrangement shown. Write a computer program which can be used to determine, for a given value of $\beta$, the magnitude and direction of the force **F** which should be exerted on the free end of the rope. Use this program to calculate $F$ and $\alpha$ for values of $\beta$ from 0 to 30° at 5° intervals.

**2.C3** Write a computer program which can be used to calculate the magnitude and direction angles of the resultant of $n$ forces which are applied at the origin and act along lines $OA_n$. For the data given below, use this program to determine the magnitude and direction of (a) $\mathbf{F}_1 + \mathbf{F}_2$, (b) $\mathbf{F}_1 + \mathbf{F}_2 + \mathbf{F}_3$, (c) $\mathbf{F}_1 + \mathbf{F}_2 + \mathbf{F}_3 + \mathbf{F}_4$.

| | $F_n$, lb | $x_n$, in. | $y_n$, in. | $z_n$, in. |
|---|---|---|---|---|
| $\mathbf{F}_1$ | 100 | 20 | 10 | 8 |
| $\mathbf{F}_2$ | 150 | 25 | 15 | −12 |
| $\mathbf{F}_3$ | 200 | −9 | −3 | 16 |
| $\mathbf{F}_4$ | 50 | 6 | 14 | −18 |

**2.C4**   A cylinder of weight $W = 650$ N is supported by two cables $AC$ and $BC$, each 4.6-m long, which are attached to the top of vertical posts. A horizontal force $\mathbf{P}$, perpendicular to the plane containing the posts, holds the cylinder in the position shown. Write a computer program and use it to calculate the elevation $h$, the magnitude of $\mathbf{P}$, and the tension in each cable for values of $a$ from 0 to 2.8 m at 0.4-m intervals.

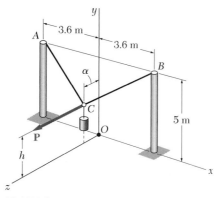

**Fig. P2.C4**

# Statics of Rigid Bodies in Two Dimensions

## 3.1. INTRODUCTION

In Chapter 2, it was assumed that each of the bodies considered could be treated as a single particle. Such a view, however, is not always possible, and a body, in general, should be treated as a combination of a large number of particles. The size of the body will have to be taken into consideration, as well as the fact that forces will act on different particles and thus will have different points of application.

Most of the bodies considered in elementary mechanics are assumed to be *rigid*, a *rigid body* being defined as one which does not deform. Actual structures and machines, however, are never absolutely rigid and deform under the loads to which they are subjected. But these deformations are usually small and do not appreciably affect the conditions of equilibrium or motion of the structure under consideration. They are important, though, as far as the resistance of the structure to failure is concerned, and are considered in the study of mechanics of materials.

In the first part of this chapter we shall study the effect of forces exerted on a rigid body and we shall learn how to replace a given system of forces by a simpler equivalent system. Our analysis will rest on the fundamental assumption that the effect of a given force on a rigid body remains unchanged if that force is moved along its line of action (*principle of transmissibility*). It follows that forces acting on a rigid body may be represented by *sliding vectors,* as indicated earlier in Sec. 2.3.

An important concept associated with the effect of a force on a rigid body is the *moment of a force about an axis* (Sec. 3.5). Another is the concept of a *couple*, i.e., the combination of two forces which have the same magnitude, parallel lines of action, and opposite sense (Sec. 3.7). As we shall see, any system of forces acting on a rigid body may be replaced by an equivalent system consisting of one force attached at a given point and one couple. This basic system is called a *force-couple system* (Sec. 3.11).

In this chapter, we shall be concerned only with two-dimensional structures subjected to coplanar forces; the study of three-dimensional bodies subjected to three-dimensional forces will be postponed until Chap. 4. We shall see in Sec. 3.11 that any system of coplanar forces may be further reduced to a single force, called the *resultant* of the system, or to a single couple, called the *resultant couple* of the system.

The second part of this chapter deals with the *equilibrium of a two-dimensional structure,* i.e., with the situation when the system of forces exerted on the structure reduces to zero (no resultant and no resultant couple). We shall see that *three equilibrium equations* may then be written (Sec. 3.13), stating, respectively, that the sum of the $x$ components of the forces, the sum of their $y$ components, and the sum of their moments about any given point $A$ must be zero.

In order to write the equations of equilibrium for a two-dimensional structure, it is essential to first identify correctly all the forces acting on that structure and draw the corresponding *free-body diagram* (Sec. 3.14). In addition to the forces *applied* to the structure, we shall consider the *reactions* exerted on the structure by its supports. We shall learn to associate a specific reaction with each type of support and determine whether the structure is properly supported, so that we may know in advance whether the equations of equilibrium may actually be solved for the unknown forces and reactions.

Two particular cases of the equilibrium of a rigid body will be given special attention: the equilibrium of a rigid body subjected to two forces (Sec. 3.18) and the equilibrium of a rigid body subjected to three forces (Sec. 3.19).

## EQUIVALENT SYSTEMS OF FORCES

### 3.2. EXTERNAL AND INTERNAL FORCES

Forces acting on rigid bodies may be separated into two groups: (1) *external forces,* (2) *internal forces.*

1. The *external forces* represent the action of other bodies on the rigid body under consideration. They are entirely responsible for the external behavior of the rigid body. They will either cause it to move or assure that it remains at rest. We shall be concerned only with external forces in this chapter and in Chaps. 4 and 5.

2. The *internal forces* are the forces which hold together the particles forming the rigid body. If the rigid body is structurally composed of several parts, the forces holding the component parts together are also defined as internal forces. Internal forces will be considered in Chaps. 6 and 7.

**Fig. 3.1**

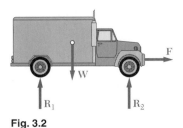

**Fig. 3.2**

As an example of external forces, we shall consider the forces acting on a disabled truck that men are pulling forward by means of a rope attached to the front bumper (Fig. 3.1). The external forces acting on the truck are shown in a *free-body diagram* (Fig. 3.2). Let us first consider the *weight* of the truck. Although it embodies the effect of the earth's pull on each of the particles forming the truck, the weight may be represented by the single force **W**. The *point of application* of this force, i.e., the point at which the force acts, is defined as the *center of gravity* of the truck. It will be seen in Chap. 5 how centers of gravity may be determined. The weight **W** tends to make the truck move vertically downward. In fact, it would actually cause the truck to move downward, i.e., to fall, if it were not for the presence of the ground. The ground opposes the downward motion of the truck by means of the reactions **R**$_1$ and **R**$_2$. These forces are exerted *by* the ground *on* the truck and must therefore be included among the external forces acting on the truck.

The men pulling on the rope exert the force **F**. The point of application of **F** is on the front bumper. The force **F** tends to make the truck move forward in a straight line and does actually make it move, since no external force opposes this motion. (Rolling resistance has been neglected here for simplicity.) This forward motion of the truck, during which all straight lines remain parallel to themselves (the floor of the truck remains horizontal, and its walls remain vertical), is known as a *translation*. Other forces might cause the truck to move differently. For example, the force exerted by a jack placed under the front axle would cause the truck to pivot about its rear axle. Such a motion is a *rotation*. It may be concluded, therefore, that each of the *external forces* acting on a *rigid body* can, if unopposed, impart to the rigid body a motion of translation or rotation, or both.

### 3.3. PRINCIPLE OF TRANSMISSIBILITY. EQUIVALENT FORCES

The *principle of transmissibility* states that the conditions of equilibrium or of motion of a rigid body will remain unchanged if a force **F** acting at a given point of the rigid body is replaced by a force **F′** of the same magnitude and same direction, but acting at a different point, *provided that the two forces have the same line of action* (Fig. 3.3). The two forces **F** and **F′** have the same effect on the rigid body and are said to be *equivalent*. This principle, which states in fact that the action of a force may be *transmitted* along its line of action, is based on experimental evidence. It *cannot* be

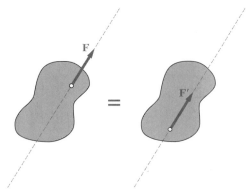

**Fig. 3.3**

derived from the properties established so far in this text and must therefore be accepted as an experimental law.†

It was indicated in Chap. 2 that the forces acting on a particle could be represented by vectors. These vectors had a well-defined point of application, namely, the particle itself, and were therefore fixed, or bound, vectors. In the case of forces acting on a rigid body, however, the point of application of the force does not matter, as long as the line of action remains unchanged. Thus forces acting on a rigid body must be represented by different kinds of vectors, known as *sliding vectors,* since these vectors may be allowed to slide along their line of action. We should note that all the properties which will be derived in the following sections for the forces acting on a rigid body will be valid more generally for any system of sliding vectors. In order to keep our presentation more intuitive, however, we shall carry it out in terms of physical forces rather than in terms of mathematical sliding vectors.

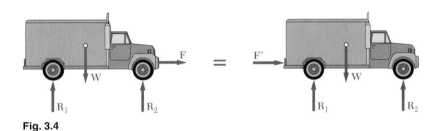

**Fig. 3.4**

Returning to the example of the truck, we first observe that the line of action of the force **F** is a horizontal line passing through both the front and the rear bumpers of the truck (Fig. 3.4). Using the principle of transmissibility, we may therefore replace **F** by an *equivalent force* **F′** acting on the rear bumper. In other words, the conditions of motion are unaffected, and all the other external forces acting on the truck (**W**, **R**$_1$, **R**$_2$) remain unchanged if the men push on the rear bumper instead of pulling on the front bumper.

The principle of transmissibility and the concept of equivalent forces have limitations, however. Consider, for example, a short bar *AB* acted upon by equal and opposite axial forces **P**$_1$ and **P**$_2$, as shown in Fig. 3.5a. According to the principle of transmissibility, the force **P**$_2$ may be replaced

---

† The principle of transmissibility may be derived from the study of the dynamics of rigid bodies, but this study requires the introduction of all Newton's three laws and of a number of other concepts as well. See F. P. Beer and E. R. Johnston, "Vector Mechanics for Engineers," 4th ed., sec. 16.5, McGraw-Hill Book Company, 1984.

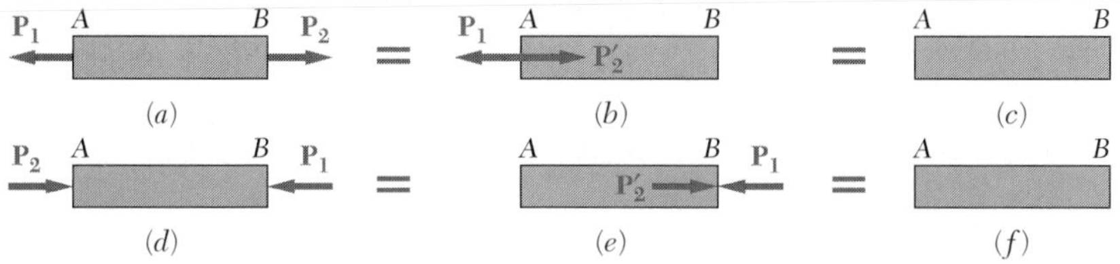

**Fig. 3.5**

by a force $\mathbf{P}'_2$ having the same magnitude, same direction, and same line of action, but acting at $A$ instead of $B$ (Fig. 3.5$b$). The forces $\mathbf{P}_1$ and $\mathbf{P}'_2$ acting on the same particle may be added according to the rules of Chap. 2, and, being equal and opposite, their sum is found equal to zero. The original system of forces shown in Fig. 3.5$a$ is thus equivalent to no force at all (Fig. 3.5$c$) from the point of view of the external behavior of the bar.

Consider now the two equal and opposite forces $\mathbf{P}_1$ and $\mathbf{P}_2$ acting on the bar $AB$ as shown in Fig. 3.5$d$. The force $\mathbf{P}_2$ may be replaced by a force $\mathbf{P}'_2$ having the same magnitude, same direction, and same line of action, but acting at $B$ instead of $A$ (Fig. 3.5$e$). The forces $\mathbf{P}_1$ and $\mathbf{P}'_2$ may then be added, and their sum is found again to be zero (Fig. 3.5$f$). From the point of view of the mechanics of rigid bodies, the systems shown in Fig. 3.5$a$ and $d$ are thus equivalent. But the *internal forces* and *deformations* produced by the two systems are clearly different. The bar of Fig. 3.5$a$ is in *tension* and, if not absolutely rigid, will increase in length slightly; the bar of Fig. 3.5$d$ is in *compression* and, if not absolutely rigid, will decrease in length slightly. Thus, while the principle of transmissibility may be used freely to determine the conditions of motion or equilibrium of rigid bodies and to compute the external forces acting on these bodies, it should be avoided, or at least used with care, in determining internal forces and deformations.

### 3.4. TWO-DIMENSIONAL STRUCTURES

The remarks made in the two preceding sections apply to rigid bodies and to all types of external forces acting on such bodies. For the rest of this chapter, however, we shall be concerned only with two-dimensional structures. By a two-dimensional structure we mean a flat structure which has length and breadth but no depth or, more generally, a structure which contains a plane of symmetry.

All forces acting on these structures will be assumed in the plane of the structure itself, or, more generally, it will be assumed that they may be reduced to forces in the plane of symmetry of the structure. Two-dimensional structures and coplanar forces may be readily represented on a sheet of paper or on a blackboard. Their analysis is therefore considerably simpler than that of three-dimensional structures and forces.

## 3.5. MOMENT OF A FORCE ABOUT AN AXIS

It was established in Sec. 3.3 that, from the point of view of the mechanics of rigid bodies, two forces **F** and **F′** are equivalent if they have the same magnitude, same direction, and same line of action (Fig. 3.6a). On the other hand, a force **F″** of the same magnitude and same direction, but having a different line of action, will not have the same effect as **F** on a given rigid body (Fig. 3.6b). While **F** and **F″** are represented by equal vectors, i.e., by vectors of the same magnitude and same direction (Sec. 2.3), these two forces are *not equivalent*. True, both forces will tend to give to the rigid body the same motion of *translation*, but they would make it *rotate* differently about an axis through A, perpendicular to the plane of the figure.

The tendency of a force to make a rigid body rotate about an axis is measured by the moment of the force about that axis. The *moment $M_A$ of the force* **F** *about an axis through A*, or, for short, the *moment of* **F** *about A*, is defined as the product of the magnitude F of the force and of the perpendicular distance d from A to the line of action of **F**:

$$M_A = Fd \tag{3.1}$$

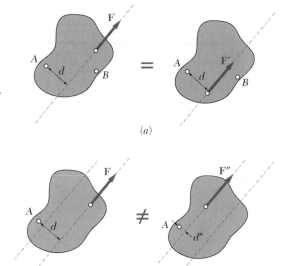

(a)

(b)

**Fig. 3.6**

In the SI system of units, where a force is expressed in newtons (N) and a distance in meters (m), the moment of a force will be expressed in newton-meters (N · m). In the U.S. customary system of units, where a force is expressed in pounds and a distance in feet or inches, the moment of a force will be expressed in lb · ft or lb · in.

The moment of a force has not only a magnitude but also a sense; depending upon the relative position of the force and of the axis, the moment of the force will be either clockwise ⤵ or counterclockwise ⤴. It is seen that the moment of **F** about A in Fig. 3.6a is counterclockwise, while its moment about B is clockwise. Moments of forces may be added algebraically as scalar quantities if a sign convention is adopted. We shall in this text consider as positive a counterclockwise moment, and as negative a clockwise moment.

Since the forces **F** and **F′** shown in Fig. 3.6a have the same magnitude, same sense, and same line of action, the moment of **F′** about A will also be Fd and its sense will also be counterclockwise. Both forces have the same moment about A. Clearly they would have *the same moment about any other axis*. It may thus be stated that *two equivalent forces have the same components $F_x$ and $F_y$ and the same moment about any axis*. On the other hand, the moment of **F″** about A in Fig. 3.6b is different; it is equal to Fd″.

A force acting on a rigid body is completely defined (except for its exact position on its line of action) if its components $F_x$ and $F_y$ and its moment $M_A$ about A are known (in magnitude and sense). The magnitude F of the force and its direction may be determined from the components $F_x$

and $F_y$ by the methods of Chap. 2, while the perpendicular distance $d$ from $A$ to the line of action may be obtained by forming the quotient of $M_A$ over $F$. The line of action itself will be drawn on one side of $A$ or the other, depending upon the sense of **F** and upon the sense of the moment.

### 3.6. VARIGNON'S THEOREM

An important theorem of statics is due to the French mathematician Varignon (1654–1722). It states that the *moment of a force about any axis is equal to the sum of the moments of its components about that axis.*

To prove this statement, consider a force **F** acting at point $A$ and the components **P** and **Q** of the force **F** in any two directions (Fig. 3.7$a$). The moment of **F** about an axis through an arbitrary point $B$ is $Fd$, where $d$ is the perpendicular distance from $B$ to the line of action of **F**. Similarly, the moments of **P** and **Q** about the axis through $B$ are $Pd_1$ and $Qd_2$, respectively, where $d_1$ and $d_2$ denote the perpendicular distances from $B$ to the lines of action of **P** and **Q**. We propose to show that

$$Fd = Pd_1 + Qd_2 \tag{3.2}$$

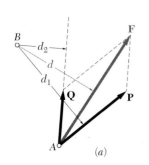

(a)

We shall attach at $A$ two rectangular axes as shown in Fig. 3.7$b$; the $x$ axis is perpendicular to $AB$, while the $y$ axis is directed along $AB$. Denoting by $\theta$ the angle that **F** forms with the $x$ axis, and observing that $\theta$ is also the angle between $AB$ and the perpendicular drawn from $B$ onto the line of action of **F**, we write

$$
\begin{aligned}
Fd = F(AB) \cos \theta &= (AB)F \cos \theta \\
&= (AB)F_x
\end{aligned} \tag{3.3}
$$

where $F_x$ denotes the $x$ component of the force **F**. Expressing in a similar way the moments $Pd_1$ and $Qd_2$ of the forces **P** and **Q**, we write

$$
\begin{aligned}
Pd_1 + Qd_2 &= (AB)P_x + (AB)Q_x \\
&= (AB)(P_x + Q_x)
\end{aligned}
$$

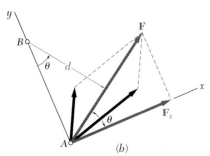

(b)

**Fig. 3.7**

But, as we saw in Sec. 2.8, the sum $P_x + Q_x$ of the $x$ components of two forces **P** and **Q** is equal to the $x$ component $F_x$ of their resultant **F**. The expression obtained for the sum $Pd_1 + Qd_2$ of the moments of **P** and **Q** is, therefore, identical with the expression obtained in (3.3) for the moment $Fd$ of **F**. The relation (3.2) is thus verified and Varignon's theorem is proved.

Varignon's theorem is very useful in the computation of moments of forces. Through a judicious choice of components, it is possible to simplify many problems. Two ways in which this may be done are indicated in Sample Probs. 3.2 and 3.3.

It should be noted that, in adding two moments, the sense (counterclockwise or clockwise) of each moment must be taken into account. As indicated in Sec. 3.5, a counterclockwise moment will be considered as positive, and a clockwise moment as negative.

# SAMPLE PROBLEM 3.1

A 100-lb vertical force is applied to the end of a lever which is attached to a shaft at $O$. Determine ($a$) the moment of the 100-lb force about $O$, ($b$) the magnitude of the horizontal force applied at $A$ which creates the same moment about $O$, ($c$) the smallest force applied at $A$ which creates the same moment about $O$, ($d$) how far from the shaft a 240-lb vertical force must act to create the same moment about $O$, ($e$) whether any one of the forces obtained in parts $b$, $c$, and $d$ is equivalent to the original force.

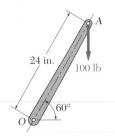

---

**$a$. Moment about $O$.** The perpendicular distance from $O$ to the line of action of the 100-lb force is

$$d = (24 \text{ in.}) \cos 60° = 12 \text{ in.}$$

The magnitude of the moment about $O$ of the 100-lb force is

$$M_O = Fd = (100 \text{ lb})(12 \text{ in.}) = 1200 \text{ lb} \cdot \text{in.}$$

Since the force tends to rotate the lever clockwise about $O$, we have

$$M_O = 1200 \text{ lb} \cdot \text{in.} \; \downarrow \quad \blacktriangleleft$$

**$b$. Horizontal Force.** In this case, we have

$$d = (24 \text{ in.}) \sin 60° = 20.8 \text{ in.}$$

Since the moment about $O$ must be 1200 lb · in., we write

$$M_O = Fd$$
$$1200 \text{ lb} \cdot \text{in.} = F(20.8 \text{ in.})$$
$$F = 57.7 \text{ lb} \qquad \mathbf{F} = 57.7 \text{ lb} \rightarrow \quad \blacktriangleleft$$

**$c$. Smallest Force.** Since $M_O = Fd$, the smallest value of $F$ occurs when $d$ is maximum. We choose the force perpendicular to $OA$ and find $d = 24$ in.; thus

$$M_O = Fd$$
$$1200 \text{ lb} \cdot \text{in.} = F(24 \text{ in.})$$
$$F = 50 \text{ lb} \qquad \mathbf{F} = 50 \text{ lb} \; \text{⦨} \; 30° \quad \blacktriangleleft$$

**$d$. 240-lb Vertical Force.** In this case $M_O = Fd$ yields

$$1200 \text{ lb} \cdot \text{in.} = (240 \text{ lb})d \qquad d = 5 \text{ in.}$$

but $\qquad OB \cos 60° = d \qquad\qquad OB = 10 \text{ in.} \quad \blacktriangleleft$

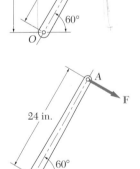

**$e$.** None of the forces considered in parts $b$, $c$, and $d$ is equivalent to the original 100-lb force. Although they have the same moment about $O$, they have different $x$ and $y$ components. In other words, although each force tends to rotate the shaft in the same manner, each causes the lever to pull on the shaft in a different way.

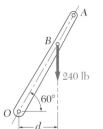

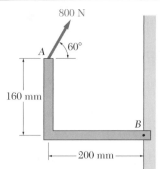

## SAMPLE PROBLEM 3.2

A force of 800 N acts on a bracket as shown. Determine the moment of the force about $B$.

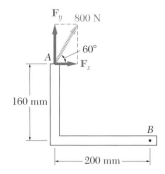

**Solution.** Resolving the force into its $x$ and $y$ components, we write

$$F_x = (800 \text{ N}) \cos 60° = 400 \text{ N}$$
$$F_y = (800 \text{ N}) \sin 60° = 693 \text{ N}$$

Observing that $\mathbf{F}_x$ tends to rotate the bracket clockwise about $B$ and recalling the sign convention introduced in Sec. 3.5, we find that the moment of $\mathbf{F}_x$ about $B$ is

$$(400 \text{ N})(0.160 \text{ m}) = 64.0 \text{ N} \cdot \text{m} \downarrow = -64.0 \text{ N} \cdot \text{m}$$

Similarly, we find that the moment of $\mathbf{F}_y$ about $B$ is

$$(693 \text{ N})(0.200 \text{ m}) = 138.6 \text{ N} \cdot \text{m} \downarrow = -138.6 \text{ N} \cdot \text{m}$$

Applying Varignon's theorem, we write

$$M_B = -64.0 \text{ N} \cdot \text{m} - 138.6 \text{ N} \cdot \text{m}$$
$$= -202.6 \text{ N} \cdot \text{m} \qquad\qquad M_B = 203 \text{ N} \cdot \text{m} \downarrow \blacktriangleleft$$

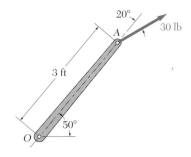

## SAMPLE PROBLEM 3.3

A 30-lb force acts on the end of the 3-ft lever as shown. Determine the moment of the force about $O$.

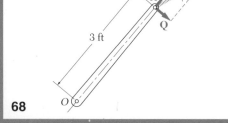

**Solution.** The force is replaced by two components, one component $\mathbf{P}$ in the direction of $OA$ and one component $\mathbf{Q}$ perpendicular to $OA$. Since $O$ is on the line of action of $\mathbf{P}$, the moment of $\mathbf{P}$ about $O$ is zero and the moment of the 30-lb force reduces to the moment of $\mathbf{Q}$, which is clockwise and, thus, negative.

$$Q = (30 \text{ lb}) \sin 20° = 10.26 \text{ lb}$$
$$M_O = -Q(3 \text{ ft}) = -(10.26 \text{ lb})(3 \text{ ft}) = -30.8 \text{ lb} \cdot \text{ft}$$
$$M_O = 30.8 \text{ lb} \cdot \text{ft} \downarrow \blacktriangleleft$$

# PROBLEMS

**3.1**  A 30-lb force is applied to the control rod $AB$ as shown. Knowing that the length of the rod is 8 in. and that $\alpha = 20°$, determine the moment of the force about $B$ by resolving the force (a) into horizontal and vertical components, (b) into components along $AB$ and in a direction perpendicular to $AB$.

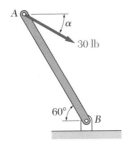

**Fig. P3.1 and P3.2**

**3.2**  A 30-lb force is applied to the control rod $AB$ as shown. Knowing that the length of the rod is 8 in. and that the moment of the force about $B$ is 160 lb · in. clockwise, determine the value of $\alpha$.

**3.3**  For the brake pedal shown, determine the magnitude and direction of the smallest force **P** which has a 130-N · m clockwise moment about $B$.

**3.4**  A force **P** is applied to the brake pedal at $A$. Knowing that $P = 450$ N and $\alpha = 30°$, determine the moment of **P** about $B$.

**3.5**  A 300-N force **P** is applied at point $A$ of the bell crank shown. (a) Compute the moment of the force **P** about $O$ by resolving it into horizontal and vertical components. (b) Using the result of part a, determine the perpendicular distance from $O$ to the line of action of **P**.

**Fig. P3.3 and P3.4**

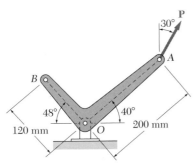

**Fig. P3.5 and P3.6**

**3.6**  A 400-N force **P** is applied at point $A$ of the bell crank shown. (a) Compute the moment of the force **P** about $O$ by resolving it into components along line $OA$ and in a direction perpendicular to that line. (b) Determine the magnitude and direction of the smallest force **Q** applied at $B$ which has the same moment as **P** about $O$.

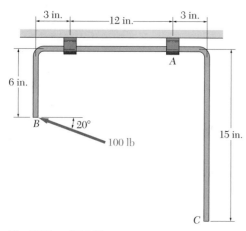

Fig. P3.7 and P3.10

**3.7 and 3.8** Compute the moment about $A$ of the force shown, $(a)$ by using the definition of the moment of a force, $(b)$ by resolving the force into horizontal and vertical components, $(c)$ by resolving the force into components along $AB$ and in the direction perpendicular to $AB$.

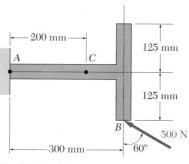

Fig. P3.8 and P3.9

**3.9 and 3.10** Determine $(a)$ the moment about $C$ of the force shown, $(b)$ the perpendicular distance from $C$ to the line of action of the force.

**3.11** A 100-N force is applied to corner $C$ of a 0.5-m square plate. Determine the moment of the force about corner $A$ $(a)$ by resolving the force into horizontal and vertical components and $(b)$ by sliding the force along its line of action to $D$.

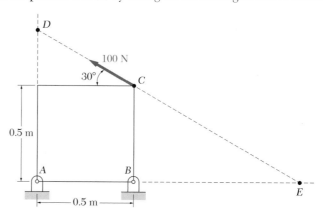

Fig. P3.11 and P3.12

**3.12** A 100-N force is applied to corner $C$ of a 0.5-m square plate. Determine the moment of the force about corner $A$ $(a)$ by sliding the force along its line of action to $E$ and $(b)$ by finding (geometrically) the perpendicular distance from the force's line of action to $A$.

**3.13** Rod $AB$ is held in place by the cord $AC$. Knowing that the tension in the cord is 250 lb and that $c = 24$ in., determine the moment about $B$ of the force exerted by the cord at point $A$ by resolving that force into horizontal and vertical components applied $(a)$ at point $A$, $(b)$ at point $C$.

**3.14** Rod $AB$ is held in place by the cord $AC$. Knowing that $c = 56$ in. and that the moment about $B$ of the force exerted by the cord at point $A$ is 280 lb · ft, determine the tension in the cord.

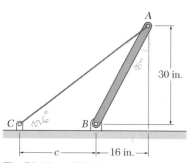

Fig. P3.13 and P3.14

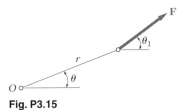

**Fig. P3.15**

3.7. Moment of a Couple **71**

**3.15** A force **F** acts at a point of coordinates $r$ and $\theta$ as shown. The force forms an angle $\theta_1$ with a line parallel to the horizontal reference axis. Show that the moment of the force about the origin of coordinates $O$ is $Fr\sin(\theta_1 - \theta)$.

**3.16** A force **F** of components $F_x$ and $F_y$ acts at a point of coordinates $x$ and $y$. Find the expression for the moment of **F** about the origin $O$ of the system of coordinates.

**3.17** Two forces **P** and **Q** have parallel lines of action and act at $A$ and $B$, respectively. The distance between $A$ and $B$ is $a$. Find the distance $x$ from $A$ to the point $C$ about which both forces have the same moment. Check the formula obtained by assuming $a = 240$ mm and (a) $P = 60$ N up, $Q = 30$ N up; (b) $P = 30$ N up, $Q = 60$ N up; (c) $P = 60$ N up, $Q = 30$ N down; (d) $P = 30$ N up, $Q = 60$ N down.

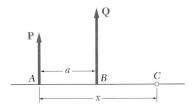

**Fig. P3.17**

**3.18** The line of action of a force **P** passes through the two points $A(x_1, y_1)$ and $B(x_2, y_2)$. If the force is directed from $A$ to $B$, determine the moment of the force about the origin.

## 3.7. MOMENT OF A COUPLE

*Two forces having the same magnitude, parallel lines of action, and opposite sense are said to form a couple.* Two such forces, **F** and **F'**, of the same magnitude $F$, are shown acting on a rigid body in Fig. 3.8.

Clearly, the sums of the $x$ components and of the $y$ components of **F** and **F'** are zero. The sum of their moments about an axis through a point $A$, however, is not zero. Thus, the effect of the couple on the rigid body is not zero. While the two forces will not move the body away, they will tend to make it turn.

Denoting by $d_1$ and $d_2$, respectively, the perpendicular distances from $A$ to the lines of action of **F** and **F'**, and recalling that counterclockwise is positive, we write the sum $M$ of the moments of the two forces about $A$ as follows:

**Fig. 3.8**

$$+\!\uparrow\ M = Fd_1 - Fd_2$$
$$= F(d_1 - d_2)$$

or, noting that $d_1 - d_2$ is equal to the distance $d$ between the lines of action of the two forces,

$$M = Fd \tag{3.4}$$

The sum *M* is called the *moment of the couple*. It may be observed *that M does not depend upon the choice of A; M* will have the same magnitude and same sense regardless of the location of *A*. This may be checked by repeating the above derivation, choosing *A* to the left of the couple, and then between the two forces forming the couple. We state therefore:

*The moment M of a couple is constant. Its magnitude is equal to the product Fd of the common magnitude F of the two forces and of the distance d between their lines of action. The sense of M (clockwise or counterclockwise) is obtained by direct observation.*

### 3.8. EQUIVALENT COUPLES

Consider the three couples shown in Fig. 3.9, which are made to act successively on the same corner plate. As seen in the preceding section, the only motion a couple may impart to a rigid body is a rotation. Each of the couples shown will cause the plate to rotate counterclockwise. Since the product *Fd* is found to be 30 N · m in each case, we may expect the three couples to have the same effect on the corner plate.

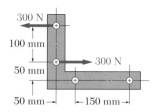

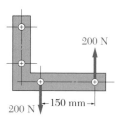

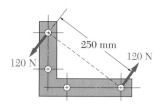

**Fig. 3.9**

As reasonable as this conclusion may appear, we should not accept it hastily. While intuitive feeling is of great help in the study of mechanics, it should not be accepted as a substitute for logical reasoning. Before stating that two systems (or groups) of forces have the same effect on a rigid body, we should prove that fact on the basis of the experimental evidence introduced so far. This evidence consists of the parallelogram law for the addition of two forces (Sec. 2.2) and of the principle of transmissibility (Sec. 3.3). Therefore, we shall state that *two systems of forces are equivalent* (i.e., they have the same effect on a rigid body) *if we can transform one of them into the other by means of one or several of the following operations:* (1) replacing two forces acting on the same particle by their resultant; (2) resolving a force into two components; (3) canceling two equal and opposite forces acting on the same particle; (4) attaching to the same particle two equal and opposite forces; (5) moving a force along its line of action. Each of these operations is easily justified on the basis of the parallelogram law or the principle of transmissibility.

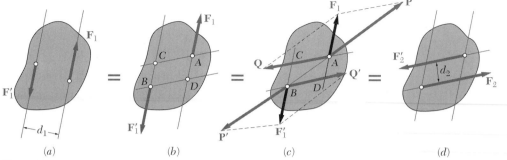

**Fig. 3.10**

Let us now prove that *two couples having the same moment (same magnitude and same sense) are equivalent*. Consider the couples shown in Fig. 3.10, consisting, respectively, of the forces $\mathbf{F}_1$ and $\mathbf{F}'_1$, of magnitude $F_1$ and at a distance $d_1$ from each other (Fig. 3.10*a*), and of the forces $\mathbf{F}_2$ and $\mathbf{F}'_2$, of magnitude $F_2$ and at a distance $d_2$ from each other (Fig. 3.10*d*). Both couples are counterclockwise, and we shall assume that they have equal moments:

$$F_1 d_1 = F_2 d_2 \qquad (3.5)$$

To prove that they are equivalent, we shall show that the couple $\mathbf{F}_1$, $\mathbf{F}'_1$ may be transformed into the couple $\mathbf{F}_2$, $\mathbf{F}'_2$ by means of the operations listed above.

Denoting by *A*, *B*, *C*, *D* the points of intersection of the lines of action of the two couples, we first slide the forces $\mathbf{F}_1$ and $\mathbf{F}'_1$ until they are attached, respectively, at *A* and *B*, as shown in Fig. 3.10*b*. The force $\mathbf{F}_1$ is then resolved into a component $\mathbf{P}$ along line *AB* and a component $\mathbf{Q}$ along *AC* (Fig. 3.10*c*); similarly, the force $\mathbf{F}'_1$ is resolved into $\mathbf{P}'$ along *AB* and $\mathbf{Q}'$ along *BD*. The forces $\mathbf{P}$ and $\mathbf{P}'$ have the same magnitude, same line of action, and opposite sense; they may be moved along their common line of action until they are applied at the same point and then canceled. Thus the couple consisting of $\mathbf{F}_1$ and $\mathbf{F}'_1$ reduces to the couple $\mathbf{Q}$, $\mathbf{Q}'$.

We shall now show that the forces $\mathbf{Q}$ and $\mathbf{Q}'$ are respectively equal to the forces $\mathbf{F}'_2$ and $\mathbf{F}_2$. The moment of the couple formed by $\mathbf{Q}$ and $\mathbf{Q}'$ may be obtained by computing the moment of $\mathbf{Q}$ about *B*; similarly, the moment of the couple formed by $\mathbf{F}_1$ and $\mathbf{F}'_1$ is the moment of $\mathbf{F}_1$ about *B*. But, by Varignon's theorem, the moment of $\mathbf{F}_1$ is equal to the sum of the moments of its components $\mathbf{P}$ and $\mathbf{Q}$. Since the moment of $\mathbf{P}$ about *B* is zero, the moment of the couple formed by $\mathbf{Q}$ and $\mathbf{Q}'$ must be equal to the moment of the couple $\mathbf{F}_1$, $\mathbf{F}'_1$. Recalling (3.5), we write

$$Q d_2 = F_1 d_1 = F_2 d_2 \qquad \text{and} \qquad Q = F_2$$

Thus the forces $\mathbf{Q}$ and $\mathbf{Q}'$ are respectively equal to the forces $\mathbf{F}'_2$ and $\mathbf{F}_2$, and the couple of Fig. 3.10*a* is equivalent to the couple of Fig. 3.10*d*.

The property we have just established is very important for the correct understanding of the mechanics of rigid bodies. It indicates that, when a couple acts on a rigid body, it does not matter where the two forces forming the couple act, or what magnitude and direction they have. The only thing which counts is the *moment* of the couple (magnitude and sense). Couples with the same moment will have the same effect on the rigid body.

### 3.9. ADDITION OF COUPLES

Consider two couples **P**, **P′** and **Q**, **Q′** acting on the same rigid body, as shown in Fig. 3.11*a*. Since a couple is completely defined by its moment, the couple formed by the forces **Q** and **Q′** may be replaced by a couple of the same moment, formed by forces **S** and **S′** which have, respectively, the same line of action as **P** and **P′** (Fig. 3.11*b*). The magnitude *S* of the new forces must be such that $Sp = Qq$.

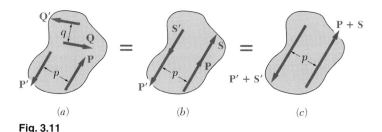

(a)          (b)          (c)

**Fig. 3.11**

Adding forces along the same line of action, we obtain a single couple consisting of the forces **P** + **S** and **P′** + **S′** (Fig. 3.11*c*). The moment of this couple is

$$M = (P + S)p = Pp + Sp = Pp + Qq$$

The moment *M* is thus found to be the algebraic sum of the moments *Pp* and *Qq* of the two original couples. It should be noted that, if the two couples have opposite senses, their moments will have opposite signs.

We state: *Two couples may be replaced by a single couple of moment equal to the algebraic sum of the moments of the given couples.*

### 3.10. RESOLUTION OF A GIVEN FORCE INTO A FORCE ACTING AT A GIVEN POINT AND A COUPLE

Consider a force **F** acting at point *B* on a rigid body, as shown in Fig. 3.12*a*. Suppose that for some reason we would rather have it acting at point *A*. We know that we can move **F** along its line of action (principle of transmissibility); but we cannot move it to a point *A* away from the original line of action without modifying the action of **F** on the rigid body.

We may attach, however, two forces at point *A*: a force **F** of the same magnitude and direction as the given force, and a force **F′** equal and opposite (Fig. 3.12*b*). As indicated in Sec. 3.8, this operation does not modify

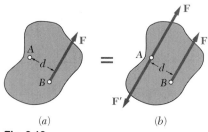

(a)          (b)

**Fig. 3.12**

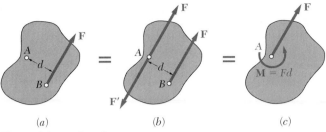

(a)    =    (b)    =    (c)    $M = Fd$

**Fig. 3.12 (completed)**

3.10. Resolution of a Given Force into a **75**
Force Acting at a Given Point and
a Couple

the action of the original force on the rigid body. We observe that the force **F** attached at $B$ and the force **F'** attached at $A$ form a couple of moment $M = Fd$. Since this couple may be replaced by any other couple of the same moment $M$ (Sec. 3.8), we may represent it by the letter **M** and the symbol ↺ placed at any convenient location.[†] The remaining force **F** attached at $A$ and the couple **M**, however, are usually grouped as shown in Fig. 3.12c.

The result we have obtained may be stated as follows: *Any force* **F** *acting on a rigid body may be moved to any given point A, provided that a couple* **M** *is added; the moment M of the couple must equal the moment of* **F** (*in its original position*) *about A*. The couple will then impart to the rigid body the same motion of rotation about $A$ that the force **F** produced before it was transferred to $A$. In the following sections we shall refer to this combination as a *force-couple system*. The location of the couple **M** is immaterial, but the force **F** should not be moved from $A$. If it were to be moved again, a new couple should be determined.

The transformation we have described may be performed in reverse. That is, *a force* **F** *acting at A and a couple* **M** *may be combined into a single resultant force* **F**. This is done by moving the force **F** until its moment about $A$ becomes equal to the moment $M$ of the couple to be eliminated. The magnitude and direction of **F** are unchanged, but its new line of action will be at a distance $d = M/F$ from $A$. An example of such a transformation is given in Fig. 3.13. Part $a$ of the figure shows the head of a bolt which is subjected to the combined action of a 600-lb · in. clockwise couple and of a 50-lb vertical downward force through the center $O$ of the head. The force-couple system may be replaced by a single 50-lb force located 12 in. to the right of $O$ (Fig. 3.13b). In this new position, the 50-lb force has a clockwise moment of 600 lb · in. about $O$. Figure 3.13a gives a good representation of the action to which the head of the bolt is subjected, while Fig. 3.13b suggests a simple way of exerting this action. The wrench was sketched to help visualize the action exerted on the bolt. This is particularly useful in Fig. 3.13b; it was not strictly necessary, however, to suggest an actual contact between the 50-lb force and the bolt.

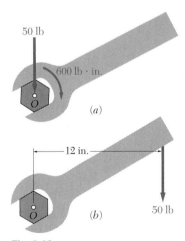

50 lb

600 lb · in.

(a)

12 in.

(b)    50 lb

**Fig. 3.13**

---

[†] The vectorial character of couples will be established in Sec. 4.4.

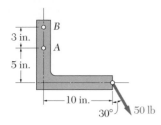

# SAMPLE PROBLEM 3.4

A 50-lb force is applied to a corner plate as shown. Determine (a) an equivalent force-couple system at A, (b) an equivalent system consisting of a 150-lb force at B and another force at A.

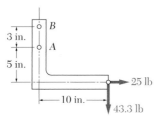

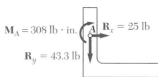

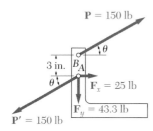

***a*. Force-Couple System at A.**  We resolve the 50-lb force into $x$ and $y$ components

$$F_x = (50 \text{ lb}) \sin 30° \qquad \mathbf{F}_x = 25.0 \text{ lb} \rightarrow \blacktriangleleft$$

$$F_y = -(50 \text{ lb}) \cos 30° \qquad \mathbf{F}_y = 43.3 \text{ lb} \downarrow \blacktriangleleft$$

These components may be moved to A if a couple is added, of moment $M_A$ equal to the moment about A of the components in their original position. Recalling that counterclockwise is positive, we obtain

$$M_A = (25.0 \text{ lb})(5 \text{ in.}) - (43.3 \text{ lb})(10 \text{ in.})$$

$$= -308 \text{ lb} \cdot \text{in.} \qquad \mathbf{M}_A = 308 \text{ lb} \cdot \text{in.} \, \downarrow \, \blacktriangleleft$$

***b*. Forces at A and B.**  We shall assume that the couple $\mathbf{M}_A$ found in part *a* consists of two 150-lb forces $\mathbf{P}$ and $\mathbf{P}'$ acting respectively at B and A. The moment of $\mathbf{P}$ about A must be equal to the moment $M_A$ of the couple; denoting by $\theta$ the angle that $\mathbf{P}$ forms with the horizontal, and applying Varignon's theorem, we write

$$M_A = -P \cos \theta \,(3 \text{ in.})$$

$$-308 \text{ lb} \cdot \text{in.} = -(150 \text{ lb}) \cos \theta \,(3 \text{ in.})$$

$$\cos \theta = \frac{308 \text{ lb} \cdot \text{in.}}{450 \text{ lb} \cdot \text{in.}} = 0.684 \qquad \theta = \pm 46.8°$$

Having found the direction of the forces $\mathbf{P}$ and $\mathbf{P}'$, we complete the solution by determining the resultant $\mathbf{Q}$ of the forces $\mathbf{F}$ and $\mathbf{P}'$ acting at A.

$$Q_x = F_x + P'_x = 25 \text{ lb} - (150 \text{ lb}) \cos \theta$$

$$Q_y = F_y + P'_y = -43.3 \text{ lb} - (150 \text{ lb}) \sin \theta$$

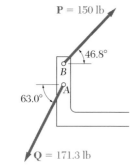

Since we have found two possible values for $\theta$, there will be two pairs of forces $\mathbf{P}$ and $\mathbf{Q}$ forming a system equivalent to the original 50-lb force:

$$\mathbf{P} = 150 \text{ lb} \, \measuredangle \, 46.8° \text{ at } B \qquad \mathbf{Q} = 171.3 \text{ lb} \, \measuredangle \, 63.0° \text{ at } A \quad \blacktriangleleft$$

or

$$\mathbf{P} = 150 \text{ lb} \, \measuredangle \, 46.8° \text{ at } B \qquad \mathbf{Q} = 102.0 \text{ lb} \, \measuredangle \, 40.4° \text{ at } A \quad \blacktriangleleft$$

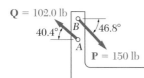

## SAMPLE PROBLEM 3.5

Replace the couple and force shown by an equivalent single force applied to the lever. Determine the distance from the shaft to the point of application of this equivalent force.

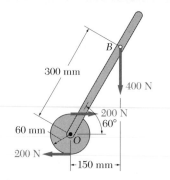

**Solution.**   First the given force and couple are replaced by an equivalent force-couple system at $O$. We move the 400-N force to $O$ and at the same time add a couple of moment equal to the moment about $O$ of the force in its original position, namely

$$M_O = -(400 \text{ N})(0.150 \text{ m}) = -60 \text{ N} \cdot \text{m} \qquad M_O = 60 \text{ N} \cdot \text{m} \downarrow$$

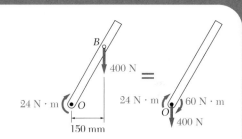

This couple is added to the 24-N·m clockwise couple formed by the two 200-N forces, and an 84-N·m clockwise couple is obtained. We now move the 400-N force to the right a distance $d$ such that the moment of the force about $O$ is 84 N·m clockwise:

$$84 \text{ N} \cdot \text{m} = (400 \text{ N})d \qquad d = 0.210 \text{ m} = 210 \text{ mm}$$

The equivalent single force, or resultant, is attached at point $C$, where its line of action intersects the lever:

$$(OC) \cos 60° = 210 \text{ mm} \qquad OC = 420 \text{ mm} \quad \blacktriangleleft$$

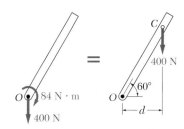

**Alternate Solution.**   Since the effect of a couple does not depend on its location, the 24-N·m clockwise couple may be moved to $B$; we thus obtain a force-couple system at $B$. We now move the 400-N force from $B$ to the right a distance $d'$ such that the moment of the force about $B$ is 24 N·m clockwise:

$$24 \text{ N} \cdot \text{m} = (400 \text{ N})d' \qquad d' = 0.060 \text{ m} = 60 \text{ mm}$$

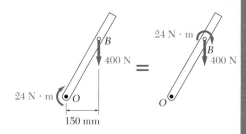

Again the point of intersection of the lever and of the line of action is determined:

$$(BC) \cos 60° = 60 \text{ mm} \qquad BC = 120 \text{ mm}$$
$$OC = OB + BC = 300 \text{ mm} + 120 \text{ mm}$$
$$OC = 420 \text{ mm} \quad \blacktriangleleft$$

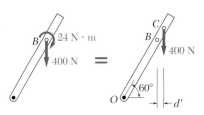

# PROBLEMS

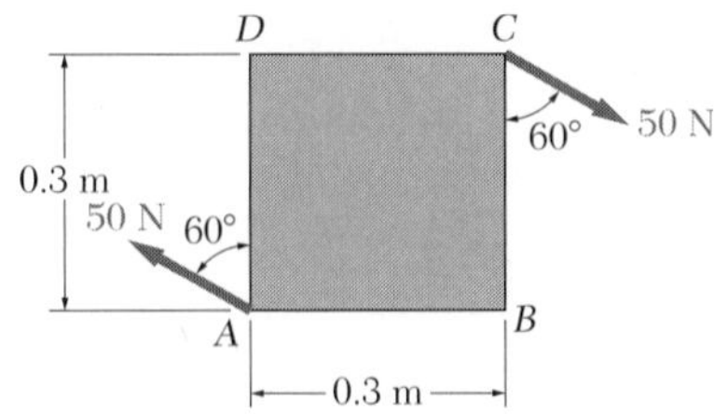

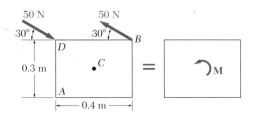

Fig. P3.19

**3.19** Two 50-N forces are applied as shown to the corners $A$ and $C$ of a 0.3-m square plate. Determine the moment of the couple formed by the two forces (a) by multiplying their magnitude by their perpendicular distance, (b) by resolving each force into horizontal and vertical components and adding the moments of the two resulting couples.

**3.20** The three pictures show equivalent couples—three different ways to apply the same couple to the same square plate. If the couple **M** is 10 N-m in the direction shown, determine the force magnitudes, (a) $F_1$ and (b) $F_2$, so that the force-pair couples are equivalent to the couple **M**.

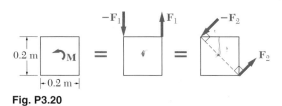

**Fig. P3.20**

**3.21** Four 1-in.-diameter pegs are attached to a board as shown. Two strings are passed around the pegs and pulled with the forces indicated. (a) Determine the resultant couple acting on the board. (b) If only one string is used, around which pegs should it pass and in what directions should it be pulled to create the same couple with the minimum tension in the string? (c) What is the value of that minimum tension?

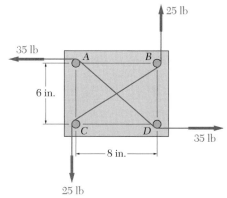

**Fig. P3.21**

**3.22** Two 50-N forces applied to the corners $B$ and $D$ of the plate form a couple equivalent to a couple **M**. Determine the moment of the 50-N force-pair couple about points (a) $A$, (b) $B$, and (c) $C$. By doing this you will "prove" that a couple is a "free vector." A couple, regardless of its point of application, produces the same moment about any point on the body.

Fig. P3.22

**3.23**  The four pictures show equivalent couples—four different ways to apply the same couple to the same rectangular plate. If the couple **M** is 12 lb-ft in the direction shown, determine the force magnitudes, (a) $F_1$, (b) $F_2$, and (c) $F_3$, so that the force-pair couples are equivalent to the couple **M**.

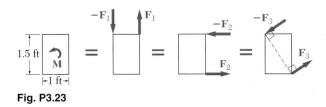

**Fig. P3.23**

**3.24**  The three pictures show equivalent couples—three different ways to apply the same couple to the same rectangular plate. If the couple **M** is 30 N-m in the direction shown, determine the force magnitudes, (a) $F_1$ and (b) $F_2$, so that the force-pair couples are equivalent to the couple **M**.

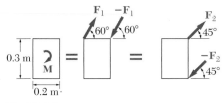

**Fig. P3.24**

**3.25**  A 30-lb vertical force **P** is applied at A to the bracket shown, which is held by screws at B and C. (a) Replace **P** by an equivalent force-couple system at B. (b) Find the two horizontal forces at B and C which are equivalent to the couple obtained in part a.

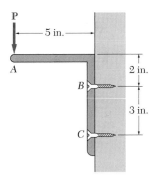

**Fig. P3.25**

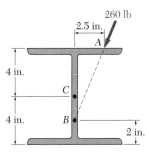

**Fig. P3.26**

**3.26**  A 260-lb force is applied at A to the rolled-steel section shown. Replace that force by an equivalent force-couple system at the center C of the section.

**3.27**  The force **P** has a magnitude of 250 N and is applied at the end C of a 500-mm rod AC attached to a bracket at A and B. Assuming $\alpha = 30°$ and $\beta = 60°$, replace **P** by (a) an equivalent force-couple system at B, (b) an equivalent system formed by two parallel forces applied at A and B.

**3.28**  Solve Prob. 3.27, assuming $\alpha = \beta = 25°$.

**3.29**  The force **P** is applied at the end C of a 500-mm rod AC attached to a bracket at A and B. Replace **P** by an equivalent system formed by two parallel forces applied at A and B and show that (a) these forces are parallel to **P**, (b) their magnitudes are independent of both $\alpha$ and $\beta$.

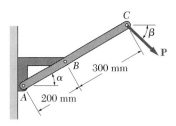

**Fig. P3.27 and P3.29**

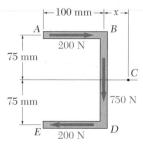

**Fig. P3.30**

**3.30** The shearing forces exerted on the cross section of a steel channel may be represented by a 750-N vertical force and two 200-N horizontal forces as shown. Replace this force and couple by a single force **F** applied at point $C$ and determine the distance $x$ from $C$ to line $BD$. (Point $C$ is defined as the *shear center* of the section.)

**3.31** A force and couple act as shown on a square plate of side $a = 20$ in. Knowing that $P = 50$ lb, $Q = 30$ lb, and $\alpha = 50°$, replace the given force and couple by a single force applied at a point located ($a$) on line $AB$, ($b$) on line $AC$. In each case determine the distance from $A$ to the point of application of the force.

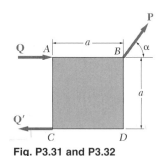

**Fig. P3.31 and P3.32**

**3.32** The force and couple shown are to be replaced by an equivalent single force. Knowing that $P = 2Q$, determine the required value of $\alpha$ if the line of action of the single equivalent force is to pass through ($a$) point $A$, ($b$) point $C$.

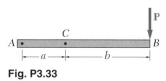

**Fig. P3.33**

**3.33** A force **P** is applied to a beam $AB$ as shown. Find the vertical forces $\mathbf{F}_A$ and $\mathbf{F}_C$ applied, respectively, at $A$ and $C$, which form a system equivalent to **P**.

**3.34** Replace the 50-lb force applied to the corner plate of Sample Prob. 3.4 by an equivalent system formed by two parallel forces at $A$ and $B$.

**3.35** Knowing that the line of action of the 600-N force passes through point $C$, replace the 600-N force by ($a$) an equivalent force-couple system at $D$, ($b$) an equivalent system formed by two parallel forces at $B$ and $D$.

**3.36** Knowing that $\alpha = 45°$, replace the 600-N force by ($a$) an equivalent force-couple system at $C$, ($b$) an equivalent system formed by two parallel forces at $B$ and $C$.

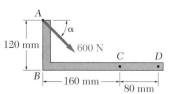

**Fig. P3.35, P3.36, and P3.37**

**3.37** Knowing that $\alpha = 30°$, replace the 600-N force by ($a$) an equivalent force-couple system at $C$, ($b$) an equivalent system consisting of a 120-N force at $B$ and another force at $C$.

## 3.11. REDUCTION OF A SYSTEM OF COPLANAR FORCES TO ONE FORCE AND ONE COUPLE. RESULTANT OF A SYSTEM OF COPLANAR FORCES

Consider a system of coplanar forces $\mathbf{F}_1$, $\mathbf{F}_2$, $\mathbf{F}_3$, etc., acting on a rigid body as shown in Fig. 3.14a. As seen in Sec. 3.10, $\mathbf{F}_1$ may be moved and attached to a given point A if a couple $\mathbf{M}_1$, of moment $M_1$ equal to the moment of $\mathbf{F}_1$ about A, is added to the original system of forces. Repeating this procedure with $\mathbf{F}_2$, $\mathbf{F}_3$, etc., we obtain the system shown in Fig. 3.14b. Since the forces act now on the same particle A, they may be added vectorially and replaced by their resultant $\mathbf{R}$ as shown in Fig. 3.14c. On the other hand, the couples may be replaced by a single couple $\mathbf{M}$ of moment $M$ equal to the algebraic sum of the moments $M_1$, $M_2$, $M_3$, etc. Any system of coplanar forces, however complex, may thus be reduced to an *equivalent force-couple system acting at a given point A.*

In the general case, when the force $\mathbf{R}$ is different from zero, the system obtained may be reduced further to a *single force* $\mathbf{R}$, called the *resultant* of the given system (Fig. 3.14d). This is done, as indicated in Sec. 3.10, by moving $\mathbf{R}$ until its moment about A becomes equal to $M$, thus eliminating the need for the couple $\mathbf{M}$. The distance from A to the line of action of $\mathbf{R}$ is $d = M/R$.

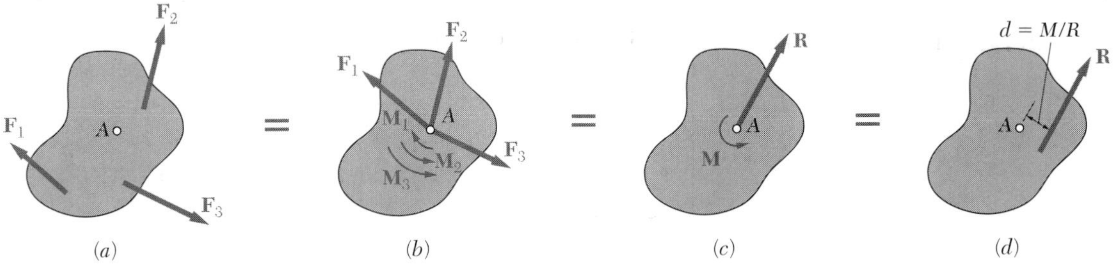

Fig. 3.14

When $\mathbf{R}$ is equal to zero, the force-couple system reduces to the couple $\mathbf{M}$. Thus, the given system of forces is reduced to a *single couple* $\mathbf{M}$, called the *resultant couple* of the system.

To summarize, *any given system of coplanar forces may be reduced to a single force or a single couple,* as the case may be. The only exception is when both $\mathbf{R}$ and $\mathbf{M}$ are zero. Then the system exerts no action on the rigid body, and the rigid body is said to be in *equilibrium.* This very particular case is of great practical interest and will be considered under Equilibrium of Rigid Bodies later in this chapter.

In practice, the reduction of a system of coplanar forces to a force $\mathbf{R}$ at A and a couple $\mathbf{M}$ will be considerably simplified if the given forces $\mathbf{F}_1$, $\mathbf{F}_2$, $\mathbf{F}_3$, etc., are resolved into their x and y components as shown in Fig. 3.15a. Adding these components, we shall obtain, respectively, the components $\mathbf{R}_x$ and $\mathbf{R}_y$ of the force $\mathbf{R}$. Besides, by using Varignon's theorem, the moments of the given forces about A may be determined more conveniently from the component forces $\mathbf{F}_x$ and $\mathbf{F}_y$ than from the forces $\mathbf{F}$ themselves; the sum $\Sigma M_A$ of these moments is equal to the moment of the couple $\mathbf{M}$. It is advisable to arrange the computations in tabular form as shown in

Sample Prob. 3.7 and to sketch the force $\mathbf{R}_x$, the force $\mathbf{R}_y$, and the couple $\mathbf{M}$ to which the given system of forces is reduced (Figure 3.15*b*). The component forces $\mathbf{R}_x$ and $\mathbf{R}_y$ may then be added vectorially, and a force-couple system at $A$ is obtained.

If the given system is to be reduced to a single force (i.e., to its resultant), the following method may be used: The forces $\mathbf{R}_x$ and $\mathbf{R}_y$ are moved to a point $B$ located on the line of action of $\mathbf{R}_x$, as shown in Fig. 3.15*c*; the displacement from $A$ to $B$ is defined by $d_x = M/R_y$. The moment of $\mathbf{R}_x$ about $A$ is still zero, but the moment of $\mathbf{R}_y$ about $A$ is now equal to $M$. Thus the couple is eliminated, and the given system reduces to the two forces $\mathbf{R}_x$ and $\mathbf{R}_y$ attached at $B$. These component forces may then be added vectorially, and the resultant $\mathbf{R}$ is obtained.

The forces $\mathbf{R}_x$ and $\mathbf{R}_y$ could also have been moved to point $C$, located on the line of action of $\mathbf{R}_y$, as shown in Fig. 3.15*d*; the displacement from $A$ to $C$ is defined by $d_y = -M/R_x$. The couple is eliminated, since the moment of $\mathbf{R}_x$ about $A$ is equal to $M$, and the given system reduces to the forces $\mathbf{R}_x$ and $\mathbf{R}_y$ at $C$, which may again be added vectorially.

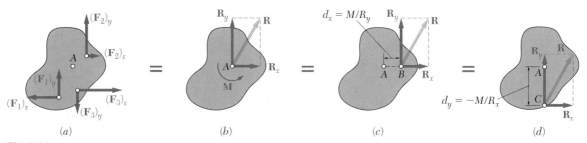

(a)          (b)          (c)          (d)

**Fig. 3.15**

## 3.12. EQUIVALENT SYSTEMS OF COPLANAR FORCES

We have seen in the preceding section that any system of coplanar forces acting on a rigid body may be reduced to a force-couple system at a given point $A$. This equivalent force-couple system characterizes completely the action exerted by the given system on the rigid body. *Two systems of coplanar forces are equivalent, therefore, if they may be reduced to the same force-couple system at a given point A.* Referring to Fig. 3.15*b* and recalling that $\mathbf{R}_x$, $\mathbf{R}_y$, and $\mathbf{M}$ are obtained, respectively, by summing the $x$ components, the $y$ components, and the moments about $A$ of the forces of the system considered, we state: *Two systems of coplanar forces are equivalent if, and only if, the sums of the x components, of the y components, and of the moments about A of their forces are, respectively, equal.*

The above statement has a simple physical significance. It means that two systems of coplanar forces are equivalent if they tend to impart to a given rigid body (1) the same translation in the $x$ direction, (2) the same translation in the $y$ direction, and (3) the same rotation about a fixed point $A$. Note that it is sufficient to establish the last property with respect to *one point only.* The property, however, holds with respect to *any point* if the two systems are equivalent.

## SAMPLE PROBLEM 3.6

A 4.80-m beam is subjected to the forces shown. Reduce the given system of forces to (a) an equivalent force-couple system at A, (b) an equivalent force-couple system at B, (c) a single force or resultant.

*Note.* Since the reactions at the supports are not included in the given system of forces, the given system will not maintain the beam in equilibrium.

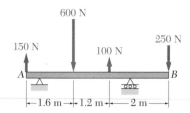

### SOLUTION

*a.* **Force-Couple System at A.** Positive senses for the force-couple system are as shown. Adding the components of the given forces and the moments about A of the given forces, we obtain

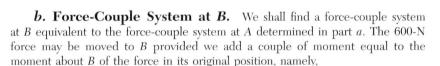

$$\xrightarrow{+} R_x = 0$$
$$+\uparrow R_y = +150 \text{ N} - 600 \text{ N} + 100 \text{ N} - 250 \text{ N} = -600 \text{ N}$$
$$+\curvearrowleft M_A = -(600 \text{ N})(1.6 \text{ m}) + (100 \text{ N})(2.8 \text{ m}) - (250 \text{ N})(4.8 \text{ m})$$
$$= -1880 \text{ N} \cdot \text{m}$$

The negative signs obtained indicate that the 600-N force acts downward and that the couple is clockwise. The equivalent force-couple system at A is thus

$$\mathbf{R} = 600 \text{ N} \downarrow \qquad \mathbf{M}_A = 1880 \text{ N} \cdot \text{m} \downarrow \quad \blacktriangleleft$$

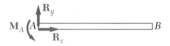

*b.* **Force-Couple System at B.** We shall find a force-couple system at B equivalent to the force-couple system at A determined in part a. The 600-N force may be moved to B provided we add a couple of moment equal to the moment about B of the force in its original position, namely,

$$+(600 \text{ N})(4.8 \text{ m}) = +2880 \text{ N} \cdot \text{m}$$

The 1880-N · m clockwise couple may be moved to B without changing its effect on the beam. Adding the two couples now at B, we obtain a couple of moment

$$+2880 \text{ N} \cdot \text{m} - 1880 \text{ N} \cdot \text{m} = +1000 \text{ N} \cdot \text{m}$$

The equivalent force-couple system at B is thus

$$\mathbf{R} = 600 \text{ N} \downarrow \qquad \mathbf{M}_B = 1000 \text{ N} \cdot \text{m} \uparrow \quad \blacktriangleleft$$

*c.* **Single Force or Resultant.** Using the results of part a, we move the 600-N force to the right through a distance x chosen so that the moment of the force about A is −1880 N · m. We write

$$-1880 \text{ N} \cdot \text{m} = -(600 \text{ N})x \qquad x = 3.13 \text{ m}$$
$$\mathbf{R} = 600 \text{ N} \downarrow \qquad x = 3.13 \text{ m} \quad \blacktriangleleft$$

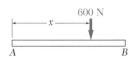

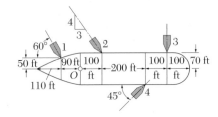

## SAMPLE PROBLEM 3.7

Four tugboats are used to bring an ocean liner to its pier. Each tugboat exerts a 5000-lb force in the direction shown. Determine (a) the equivalent force-couple system at the foremast O, (b) the point on the hull where a single, more powerful tugboat should push to produce the same effect as the original four tugboats.

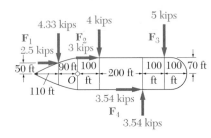

**a. Force-Couple System at O.**   Each of the given forces is resolved into components in the diagram shown (kip units are used). For convenience the components and their moments about O are recorded in a table, from which $R_x$, $R_y$, and M are then obtained.

| Force | $F_x \xrightarrow{}$, kips | $F_y + \uparrow$ kips | $M_O + \uparrow$ kip · ft |
|---|---|---|---|
| $F_1$ | +2.50 | | −125 |
| | | −4.33 | +390 |
| $F_2$ | +3.00 | | −210 |
| | | −4.00 | −400 |
| $F_3$ | 0 | | 0 |
| | | −5.00 | −2000 |
| $F_4$ | +3.54 | | +248 |
| | | +3.54 | +1062 |
| | $R_x = \Sigma F_x$ $= +9.04$ $= 9.04 \rightarrow$ | $R_y = \Sigma F_y$ $= -9.79$ $= 9.79\downarrow$ | $M = \Sigma M_O$ $= -1035$ $= 1035 \downarrow$ |

The force components are combined to yield a force-couple system equivalent to the original system.

$$\mathbf{R} = 13.33 \text{ kips} \searrow 47.3° \qquad \mathbf{M} = 1035 \text{ kip·ft} \downarrow \quad \blacktriangleleft$$

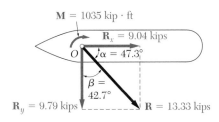

**b. Single Tugboat.**   The force exerted by a single tugboat must be equivalent to the force-couple system found in part a and thus have a magnitude of 13.33 kips. Since this force acts on the hull at a point A, we resolve it into components at A. The distance x must be such that the total moment about O is −1035 kip · ft.

$$-1035 \text{ kip} \cdot \text{ft} = -(9.04 \text{ kips})(70 \text{ ft}) - (9.79 \text{ kips})x$$
$$x = 41.1 \text{ ft}$$
$$\mathbf{R} = 13.33 \text{ kips} \searrow 47.3° \qquad x = 41.1 \text{ ft} \quad \blacktriangleleft$$

# PROBLEMS

**3.38** A 3-m beam is loaded in the various ways represented in the figure. Find two loadings which are equivalent.

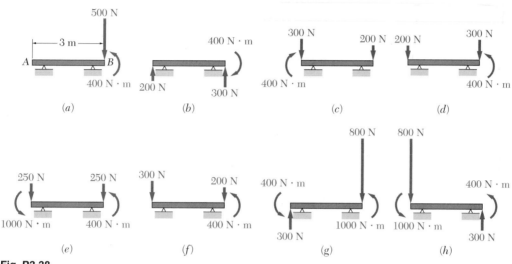

Fig. P3.38

**3.39** A 3-m beam is loaded as shown. Determine the loading of Prob. 3.38 which is equivalent to this loading.

**3.40** Determine the resultant of the loads and the distance from point $A$ to its line of action for the beam and loading of (a) Prob. 3.38a, (b) Prob. 3.38b, (c) Prob. 3.39.

**3.41** Three horizontal forces are applied as shown to a vertical cast-iron arm. Determine the resultant of the forces and the distance from the ground to its line of action when (a) $P = 200$ N, (b) $P = 2400$ N, (c) $P = 1000$ N.

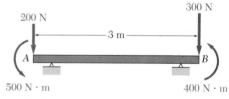

Fig. P3.39

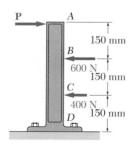

Fig. P3.41

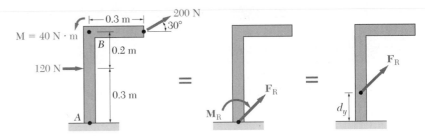

**Fig. P3.42**

**3.42** A system of forces and a couple are applied to the inverted L-shaped member. Perform two stages of simplification of this system. (*a*) First, replace this system with a resultant force $\mathbf{F}_R$ and a resultant couple, $\mathbf{M}_R$, at *A*. (*b*) Next, find the distance $d_y$ that $\mathbf{F}_R$ may be moved vertically along *AB* so that no couple is necessary. What you have demonstrated here is that a general planar force-couple system can be replaced by a single force, $\mathbf{F}_R$, properly located.

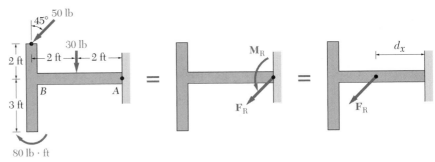

**Fig. P3.43**

**3.43** A system of forces and a couple are applied to the T-shaped member. Perform two stages of simplification of this system. (*a*) First, replace this system with a resultant force $\mathbf{F}_R$ and a resultant couple, $\mathbf{M}_R$, at *A*. (*b*) Next, find the distance $d_x$ that $\mathbf{F}_R$ may be moved horizontally along *AB* so that no couple is necessary. What you have demonstrated here is that a general planar force-couple system can be replaced by a single force, $\mathbf{F}_R$, properly located.

**3.44** Four forces act on a 28 × 15 in. plate as shown. (*a*) Find the resultant of these forces. (*b*) Locate the two points where the line of action of the resultant intersects the edge of the plate.

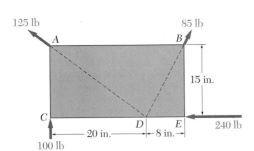

**Fig. P3.44**

**3.45** Solve Prob. 3.44, assuming that the 240-lb force is directed to the right.

**3.46** The three forces shown and a couple of moment $M = 6$ N·m are applied to an angle bracket. (*a*) Find the resultant of this system of forces. (*b*) Locate the points where the line of action of the resultant intersects line *AB* and line *BC*.

**3.47** The three forces shown and a couple $\mathbf{M}$ are applied to an angle bracket. Find the moment of the couple if the line of action of the resultant of the force system is to pass through (*a*) point *A*, (*b*) point *B*, (*c*) point *C*.

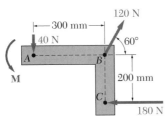

**Fig. P3.46 and P3.47**

**3.48** A bracket is subjected to the forces and couples shown. Find the moment of the couple **M** if the line of action of the resultant of this force system is to pass through (a) point A, (b) point B, (c) point C, (d) point D.

**3.49** A bracket is subjected to the forces and couples shown. Knowing that the moment of the couple **M** is 200 lb · in. clockwise, determine (a) the resultant of this force system, (b) the points where the line of action of the resultant intersects the centerlines of portions AB, BC, and CD.

**3.50** Two cables exert forces of 90 kN each on a truss of weight $W = 200$ kN. Find the resultant force acting on the truss and the point of intersection of its line of action with line AB.

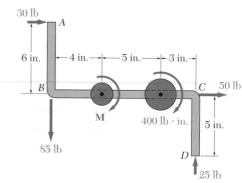

**Fig. P3.48 and P3.49**

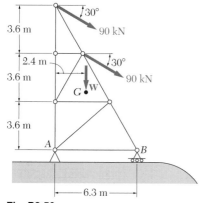

**Fig. P3.50**

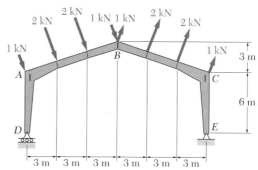

**Fig. P3.51**

**3.51** The roof of a building frame is subjected to the wind loading shown. Determine (a) the equivalent force-couple system at D, (b) the resultant of the loading and its line of action.

**3.52** Two forces **P** and **Q** are applied as shown to the corners A and B of a square plate. Determine their magnitudes P and Q and the angle β, knowing that their resultant **R** has a magnitude $R = 120$ N and a line of action passing through O and forming an angle $\theta = 30°$ with the x axis.

**3.53** Two forces **P** and **Q**, of magnitude P and Q, respectively, are applied as shown to the corners A and B of a square plate. Knowing that their resultant **R** has a magnitude $R = Q\sqrt{2}$ and a line of action passing through O, determine (a) the two possible values of the angle β, (b) the corresponding values of the ratio $P/Q$ of the magnitudes of the two forces, (c) the corresponding values of the angle θ that the resultant **R** forms with the x axis.

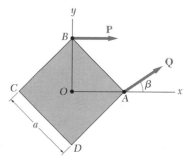

**Fig. P3.52 and P3.53**

**\*3.54** Solve Prob. 3.53 assuming that the resultant **R** of the two forces has a magnitude $R = 3Q/2$ and a line of action passing through O.

## EQUILIBRIUM OF RIGID BODIES

### 3.13. RIGID BODY IN EQUILIBRIUM

A *rigid body is said to be in equilibrium when the external forces acting on it form a system of forces equivalent to zero*, i.e., a system which has no resultant and no resultant couple. The necessary and sufficient conditions for the equilibrium of a rigid body in two dimensions may thus be expressed analytically by writing

$$\Sigma F_x = 0 \qquad \Sigma F_y = 0 \qquad \Sigma M_A = 0 \tag{3.6}$$

Since the direction of the axes of coordinates and the location of point $A$ may be chosen arbitrarily, Eqs. (3.6) indicate that the external forces acting on the rigid body do not impart to the body any motion of translation and do not make it rotate about any point. The action of any individual external force is thus canceled by the action of the other forces of the system, and the external forces are said to be *balanced*.

### 3.14. FREE-BODY DIAGRAM

In solving a problem concerning the equilibrium of a rigid body, it is essential to consider *all* the forces acting on the body; it is equally important to exclude any force which is not directly applied on the body. Omitting a force or adding an extraneous one would destroy the conditions of equilibrium. Therefore, the first step in the solution of the problem should consist in drawing a *free-body diagram* of the rigid body under consideration. Free-body diagrams have already been used on many occasions in Chap. 2. However, in view of their importance to the solution of equilibrium problems, we shall summarize here the various steps which must be followed in drawing a free-body diagram.

First, a clear decision is made regarding the choice of the free body to be used. This body is then detached from the ground and separated from any other body. The contour of the body thus isolated is sketched.

All external forces are then indicated. These forces represent the action exerted *on* the free body *by* the ground and the bodies which have been detached; they should be applied at the various points where the free body was supported by the ground or connected to the other bodies. The *weight* of the free body should also be included among the external forces, since it represents the attraction exerted by the earth on the various particles forming the free body. As will be seen in Chap. 5, the weight should be applied at the center of gravity of the body. When the free body is made of several parts, the forces the various parts exert on each other should *not* be included among the external forces. These forces are internal forces as far as the free body is concerned.

The magnitude and direction of the *known external forces* should be clearly marked on the free-body diagram. Care should be taken to indicate the sense of the force exerted *on* the free body, not that of the force exerted *by* the free body. Known external forces generally include the *weight* of the free body and *forces applied* for a given purpose.

3.15. Reactions at Supports and **89**
Connections for a Two-Dimensional
Structure

*Unknown external forces* usually consist of the *reactions*—sometimes called *constraining forces*—through which the ground and other bodies oppose a possible motion of the free body and thus constrain it to remain in the same position. Reactions are exerted at the points where the free body is *supported* or *connected* to other bodies. They will be discussed in detail in the next section.

The free-body diagram should also include dimensions, since these may be needed in the computation of moments of forces. Any other detail, however, should be omitted.

## 3.15. REACTIONS AT SUPPORTS AND CONNECTIONS FOR A TWO-DIMENSIONAL STRUCTURE

The reactions exerted on a rigid two-dimensional structure may be divided into three groups, corresponding to three types of *supports*, or *connections:*

1. *Reactions Equivalent to a Force with Known Line of Action.* Supports and connections causing reactions of this group include *rollers, rockers, frictionless surfaces, short links and cables, collars on frictionless rods,* and *frictionless pins in slots.* Each of these supports and connections can prevent motion in one direction only. They are shown in Fig. 3.16, together with the reaction they produce. Reactions of this group involve *one unknown,* namely, the magnitude of the reaction; this magnitude should be denoted by an appropriate letter. The line of action of the reaction is known and should be indicated clearly in the free-body diagram. The sense of the reaction must be as shown in Fig. 3.16 in the case of a frictionless surface (away from the surface) or of a cable (tension in the direction of the cable). The reaction may be directed either way in the case of double-track rollers, links, collars on rods, and pins in slots. Single-track rollers and rockers are generally assumed to be reversible, and thus the corresponding reactions may also be directed either way.

2. *Reactions Equivalent to a Force of Unknown Direction.* Supports and connections causing reactions of this group include *frictionless pins in fitted holes, hinges,* and *rough surfaces.* They can prevent translation of the free body in all directions, but they cannot prevent the body from rotating about the connection. Reactions of this group involve *two unknowns* and are usually represented by their $x$ and $y$ components. In the case of a rough surface, the component normal to the surface must be directed away from the surface.

3. *Reactions Equivalent to a Force and a Couple.* These reactions are caused by *fixed supports* which oppose any motion of the free body and thus constrain it completely. Fixed supports actually produce forces over the entire surface of contact; these forces, however, form a system which may be reduced to a force and a couple. Reactions of this group involve *three unknowns,* consisting usually of the two components of the force and the moment of the couple.

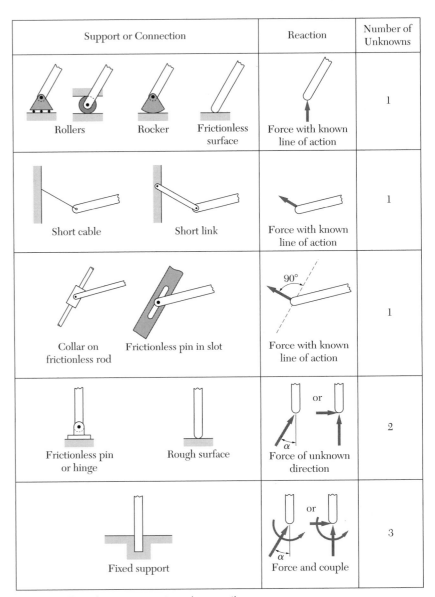

**Fig. 3.16** Reactions at supports and connections.

When the sense of an unknown force or couple is not clearly apparent, no attempt should be made to determine it. Instead, the sense of the force or couple should be arbitrarily assumed; the sign of the answer obtained will indicate whether the assumption is correct or not.

## 3.16. PROBLEMS INVOLVING THE EQUILIBRIUM OF A RIGID BODY IN TWO DIMENSIONS

We saw in Sec. 3.13 that the conditions for the equilibrium of a two-dimensional structure may be expressed by writing the three equations

$$\Sigma F_x = 0 \qquad \Sigma F_y = 0 \qquad \Sigma M_A = 0 \tag{3.6}$$

where $A$ is any point in the plane of the structure. The three equations obtained may be solved for no more than *three unknowns.*

We saw in the preceding section that unknown forces usually consist of reactions, and that the number of unknowns corresponding to a given reaction depends upon the type of support or connection causing that reaction. Referring to Sec. 3.15, we check that the equilibrium equations (3.6) may be used to determine the reactions of two rollers and one cable, or of one fixed support, or of one roller and one pin in a fitted hole, etc.

Consider, for instance, the truss shown in Fig. 3.17a, which is subjected to the given forces **P**, **Q**, and **S**. The truss is held in place by a pin at $A$ and a roller at $B$. The pin prevents point $A$ from moving by exerting on the truss a force which may be resolved into the components $\mathbf{A}_x$ and $\mathbf{A}_y$; the roller keeps the truss from rotating about $A$ by exerting the vertical force **B**. The free-body diagram of the truss is shown in Fig. 3.17b; it includes the reactions $\mathbf{A}_x$, $\mathbf{A}_y$, and **B** as well as the applied forces **P**, **Q**, **S**, and the weight **W** of the truss. Expressing that the sum of the moments about $A$ of all the forces shown in Fig. 3.17b is zero, we write the equation $\Sigma M_A = 0$, which may be solved for the magnitude $B$ since it does not contain $A_x$ or $A_y$. Expressing, then, that the sum of the $x$ components and the sum of the $y$ components of the forces are zero, we write the equations $\Sigma F_x = 0$ and $\Sigma F_y = 0$, which may be solved for the components $A_x$ and $A_y$, respectively.

Additional equations could be obtained by expressing that the sum of the moments of the external forces about points other than $A$ is zero. We could write, for instance, $\Sigma M_B = 0$. Such a statement, however, does not contain any new information, since it has already been established that the system of the forces shown in Fig. 3.17b is equivalent to zero. The additional equation *is not independent* and cannot be used to determine a fourth unknown. It will be useful, however, for checking the solution obtained from the original three equations of equilibrium.

While the three equations of equilibrium cannot be *augmented* by additional equations, any of them may be *replaced* by another equation. Therefore, an alternate system of equations of equilibrium is

$$\Sigma F_x = 0 \qquad \Sigma M_A = 0 \qquad \Sigma M_B = 0 \tag{3.7}$$

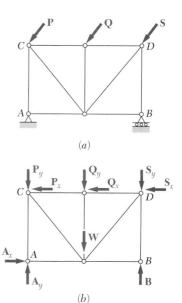

(a)

(b)

**Fig. 3.17**

where the line $AB$ is chosen in a direction different from the $y$ direction (Fig. 3.17$b$). These equations are sufficient conditions for the equilibrium of the truss. The first two equations indicate that the external forces must reduce to a single vertical force at $A$. Since the third equation requires that the moment of this force be zero about a point $B$ which is not on its line of action, the force must be zero and the rigid body is in equilibrium.

A third possible set of equations of equilibrium is

$$\Sigma M_A = 0 \qquad \Sigma M_B = 0 \qquad \Sigma M_C = 0 \qquad (3.8)$$

where the points $A$, $B$, and $C$ are not in a straight line (Fig. 3.17$b$). The first equation requires that the external forces reduce to a single force at $A$; the second equation requires that this force pass through $B$; the third, that it pass through $C$. Since the points $A$, $B$, $C$ are not in a straight line, the force must be zero, and the rigid body is in equilibrium.

The equation $\Sigma M_A = 0$, which expresses that the sum of the moments of the forces about pin $A$ is zero, possesses a more definite physical meaning than either of the other two equations (3.8). These two equations express a similar idea of balance, but with respect to points about which the rigid body is not actually hinged. They are, however, as useful as the first equation, and our choice of equilibrium equations should not be unduly influenced by the physical meaning of these equations. Indeed, it will be desirable in practice to choose equations of equilibrium containing only one unknown, since this eliminates the necessity of solving simultaneous equations. Equations containing only one unknown may be obtained by summing moments about the point of intersection of the lines of action of two unknown forces or, if these forces are parallel, by summing components in a direction perpendicular to their common direction. In the case of the truss of Fig. 3.18, for example, which is held by rollers at $A$ and $B$ and a short link at $D$, the reactions at $A$ and $B$ may be eliminated by summing $x$ components. The reactions at $A$ and $D$ will be eliminated by summing moments about $C$ and the reactions at $B$ and $D$ by summing moments about $D$. The equations obtained are

$$\Sigma F_x = 0 \qquad \Sigma M_C = 0 \qquad \Sigma M_D = 0$$

Each of these equations contains only one unknown.

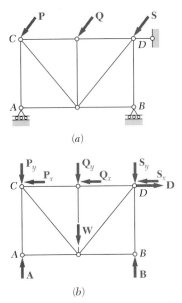

(a)

(b)

**Fig. 3.18**

## 3.17. STATICALLY INDETERMINATE REACTIONS. PARTIAL CONSTRAINTS

In each of the two examples considered in the preceding section (Figs. 3.17 and 3.18), the types of supports used were such that the rigid body could not possibly move under the given loads or under any other loading conditions. In such cases, the rigid body is said to be *completely constrained*. We also recall that the reactions corresponding to these supports involved *three unknowns* and could be determined by solving the three equations of equilibrium. When such a situation exists, the reactions are said to be *statically determinate*.

Consider now the truss shown in Fig. 3.19*a*, which is held by pins at *A* and *B*. These supports provide more constraints than are necessary to keep the truss from moving under the given loads or under any other loading conditions. We also note from the free-body diagram of Fig. 3.19*b* that the corresponding reactions involve *four unknowns*. Since, as was pointed out in Sec. 3.16, only three independent equilibrium equations are available, there are *more unknowns than equations*, and all the unknowns cannot be determined. While the equations $\Sigma M_A = 0$ and $\Sigma M_B = 0$ yield the vertical components $B_y$ and $A_y$, respectively, the equation $\Sigma F_x = 0$ gives only the sum $A_x + B_x$ of the horizontal components of the reactions at *A* and *B*. The components $A_x$ and $B_x$ are said to be *statically indeterminate*. They could be determined by considering the deformations produced in the truss by the given loading, but this method is beyond the scope of statics and belongs to the study of mechanics of materials.

The supports used to hold the truss shown in Fig. 3.20*a* consist of rollers at *A* and *B*. Clearly, the constraints provided by these supports are not sufficient to keep the truss from moving. While any vertical motion is prevented, the truss is free to move horizontally. The truss is said to be *partially constrained.*† Turning our attention to Fig. 3.20*b*, we note that the reactions at *A* and *B* involve only *two unknowns*. Since three equations of equilibrium must still be satisfied, there are *fewer unknowns than equations*, and one of the equilibrium equations will not be satisfied. While the equations $\Sigma M_A = 0$ and $\Sigma M_B = 0$ can be satisfied by a proper choice of reactions at *A* and *B*, the equation $\Sigma F_x = 0$ will not be satisfied unless the sum of the horizontal components of the applied forces happens to be zero. We thus check that the equilibrium of the truss of Fig. 3.20 cannot be maintained under general loading conditions.

It appears from the above that if a rigid body is to be completely constrained and if the reactions at its supports are to be statically determinate, *there must be as many unknowns as there are equations of equilibrium.* When this condition is *not* satisfied, we may be sure that the rigid body is not completely constrained, or that the reactions at its supports are not statically determinate, or both.

We should note, however, that, while *necessary*, the above condition is *not sufficient*. In other words, the fact that the number of unknowns is equal to the number of equations is no guarantee that the body is completely constrained or that the reactions at its supports are statically determinate. Consider the truss shown in Fig. 3.21*a*, which is held by rollers at *A*, *B*, and *E*. While there are three unknown reactions, **A**, **B**, and **E** (Fig. 3.21*b*), the equation $\Sigma F_x = 0$ will not be satisfied unless the sum of

† Partially constrained bodies are often referred to as *unstable*. However, to avoid confusion between this type of instability, due to insufficient constraints, and the type of instability considered in Chap. 10, which relates to the behavior of a rigid body when its equilibrium is disturbed, we shall restrict the use of the words *stable* and *unstable* to the latter case.

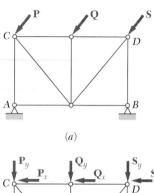

*(a)*

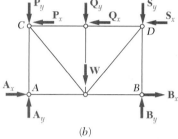

*(b)*

**Fig. 3.19** Statically indeterminate reactions.

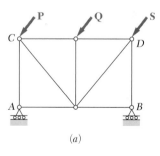

*(a)*

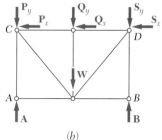

*(b)*

**Fig. 3.20** Partial constraints.

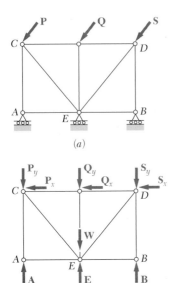

(a)

(b)

**Fig. 3.21**   Improper constraints.

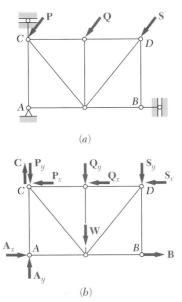

(a)

(b)

**Fig. 3.22**   Improper constraints.

the horizontal components of the applied forces happens to be zero. There is a sufficient number of constraints, but these constraints are not properly arranged, and the truss is free to move horizontally. We say that the truss is *improperly constrained.* Since only two equilibrium equations are left for determining three unknowns, the reactions will be statically indeterminate. Thus, improper constraints also produce static indeterminacy.

Another example of improper constraints—and of static indeterminacy—is provided by the truss shown in Fig. 3.22. This truss is held by a pin at $A$ and by rollers at $B$ and $C$, which altogether involve four unknowns. Since only three independent equilibrium equations are available, the reactions at the supports are statically indeterminate. On the other hand, we note that the equation $\Sigma M_A = 0$ cannot be satisfied under general loading conditions, since the lines of action of the reactions $\mathbf{B}$ and $\mathbf{C}$ are required to pass through $A$. We conclude that the truss can rotate about $A$ and that it is improperly constrained.†

The examples of Figs. 3.21 and 3.22 lead us to conclude that *a rigid body is improperly constrained whenever the supports,* even though they may provide a sufficient number of reactions, *are arranged in such a way that the reactions must be either concurrent or parallel.*‡

In summary, to be sure that a two-dimensional rigid body is completely constrained and that the reactions at its supports are statically determinate, we should check that the reactions involve three—and only three—unknowns, and that the supports are arranged in such a way that they do not require the reactions to be either concurrent or parallel.

Supports involving statically indeterminate reactions should be used with care in the *design* of structures, and only with a full knowledge of the problems they may cause. On the other hand, the *analysis* of structures possessing statically indeterminate reactions often may be partially carried out by the methods of statics. In the case of the truss of Fig. 3.19, for example, the vertical components of the reactions at $A$ and $B$ were obtained from the equilibrium equations.

For obvious reasons, supports producing partial or improper constraints should be avoided in the design of stationary structures. However, a partially or improperly constrained structure will not necessarily collapse; under particular loading conditions, equilibrium may be maintained. For example, the trusses of Figs. 3.20 and 3.21 will be in equilibrium if the applied forces $\mathbf{P}$, $\mathbf{Q}$, and $\mathbf{S}$ are vertical. Besides, structures which are designed to move *should* be only partially constrained. A railroad car, for instance, would be of little use if it were completely constrained by having its brakes applied permanently.

---

† Rotation of the truss about $A$ requires some "play" in the supports at $B$ and $C$. In practice such play will always exist. Besides, we may note that if the play is kept small, the displacements of the rollers $B$ and $C$ and, thus, the distances from $A$ to the lines of action of the reactions $\mathbf{B}$ and $\mathbf{C}$ will also be small. The equation $\Sigma M_A = 0$ will then require that $\mathbf{B}$ and $\mathbf{C}$ be very large, a situation which may well result in the failure of the supports at $B$ and $C$.

‡ Because this situation arises from an inadequate arrangement or *geometry* of the supports, it is often referred to as *geometric instability.*

# SAMPLE PROBLEM 3.8

A fixed crane has a mass of 1000 kg and is used to lift a 2400-kg crate. It is held in place by a pin at $A$ and a rocker at $B$. The center of gravity of the crane is located at $G$. Determine the components of the reactions at $A$ and $B$.

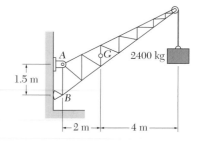

**Solution.** A free-body diagram of the crane is drawn. Multiplying the masses of the crane and of the crate by $g = 9.81$ m/s$^2$, we obtain the corresponding weights, that is, 9810 N or 9.81 kN, and 23 500 N or 23.5 kN. The reaction at pin $A$ is a force of unknown direction, represented by its components $\mathbf{A}_x$ and $\mathbf{A}_y$. The reaction at the rocker $B$ is perpendicular to the rocker surface; thus it is horizontal. We assume that $\mathbf{A}_x$, $\mathbf{A}_y$, and $\mathbf{B}$ act in the directions shown.

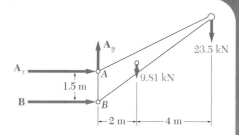

**Determination of B.** We express that the sum of the moments of all external forces about point $A$ is zero. The equation obtained will contain neither $A_x$ nor $A_y$ since the moments of $\mathbf{A}_x$ and $\mathbf{A}_y$ about $A$ are zero. Multiplying the magnitude of each force by its perpendicular distance from $B$, we write

$+\circlearrowleft \Sigma M_A = 0$:     $+B(1.5 \text{ m}) - (9.81 \text{ kN})(2 \text{ m}) - (23.5 \text{ kN})(6 \text{ m}) = 0$
                 $B = +107.1 \text{ kN}$                   $\mathbf{B} = 107.1 \text{ kN} \rightarrow$   ◀

Since the result is positive, the reaction is directed as assumed.

**Determination of $A_x$.** The magnitude $A_x$ is determined by expressing that the sum of the horizontal components of all external forces is zero.

$\xrightarrow{+} \Sigma F_x = 0$:     $A_x + B = 0$
             $A_x + 107.1 \text{ kN} = 0$
             $A_x = -107.1 \text{ kN}$         $\mathbf{A}_x = 107.1 \text{ kN} \leftarrow$   ◀

Since the result is negative, the sense of $\mathbf{A}_x$ is opposite to that assumed originally.

**Determination of $A_y$.** The sum of the vertical components must also equal zero.

$+\uparrow \Sigma F_y = 0$:     $A_y - 9.81 \text{ kN} - 23.5 \text{ kN} = 0$
             $A_y = +33.3 \text{ kN}$              $\mathbf{A}_y = 33.3 \text{ kN} \uparrow$   ◀

Adding vectorially the components $\mathbf{A}_x$ and $\mathbf{A}_y$, we find that the reaction at $A$ is 112.2 kN $\searsimeq$ 17.3°.

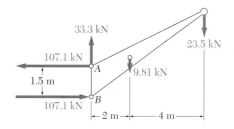

**Check.** The values obtained for the reactions may be checked by recalling that the sum of the moments of all external forces about any point must be zero. For example, considering point $B$, we write

    $+\circlearrowleft \Sigma M_B = -(9.81 \text{ kN})(2 \text{ m}) - (23.5 \text{ kN})(6 \text{ m}) + (107.1 \text{ kN})(1.5 \text{ m}) = 0$

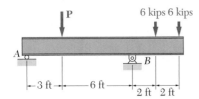

6 kips 6 kips

# SAMPLE PROBLEM 3.9

Three loads are applied to a beam as shown. The beam is supported by a roller at $A$ and by a pin at $B$. Neglecting the weight of the beam, determine the reactions at $A$ and $B$ when $P = 15$ kips.

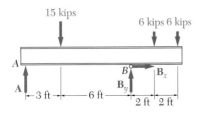

**Solution.** A free-body diagram of the beam is drawn. The reaction at $A$ is vertical and is denoted by **A**. The reaction at $B$ is represented by components $\mathbf{B}_x$ and $\mathbf{B}_y$. Each component is assumed to act in the direction shown.

*Equilibrium Equations.* We write the following three equilibrium equations and solve for the reactions indicated:

$$\xrightarrow{+} \Sigma F_x = 0: \qquad\qquad B_x = 0 \qquad\qquad \mathbf{B}_x = 0 \blacktriangleleft$$

$$+\uparrow \Sigma M_A = 0:$$
$$-(15 \text{ kips})(3 \text{ ft}) + B_y(9 \text{ ft}) - (6 \text{ kips})(11 \text{ ft}) - (6 \text{ kips})(13 \text{ ft}) = 0$$
$$B_y = +21.0 \text{ kips} \qquad\qquad \mathbf{B}_y = 21.0 \text{ kips} \uparrow \blacktriangleleft$$

$$+\uparrow \Sigma M_B = 0:$$
$$-A(9 \text{ ft}) + (15 \text{ kips})(6 \text{ ft}) - (6 \text{ kips})(2 \text{ ft}) - (6 \text{ kips})(4 \text{ ft}) = 0$$
$$A = +6.00 \text{ kips} \qquad\qquad \mathbf{A} = 6.00 \text{ kips} \uparrow \blacktriangleleft$$

**Check.** The results are checked by adding the vertical components of all the external forces:

$$+\uparrow \Sigma F_y = +6.00 \text{ kips} - 15 \text{ kips} + 21.0 \text{ kips} - 6 \text{ kips} - 6 \text{ kips} = 0$$

**Remark.** In this problem the reactions at both $A$ and $B$ are vertical; however, these reactions are vertical for different reasons. At $A$, the beam is supported by a roller; hence the reaction cannot have any horizontal component. At $B$, the horizontal component of the reaction is zero because it must satisfy the equilibrium equation $\Sigma F_x = 0$ and because none of the other forces acting on the beam has a horizontal component.

We could have noticed at first glance that the reaction at $B$ was vertical and dispensed with the horizontal component $\mathbf{B}_x$. This, however, is a bad practice. In following it, we would run the risk of forgetting the component $\mathbf{B}_x$ when the loading conditions require such a component (i.e., when a horizontal load is included). Also, the component $\mathbf{B}_x$ was found to be zero by using and solving an equilibrium equation, $\Sigma F_x = 0$. By setting $\mathbf{B}_x$ equal to zero immediately, we might not realize that we actually make use of this equation and thus might lose track of the number of equations available for solving the problem.

# SAMPLE PROBLEM 3.10

A loading car is at rest on a track forming an angle of 25° with the vertical. The gross weight of the car and its load is 5500 lb, and it is applied at a point 30 in. from the track, halfway between the two axles. The car is held by a cable attached 24 in. from the track. Determine the tension in the cable and the reaction at each pair of wheels.

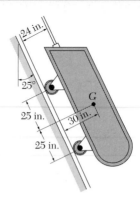

**Solution.** A free-body diagram of the car is drawn. The reaction at each wheel is perpendicular to the track, and the tension force **T** is parallel to the track. For convenience, we choose the $x$ axis parallel to the track and the $y$ axis perpendicular to the track. The 5500-lb weight is then resolved into $x$ and $y$ components.

$$W_x = +(5500 \text{ lb}) \cos 25° = +4980 \text{ lb}$$
$$W_y = -(5500 \text{ lb}) \sin 25° = -2320 \text{ lb}$$

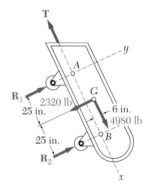

**Equilibrium Equations.** We take moments about $A$ to eliminate **T** and $R_1$ from the computation.

$+\uparrow \Sigma M_A = 0$:   $-(2320 \text{ lb})(25 \text{ in.}) - (4980 \text{ lb})(6 \text{ in.}) + R_2(50 \text{ in.}) = 0$
$R_2 = +1758 \text{ lb}$                     $R_2 = 1758 \text{ lb} \nearrow$ ◀

Now, taking moments about $B$ to eliminate **T** and $R_2$ from the computation, we write

$+\uparrow \Sigma M_B = 0$:   $(2320 \text{ lb})(25 \text{ in.}) - (4980 \text{ lb})(6 \text{ in.}) - R_1(50 \text{ in.}) = 0$
$R_1 = +562 \text{ lb}$                     $R_1 = 562 \text{ lb} \nearrow$ ◀

The value of $T$ is found by writing

$\searrow +\Sigma F_x = 0$:   $+4980 \text{ lb} - T = 0$
$T = +4980 \text{ lb}$                     $T = 4980 \text{ lb} \nwarrow$ ◀

The computed values of the reactions are shown in the adjacent sketch.

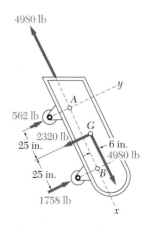

**Check.** The computations are verified by writing

$$\nearrow +\Sigma F_y = +562 \text{ lb} + 1758 \text{ lb} - 2320 \text{ lb} = 0$$

A check could also have been obtained by computing moments about any point except $A$ or $B$.

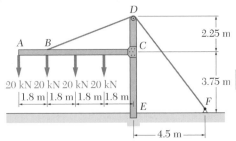

## SAMPLE PROBLEM 3.11

The frame shown supports part of the roof of a small building. Knowing that the tension in the cable is 150 kN, determine the reaction at the fixed end $E$.

**Solution.** A free-body diagram of the frame and of the cable $BDF$ is drawn. The reaction at the fixed end $E$ is represented by the force components $\mathbf{E}_x$ and $\mathbf{E}_y$ and the couple $\mathbf{M}_E$. The other forces acting on the free body are the four 20-kN loads and the 150-kN force exerted at end $F$ of the cable.

**Equilibrium Equations.** Noting that $DF = \sqrt{(4.5 \text{ m})^2 + (6 \text{ m})^2} = 7.5 \text{ m}$, we write

$$\xrightarrow{+} \Sigma F_x = 0: \qquad E_x + \frac{4.5}{7.5}(150 \text{ kN}) = 0$$

$$E_x = -90.0 \text{ kN} \qquad \mathbf{E}_x = 90.0 \text{ kN} \leftarrow \blacktriangleleft$$

$$+\uparrow \Sigma F_y = 0: \qquad E_y - 4(20 \text{ kN}) - \frac{6}{7.5}(150 \text{ kN}) = 0$$

$$E_y = +200 \text{ kN} \qquad \mathbf{E}_y = 200 \text{ kN} \uparrow \blacktriangleleft$$

$$+\uparrow \Sigma M_E = 0: \quad (20 \text{ kN})(7.2 \text{ m}) + (20 \text{ kN})(5.4 \text{ m}) + (20 \text{ kN})(3.6 \text{ m})$$

$$+ (20 \text{ kN})(1.8 \text{ m}) - \frac{6}{7.5}(150 \text{ kN})(4.5 \text{ m}) + M_E = 0$$

$$M_E = +180.0 \text{ kN} \cdot \text{m} \qquad \mathbf{M}_E = 180.0 \text{ kN} \cdot \text{m} \uparrow \blacktriangleleft$$

## SAMPLE PROBLEM 3.12

A 400-lb weight is attached to the lever $AO$ as shown. The constant of the spring $BC$ is $k = 250$ lb/in., and the spring is unstretched when $\theta = 0$. Determine the position of equilibrium.

**Solution.** *Force Exerted by Spring.* Denoting by $s$ the deflection of the spring from its undeformed position, and noting that $s = r\theta$, we write

$$F = ks = kr\theta$$

**Equilibrium Equation.** Summing the moments of $\mathbf{W}$ and $\mathbf{F}$ about $O$,

$$+\uparrow \Sigma M_O = 0: \qquad Wl \sin \theta - r(kr\theta) = 0 \qquad \sin \theta = \frac{kr^2}{Wl}\theta$$

Substituting the given data, we obtain

$$\sin \theta = \frac{(250 \text{ lb/in.})(3 \text{ in.})^2}{(400 \text{ lb})(8 \text{ in.})}\theta \qquad \sin \theta = 0.703\theta$$

Solving by trial, we find

$$\theta = 0 \qquad \theta = 80.3° \blacktriangleleft$$

# PROBLEMS

**3.55** A 3000-kg (acts at $G_2$) forklift truck is used to lift a 500-kg (acts at $G_1$) crate. Determine the reaction at each of the two (a) front wheels A, (b) rear wheels B.

**3.56** A hand truck is used to move a nitrogen cylinder. Knowing that the combined weight of the truck and cylinder is 180 lb, determine (a) the vertical force **P** which should be applied to the handle to maintain the cylinder in the position shown, (b) the corresponding reaction at each of the two wheels.

**3.57** A sports car is parked on a 20° slope, and the parking brake is set. Assume that the upslope-acting friction forces keeping the car in place act only on the rear wheels at B. (Mechanical parking brakes act on the rear wheels only on most cars.) Determine the normal reactions at each of the (a) front tires A, (b) rear wheels B, and (c) the upslope-acting friction forces at each of the rear wheels B. (Most cars have a mass center that is somewhat closer to the front wheels than to the rear. This car, being well designed, has a mass center exactly at the car's midpoint. This is desirable for predictable performance in road curves and in possible lateral slip situations.)

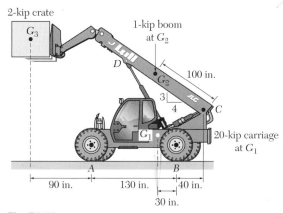

**Fig. P3.57**

**3.58** A JLG Lull Telehandler lifts a 2-kip load. Its large tires enable it to operate on rough terrain. At the boom position shown, determine the reaction at each of the two (a) front wheels A, (b) rear wheels B.

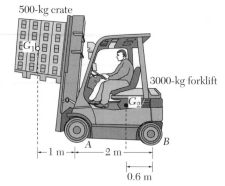

**Fig. P3.55**

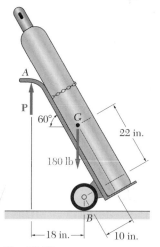

**Fig. P3.56**

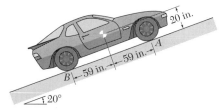

**Fig. P3.58**

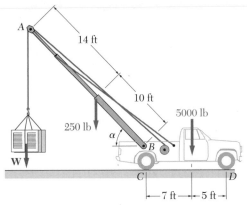

**Fig. P3.59**

**3.59** A truck-mounted crane is used to lift a 600-lb compressor. The weights of the boom $AB$ and of the truck are as shown, and the angle the boom forms with the horizontal is $\alpha = 45°$. Determine the reaction at each of the two (*a*) rear wheels $C$, (*b*) front wheels $D$.

**3.60** For the truck-mounted crane of Prob. 3.59, determine the smallest allowable value of $\alpha$ if the truck is not to tip over when a 3000-lb load is lifted.

**3.61** An adventurous boy pulls a wagon filled with rocks up a 10° slope at a constant speed. The combined weight of the wagon plus the load is 50 lb, acting at $G$. Determine (*a*) the force **P** the boy exerts at the handle, (*b*) the normal reactions at each front wheel at $A$, and (*c*) the normal reactions at each rear wheel at $B$. Neglect rolling friction (only normal forces act at the wheels).

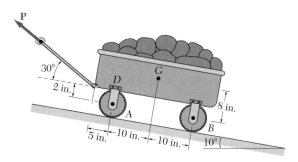

**Fig. P3.61**

**3.62** Three loads are applied as shown to a light beam supported by cables attached at $B$ and $D$. Neglecting the weight of the beam, determine the range of values of $Q$ for which neither cable becomes slack when $P = 0$.

**3.63** Three loads are applied as shown to a light beam supported by cables attached at $B$ and $D$. Knowing that the maximum allowable tension in each cable is 4 kN and neglecting the weight of the beam, determine the range of values of $Q$ for which the loading is safe when $P = 0$.

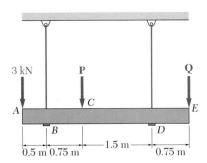

**Fig. P3.62 and P3.63**

**3.64** For the beam of Prob. 3.63, determine the range of values of $Q$ for which the loading is safe when $P = 1$ kN.

**3.65** A bent steel member is firmly embedded into the wall at $A$. Two forces and a 40 N-m couple are applied to the member. Determine the support reactions at the wall at $A$ necessary for equilibrium. Neglect the weight of the member itself.

**3.66** A T-shaped bracket supports a 200-N load as shown. Determine the reactions at $A$ and $C$.

**3.67** Two links $AB$ and $DE$ are connected by a bell crank as shown. Knowing that the tension in link $AB$ is 150 lb, determine (*a*) the tension in link $DE$, (*b*) the reaction at $C$.

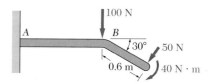

**Fig. P3.65**

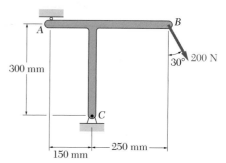

**Fig. P3.66**

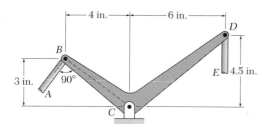

**Fig. P3.67 and P3.68**

**3.68** Two links $AB$ and $DE$ are connected by a bell crank as shown. Determine the maximum force which may be safely exerted by link $AB$ on the bell crank if the maximum allowable value for the reaction at $C$ is 400 lb.

**3.69** The lever $AB$ is hinged at $C$ and attached to a control cable at $A$. If the lever is subjected at $B$ to a 400-N horizontal force, determine (*a*) the tension in the cable, (*b*) the reaction at $C$.

**3.70** A truss may be supported in three different ways as shown. In each case, determine the reactions at the supports.

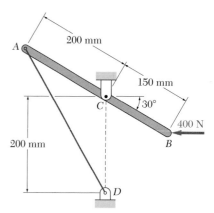

**Fig. P3.69**

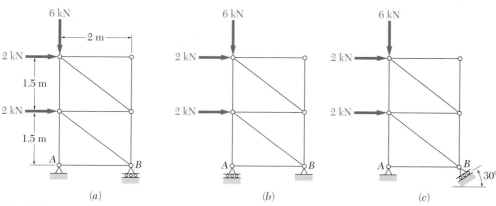

(*a*)          (*b*)          (*c*)

**Fig. P3.70**

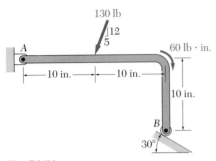

130 lb

12
5

60 lb · in.

A

—10 in.— —10 in.—

10 in.

B

30°

**Fig. P3.71**

**3.71** A force and a couple are applied to the inverted L-shaped member. Neglect the weight of the member. Determine the reactions at (a) the pin at A, (b) the roller at B necessary for equilibrium.

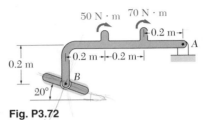

50 N · m   70 N · m

0.2 m

0.2 m + 0.2 m

0.2 m

A

B

20°

**Fig. P3.72**

**3.72** The odd-shaped member is loaded by two couples and supported by a pin at A and a roller in a smooth slot at B. Neglect the weight of the member. Determine the reactions at (a) the pin at A, (b) the roller at B necessary for equilibrium.

**3.73** Knowing that the tension in all portions of the belt is 800 N, determine the reactions at the supports A, B of the plate when (a) $\alpha = 0$, (b) $\alpha = 90°$, (c) $\alpha = 30°$.

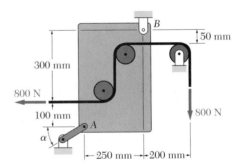

B

50 mm

300 mm

800 N

100 mm

800 N

A

$\alpha$

— 250 mm — 200 mm —

**Fig. P3.73**

**3.74** A light bar AD is suspended from a cable BE and supports a 20-kg block at C. The extremities A and D of the bar are in contact with frictionless, vertical walls. Determine the tension in cable BE and the reactions at A and D.

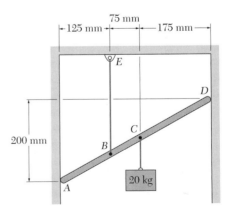

75 mm

— 125 mm — — 175 mm —

E

D

200 mm

C

B

20 kg

A

**Fig. P3.74**

**3.75**   A light bar *AB* supports a 15-kg block at its midpoint *C*. Rollers at *A* and *B* rest against frictionless surfaces, and a horizontal cable *AD* is attached at *A*. Determine the tension in cable *AD* and the reactions at *A* and *B*.

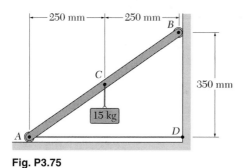

250 mm · 250 mm · 350 mm · 15 kg

**Fig. P3.75**

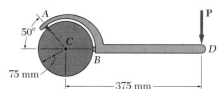

8 in. · 8 in. · 8 in. · 30° · 120 lb · α

**Fig. P3.76**

**3.76**   A light rod *AD* is supported by frictionless pegs at *B* and *C* and rests against a frictionless wall at *A*. A vertical 120-lb force (α = 0) is applied at *D*. Determine the reactions at *A*, *B*, and *C*.

**3.77**   Solve Prob. 3.76, assuming that the 120-lb force is applied in a direction perpendicular to the rod (α = 30°).

**3.78**   The spanner shown is used to rotate a shaft. A pin fits in a hole at *A*, while a flat, frictionless surface rests against the shaft at *B*. If a 250-N force **P** is exerted on the spanner at *D*, find (*a*) the reaction at *B*, (*b*) the component of the reaction at *A* in a direction perpendicular to *AC*.

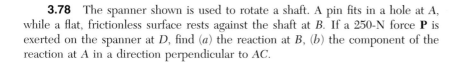

50° · **P** · *C* · *B* · *D* · 75 mm · 375 mm

**Fig. P3.78 and P3.79**

**3.79**   The spanner shown is used to rotate a shaft. A pin fits in a hole at *A*, while a flat, frictionless surface rests against the shaft at *B*. If the moment about *C* of the force exerted on the shaft at *A* is to be 75 N · m, find (*a*) the force **P** which should be exerted on the spanner at *D*, (*b*) the corresponding value of the force exerted on the spanner at *B*.

**3.80**   A 160-lb overhead garage door consists of a uniform rectangular panel *AC*, 84 in. high, supported by the cable *AE* attached at the middle of the upper edge of the door and by two sets of frictionless rollers at *A* and *B*. Each set consists of two rollers located on either side of the door. The rollers *A* are free to move in horizontal channels, while the rollers *B* are guided by vertical channels. If the door is held in the position for which *BD* = 42 in., determine (*a*) the tension in cable *AE*, (*b*) the reaction at each of the four rollers.

**3.81**   In Prob. 3.80, determine the distance *BD* for which the tension in cable *AE* is equal to 900 lb.

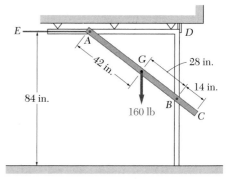

E · A · D · 42 in. · 28 in. · 14 in. · 84 in. · G · 160 lb · B · C

**Fig. P3.80**

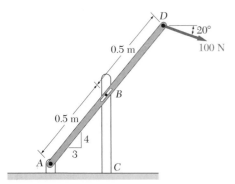

**Fig. P3.82**

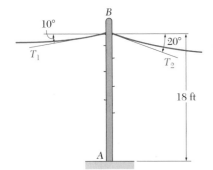

**Fig. P3.83**

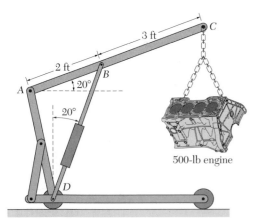

**Fig. P3.85**

**3.82**  Member $ABD$ is supported by a pin at $A$ and a pin in a smooth slot at $B$. Member $BC$ serves as a support post. A 100-N force is applied at $D$. Determine the reactions at (a) the pin at $A$, (b) the slot at $B$ necessary for equilibrium of $ABD$.

**3.83**  An 18-ft telephone pole weighing 320 lb is used to support the ends of two wires. The wires form the angles shown with the horizontal and the tensions in the wires are, respectively, $T_1 = 120$ lb and $T_2 = 75$ lb. Determine the reaction at the fixed end $A$.

**3.84**  A waterslide at an aquatic center is supported at many places along its twisting, turning route by simple frames like that shown here. This one consists of a 1.1-m horizontal member $ABD$ supported by an angled brace $BC$. This simple frame carries the 800-N weight of a 4-m section of the waterslide plus the flowing water. Determine the forces at pins $A$ and $B$ on member $ABD$ required for equilibrium. Neglect friction forces from flowing water.

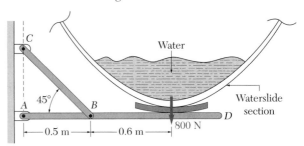

**Fig. P3.84**

**3.85**  An engine hoist is used to support a 500-lb engine. Determine the forces at $A$ and $B$ required for equilibrium of member $ABC$. Neglect the weight of $ABC$.

**3.86**  Rod $AD$ is acted upon by a vertical force $\mathbf{P}$ at end $A$ and by two equal and opposite horizontal forces of magnitude $Q$ at points $B$ and $C$. Neglecting the weight of the rod, express the angle $\theta$ corresponding to the equilibrium position in terms of $P$ and $Q$.

**3.87**  A uniform rod $AB$ of length $l$ and weight $W$ is suspended from two cords $AC$ and $BC$ of equal length. Determine the angle $\theta$ corresponding to the equilibrium position when a couple $\mathbf{M}$ is applied to the rod.

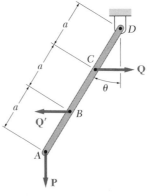

**Fig. P3.86**

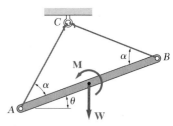

**Fig. P3.87**

**3.88 and 3.89** A vertical load **P** is applied at end $B$ of rod $BC$. (a) Neglecting the weight of the rod, express the angle $\theta$ corresponding to the equilibrium position in terms of $P$, $l$, and the counterweight $W$. (b) Determine the value of $\theta$ corresponding to equilibrium if $P = 2W$.

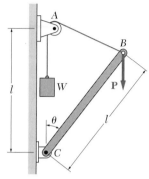

**Fig. P3.88**

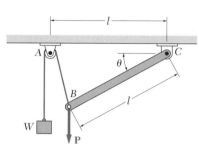

**Fig. P3.89**

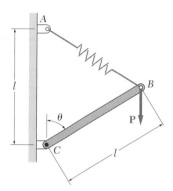

**Fig. P3.90**

**3.90** A vertical load **P** is applied at end $B$ of rod $BC$. The constant of the spring is $k$ and the spring is unstretched when $\theta = 60°$. (a) Neglecting the weight of the rod, express the angle $\theta$ corresponding to the equilibrium position in terms of $P$, $k$, and $l$. (b) Determine the value of $\theta$ corresponding to equilibrium if $P = \frac{1}{4}kl$.

**3.91** A vertical load **P** is applied at end $B$ of rod $BC$. The constant of the spring is $k$ and the spring is unstretched when $\theta = 0$. (a) Neglecting the weight of the rod, express the angle $\theta$ corresponding to the equilibrium position in terms of $P$, $k$, and $l$. (b) Determine the value of $\theta$ corresponding to equilibrium if $P = 2kl$.

**3.92** A collar $B$, of weight $W$, may move freely along the vertical rod shown. The constant of the spring is $k$ and the spring is unstretched when $y = 0$. (a) Derive an equation in $y$, $W$, $a$, and $k$ which must be satisfied when the collar is in equilibrium. (b) Find the value of $k$, knowing that $W = 60$ N, $a = 400$ mm, and that the collar is in equilibrium when $y = 300$ mm.

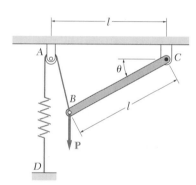

**Fig. P3.91**

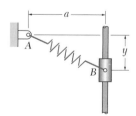

**Fig. P3.92**

**3.93** A slender rod $AB$, of weight $W$, is attached to blocks $A$ and $B$ which may move freely in the guides shown. The constant of the spring is $k$ and the spring is unstretched when $AB$ is horizontal. (a) Neglecting the weight of the blocks, derive an equation in $\theta$, $W$, $l$, and $k$ which must be satisfied when the rod is in equilibrium. (b) Find the value of $k$, knowing that $W = 30$ N, $l = 0.5$ m, and that the rod is in equilibrium when $\theta = 30°$.

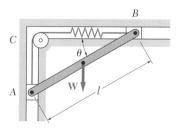

**Fig. P3.93**

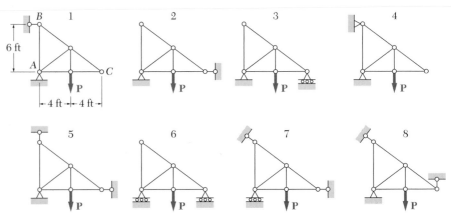

**Fig. P3.94**

**3.94** A small truss is supported in eight different ways as shown. All connections consist of frictionless pins, rollers, and short links. In each case, determine whether (*a*) the truss is completely, partially, or improperly constrained, (*b*) the reactions are statically determinate or indeterminate, (*c*) the equilibrium of the truss is maintained in the position shown. Also, wherever possible, compute the reactions, assuming that the magnitude of the force **P** is 10 kips.

**3.95** Nine identical rectangular plates, 600 × 900 mm, and each of mass *m* = 50 kg, are held in a vertical plane as shown. All connections consist of frictionless pins, rollers, or short links. For each case, answer the questions listed in Prob. 3.94, and, wherever possible, compute the reactions.

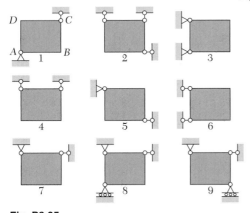

**Fig. P3.95**

## 3.18. EQUILIBRIUM OF A TWO-FORCE BODY

A particular case of equilibrium which is of considerable interest is that of a rigid body subjected to two forces. Such a body is commonly called a *two-force body*. We shall show that *if a two-force body is in equilibrium, the two forces must have the same magnitude, same line of action, and opposite sense.*

Consider a corner plate subjected to two forces $\mathbf{F}_1$ and $\mathbf{F}_2$ acting at $A$ and $B$, respectively (Fig. 3.23*a*). If the plate is to be in equilibrium, the sum of the moments of $\mathbf{F}_1$ and $\mathbf{F}_2$ about any axis must be zero. First, we sum moments about $A$: Since the moment of $\mathbf{F}_1$ is obviously zero, the moment of $\mathbf{F}_2$ must also be zero and the line of action of $\mathbf{F}_2$ must pass through $A$ (Fig. 3.23*b*). Summing moments about $B$, we prove similarly that the line

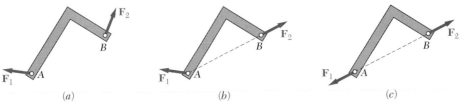

**Fig. 3.23** Note that a two-force body is assumed to be massless. If it had a weight vector, it would no longer be a two-force body.

of action of $\mathbf{F}_1$ must pass through $B$ (Fig. 3.23c). Both forces have the same line of action (line $AB$). From the equation $\Sigma F_x = 0$ or $\Sigma F_y = 0$, it is seen that they must have also the same magnitude but opposite sense.

If several forces act at two points $A$ and $B$, the forces acting at $A$ may be replaced by their resultant $\mathbf{F}_1$ and those acting at $B$ by their resultant $\mathbf{F}_2$. Thus a two-force body may be more generally defined as *a rigid body subjected to forces acting at only two points*. The resultants $\mathbf{F}_1$ and $\mathbf{F}_2$ then must have the same line of action, same magnitude, and opposite sense.

Students must learn to recognize two-force bodies. Seeing these greatly simplifies many statics problems. Hydraulic cylinders, common in machinery, may be considered to be two-force bodies (Fig. 3.24).

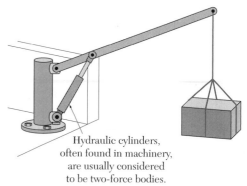

Hydraulic cylinders, often found in machinery, are usually considered to be two-force bodies.

**Fig. 3.24**

## 3.19. EQUILIBRIUM OF A THREE-FORCE BODY

Another case of equilibrium that is of interest is that of a *three-force body*, i.e., a rigid body subjected to three forces or, more generally, *a rigid body subjected to forces acting at only three points*. Consider a rigid body subjected to a system of forces which may be reduced to three forces $\mathbf{F}_1$, $\mathbf{F}_2$, and $\mathbf{F}_3$ acting at $A$, $B$, and $C$, respectively (Fig. 3.25a). We shall show that if the body is in equilibrium, *the lines of action of the three forces must be either concurrent or parallel*.

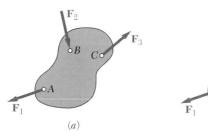

(a)           (b)           (c)

**Fig. 3.25**

Since the rigid body is in equilibrium, the sum of the moments of $\mathbf{F}_1$, $\mathbf{F}_2$, and $\mathbf{F}_3$ about any axis must be zero. Assuming that the lines of action of $\mathbf{F}_1$ and $\mathbf{F}_2$ intersect, and denoting their point of intersection by $D$, we sum moments about $D$ (Fig. 3.25b); since the moments of $\mathbf{F}_1$ and $\mathbf{F}_2$ about $D$ are zero, the moment of $\mathbf{F}_3$ about $D$ must also be zero and the line of action of $\mathbf{F}_3$ must pass through $D$ (Fig. 3.25c). The three lines of action are concurrent. The only exception occurs when none of the lines intersect; the lines of action are then parallel.

Although problems concerning three-force bodies may be solved by the general methods of Secs. 3.13 to 3.17, the property just established may be used to solve them either graphically, or mathematically from simple trigonometric or geometric relations.

Identifying three-force bodies is a useful, but not an essential, skill. Problems involving three-force bodies may be easily solved by standard statics methods. Identifying two-force bodies, in comparison, is absolutely essential.

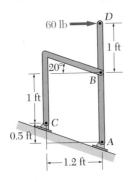

# SAMPLE PROBLEM 3.13

A 60-lb force acts at point $D$ on the frame shown at left. Determine the reactions at points $A$ and $B$ on member $ABD$ necessary for equilibrium. Consider both $ABD$ and $BC$ to be weightless.

**Free-body diagram of *ABD*:**

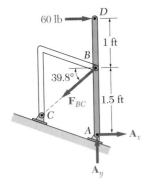

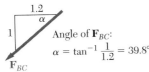

Angle of $\mathbf{F}_{BC}$:
$\alpha = \tan^{-1} \dfrac{1}{1.2} = 39.8°$

**Free-body diagram of *ABD*:**
**(not recognizing that FBC is a two-force)**
**Four unknowns! Cannot find all four.**

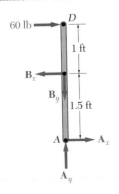

**Solution.** Draw a free-body diagram of member $ABD$. Recognize that **member *BC* is a two-force body,** thus a line drawn from pin $B$ to $C$ defines the line of action of $\mathbf{F}_{BC}$. The fact that $BC$ is an odd-shaped member is immaterial. The $B$ and $C$ pin locations—not the $BC$ shape—define the force direction.

*Write the Equilibrium Equations.* **Note there are three unknowns ($\mathbf{A_x}$, $\mathbf{A_y}$, and $\mathbf{F_{BC}}$), so the three static equilibrium equations are sufficient to determine them.**

$$+\!\downarrow \Sigma M_A = 0: \qquad F_{BC}(\cos 39.8)(1.5) - 60(2.5) = 0$$

$$F_{BC} = 130.2 \text{ lb} \quad 39.8° \nearrow \quad \blacktriangleleft$$

$$+\!\uparrow\!\Sigma F_y = 0: \qquad -F_{BC}(\sin 39.8) + A_y = 0$$

$$A_y = 83.3 \text{ lb}\!\uparrow \quad \blacktriangleleft$$

$$\overset{+}{\rightarrow}\Sigma F_x = 0: \qquad -F_{BC}(\cos 39.8) + 60 + A_x = 0$$

$$A_x = 40 \text{ lb} \rightarrow \quad \blacktriangleleft$$

**Concluding Remarks.** Recognizing that $BC$ is a two-force body gives the line of action of $\mathbf{F}_{BC}$, so the single unknown at $B$ is the *magnitude* $F_{BC}$. Finding three unknowns ($F_{BC}$, $A_x$, $A_y$) is easy. Not recognizing that $BC$ is a two-force body results in four unknowns on the FBD. You can manage to solve for $A_x$ and $B_x$, but not $A_y$ and $B_y$. It is good practice in statics to pause a moment after drawing the FBD and count unknowns. If you have more than three unknowns, check the support reactions and check for two-force bodies. In Chap. 6 you will study frames, which consist of multiple connected members. You will learn to draw multiple FBD's—one per member—which will give you more equilibrium equations—three per member—to use. But still, on frames, correctly finding two-force bodies leads to easier solutions.

# SAMPLE PROBLEM 3.14

A man raises a 10-kg joist, of length 4 m, by pulling on a rope. Find the tension $T$ in the rope and the reaction at $A$.

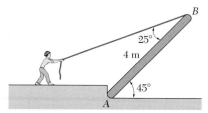

**Solution.** The joist is a three-force body since it is acted upon by three forces: its weight $\mathbf{W}$, the force $\mathbf{T}$ exerted by the rope, and the reaction $\mathbf{R}$ of the ground at $A$. We note that

$$W = mg = (10 \text{ kg})(9.81 \text{ m/s}^2) = 98.1 \text{ N}$$

*Three-Force Body.* Since the joist is a three-force body, the forces acting on it must be concurrent. The reaction $\mathbf{R}$, therefore, will pass through the point of intersection $C$ of the lines of action of the weight $\mathbf{W}$ and of the tension force $\mathbf{T}$. This fact will be used to determine the angle $\alpha$ that $\mathbf{R}$ forms with the horizontal.

Drawing the vertical $BF$ through $B$ and the horizontal $CD$ through $C$, we note that

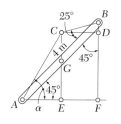

$$AF = BF = (AB)\cos 45° = (4 \text{ m})\cos 45° = 2.828 \text{ m}$$
$$CD = EF = AE = \tfrac{1}{2}(AF) = 1.414 \text{ m}$$
$$BD = (CD)\cot(45° + 25°) = (1.414 \text{ m})\tan 20° = 0.515 \text{ m}$$
$$CE = DF = BF - BD = 2.828 \text{ m} - 0.515 \text{ m} = 2.313 \text{ m}$$

We write

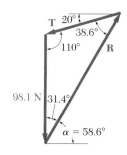

$$\tan \alpha = \frac{CE}{AE} = \frac{2.313 \text{ m}}{1.414 \text{ m}} = 1.636$$

$$\alpha = 58.6° \quad \blacktriangleleft$$

We now know the direction of all the forces acting on the joist.

*Force Triangle.* A force triangle is drawn as shown, and its interior angles are computed from the known directions of the forces. Using the law of sines, we write

$$\frac{T}{\sin 31.4°} = \frac{R}{\sin 110°} = \frac{98.1 \text{ N}}{\sin 38.6°}$$

$$T = 81.9 \text{ N} \quad \blacktriangleleft$$
$$\mathbf{R} = 147.8 \text{ N} \;\measuredangle\; 58.6° \quad \blacktriangleleft$$

# PROBLEMS

**3.96**   The spanner shown is used to rotate a shaft. A pin fits in a hole at *A*, while a flat, frictionless surface rests against the shaft at *B*. If a 300-N force **P** is exerted on the spanner at *D*, find the reactions at *A* and *B*.

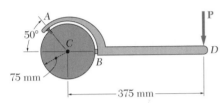

**Fig. P3.96**

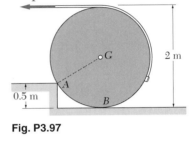

**Fig. P3.97**

**3.97**   A 250-kg cylindrical tank, 2 m in diameter, is to be raised over a 0.5-m obstruction. A cable is wrapped around the tank and pulled horizontally as shown. Knowing that the corner of the obstruction at *A* is rough, find the required tension in the cable and the reaction at *A*.

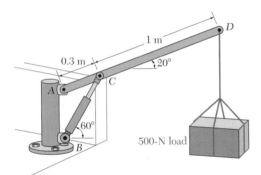

**Fig. P3.98**

**3.98**   A lift mounted on the bed of a pickup truck consists of a main support post *BA*, a boom *ACD*, and a hydraulic cylinder *BC*. If the lift supports a 500-N load, determine the reactions at points *A* and *C* on *ACD* necessary for equilibrium.

**3.99**   A simple frame supports a 40 N-m couple and a 100-N force. Determine the reactions at points *A* and *B* on member *AB* necessary for equilibrium.

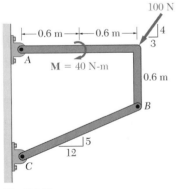

**Fig. P3.99**

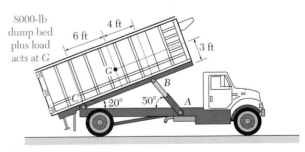

**Fig. P3.100**

**3.100**   A dump truck designed to carry grain is shown. The dump bed is actuated by a hydraulic cylinder *AB*. If the combined weight of the dump bed plus its load is 8000 lb, and the bed is at the position shown, determine the forces acting on the bed at *B* and *C* necessary for equilibrium.

**3.101**  One end of rod *AB* rests in the corner *A* and the other is attached to the cord *BD*. If the rod supports a 180-N load at its midpoint *C*, find the reaction at *A* and the tension in the cord.

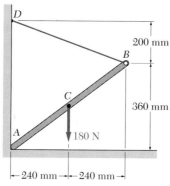

**Fig. P3.101**

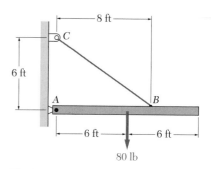

**Fig. P3.102**

**3.102**  A 12-ft wooden beam weighing 80 lb is supported by a pin and bracket at *A* and by the cable *BC*. Find the reaction at *A* and the tension in the cable.

**3.103**  A small skidsteer loader, common on construction sites, carries a load with bucket *F*, supported by an inverted L-shaped arm *CBDE* which pivots about *C*. The position of arm *CBDE* is controlled by a hydraulic cylinder *AB*. When arm *CBDE* is at the position shown, determine the reactions at *B* and *C* necessary for equilibrium. (The 300-lb weight shown is *half* of the total bucket plus load weight. Another arm identical to *CBDE* on the other side of the loader carries the other half.)

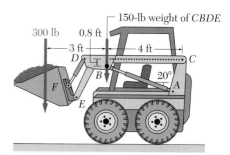

**Fig. P3.103**

**3.104**  A frame consisting of an L-shaped member *AB* and a quarter-circle link *BC* supports a 120 lb-ft couple. Determine the reactions at *A* and *B* on *AB* necessary for equilibrium.

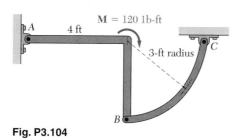

**Fig. P3.104**

**3.105**  A frame consisting of an L-shaped member *AB* and an elbow-shaped link *BC* supports a 150-N force. Determine the reactions at *A* and *B* on *AB* necessary for equilibrium.

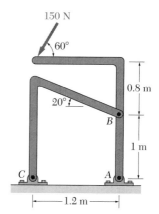

**Fig. P3.105**

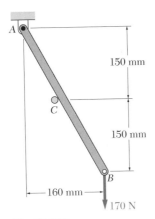

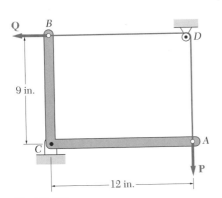

Fig. P3.106

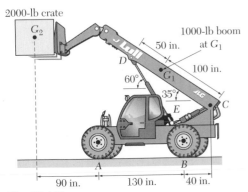

Fig. P3.108

Fig. P3.111

**3.106** Rod $AB$ is supported by a pin and bracket at $A$ and rests against a frictionless peg at $C$. Determine the reactions at $A$ and $C$ when a 170-N vertical force is applied at $B$.

**3.107** A frame consisting of a horizontal member $ABD$ and a link $BC$ supports a 400-N force. Determine the reactions at $A$ and $B$ on $ABD$ necessary for equilibrium.

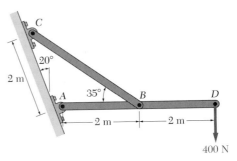

Fig. P3.107

**3.108** The L-shaped member $ACB$ is supported by a pin and bracket at $C$ and by an inextensible cord attached at $A$ and $B$ and passing over a frictionless pulley at $D$. The tension may be assumed to be the same in portions $AD$ and $BD$ of the cord. If the magnitudes of the forces applied at $A$ and $B$ are, respectively, $P = 30$ lb and $Q = 0$, determine (a) the tension in the cord, (b) the reaction at $C$.

**3.109** For the L-shaped member of Prob. 3.108, (a) express the tension $T$ in the cord in terms of the magnitudes $P$ and $Q$ of the forces applied at $A$ and $B$, (b) assuming $P = 30$ lb, find the largest allowable value of $Q$ if the equilibrium is to be maintained.

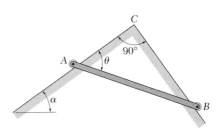

Fig. P3.110

**3.110** The uniform rod $AB$ lies in a vertical plane. Its ends are connected to rollers which rest against frictionless surfaces. Determine the relation between the angles $\theta$ and $\alpha$ when the rod is in equilibrium.

**3.111** A rough-terrain construction lift supports a 2000-lb crate. The boom $CD$ weighs 1000 lb, acting at $G_1$. (Assume $G_1$ acts along a line $CD$ [not shown] drawn from $C$ to $D$.) The boom $CD$ is supported by a hydraulic cylinder $DE$. When the boom is in the position shown, determine the reactions at $C$ and $D$ on the boom necessary for equilibrium. (Ignore the small hydraulic cylinder located near $C$.)

**3.112**  A thin ring of mass 2 kg and radius $r = 140$ mm is held against a frictionless wall by a 125-mm string $AB$. Determine (a) the distance $d$, (b) the tension in the string, (c) the reaction at $C$.

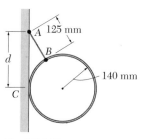

**Fig. P3.112**

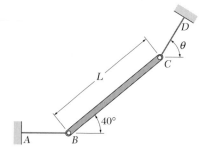

**Fig. P3.113**

**3.113**  A slender rod $BC$ of length $L$ and weight $W$ is held by two cables as shown. Knowing that cable $AB$ is horizontal and that the rod forms an angle of $40°$ with the horizontal, determine (a) the angle $\theta$ that cable $CD$ forms with the horizontal, (b) the tension in each cable.

**\*3.114**  A slender rod of length $2r$ and weight $W$ is attached to a collar at $B$ and rests on a circular cylinder of radius $r$. Knowing that the collar may slide freely along a vertical guide and neglecting friction, determine the value of $\theta$ corresponding to equilibrium.

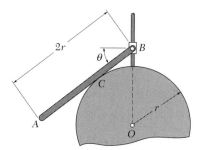

**Fig. P3.114**

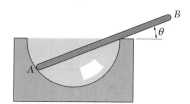

**Fig. P3.115**

**\*3.115**  A uniform rod $AB$ of length $3R$ and weight $W$ rests inside a hemispherical bowl of radius $R$ as shown. Neglecting friction, determine the angle $\theta$ corresponding to equilibrium.

# REVIEW AND SUMMARY
# FOR CHAPTER 3

In the first part of this chapter we studied *the effect of coplanar forces on rigid, two-dimensional structures.* We first learned to distinguish between external forces and internal forces [Sec. 3.2] and saw that, according to the *principle of transmissibility,* the effect of an external force on a rigid body remains unchanged if that force is moved along its line of action [Sec. 3.3]. In other words, two forces **F** and **F′** acting on a rigid body at two different points have the same effect on that body if they have the same magnitude, same direction, and same line of action (Fig. 3.26). Two such forces are said to be *equivalent.*

**Principle of transmissibility**

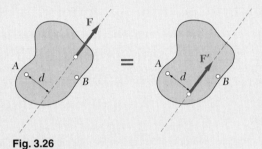

**Fig. 3.26**

**Moment of a force**

The *moment $M_A$ of the force* **F** about a point $A$—or more precisely about an axis through $A$ perpendicular to the plane of the figure—was defined [Sec. 3.4] as the product of the magnitude $F$ of the force and the perpendicular distance $d$ from $A$ to the line of action of **F** (Fig. 3.26):

$$M_A = Fd \tag{3.1}$$

The moment of a force about an axis measures the tendency of the force to rotate a rigid body about that axis. It has not only a magnitude but also a sense; in Fig. 3.26 the moment of **F** about $A$ is seen to be counterclockwise ↱ and its moment about $B$ clockwise ↓. Moments of forces may be added algebraically as scalar quantities if a sign convention is adopted; in this text we consider ↱ as positive and ↓ as negative.

It follows from the above definition that the moment of a force remains unchanged when the force is moved along its line of action. Thus, two equivalent forces **F** and **F′** (Fig. 3.26) have the same components $F_x$ and $F_y$ and the same moment $M_A$ about any given axis.

The application of *Varignon's theorem* [Sec. 3.6] may facilitate the computation of the moment of a force **F** about an axis through a given point $A$: The force is first resolved into $x$ and $y$ components (or some other appropriate components); the moments of the components are then computed about $A$ and added to obtain the moment of **F** about $A$ [Sample Probs. 3.2 and 3.3].

Two forces **F** and **F′** having the same magnitude $F$, parallel lines of action, and opposite sense are said to form a *couple* (Fig. 3.27). It was shown in Sec. 3.7 that the sum $M$ of the moments of **F** and **F′** about a point $A$ is independent of the choice of $A$ and equal to the product of the common magnitude $F$ of the forces and the distance $d$ between their lines of action:

$$M = Fd \qquad (3.4)$$

It was further shown [Sec. 3.8] that the effect of a couple on a rigid body is completely defined by its moment (magnitude and sense). *Two couples are equivalent* (i.e., they have the same effect on a rigid body), therefore, *if they have the same moment;* it does not matter where the forces forming each of the two couples act, or what their magnitude and direction are. It also follows that two couples may be replaced by a single couple of moment equal to the algebraic sum of the moments of the given couples [Sec. 3.9].

Any force **F** acting on a rigid body may be moved to a given point $A$, provided that a couple **M** is added, of moment $M$ equal to the moment of **F** (in its original position) about $A$ (Fig. 3.28); the force and couple thus obtained are referred to as a *force-couple system* [Sec. 3.10].

**Moment of a couple**

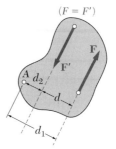

**Fig. 3.27**

**Force-couple system**

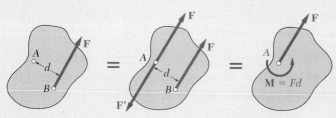

**Fig. 3.28**

*Any system of coplanar forces may be reduced to a force-couple system at an arbitrary point $A$* by replacing each of the forces of the system by an equivalent force-couple system at $A$ and adding all the forces and all the couples obtained in this manner [Sec. 3.11]. The system may be further reduced to a single equivalent force **R** called the *resultant of the system.* In the special case where **R** = 0, the system reduces to a single couple **M** called the *resultant couple* of the system. The reduction of a system of coplanar forces to a single force or couple may usually be facilitated by resolving each of the original forces into its $x$ and $y$ components [Sample Prob. 3.7].

Two systems of coplanar forces are *equivalent* if they may be reduced to the same force-couple system at a given point $A$. It follows that two systems of coplanar forces are equivalent if the sums of the $x$ components, of the $y$ components, and of the moments about $A$ of their forces are, respectively, equal [Sec. 3.12].

**Resultant of a system of forces**

**Equivalent systems of forces**

**Equilibrium equations**

In the second part of the chapter we considered the *equilibrium of a two-dimensional structure*, i.e., the situation when the external forces exerted on the structure form a system equivalent to zero. It follows from the preceding paragraph that the conditions for the equilibrium of a rigid two-dimensional structure may be expressed by the *three equilibrium equations*

$$\Sigma F_x = 0 \qquad \Sigma F_y = 0 \qquad \Sigma M_A = 0 \tag{3.6}$$

where $A$ is an arbitrary point in the plane of the structure [Sec. 3.13]. These equations may be solved for three unknowns.

Before applying the equilibrium equations (3.6) to the solution of a given problem, it is necessary to identify correctly all the forces acting on the structure and draw the corresponding *free-body diagram* [Sec. 3.14]. In addition to the forces *applied* to the structure, we had to consider the *reactions* exerted on the structure by its supports [Sec. 3.15]. Depending upon the type of support, a reaction could involve one, two, or three unknowns, as shown in Fig. 3.16.

While the three equilibrium equations (3.6) cannot be *augmented* by additional equations, any of them may be *replaced* by another equation [Sec. 3.16]. Therefore, we may write alternative sets of equilibrium equations, such as

$$\Sigma F_x = 0 \qquad \Sigma M_A = 0 \qquad \Sigma M_B = 0 \tag{3.7}$$

where the line $AB$ is chosen in a direction different from the $y$ direction, or

$$\Sigma M_A = 0 \qquad \Sigma M_B = 0 \qquad \Sigma M_C = 0 \tag{3.8}$$

where the points $A$, $B$, and $C$ are not in a straight line.

**Statical indeterminacy**

**Partial constraints**

**Improper constraints**

Since any set of equilibrium equations may be solved for only three unknowns, the reactions at the supports of a rigid two-dimensional structure may not be completely determined if they involve *more than three unknowns;* they are said to be *statically indeterminate* [Sec. 3.17]. On the other hand, if the reactions involve *fewer than three unknowns,* equilibrium will not be maintained under general loading conditions; the structure is said to be *partially constrained.* The fact that the reactions involve exactly three unknowns is no guarantee that the equilibrium equations can be solved for all three unknowns. If the supports are arranged in such a way that the reactions *must be either concurrent or parallel,* the reactions are statically indeterminate and the structure is said to be *improperly constrained.*

**Two-force body**

Two particular cases of equilibrium of a rigid body were considered at the end of the chapter. In Sec. 3.18, a *two-force body* was defined as a rigid body subjected to forces at only two points, and it was shown that the resultants $\mathbf{F}_1$ and $\mathbf{F}_2$ of these forces must have the *same magnitude, same line of action, and opposite sense* (Fig. 3.29), a property which will simplify the solution of certain

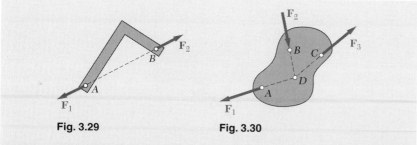

**Fig. 3.29**    **Fig. 3.30**

problems [Sample Prob. 3.13]. In Sec. 3.19, a *three-force body* was defined as a rigid body subjected to forces at only three points, and it was shown that the resultants $\mathbf{F}_1$, $\mathbf{F}_2$, and $\mathbf{F}_3$ of these forces must be *either concurrent* (Fig. 3.30) *or parallel.* This property provides us with an alternative approach to the solution of problems involving a three-force body [Sample Prob. 3.14].

Three-force body

# REVIEW PROBLEMS

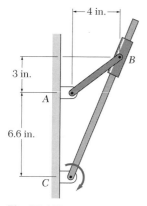

**Fig. P3.116**

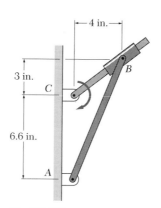

**Fig. P3.117**

**3.116 and 3.117**    It is known that the slider $B$ exerts on rod $AB$ a 130-lb force perpendicular to rod $BC$ and directed down and to the right. Determine the moment of that force about $A$.

**3.118**    A 500-N force is applied to a bent plate as shown. Determine (*a*) an equivalent force-couple system at $A$, (*b*) an equivalent system formed by parallel forces acting at $A$ and $B$.

**3.119**    A 500-N force is applied to a bent plate as shown. Determine (*a*) an equivalent force-couple system at $B$, (*b*) an equivalent system formed by a vertical force at $A$ and a force at $B$.

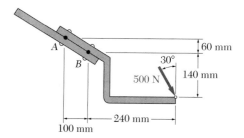

**Fig. P3.118 and P3.119**

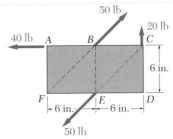

**Fig. P3.120**

**3.120** A 6 × 12 in. plate is subjected to four loads. Find the resultant of the four loads and the two points at which the line of action of the resultant intersects the edge of the plate.

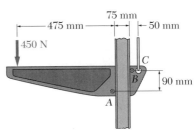

**Fig. P3.121**

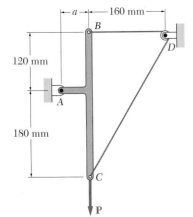

**Fig. P3.122**

**3.121** A movable bracket is held at rest by a cable attached at $C$ and by frictionless rollers at $A$ and $B$. For the loading shown, determine (*a*) the tension in the cable, (*b*) the reactions at $A$ and $B$.

**3.122** A force **P** of magnitude 450 N is applied to member $ABC$ which is supported by a frictionless pin at $A$ and by the cable $BDC$. Since the cable passes over a pulley at $D$, the tension may be assumed to be the same in the portions $BD$ and $CD$ of the cable. Determine (*a*) the tension in the cable, (*b*) the reaction at $A$ when $a = 60$ mm.

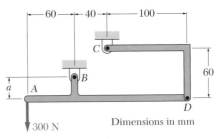

**Fig. P3.123**

**3.123** Determine the reactions at $B$ and $C$ when $a = 25$ mm.

**3.124** A light rod $AD$ supports a vertical load **P** and is attached to collars $B$ and $C$ which may slide freely on the rods shown. Knowing that the wire attached at $A$ forms an angle $a = 30°$ with the horizontal, determine (a) the tension in the wire, (b) the reactions at $B$ and $C$.

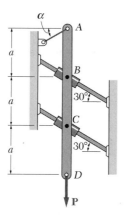

**Fig. P3.124**

**3.125** A force **P** of magnitude 60 lb is applied to end $E$ of cable $CDE$ which passes under pulley $D$ and is attached to the mechanism at $C$. Neglecting the weight of the mechanism and the radius of the pulley, determine the value of $\theta$ corresponding to equilibrium. The constant of the spring is $k = 20$ lb/in., and the spring is unstretched when $\theta = 90°$.

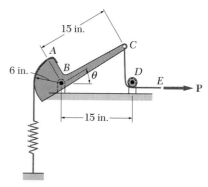

**Fig. P3.125**

**3.126** A force **P** is applied to a bent rod $AD$ which may be supported in four different ways as shown. In each case determine the reactions at the supports.

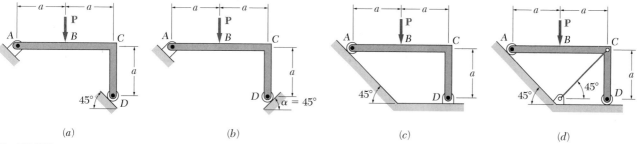

(a)  (b)  (c)  (d)

**Fig. P3.126**

**\*3.127** In the problems listed below, the rigid bodies considered were completely constrained, and the reactions were statically determinate. For each of these rigid bodies it is possible to create an improper set of constraints by changing either a dimension of the body or the direction of a reaction. In each problem determine the value of $\alpha$ or of $a$ which results in improper constraints. (a) Prob. 3.122, (b) Prob. 3.123, (c) Prob. 3.124, (d) Prob. 3.126b.

# COMPUTER PROBLEMS

**3.C1**  A beam $AB$ is subjected to several vertical forces as shown. Write a computer program which can be used to determine the magnitude of the resultant and the point where its line of action intersects $AB$. Use this program to solve (a) Sample Prob. 3.6c, (b) Prob. 3.42.

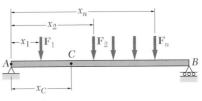

**Fig. P3.C1**

**3.C2**  Two forces **P** and **Q**, of magnitude $P = 50$ N and $Q = 120$ N, are applied as shown to corners $A$ and $B$ of a square plate of side $a = 500$ mm. Write a computer program which can be used to calculate the equivalent force-couple system at $D$. Use this program to determine the magnitude and the direction of **R** and the moment of $\mathbf{M}_D$ for values of $\beta$ from 0 to 360° at 30° intervals.

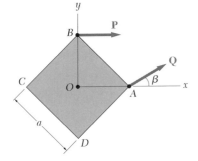

**Fig. P3.C2**

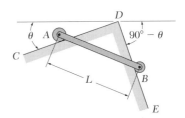

**Fig. P3.C3**

**3.C3**  Two wheels $A$ and $B$, of weight $W_A$ and $W_B$, respectively, are connected by a rod of negligible weight and are free to roll on the surface shown. Write a computer program which can be used to determine the angle $\alpha$ that the rod forms with the horizontal in its equilibrium position. Use this program to calculate $\alpha$ for values of $\theta$ from 0 to 90° at 5° intervals. Consider the cases when (a) $W_A = W_B$, (b) $W_A = 2W_B$.

**3.C4**  A vertical load **P** is applied to the end of rod $BC$. Write a computer program which can be used to calculate the magnitude $P$ of the load required for the equilibrium of the system. For $W = 10$ lb and $l = 25$ in., use this program to calculate $P$ for values of $\theta$ from 0 to 90° at 5° intervals.

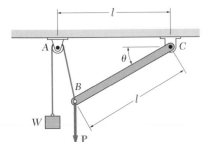

**Fig. P3.C4**

# CHAPTER 4

# Statics of Rigid Bodies in Three Dimensions

## 4.1. INTRODUCTION

In the first part of this chapter we shall study the effect of forces on a three-dimensional rigid body and we shall learn to replace a given system of forces by a simpler equivalent system. While our approach will be essentially the same as that used in Chap. 3 for two-dimensional structures and will again be based on the application of the principle of transmissibility, it will be necessary to extend some of the concepts introduced earlier.

For example, in Chap. 3 when we defined the moment of a force about an axis, we assumed that the axis was perpendicular to the plane containing the force. In three-dimensional space, this will generally not be the case and a new definition of the moment of a force about an axis, based on the projection of the force on a plane perpendicular to the axis, will be required (Sec. 4.2). We shall see that the computation of the moments $M_x$, $M_y$, and $M_z$ of a force $\mathbf{F}$ about each of the coordinate axes is greatly facilitated by drawing the projections of $\mathbf{F}$ on each of the coordinate planes.

When we introduced the concept of a couple in Chap. 3, we found that the effect of a couple on a two-dimensional structure was completely defined by the moment $M$ of the couple (magnitude and sense). We shall see in this chapter that the effect of a couple on a three-dimensional body also depends upon the orientation of the plane in which it acts, and that a couple may be represented by a vector, called a *couple vector*, perpendicular to the plane of the couple and of magnitude equal to the moment $M$ of the couple (Secs. 4.3 and 4.4).

As we shall see, any system of forces can again be reduced to a *force-couple system* at a given point. However, in the general case of three-dimensional forces, the system cannot be reduced further to a single resultant force or couple; yet it may be reduced to a special force-couple system called a *wrench* (Sec. 4.6).

The second part of this chapter deals with the *equilibrium of a three-dimensional rigid body*, i.e., with the situation when the system of forces exerted on the body reduces to zero. We shall see that *six equilibrium equations* may then be written (Sec. 4.7), stating, respectively, that the sum of the components of the forces in the $x$, $y$, and $z$ directions and the sum of their moments about the $x$, $y$, and $z$ axes must be zero.

In order to write the equations of equilibrium for a rigid body, it is necessary to first identify all the forces acting on that body and draw the corresponding *free-body diagram*. In addition to the forces *applied* to the body, we shall consider the *reactions* exerted on the body by its supports. We shall learn to associate a specific reaction with each type of support and determine whether the body is properly supported, so that we may know in advance whether the equations of equilibrium may actually be solved for the unknown forces and reactions (Sec. 4.8).

## EQUIVALENT SYSTEMS OF FORCES

### 4.2. MOMENT OF A FORCE ABOUT AN AXIS

When the moment of a force about an axis was defined in Chap. 3, all the forces considered were contained in the plane of the figure and their moments were computed about axes perpendicular to that plane. In space, however, the forces generally are not contained in a plane perpendicular to the axis about which their moment is to be determined. The definition of the moment of a force given in Sec. 3.5 needs therefore to be extended.

Consider a force $\mathbf{F}$ applied at $B$ and an axis $AA'$ (Fig. 4.1). We resolve $\mathbf{F}$ into two perpendicular components $\mathbf{F}_1$ and $\mathbf{F}_2$; the component $\mathbf{F}_1$ is chosen in a direction parallel to $AA'$, and the component $\mathbf{F}_2$ lies in a plane $P$ perpendicular to $AA'$, and passing through $B$. (We say that $\mathbf{F}_2$ is the *projection* of $\mathbf{F}$ on $P$.) If $\mathbf{F}$ acted on a rigid body, only its component $\mathbf{F}_2$ would tend to make the body rotate about the axis $AA'$. Therefore, *the moment of $\mathbf{F}$ about the axis $AA'$ is defined as the moment of $\mathbf{F}_2$ about that axis*. It is equal to the product $F_2 d$, where $d$ is the perpendicular distance from the axis $AA'$ to the line of action of $\mathbf{F}_2$. The sense of the moment (clockwise or counterclockwise) is determined from the point of view of an observer located at $A'$ and looking toward $A$. The moment of $\mathbf{F}$ about the axis $AA'$ (directed from $A$ to $A'$) shown in Fig. 4.1 is thus counterclockwise.

It should be noted that the moment of a force $\mathbf{F}$ about an axis $AA'$ is zero if the line of action of $\mathbf{F}$ either intersects the axis or is parallel to it. In the first case, the line of action of the projection $\mathbf{F}_2$ of $\mathbf{F}$ also intersects the axis (Fig. 4.2), and, in the second case, the magnitude of $\mathbf{F}_2$ is zero (Fig. 4.3). In both cases the moment of $\mathbf{F}_2$, and thus the moment of $\mathbf{F}$, is zero.

Consider now a force $\mathbf{F}$ acting on the corner $B$ of a rectangular box of sides $a$, $b$, and $c$ (Fig. 4.4). The force may impart to the box a motion of translation in the $x$, $y$, or $z$ directions, or a motion of rotation about the $x$, $y$, or $z$ axes, or a combination of all these motions. The ability of the force

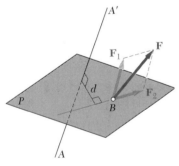

**Fig. 4.1**

to produce a translation is measured by its rectangular components $F_x$, $F_y$, $F_z$; its ability to produce a rotation about any of the coordinate axes is measured by its moment about the axis. The moments of $\mathbf{F}$ about the $x$, $y$, and $z$ axes are denoted, respectively, by $M_x$, $M_y$, and $M_z$.

According to the definition given above for the moment of a force, the moment $M_x$ of $\mathbf{F}$ about the $x$ axis will be obtained by projecting $\mathbf{F}$ on the $yz$ plane and computing the moment of its projection about the $x$ axis. This is done in Fig. 4.5$a$, which shows the $yz$ plane as an observer located along the positive half of the $x$ axis and looking toward $O$ would see it. It will generally be found convenient to carry out the actual computation of $M_x$ by using Varignon's theorem and computing the moments of the component forces $\mathbf{F}_y$ and $\mathbf{F}_z$ about $O$. Figure 4.5$b$ shows the projection of $\mathbf{F}$ on the $zx$ plane, as seen by an observer located along the positive half of the $y$ axis and looking toward $O$. The moment of this projection about the $y$ axis is the moment $M_y$ of $\mathbf{F}$ about that axis. Figure 4.5$c$ shows the projection of $\mathbf{F}$ on the $xy$ plane, as seen by an observer located along the positive half of the $z$ axis and looking toward $O$. The moment of this projection about the $z$ axis is the moment $M_z$ of $\mathbf{F}$ about that axis. In the case considered here, the moments $M_x$, $M_y$, and $M_z$ are, respectively, clockwise, clockwise, and counterclockwise. Recalling the sign convention adopted in Sec. 3.5, we may also state that the moments $M_x$, $M_y$, and $M_z$ are, respectively, negative, negative, and positive.

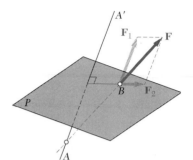

**Fig. 4.2**

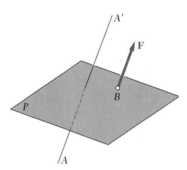

**Fig. 4.3**

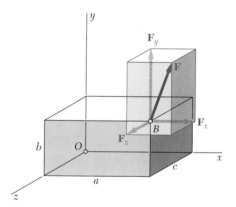

**Fig. 4.4**

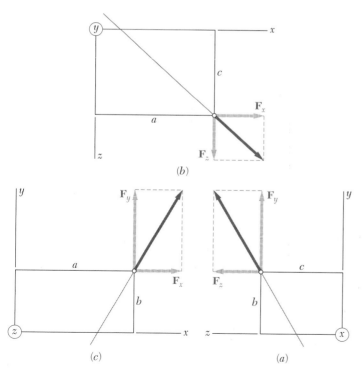

**Fig. 4.5**

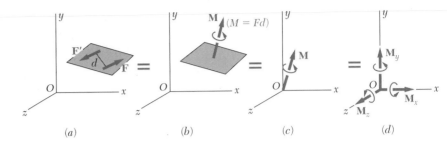

**Fig. 4.6**      (*a*)      (*b*)      (*c*)      (*d*)

## 4.3. COUPLES IN SPACE

A *couple* was defined in Sec. 3.7 as the combination of two forces **F** and **F′**
having the same magnitude $F$, parallel lines of action, and opposite sense. It
was shown in Sec. 3.8 that the action a couple exerts on a rigid body depends
only upon the moment $M$ of the couple, which is equal to the product $Fd$
of the magnitude $F$ of the forces and the distance $d$ between their lines of
action. In the case of a couple in space, however, *the plane in which the couple acts* should be specified (Fig. 4.6*a*). Since the couple may be replaced
by any other couple of the same moment acting in the same plane, it will be
found convenient to represent a couple in space by a *vector* perpendicular
to the plane of the couple and of magnitude equal to the moment $M$ of the
couple (Fig. 4.6*b*). This vector, denoted by M, is called a *couple vector,* and
its sense is such that an observer located at the tip of **M** will see the couple
acting counterclockwise. The sense of the couple vector may also be found
by applying the *right-hand rule:* Close your right hand, and hold it so that
your fingers are curled in the sense defined by the couple; your thumb will
indicate the sense of the couple vector. Note that the symbol ↑ is added to
the couple vector to distinguish it from vectors representing forces.

The point of application of the couple vector may be chosen anywhere in the plane of the couple. In fact, it may be chosen anywhere in
space since, as will be seen below, only the *orientation* of the plane of the
couple matters. (In other words, two couples of the same moment and
acting in parallel planes are equivalent.) Because of this fact, couple vectors are *free vectors,* as opposed to bound vectors, which have well-defined
points of application, and sliding vectors, which have well-defined lines of
action. The point of application of the couple vector, therefore, may be
chosen at the origin of the system of coordinates (Fig. 4.6*c*). Furthermore,
the couple vector **M** may be resolved into component vectors $\mathbf{M}_x$, $\mathbf{M}_y$, $\mathbf{M}_z$,
directed along the coordinate axes (Fig. 4.6*d*). The vector $\mathbf{M}_x$, of magnitude $M_x$, directed along the $x$ axis, represents a couple of moment $M_x$ acting
in the $yz$ plane; similarly, the other two vectors represent couples of moment $M_y$ and $M_z$ acting, respectively, in the $zx$ and $xy$ planes.

## 4.4. COUPLES ARE VECTORS

The representation of a couple by means of a vector is convenient. This
representation, however, should be *justified,* i.e., we should show that couples possess all the characteristics of vectors (see Sec. 2.3). We already
know that couples have magnitude and direction; we should also show that
couples may be added by adding the corresponding couple vectors according to the parallelogram law.

Consider two couples of moment $M_1$ and $M_2$ acting, respectively, in planes $P_1$ and $P_2$ (Fig. 4.7a). Each of these couples may be represented by two forces $\mathbf{F}$ and $\mathbf{F}'$ of arbitrary magnitude $F$. The distance $d_1$ between the forces forming the first couple and the distance $d_2$ between the forces forming the second couple should satisfy the equations

$$M_1 = Fd_1 \qquad M_2 = Fd_2 \qquad (4.1)$$

One force in each couple may be chosen along the line of intersection $AA'$ of the two planes, one in the sense $AA'$, the other in the opposite sense. The forces along $AA'$ cancel each other, and the given couples thus reduce to the forces along $BB'$ and $CC'$. These two forces form a couple of moment $M = Fd$ acting in the plane $P$ defined by $BB'$ and $CC'$ (Fig. 4.7b).

Let us now represent the two given couples of Fig. 4.7a by arrows $\mathbf{M}_1$ and $\mathbf{M}_2$, of magnitude $M_1$ and $M_2$, respectively perpendicular to the planes $P_1$ and $P_2$ (Fig. 4.8a). Drawing $\mathbf{M}_1$ and $\mathbf{M}_2$ in tip-to-tail fashion, we define a plane parallel to plane $ABC$ of Fig. 4.8c. Connecting the tail of $\mathbf{M}_1$ to the tip of $\mathbf{M}_2$ (Fig. 4.8b), we obtain a triangle similar to triangle $ABC$ since the sides $\mathbf{M}_1$ and $\mathbf{M}_2$ are, respectively, perpendicular to the sides $AB$ and $AC$ and since, according to (4.1), they are proportional to $d_1$ and $d_2$. The third side of the triangle of Fig. 4.8b must therefore be perpendicular to $BC$ and be proportional to $d$; its magnitude is therefore $Fd = M$. Thus, the third side of the triangle constructed on $\mathbf{M}_1$ and $\mathbf{M}_2$ is equal to the arrow $\mathbf{M}$ which represents the resultant couple (Fig. 4.8c). We conclude that $\mathbf{M}$ may be obtained by adding $\mathbf{M}_1$ and $\mathbf{M}_2$ vectorially and, thus, that the arrows $\mathbf{M}_1$, $\mathbf{M}_2$, and $\mathbf{M}$ representing the couples *are truly vectors.*

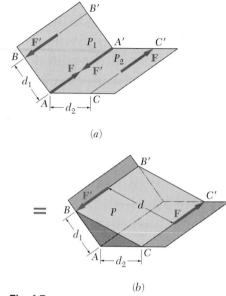

**Fig. 4.7**

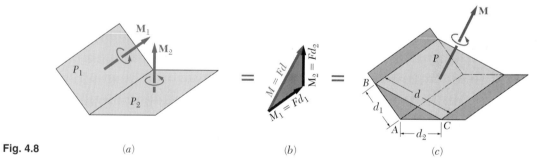

**Fig. 4.8**  (a)  (b)  (c)

We note that, if forces $\mathbf{Q}$ and $\mathbf{Q}'$, of magnitude $Q$ different from $F$, are used to represent the couples in Fig. 4.7a, new distances $d_1'$ and $d_2'$ are obtained and the resultant couple, which has the same moment $M$, is found to act in a plane $P'$ parallel to the plane $P$ of Fig. 4.7b. Clearly, this new couple is equivalent to the couple of Fig. 4.7b. Since by an appropriate choice of the forces $\mathbf{Q}$ and $\mathbf{Q}'$, the plane $P'$ may be chosen at any distance from $P$, we conclude that *couples of the same moment and acting in parallel planes are equivalent.* Couples acting in parallel planes may thus be moved freely from one plane to another and added as if they were coplanar.

Summarizing the results obtained, we conclude that *couples in space may be truly represented by vectors.* These vectors, called couple vectors, may be added or resolved according to the parallelogram law; they are *free vectors* and may be applied at any point.

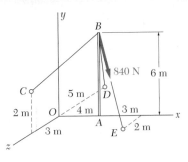

## SAMPLE PROBLEM 4.1

A pole $AB$, 6 m long, is held by three guy wires as shown. Determine the moment about each of the coordinate axes of the force exerted by wire $BE$ on point $B$. The tension $T$ in wire $BE$ is known to be 840 N.

**Solution.** The force $\mathbf{T}$ exerted by $BE$ is first resolved into components. The components and magnitude of the vector $\overrightarrow{BE}$, which joins points $B$ and $E$, are

$$d_x = +3\text{ m} \qquad d_y = -6\text{ m} \qquad d_z = +2\text{ m} \qquad d = 7\text{ m}$$

Therefore,

$$\frac{T_x}{+3\text{ m}} = \frac{T_y}{-6\text{ m}} = \frac{T_z}{+2\text{ m}} = \frac{840\text{ N}}{7\text{ m}}$$

$$T_x = +360\text{ N} \qquad T_y = -720\text{ N} \qquad T_z = +240\text{ N}$$

The projections of $\mathbf{T}$ on the three coordinate planes are now sketched. Using Varignon's theorem, we write

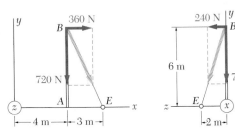

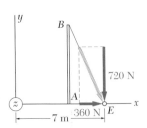

$$+\!\uparrow M_x = +(240\text{ N})(6\text{ m}) = +1440\text{ N}\cdot\text{m} \qquad \blacktriangleleft$$
$$+\!\uparrow M_y = -(240\text{ N})(4\text{ m}) = -960\text{ N}\cdot\text{m} \qquad \blacktriangleleft$$
$$+\!\uparrow M_z = -(720\text{ N})(4\text{ m}) - (360\text{ N})(6\text{ m}) = -5040\text{ N}\cdot\text{m} \qquad \blacktriangleleft$$

**Alternate Solution.** The computation of $M_z$ may be simplified by moving the projection of $\mathbf{T}$ along its line of action to the point where it intersects the $x$ axis; the moment of $\mathbf{T}_x$ about the $z$ axis is now zero. We obtain

$$+\!\uparrow M_z = -(720\text{ N})(7\text{ m}) = -5040\text{ N}\cdot\text{m} \qquad \blacktriangleleft$$

# SAMPLE PROBLEM 4.2

Determine the components of the single couple equivalent to the two couples shown.

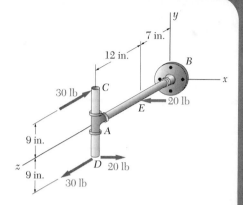

**Solution.** Our computations will be simplified if we attach two equal and opposite 20-lb forces at $A$. This enables us to replace the original 20-lb-force couple by two new 20-lb-force couples, one of which lies in the $xz$ plane and the other in a plane parallel to the $xy$ plane. The three couples shown in the adjoining sketch may be represented by three couple vectors $\mathbf{M}_x$, $\mathbf{M}_y$, and $\mathbf{M}_z$ directed along the coordinate axes. The corresponding moments are

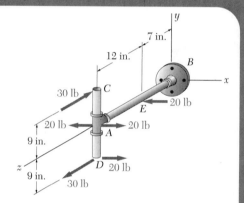

$$M_x = -(30 \text{ lb})(18 \text{ in.}) = -540 \text{ lb} \cdot \text{in.}$$
$$M_y = +(20 \text{ lb})(12 \text{ in.}) = +240 \text{ lb} \cdot \text{in.}$$
$$M_z = +(20 \text{ lb})(9 \text{ in.}) = +180 \text{ lb} \cdot \text{in.}$$

The three couple vectors $\mathbf{M}_x$, $\mathbf{M}_y$, $\mathbf{M}_z$ represent the vector components of the single couple $\mathbf{M}$ equivalent to the two given couples. The scalar components of $\mathbf{M}$, therefore, are

$$M_x = -540 \text{ lb} \cdot \text{in.} \qquad M_y = +240 \text{ lb} \cdot \text{in.} \qquad M_z = +180 \text{ lb} \cdot \text{in.} \blacktriangleleft$$

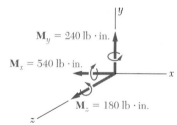

**Alternate Solution.** The components of the equivalent single couple $\mathbf{M}$ may also be obtained by computing the sum of the moments of the four given forces about each of the coordinate axes:

$$M_x = -(30 \text{ lb})(9 \text{ in.}) - (30 \text{ lb})(9 \text{ in.}) = -540 \text{ lb} \cdot \text{in.}$$
$$M_y = +(20 \text{ lb})(19 \text{ in.}) - (20 \text{ lb})(7 \text{ in.}) = +240 \text{ lb} \cdot \text{in.}$$
$$M_z = +(20 \text{ lb})(9 \text{ in.}) = +180 \text{ lb} \cdot \text{in.}$$

The scalar components of $\mathbf{M}$, therefore, are

$$M_x = -540 \text{ lb} \cdot \text{in.} \qquad M_y = +240 \text{ lb} \cdot \text{in.} \qquad M_z = +180 \text{ lb} \cdot \text{in.} \blacktriangleleft$$

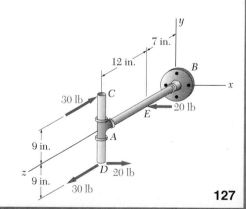

**127**

# PROBLEMS

**4.1**  Three forces act on a rectangular box as shown. Determine the moment of each force about each of the coordinate axes.

**4.2**  Three forces act on a rectangular box as shown. Determine the moment of each force about (*a*) edge *AD*, (*b*) edge *GF*, (*c*) edge *BC*, (*d*) a line joining corners *A* and *C*.

**4.3**  The wire *AE* is stretched between the corners *A* and *E* of a bent plate. Knowing that the tension in the wire is 435 N, determine the moment about each of the coordinate axes of the force exerted by the wire on corner *A*.

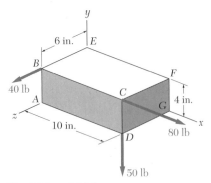

**Fig. P4.1 and P4.2**

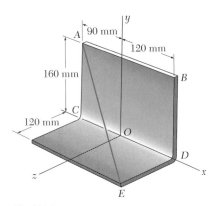

**Fig. P4.3**

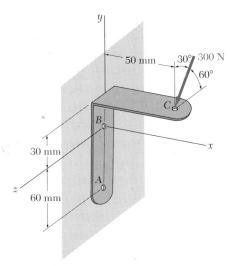

**Fig. P4.4**

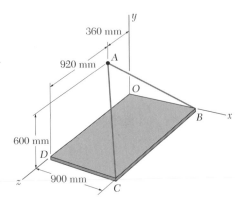

**Fig. P4.5 and P4.6**

**4.4**  A 300-N force is applied as shown to the bracket *ABC*. Determine the moment of the force about each of the coordinate axes.

**4.5**  Knowing that the tension in cable *AB* is 190 N, determine the moment about each of the coordinate axes of the force exerted on the plate at *B*.

**4.6**  Knowing that the tension in cable *AC* is 355 N, determine the moment about each of the coordinate axes of the force exerted on the plate at *C*.

**4.7** The jib crane is oriented so that the boom $DA$ is parallel to the $x$ axis. At the instant shown the tension in cable $AB$ is 13 kN. Determine the moment about each of the coordinate axes of the force exerted on $A$ by the cable $AB$.

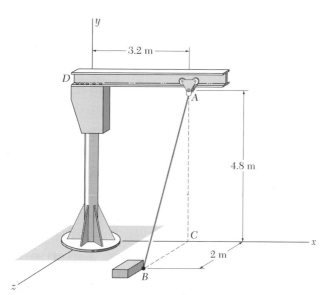

**Fig. P4.7 and P4.8**

**4.8** The jib crane is oriented so that the boom $DA$ is parallel to the $x$ axis. Determine the maximum permissible tension in the cable $AB$ if the absolute value of the moments about the coordinate axes of the force exerted on $A$ must be as follows: $M_x \leq 10$ kN · m, $M_y \leq 6$ kN · m, $M_z \leq 16$ kN · m.

**4.9** A small boat hangs from two davits, one of which is shown in the figure. It is known that the moment about the $z$ axis of the resultant force $\mathbf{R}_A$ exerted on the davit at $A$ must not exceed 160 lb · ft in absolute value. Determine the largest allowable tension in line $ABAD$ when $x = 4.8$ ft.

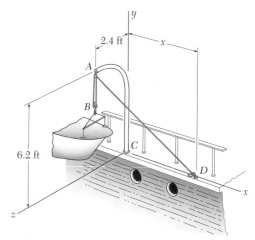

**Fig. P4.9**

**4.10** For the davit of Prob. 4.9, determine the largest allowable distance $x$ when the tension in line $ABAD$ is 48 lb.

**4.11** A single force $\mathbf{P}$ acts at $C$ in a direction perpendicular to the handle $BC$ of the crank shown. Knowing that $M_x = +20$ N · m, $M_y = -8.75$ N · m, and $M_z = -30$ N · m, determine the magnitude of $\mathbf{P}$ and the values of $\phi$ and $\theta$.

**4.12** A single force $\mathbf{P}$ acts at $C$ in a direction perpendicular to the handle $BC$ of the crank shown. Determine the moment $M_x$ of $\mathbf{P}$ about the $x$ axis when $\theta = 70°$, knowing that $M_y = -20$ N · m and $M_z = -37.5$ N · m.

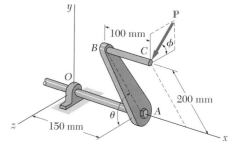

**Fig. P4.11 and P4.12**

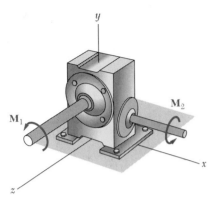

**Fig. P4.15**

**4.13** A force **F** of components $F_x$, $F_y$, and $F_z$ acts at a point of coordinates $x$, $y$, and $z$. Find expressions for the moment of **F** about each of the coordinate axes.

**4.14** A single force **F** acts at a point $A$ of coordinates $x = y = z = a$. Show that $M_x + M_y + M_z = 0$; that is, show that the *algebraic* sum of the moments of **F** about the coordinate axes is zero.

**4.15** The two shafts of a speed-reducer unit are subjected to couples of magnitude $M_1 = 15$ lb · ft and $M_2 = 3$ lb · ft, respectively. Replace the two couples by a single equivalent couple, specifying its moment and the direction of its axis.

**4.16** The two couples shown are to be replaced by a single equivalent couple. Determine (*a*) the couple vector representing the equivalent couple, (*b*) the two forces acting at $B$ and $C$ which may be used to form that couple.

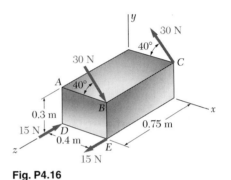

**Fig. P4.16**

**4.17** Solve part *a* of Prob. 4.16, assuming that two 25-N vertical forces have been added, one acting upward at $A$ and the other downward at $C$.

**4.18** Shafts $A$ and $B$ connect the gear box to the wheel assemblies of a tractor, and shaft $C$ connects it to the engine. Shafts $A$ and $B$ lie in the vertical $yz$ plane, while shaft $C$ is directed along the $x$ axis. Replace the couples applied to the shafts by a single equivalent couple, specifying its moment and the direction of its axis.

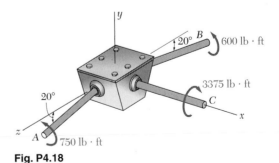

**Fig. P4.18**

## 4.5. RESOLUTION OF A GIVEN FORCE INTO A FORCE AT *O* AND A COUPLE

Consider a force **F** applied at the corner *A* of a rectangular box. Since a force may always be resolved into components parallel to the coordinate axes, we shall assume that **F** is parallel to one of the axes, say, the *z* axis (Fig. 4.9*a*).

Operating in plane *ABCD* according to the rules established in Chap. 3, we replace the given force by a force-couple system at *B*. The force, of magnitude *F*, is parallel to the *z* axis; the couple, of moment equal to the moment $M_y$ of **F** (in its original position) about the *y* axis, is contained in plane *ABCD* and may therefore be represented by a vector $\mathbf{M}_y$ directed along the *y* axis (Fig. 4.9*b*). Operating now in plane *BCEO*, we move the couple vector to *O* and replace the force **F** acting at *B* by a force-couple system at *O*; the new couple, of moment equal to the moment $M_x$ of **F** (acting at *B* or at *A*) about the *x* axis, may be represented by a vector $\mathbf{M}_x$ directed along the *x* axis (Fig. 4.9*c*). We conclude therefore that *the force* **F** *acting at A may be moved to O, provided that couple vectors are added along the coordinate axes; the magnitude of each couple vector must equal the moment of* **F** *(in its original position) about the corresponding axis.* Since, in the case considered, **F** has zero moment about the *z* axis, there is no couple vector along that axis.

The couple vectors may be replaced by their resultant. In the case considered, the resultant of $\mathbf{M}_x$ and $\mathbf{M}_y$ is a couple vector **M** lying in the *xy* plane (Fig. 4.9*d*). Thus, *the couple vector* **M** *is perpendicular to the force* **F**. Note that the vector **M** represents a couple acting in plane *ADEO* and could have been obtained more directly by moving the force from *A* to *O* within that plane. While the representation given in Fig. 4.9*d* is more compact than that given in Fig. 4.9*c*, the representation of Fig. 4.9*c* is usually preferred since it shows more clearly the action of the original force on the rigid body.

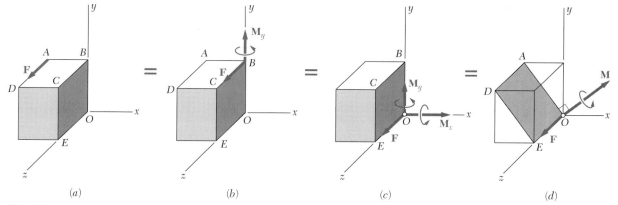

(*a*)  (*b*)  (*c*)  (*d*)

**Fig. 4.9**

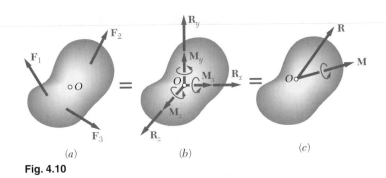

**Fig. 4.10**

## 4.6. REDUCTION OF A SYSTEM OF FORCES TO ONE FORCE AND ONE COUPLE

Consider a system of forces $\mathbf{F}_1$, $\mathbf{F}_2$, $\mathbf{F}_3$, etc., acting on a rigid body (Fig. 4.10$a$). We first resolve each force into rectangular components; according to Sec. 4.5, these components may be replaced by equivalent forces and couples at $O$. Adding all $x$, $y$, and $z$ components, we obtain, respectively, the forces $\mathbf{R}_x$, $\mathbf{R}_y$, and $\mathbf{R}_z$, directed along the coordinate axes; similarly, adding all moments about the $x$, $y$, and $z$ axes, we obtain, respectively, the couple vectors $\mathbf{M}_x$, $\mathbf{M}_y$, and $\mathbf{M}_z$ (Fig. 4.10$b$). It is advisable to arrange the computations in tabular form as shown in Sample Probs. 4.3 and 4.4. The forces and couple vectors obtained are together equivalent to the given system; they give a clear representation of the action exerted by that system on the rigid body. Note that different couple vectors would be obtained if a different origin $O$ were chosen.

We shall observe that, to be equivalent, two given systems of forces should reduce to the same forces and couple vectors at $O$. The sums of their $x$, $y$, and $z$ components and the sums of their moments about the $x$, $y$, and $z$ axes should therefore be, respectively, equal.

If desired, the three forces $\mathbf{R}_x$, $\mathbf{R}_y$, $\mathbf{R}_z$ of Fig. 4.10$b$ may be replaced by their resultant $\mathbf{R}$; similarly, the couple vectors $\mathbf{M}_x$, $\mathbf{M}_y$, $\mathbf{M}_z$ may be replaced by a single couple vector $\mathbf{M}$ (Fig. 4.10$c$). The magnitude and direction of the force $\mathbf{R}$ and of the couple vector $\mathbf{M}$ may be obtained by the method of Sec. 2.12. Denoting by $\theta_x$, $\theta_y$, $\theta_z$ and $\phi_x$, $\phi_y$, $\phi_z$ the angles formed, respectively, by $\mathbf{R}$ and $\mathbf{M}$ with the coordinate axes, we write

$$R = \sqrt{R_x^2 + R_y^2 + R_z^2}$$

$$\cos \theta_x = \frac{R_x}{R} \qquad \cos \theta_y = \frac{R_y}{R} \qquad \cos \theta_z = \frac{R_z}{R} \tag{4.2}$$

$$M = \sqrt{M_x^2 + M_y^2 + M_z^2}$$

$$\cos \phi_x = \frac{M_x}{M} \qquad \cos \phi_y = \frac{M_y}{M} \qquad \cos \phi_z = \frac{M_z}{M} \tag{4.3}$$

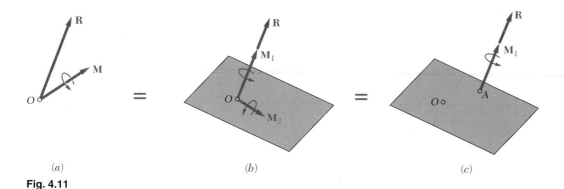

(a)   =   (b)   =   (c)

**Fig. 4.11**

An important difference between the reduction of a system of coplanar forces and the reduction of a system of space forces should be noted. It
was shown in Sec. 3.11 that the force-couple system obtained by reducing a
system of coplanar forces may be further reduced to a single force. In the
case of a system of space forces, this will generally not be possible. A
force-couple system may be reduced to a single force only if the force and
the couple act in the same plane, i.e., if the force and the couple vector are
mutually perpendicular (see Sec. 4.5).

If, as is generally the case, the force and couple vector are not at a
right angle (Fig. 4.11a), the couple vector may be replaced by two other
couple vectors obtained by resolving $\mathbf{M}$ into a component $\mathbf{M}_1$ along $\mathbf{R}$ and
a component $\mathbf{M}_2$ in a direction perpendicular to $\mathbf{R}$ (Fig. 4.11b). The couple
vector $\mathbf{M}_2$ and the force $\mathbf{R}$ may then be replaced by a single force $\mathbf{R}$ acting
along a new line of action; the original system of forces reduces to $\mathbf{R}$ and to
a couple of moment $M_1$ acting in a plane perpendicular to $\mathbf{R}$ (Fig. 4.11c).
This particular force-couple combination is called a *wrench*. The force $\mathbf{R}$
and the couple of moment $M_1$ tend to move the rigid body along the line of
action of $\mathbf{R}$ and, at the same time, tend to rotate it about that line. The line
of action of $\mathbf{R}$ is known as the *axis of the wrench* and the ratio $M_1/R$ is
called the *pitch of the wrench*.

Thus, it is seen that, while the action of a system of coplanar forces
may always be represented by that of a single force (or of a single couple),
it is generally necessary to use a wrench, i.e., the combination of a force
and a couple, to represent the action of space forces. An important exception to this rule is the case of a system of parallel forces; such a system may
be reduced to a single force (see Sample Prob. 4.4).

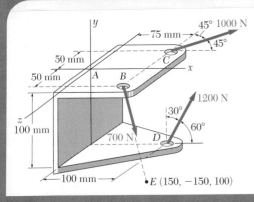

## SAMPLE PROBLEM 4.3

Three cables are attached to a bracket as shown. Replace the forces exerted by the cables by three forces at $A$ and three couples directed along the coordinate axes.

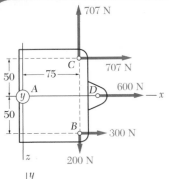

**Solution.** Because of the many computations involved, this problem is most efficiently solved in tabular fashion. First the components of the forces are determined and their values entered in the table. We note that the components and magnitude of the vector $\vec{BE}$ are, respectively,

$$d_x = 150 - 75 = 75 \text{ mm} \qquad d_y = -150 \text{ mm} \qquad d_z = 100 - 50 = 50 \text{ mm}$$

$$d = \sqrt{(75 \text{ mm})^2 + (-150 \text{ mm})^2 + (50 \text{ mm})^2} = 175 \text{ mm}$$

Recalling Eq. (2.18), we obtain the components of the force $\mathbf{B}$ from the proportion

$$\frac{B_x}{+75 \text{ mm}} = \frac{B_y}{-150 \text{ mm}} = \frac{B_z}{+50 \text{ mm}} = \frac{700 \text{ N}}{175 \text{ mm}}$$

The components are also plotted on the three projection planes, and their moments are computed about each axis and entered in the table. The components and moments are then added.

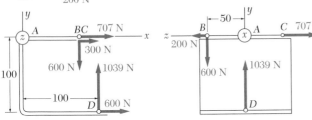

| Force | $F_x$, N | $F_y$, N | $F_z$, N | $M_x$, N·m | $M_y$, N·m | $M_z$, N·m |
|-------|----------|----------|----------|------------|------------|------------|
| **B** | +300 | −600 | +200 | +30 | +15<br>−15 | −45 |
| **C** | +707 | 0 | −707 | 0 | +53.03<br>−35.35 | 0 |
| **D** | +600 | +1039 | 0 | 0 | 0 | +103.9<br>+60 |
| | $R_x =$<br>+1607 | $R_y =$<br>+439 | $R_z =$<br>−507 | $M_x =$<br>+30 | $M_y =$<br>+17.68 | $M_z =$<br>+118.9 |

The given system may thus be replaced by three forces and three couple vectors, as shown. The three forces could be replaced by a single force $\mathbf{R}$ of components $R_x$, $R_y$, and $R_z$ and the three couple vectors by a single couple vector $\mathbf{M}$ of components $M_x$, $M_y$, and $M_z$.

# SAMPLE PROBLEM 4.4

A square foundation mat supports the four columns shown. Determine the magnitude and point of application of the resultant of the four loads.

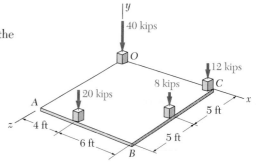

## Solution.
The $x$ and $z$ components of the given forces and their moments about the $y$ axis are obviously zero. Therefore only the $y$ components and the moments $M_x$ and $M_z$ are entered in the table.

| $F_y$, kips | $M_x$, kip · ft | $M_z$, kip · ft |
|:---:|:---:|:---:|
| −40 | | |
| −12 | | $-(12)(10) = -120$ |
| −8 | $+(8)(5) = +40$ | $-(8)(10) = -80$ |
| −20 | $+(20)(10) = +200$ | $-(20)(4) = -80$ |
| $R_y = -80$ | $M_x = +240$ | $M_z = -280$ |

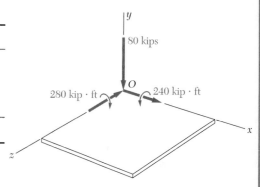

From the table, it is seen that the given system of forces may be replaced by an 80-kip force acting at $O$ and two couples represented by couple vectors located along the $x$ and $z$ axes.

The two couple vectors are perpendicular to the force and thus may be eliminated by moving the force to a new point of application. The coordinates of this point are

$$x = \frac{280 \text{ kip · ft}}{80 \text{ kips}} = 3.50 \text{ ft} \qquad z = \frac{240 \text{ kip · ft}}{80 \text{ kips}} = 3.00 \text{ ft}$$

We conclude that the resultant of the given system of forces is

$$\mathbf{R} = 80 \text{ kips}\downarrow \qquad \text{at } x = 3.50 \text{ ft, } z = 3.00 \text{ ft} \quad \blacktriangleleft$$

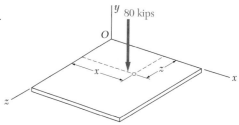

# PROBLEMS

**4.19**  Replace the 800-N force **P** by an equivalent force-couple system at *G*.

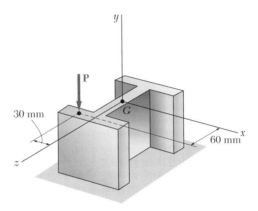

**Fig. P4.19**

**4.20**  The 15-ft boom *AB* has a fixed end *A* and the tension in cable *BC* is 190 lb. Replace the force that the cable exerts at *B* by an equivalent force-couple system at *A*.

**4.21**  A 520-lb force is applied at point *D* of the cast-iron post shown. Replace that force by an equivalent force-couple system at the center *A* of the base section.

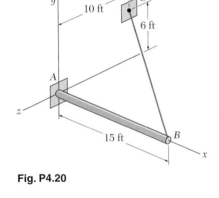

**Fig. P4.20**

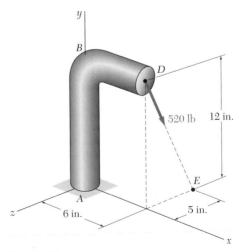

**Fig. P4.21**

**4.22** Replace the 150-N force by an equivalent force-couple system at $A$.

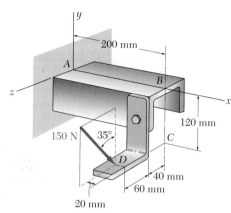

**Fig. P4.22**

**4.23** The speed-reducer unit shown weighs 60 lb and its center of gravity is located on the $y$ axis. Show that the weight of the unit and the two couples acting on it, of moment $M_1 = 15$ lb·ft and $M_2 = 3$ lb·ft, respectively, may be replaced by a single equivalent force and determine ($a$) the magnitude and direction of that force, ($b$) the point where its line of action intersects the floor.

**4.24** Five separate force-couple systems act at the corners of a bent plate as shown. Find two force-couple systems which are equivalent.

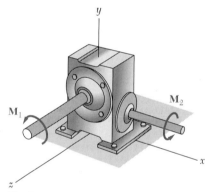

**Fig. P4.23**

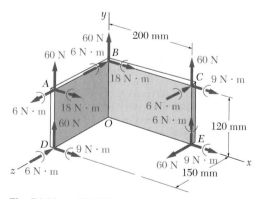

**Fig. P4.24 and P4.25**

**4.25** Five separate force-couple systems act at the corners of a bent plate as shown. Determine which of these systems is equivalent to a force $\mathbf{F}$ applied at $O$, of components $F_x = 0$, $F_y = +60$ N, $F_z = 0$, and a couple $\mathbf{M}$ of moments $M_x = M_y = 0$, $M_z = -6$ N·m.

**4.26** A blade held in a brace is used to tighten a screw at $A$. ($a$) Determine the horizontal force exerted at $B$ and the components of the force exerted at $C$, knowing that these forces are equivalent to a force-couple system at $A$ consisting of a force $\mathbf{R}$ with $R_x = -30$ N and $R_y$ and $R_z$ unspecified, and a couple $\mathbf{M}$ of moments $M_x = -12$ N·m, $M_y = M_z = 0$. ($b$) Find the corresponding values of $R_y$ and $R_z$. ($c$) What is the orientation of the slot in the head of the screw for which the blade is least likely to slip when the brace is in the position shown?

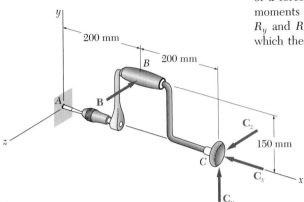

**Fig. P4.26**

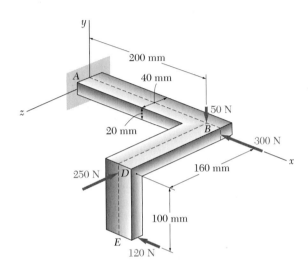

**Fig. P4.27**

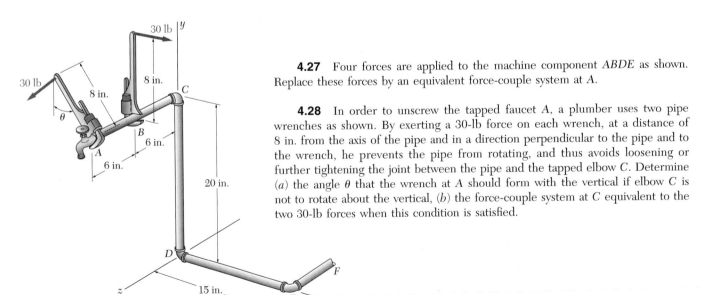

**4.27** Four forces are applied to the machine component $ABDE$ as shown. Replace these forces by an equivalent force-couple system at $A$.

**4.28** In order to unscrew the tapped faucet $A$, a plumber uses two pipe wrenches as shown. By exerting a 30-lb force on each wrench, at a distance of 8 in. from the axis of the pipe and in a direction perpendicular to the pipe and to the wrench, he prevents the pipe from rotating, and thus avoids loosening or further tightening the joint between the pipe and the tapped elbow $C$. Determine ($a$) the angle $\theta$ that the wrench at $A$ should form with the vertical if elbow $C$ is not to rotate about the vertical, ($b$) the force-couple system at $C$ equivalent to the two 30-lb forces when this condition is satisfied.

**Fig. P4.28**

**4.29** Assuming $\theta = 60°$ in Prob. 4.28, replace the two 30-lb forces by an equivalent force-couple system at $D$ and determine whether the plumber's action tends to tighten or loosen the joint between (a) pipe $CD$ and elbow $D$, (b) elbow $D$ and pipe $DE$. Assume all threads to be right-handed.

**4.30** Assuming $\theta = 60°$ in Prob. 4.28, replace the two 30-lb forces by an equivalent force-couple system at $E$ and determine whether the plumber's action tends to tighten or loosen the joint between (a) pipe $DE$ and elbow $E$. (b) elbow $E$ and pipe $EF$. Assume all threads to be right-handed.

**4.31** A rectangular concrete foundation mat supports four column loads as shown. Determine the magnitude and point of application of the resultant of the four loads.

**4.32** A concrete foundation mat in the shape of a regular hexagon of side 12 ft supports four column loads as shown. Determine the magnitude and point of application of the resultant of the four loads.

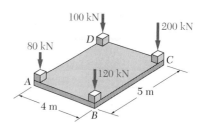

**Fig. P4.31**

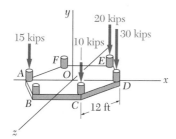

**Fig. P4.32 and P4.33**

**4.33** Determine the magnitudes of the additional loads which must be applied at $B$ and $F$ if the resultant of all six loads is to pass through the center of the mat.

**4.34** In Prob. 4.31, determine the magnitude and point of application of the smallest additional load which must be applied to the foundation mat if the resultant of the five loads is to pass through the center of the mat.

**\*4.35** A rectangular block is acted upon by the four forces shown, which are directed along its edges. Reduce this system of forces to (a) a force-couple system at the origin, (b) a wrench. (Specify the pitch and axis of the wrench.)

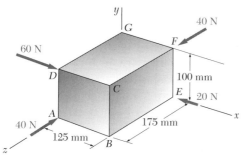

**Fig. P4.35**

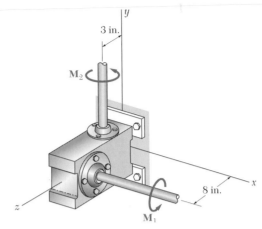

**Fig. P4.36**

**\*4.36** The two shafts of a speed-reducer unit are subjected to couples of moment $M_1 = 15$ lb·ft and $M_2 = 3$ lb·ft, respectively. The unit weighs 60 lb and its center of gravity is located on the $z$ axis at $z = 6$ in. Replace the weight and the two couples by an equivalent wrench and determine (a) the resultant force **R**, (b) the pitch of the wrench, (c) the point where the axis of the wrench intersects the $xz$ plane.

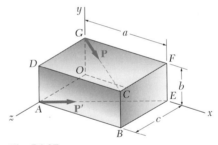

**Fig. P4.37**

**\*4.37** Two forces of magnitude $P$ act as shown along the diagonals $GC$ and $AE$ of the horizontal faces of a rectangular parallelepiped. Replace the two forces by an equivalent system consisting of (a) a force and couple at $O$, (b) a wrench. (Specify the pitch and axis of the wrench.)

**\*4.38** In an automated manufacturing process three holes are drilled simultaneously as shown in an aluminum block. Each drill exerts a 40-N force and a 0.100-N·m couple on the block. Knowing that drill $A$ rotates counterclockwise, and drills $B$ and $C$ clockwise (as observed from each drill), reduce the forces and couples exerted by the drills on the block to an equivalent wrench and determine (a) the resultant force **R**, (b) the pitch of the wrench, (c) the point where the axis of the wrench intersects the $xz$ plane.

**\*4.39** If only drills $A$ and $B$ are used in the manufacturing process of Prob. 4.38, show that the forces and couples exerted by the two drills on the aluminum block may be reduced to a single force **R** and determine (a) the magnitude and direction of **R**, (b) the line of action of **R**.

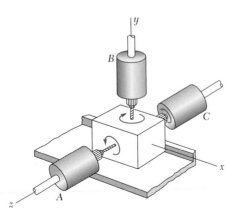

**Fig. P4.38**

EQUILIBRIUM OF RIGID BODIES

4.8. Reactions at Supports and
Connections for a Three-Dimensional
Structure    **141**

## 4.7. EQUILIBRIUM OF A RIGID BODY IN SPACE

As was indicated in Sec. 3.13, *a rigid body is said to be in equilibrium when the external forces acting on it form a system of forces equivalent to zero.* In the case of space forces, a system of forces is equivalent to zero when the three forces and three couple vectors shown in Fig. 4.10*b* are all zero. The necessary and sufficient conditions for the equilibrium of a rigid body in space may thus be expressed analytically by writing the six equations

$$\Sigma F_x = 0 \qquad \Sigma F_y = 0 \qquad \Sigma F_z = 0 \qquad (4.4)$$

$$\Sigma M_x = 0 \qquad \Sigma M_y = 0 \qquad \Sigma M_z = 0 \qquad (4.5)$$

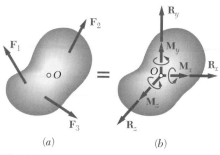

(*a*)                (*b*)

**Fig. 4.10*a* and *b* (*repeated*)**

Equations (4.4) express the fact that the components of the external forces in the $x$, $y$, and $z$ directions are balanced; Eqs. (4.5) express the fact that the moments of the external forces about the $x$, $y$, and $z$ axes are balanced. The system of the external forces, therefore, will impart no motion of translation or rotation to the rigid body considered.

After a free-body diagram has been drawn, showing all the external forces acting on the body, the equations of equilibrium (4.4) and (4.5) may be solved for no more than *six unknowns*. These unknowns generally represent reactions at supports and connections. In practice, it will be found convenient to solve the moment equations (4.5) first, choosing axes which either intersect or are parallel to the lines of action of several unknown forces or force components. In this way, equations containing only two unknowns, possibly one, will be obtained. Once these unknowns have been determined, their values are carried into Eqs. (4.4) in order to reduce the number of unknowns in these equations.

While the six equations of equilibrium cannot be augmented by additional equations, any of them may be replaced by another equation. Thus, moment equations, written with respect to lines other than the axes of coordinates, may sometimes be substituted advantageously for one or several of Eqs. (4.4) and (4.5).

## 4.8. REACTIONS AT SUPPORTS AND CONNECTIONS FOR A THREE-DIMENSIONAL STRUCTURE

The reactions on a three-dimensional structure range from the single force of known direction exerted by a frictionless surface to the force-couple system exerted by a fixed support. Consequently, the number of unknowns associated with the reaction at a support or connection may vary from one to six in problems involving the equilibrium of a three-dimensional structure. Various types of supports and connections are shown in Fig. 4.12 with the corresponding reactions. A simple way of determining the type of

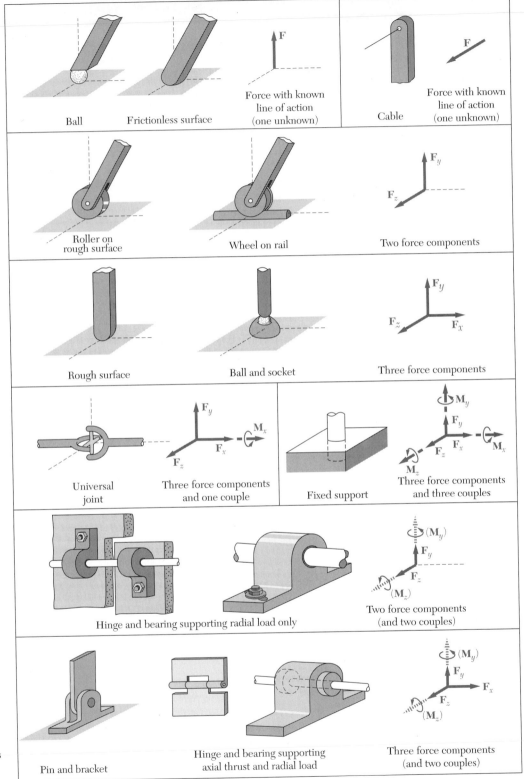

**Fig. 4.12** Reactions at supports and connections.

Ball   Frictionless surface   Force with known line of action (one unknown)

Cable   Force with known line of action (one unknown)

Roller on rough surface   Wheel on rail   Two force components

Rough surface   Ball and socket   Three force components

Universal joint   Three force components and one couple

Fixed support   Three force components and three couples

Hinge and bearing supporting radial load only   Two force components (and two couples)

Pin and bracket   Hinge and bearing supporting axial thrust and radial load   Three force components (and two couples)

4.8. Reactions at Supports and **143**
Connections for a Three-Dimensional
Structure

reaction corresponding to a given support or connection and the number of unknowns involved is to find which of the six fundamental motions (translation in $x$, $y$, and $z$ directions, rotation about the $x$, $y$, and $z$ axes) are allowed and which motions are prevented.

Ball supports, frictionless surfaces, and cables, for example, prevent translation in one direction only and thus exert a single force of known line of action; they each involve one unknown, namely, the magnitude of the reaction. Rollers on rough surfaces and wheels on rails prevent translation in two directions; the corresponding reactions consist of two unknown force components. Rough surfaces in direct contact and ball-and-socket supports prevent translation in three directions; these supports involve three unknown force components.

Some supports and connections may prevent rotation as well as translation; the corresponding reactions include, then, couples as well as forces. The reaction at a fixed support, for example, which prevents any motion (rotation as well as translation), consists of three unknown forces and three unknown couples. A universal joint, which is designed to allow rotation about two axes, will exert a reaction consisting of three unknown force components and one unknown couple.

Other supports and connections are primarily intended to prevent translation; their design, however, is such that they also prevent some rotations. The corresponding reactions consist essentially of force components but *may* also include couples. One group of supports of this type includes hinges and bearings designed to support radial loads only (for example, journal bearings, roller bearings). The corresponding reactions consist of two force components but may also include two couples. Another group includes pin-and-bracket supports, hinges, and bearings designed to support an axial thrust as well as a radial load (for example, ball bearings). The corresponding reactions consist of three force components but may include two couples. However, these supports will not exert any appreciable couples under normal conditions of use. Therefore, *only* force components should be included in their analysis, *unless* it is found that couples are necessary to maintain the equilibrium of the rigid body, or unless the support is known to have been specifically designed to exert a couple (see Probs. 4.64 through 4.67).

If the reactions involve more than six unknowns, there are more unknowns than equations, and some of the reactions are *statically indeterminate*. If the reactions involve less than six unknowns, there are more equations than unknowns, and some of the equations of equilibrium cannot be satisfied under general loading conditions; the rigid body is only *partially constrained*. Under the particular loading conditions corresponding to a given problem, however, the extra equations often reduce to trivial identities such as $0 = 0$ and may be disregarded; although only partially constrained, the rigid body remains in equilibrium (see Sample Probs. 4.5 and 4.6). Even with six or more unknowns, it is possible that some equations of equilibrium will not be satisfied. This may occur when the supports are such that the reactions are forces which must either be parallel or intersect the same line; the rigid body is then *improperly constrained*.

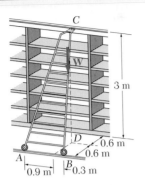

# SAMPLE PROBLEM 4.5

A 20-kg ladder used to reach high shelves in a storeroom is supported by two flanged wheels $A$ and $B$ mounted on a rail and by an unflanged wheel $C$ resting against a rail fixed to the wall. An 80-kg man stands on the ladder and leans to the right. The line of action of the combined weight $\mathbf{W}$ of the man and ladder intersects the floor at point $D$. Determine the reactions at $A$, $B$, and $C$.

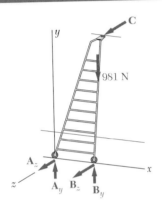

**Solution.** A free-body diagram of the ladder is drawn. The forces involved are the combined weight of the man and ladder, of magnitude

$$W = mg = (80 \text{ kg} + 20 \text{ kg})(9.81 \text{ m/s}^2) = 981 \text{ N}$$

and five unknown reaction components, two at each flanged wheel and one at the unflanged wheel. The ladder is thus only partially constrained; it is free to roll along the rails. It is, however, in equilibrium under the given load since the equation $\Sigma F_x = 0$ is satisfied.

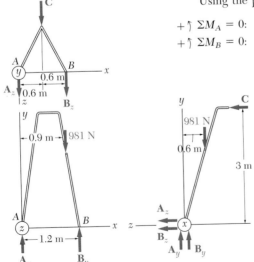

**Equilibrium Equations.** Using the projection on the $yz$ plane, we write

$$+\!\uparrow \ \Sigma M_x = 0: \qquad -(981 \text{ N})(0.6 \text{ m}) + C(3 \text{ m}) = 0 \qquad C = +196.2 \text{ N} \ \blacktriangleleft$$

Using the projection on the $xy$ plane, we write

$$+\!\uparrow \ \Sigma M_A = 0: \qquad -(981 \text{ N})(0.9 \text{ m}) + B_y(1.2 \text{ m}) = 0 \qquad B_y = +736 \text{ N} \ \blacktriangleleft$$

$$+\!\uparrow \ \Sigma M_B = 0: \qquad +(981 \text{ N})(0.3 \text{ m}) - A_y(1.2 \text{ m}) = 0 \qquad A_y = +245 \text{ N} \ \blacktriangleleft$$

Using the projection on the $xz$ plane and the computed value of $C$, we write

$$+\!\uparrow \ \Sigma M_A = 0: \quad -(196.2 \text{ N})(0.6 \text{ m}) - B_z(1.2 \text{ m}) = 0$$
$$B_z = -98.1 \text{ N} \ \blacktriangleleft$$

$$+\!\uparrow \ \Sigma M_B = 0: \quad +(196.2 \text{ N})(0.6 \text{ m}) + A_z(1.2 \text{ m}) = 0$$
$$A_z = -98.1 \text{ N} \ \blacktriangleleft$$

**Check.** We may check the results by verifying that equilibrium equations not used in the solution are satisfied; for example,

$$\Sigma F_y = A_y + B_y - 981 \text{ N}$$
$$= +245 \text{ N} + 736 \text{ N} - 981 \text{ N} = 0$$
$$\Sigma F_z = A_z + B_z + C$$
$$= -98.1 \text{ N} - 98.1 \text{ N} + 196.2 \text{ N} = 0$$

# SAMPLE PROBLEM 4.6

A 5 × 8 ft sign of uniform density weighs 270 lb and is supported by a ball and socket at $A$ and by two cables. Determine the tension in each cable and the reaction at $A$.

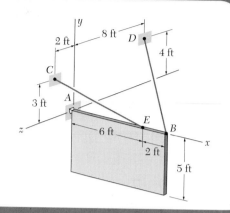

## Solution.
A free-body diagram of the sign is drawn. The forces acting on the free body consist of the 270-lb weight and of the reactions at $A$, $B$, and $E$. The reaction at $A$ is a force of unknown direction and is represented by three unknown components. Since the directions of the forces exerted by the cables are known, these forces involve only one unknown each, namely the magnitudes $T_{BD}$ and $T_{EC}$. Since there are only five unknowns, the sign is partially constrained. It may rotate freely about the $x$ axis; it is, however, in equilibrium under the given loading since the equation $\Sigma M_x = 0$ is satisfied.

Recalling that the components and magnitude of the forces $\mathbf{T}_{BD}$ and $\mathbf{T}_{EC}$ are respectively proportional to the components and magnitude of the vectors $\vec{BD}$ and $\vec{EC}$, we write

$$\frac{(T_{BD})_x}{-8 \text{ ft}} = \frac{(T_{BD})_y}{4 \text{ ft}} = \frac{(T_{BD})_z}{-8 \text{ ft}} = \frac{T_{BD}}{12 \text{ ft}}$$

$$\frac{(T_{EC})_x}{-6 \text{ ft}} = \frac{(T_{EC})_y}{3 \text{ ft}} = \frac{(T_{EC})_z}{2 \text{ ft}} = \frac{T_{EC}}{7 \text{ ft}}$$

and express the components of these forces in terms of their magnitudes $T_{BD}$ and $T_{EC}$ as shown on the two projections.

*Equilibrium Equations.* Using the projections shown, we write

$+\uparrow \Sigma M_y = 0:$      $(\tfrac{2}{3}T_{BD})(8 \text{ ft}) - (\tfrac{6}{7}T_{EC})(6 \text{ ft}) = 0$

$+\uparrow \Sigma M_z = 0:$      $(\tfrac{1}{3}T_{BD})(8 \text{ ft}) + (\tfrac{3}{7}T_{EC})(6 \text{ ft}) - (270 \text{ lb})(4 \text{ ft}) = 0$

Solving these two equations simultaneously, we obtain

$$T_{BD} = 101.3 \text{ lb} \qquad T_{EC} = 315 \text{ lb} \quad \blacktriangleleft$$

Using the values found for $T_{BD}$ and $T_{EC}$, we write

$\Sigma F_x = 0:$    $A_x - \tfrac{2}{3}T_{BD} - \tfrac{6}{7}T_{EC} = 0$

         $A_x - \tfrac{2}{3}(101.3 \text{ lb}) - \tfrac{6}{7}(315 \text{ lb}) = 0$      $A_x = 338 \text{ lb} \quad \blacktriangleleft$

$\Sigma F_y = 0:$    $A_y + \tfrac{1}{3}T_{BD} + \tfrac{3}{7}T_{EC} - 270 \text{ lb} = 0$

         $A_y + \tfrac{1}{3}(101.3 \text{ lb}) + \tfrac{3}{7}(315 \text{ lb}) - 270 \text{ lb} = 0$

                               $A_y = 101.2 \text{ lb} \quad \blacktriangleleft$

$\Sigma F_z = 0:$    $A_z - \tfrac{2}{3}T_{BD} + \tfrac{2}{7}T_{EC} = 0$

         $A_z - \tfrac{2}{3}(101.3 \text{ lb}) + \tfrac{2}{7}(315 \text{ lb}) = 0$      $A_z = -22.5 \text{ lb} \quad \blacktriangleleft$

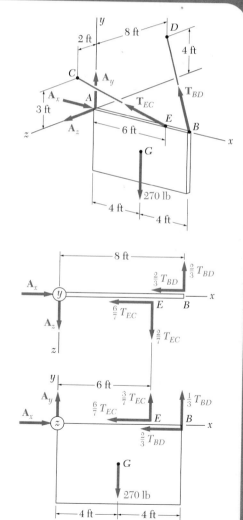

**145**

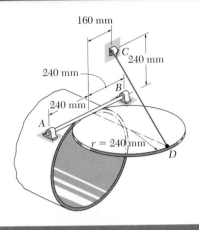

160 mm

C

240 mm

240 mm

B

240 mm

A

$r = 240$ mm

D

# SAMPLE PROBLEM 4.7

A uniform pipe cover of radius $r = 240$ mm and mass 30 kg is held in a horizontal position by the cable $CD$. Assuming that the bearing at $B$ does not exert any axial thrust, determine the tension in the cable and the reactions at $A$ and $B$.

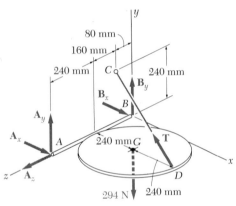

**Solution.**  A free-body diagram is drawn with the coordinate axes shown. The forces acting on the free body are the weight of the cover, of magnitude

$$W = mg = (30 \text{ kg})(9.81 \text{ m/s}^2) = 294 \text{ N}$$

and reactions involving six unknowns, namely, the magnitude of the force $\mathbf{T}$ exerted by the cable, three force components at hinge $A$, and two at hinge $B$. The components of $\mathbf{T}$ are expressed in terms of the unknown magnitude $T$ by writing

$$\frac{T_x}{-480 \text{ mm}} = \frac{T_y}{+240 \text{ mm}} = \frac{T_z}{-160 \text{ mm}} = \frac{T}{560 \text{ mm}}$$

$$T_x = -\tfrac{6}{7}T \qquad T_y = +\tfrac{3}{7}T \qquad T_z = -\tfrac{2}{7}T$$

***Equilibrium Equations.***  Using the projections shown, we write

| | | |
|---|---|---|
| $+\uparrow \Sigma M_z = 0$: | $(\tfrac{3}{7}T)(2r) - (294 \text{ N})r = 0$ | $T = 343 \text{ N}$  ◀ |
| $+\uparrow \Sigma M_x = 0$: | $(294 \text{ N})r - (\tfrac{3}{7}T)r - A_y(2r) = 0$ | |
| | $2A_y = 294 \text{ N} - \tfrac{3}{7}(343 \text{ N})$ | $A_y = +73.5 \text{ N}$  ◀ |
| $+\uparrow \Sigma M_y = 0$: | $(\tfrac{2}{7}T)(2r) - (\tfrac{6}{7}T)r + A_x(2r) = 0$ | |
| | $A_x = -\tfrac{2}{7}T + \tfrac{3}{7}T = \tfrac{1}{7}(343 \text{ N})$ | $A_x = +49.0 \text{ N}$  ◀ |
| $\Sigma F_x = 0$: | $A_x + B_x - \tfrac{6}{7}T = 0$ | |
| | $B_x = \tfrac{6}{7}(343 \text{ N}) - 49 \text{ N}$ | $B_x = +245 \text{ N}$  ◀ |
| $\Sigma F_y = 0$: | $A_y + B_y + \tfrac{3}{7}T - 294 \text{ N} = 0$ | |
| | $B_y = 294 \text{ N} - 73.5 \text{ N} - \tfrac{3}{7}(343 \text{ N})$ | $B_y = +73.5 \text{ N}$  ◀ |
| $\Sigma F_z = 0$: | $A_z - \tfrac{2}{7}T = 0 \qquad A_z = \tfrac{2}{7}(343 \text{ N})$ | $A_z = +98.0 \text{ N}$  ◀ |

# PROBLEMS

**4.40** A 20-kg circular porthole cover is supported by a single hinge at $A$ and a cable $BC$. Determine the reactions at $A$ and $B$ necessary for equilibrium.

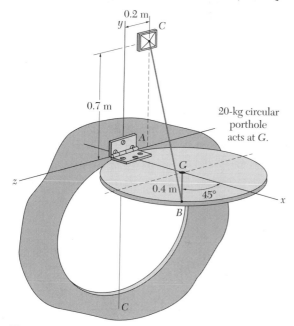

**Fig. P4.40**

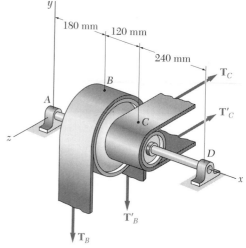

**Fig. P4.41**

**4.41** Two transmission belts pass over a double-sheaved pulley which is attached to an axle supported by bearings at $A$ and $D$. The radius of the inner sheave is 150 mm and the radius of the outer sheave is 300 mm. Knowing that when the system is at rest, the tension is 75 N in both portions of belt $B$ and 135 N in both portions of belt $C$, determine the reactions at $A$ and $D$. Assume that the bearing at $D$ does not exert any axial thrust.

**4.42** Solve Prob. 4.41, assuming that the pulley rotates at a constant rate and that $T_B = 87$ N, $T'_B = 69$ N, and $T_C = 162$ N.

**4.43** A small winch is used to raise a 150-lb load as shown. Find (*a*) the magnitude of the horizontal force **P** which should be applied at $C$ to maintain equilibrium, (*b*) the reactions at $A$ and $B$, assuming that the bearing at $B$ does not exert any axial thrust.

**4.44** A 250-mm lever and a 300-mm-diameter pulley are welded to the axle $BE$ which is supported by bearings at $C$ and $D$. If a 450-N vertical load is applied at $A$ when the lever is horizontal, determine (*a*) the tension in the cord, (*b*) the reactions at $C$ and $D$. Assume that the bearing at $D$ does not exert any axial thrust.

**4.45** Solve Prob. 4.44, assuming that the axle has been rotated clockwise in its bearings by 60°.

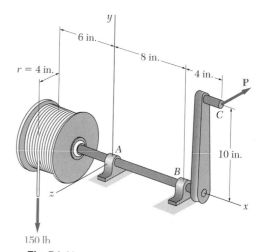

**Fig. P4.43**

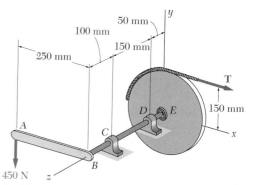

**Fig. P4.44**

**4.46** The T-shaped member is supported by a hinge at $A$ and a cable $BC$. If the member is massless, and a force $\mathbf{F}_D$ is applied at $D$, determine the reactions at $A$ and $B$ necessary for equilibrium.

**4.47** The rectangular plate shown weighs 60 lb and is supported by three wires. Determine the tension in each wire.

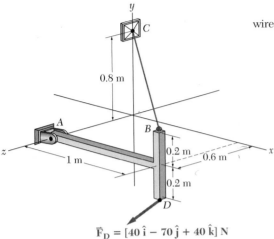

$$\bar{\mathbf{F}}_D = [40\,\hat{\mathbf{i}} - 70\,\hat{\mathbf{j}} + 40\,\hat{\mathbf{k}}]\ \text{N}$$

**Fig. P4.46**

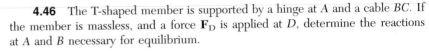

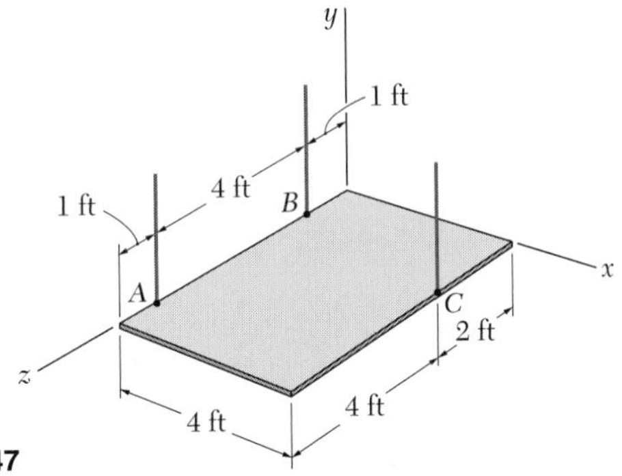

**Fig. P4.47**

**4.48** A load $W$ is to be placed on the 60-lb plate of Prob. 4.47. Determine the magnitude of $W$ and the point where it should be placed if the tension is to be 50 lb in each of the three wires.

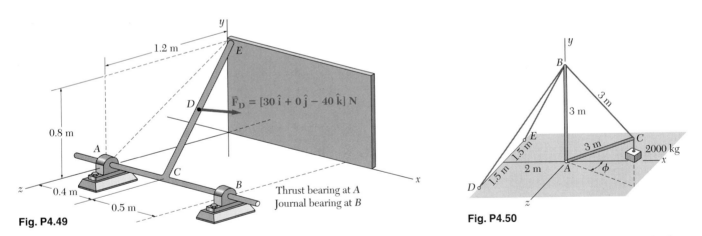

**Fig. P4.49**

$$\bar{\mathbf{F}}_D = [30\,\hat{\mathbf{i}} + 0\,\hat{\mathbf{j}} - 40\,\hat{\mathbf{k}}]\ \text{N}$$

Thrust bearing at $A$
Journal bearing at $B$

**Fig. P4.50**

**4.49** The angled T-shaped member is supported by two bearings $A$ and $B$, properly aligned. End $E$ of the $T$ member rests against a frictionless wall. Support $A$ is a thrust bearing, and $B$ is a journal bearing. If the member is massless and a force $\mathbf{F}_D$ is applied at $D$, determine the reactions at $A$, $B$, and $E$ necessary for equilibrium. ($D$ is the midpoint of $CE$.)

**4.50** The derrick shown supports a 2000-kg crate. It is held by a ball and socket at $A$ and by two cables attached at $D$ and $E$. Knowing that the derrick stands in the vertical $xy$ plane ($\phi = 0$), determine (a) the tension in each cable, (b) the reaction at $A$.

**4.51** Solve Prob. 4.50, assuming that the derrick stands in a vertical plane forming an angle $\phi = 20°$ with the $xy$ plane.

**4.52** The 12-ft boom $AB$ is acted upon by the 850-lb load shown. Determine (a) the tension in each cable, (b) the reaction at the ball and socket at $A$.

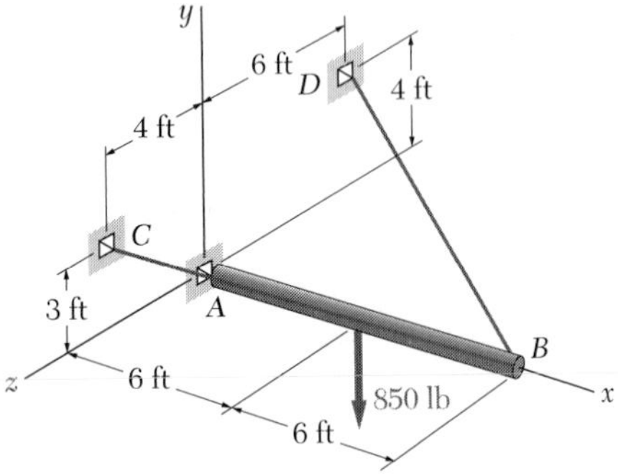

**Fig. P4.52**

**4.53** Solve Prob. 4.52, assuming that the 850-lb load is applied at point *B*.

**4.54** The angled member *AB* is supported by a thrust bearing (capable of developing an axial force) at *A* and a cable *BC*. If the member is massless and a 50 lb-in. couple is applied as shown, determine the reactions at *A* and *B* necessary for equilibrium.

**4.55** In Prob. 4.50, determine the angle $\phi$ for which the tension in *BE* is maximum (*a*) if neither cable may become slack, (*b*) if cables *BE* and *BD* are replaced by rods which can withstand either a tension or a compression force.

**4.56** The 40-lb attic access hatch is supported by two hinges *A* and *B* and a cable *CD*. Hinge *A* is capable of developing an axial force along the *z* axis. Determine the reactions at *A*, *B*, and *C* necessary for equilibrium of the hatch.

**4.57** The 200-N sign is supported by a ball joint at *A* and two cables *BC* and *BD*. Determine the reactions at *A* and the cable tensions necessary for equilibrium.

**4.58** A 3 × 5 ft storm window weighing 15 lb is held by hinges at *A* and *B*. In the position shown, it is held away from the side of the house by a 2-ft stick *CD*. Assuming that the hinge at *A* does not exert any axial thrust, determine (*a*) the magnitude of the force exerted by the stick, (*b*) the reactions at *A* and *B*.

**4.59** A 1.2 × 2.4 m sheet of plywood is temporarily held by nails at *D* and *E* and by two wooden braces nailed at *A*, *B*, and *C*. Wind is blowing on the hidden face of the plywood sheet and it is assumed that its effect may be represented by a force **P** applied at the center of the sheet in a direction perpendicular to the sheet. Knowing that each brace becomes unsafe with respect to buckling when subjected to a 1.5-kN axial force, determine (*a*) the maximum allowable value of the magnitude *P* of the wind force, (*b*) the corresponding value of the *z* component of the reaction at *E*. Assume that the nails are loose and do not exert any couple.

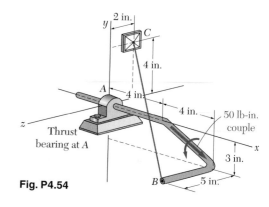

**Fig. P4.54**

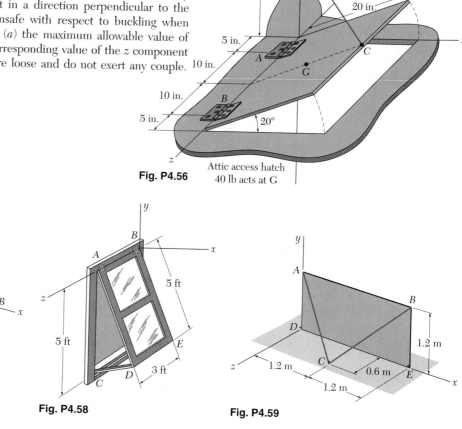

**Fig. P4.56**

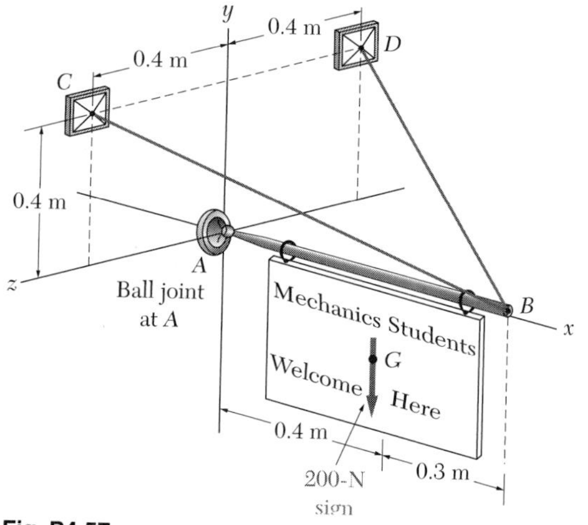

Attic access hatch
40 lb acts at *G*

**Fig. P4.57**

**Fig. P4.58**

**Fig. P4.59**

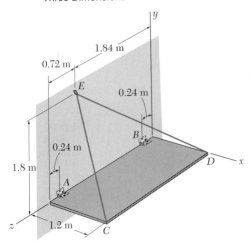

**Fig. P4.60**

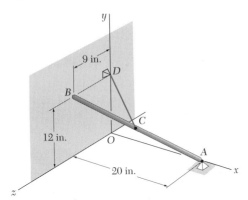

**Fig. P4.68**

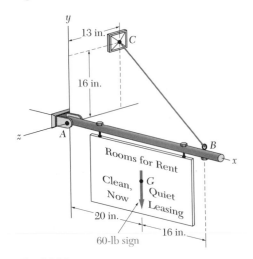

**Fig. P4.69**

**4.60** A 200-kg uniform rectangular platform, 2.56 m long and 1.20 m wide, is supported by hinges at $A$ and $B$ and by a cable attached to the corners $C$ and $D$ of the platform and passing over a frictionless hook $E$. Assuming that the tension is the same in both parts of the cable and that the hinge at $A$ does not exert any axial thrust, determine (a) the tension in the cable, (b) the reactions at $A$ and $B$.

**4.61** Solve Prob. 4.60, assuming that cable $CED$ is replaced by a cable attached to corner $C$ and to hook $E$.

**4.62** Solve Prob. 4.60, assuming that cable $CED$ is replaced by a cable attached to corner $D$ and to hook $E$.

**4.63** The 50-lb sign is supported by a ball joint at $A$ and two cables, $BC$ and $BD$. Determine the reactions at $A$ and the cable tensions necessary for equilibrium.

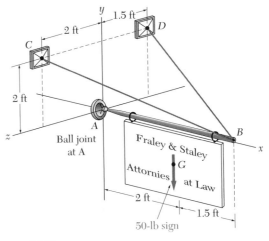

**Fig. P4.63**

**4.64** Solve Prob. 4.58, assuming that the hinge at $A$ has been removed and that the hinge at $B$ can exert couples about axes parallel to the $x$ and $y$ axes, respectively.

**4.65** Solve Prob. 4.58, assuming that the hinge at $B$ has been removed and that the hinge at $A$ can exert couples about axes parallel to the $x$ and $y$ axes, respectively.

**4.66** Solve Prob. 4.60, assuming that the hinge at $A$ has been removed and that the hinge at $B$ can exert couples about axes parallel to the $x$ and $y$ axes, respectively.

**4.67** Solve Prob. 4.60, assuming that the hinge at $B$ has been removed and that the hinge at $A$ can exert an axial thrust, as well as couples about axes parallel to the $x$ and $y$ axes, respectively.

**4.68** The uniform rod $AB$ weighs 12 lb. It is supported by a ball and socket at $A$ and by the cord $CD$ which is attached to the midpoint $C$ of the rod. Knowing that the rod leans against a frictionless vertical wall at $B$, determine (a) the tension in the cord, (b) the reactions at $A$ and $B$.

**4.69** The 60-lb sign is supported by a hinge at $A$ and a single cable $BC$. Determine the reactions at $A$ and the cable tension necessary for equilibrium.

**4.70**  A 200-kg catamaran is sailing in a straight course and at a constant speed. The effect of the wind on the sail may be represented by the force **F** of components $F_x = 450$ N, $F_y = 0$, $F_z = -600$ N, applied at the point of coordinates $x = -0.48$ m, $y = 3$ m, $z = -0.36$ m. Due to the efforts of the crew, the mast is maintained in a vertical position and the twin hulls are subjected to equal forces, consisting of vertical buoyant forces **B** along the lines of action $x = -0.6$ m, $z = \pm1.2$ m, drag forces **D** directed toward the negative $x$ axis and acting along the lines $y = -0.6$ m, $z = \pm1.2$ m, and side forces **S** directed toward the positive $z$ axis and acting along the line $x = -d$, $y = -0.6$ m. Determine (*a*) the magnitude $D$ of the drag on each of the hulls, (*b*) the magnitude $S$ of each of the side forces, and the distance $d$ defining their line of action.

**4.71**  The catamaran of Prob. 4.70 is manned by two sailors, one of mass 80 kg and the other 65 kg. The center of gravity of the boat is located at $x = -0.9$ m, $z = 0$, and the 80-kg sailor is sitting at $x = -1.5$ m, $z = +1.2$ m, while the 65-kg sailor leans overboard to maintain the mast in a vertical position. Determine the $x$ and $z$ coordinates of the center of gravity of the second sailor.

**4.72**  In order to clean the clogged drainpipe *AE*, a plumber has disconnected both ends of the pipe and introduced a power snake through the opening at *A*. The cutting head of the snake is connected by a heavy cable to an electric motor which keeps it rotating at a constant speed while the plumber forces the cable into the pipe. The forces exerted by the plumber and the motor on the end of the cable may be represented by a wrench consisting of a 6.5-lb force **F** and a 97.5-lb·ft couple vector **M**, both directed vertically downward. Determine the additional reactions at *B*, *C*, and *D* caused by the cleaning operation. Assume that the reaction at each support consists of two force components perpendicular to the pipe.

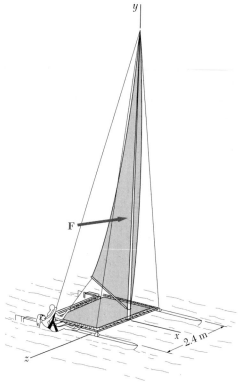

**Fig. P4.70**

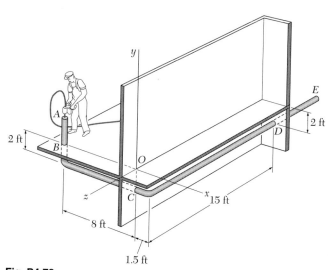

**Fig. P4.72**

**4.73**  Solve Prob. 4.72, assuming that the plumber exerts no force on the end of the cable and that the motor exerts a 97.5-lb·ft couple represented by a vector directed vertically downward.

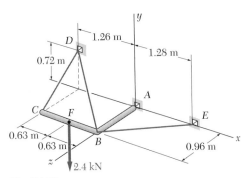

**Fig. P4.74**

**4.74** The rigid L-shaped member $ABC$ is supported by a ball and socket at $A$ and by three cables. If a 2.4-kN load is applied at $F$, determine the tension in each cable.

**4.75** Solve Prob. 4.74, assuming that the 2.4-kN load is applied at $C$.

# REVIEW AND SUMMARY
# FOR CHAPTER 4

Moment of a force about an axis

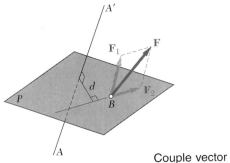

**Fig. 4.13**

Couple vector

In the first part of this chapter we studied the effect of three-dimensional forces on rigid bodies. We first extended the definition of the *moment of a force* about an axis by defining the moment $M$ of a force $\mathbf{F}$ about an arbitrary axis $AA'$ as the moment about $AA'$ of the projection $\mathbf{F}_2$ of $\mathbf{F}$ on a plane perpendicular to the axis $AA'$ (Fig. 4.13): $M = F_2 d$ [Sec. 4.2]. The moments computed most frequently are the moments $M_x$, $M_y$, $M_z$ of $\mathbf{F}$ about the coordinate axes, and its moments about axes parallel to the coordinate axes. All these may be readily obtained from the projections of $\mathbf{F}$ on the three coordinate planes $yz$, $zx$, and $xy$ [Sample Prob. 4.1].

We saw that the effect of a *couple* on a three-dimensional body depends upon the moment $M = Fd$ of the couple and the orientation of the plane in which it acts (Fig. 4.14a). A couple in space may thus be represented by a vector $\mathbf{M}$ (Fig. 4.14b), called a *couple vector,* perpendicular to the plane of the couple, of magnitude equal to the moment $M$ of the couple and of sense defined by the right-hand rule [Sec. 4.3]. It was further established that couple vectors add according to the parallelogram law and, thus, are truly vectors [Sec. 4.4]. Also, since two couples with the same moment and acting in parallel planes are equivalent, the couple vector representing them is a *free vector,* which may be attached at the origin $O$ (Fig. 4.14c) and resolved into couple vectors along the coordinate axes if so desired (Fig. 4.14d).

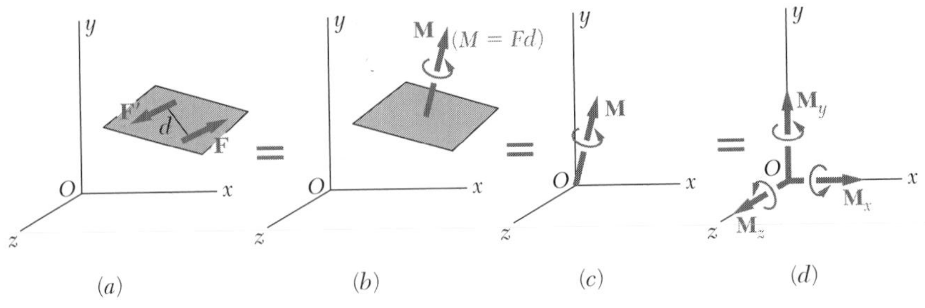

(a)          (b)          (c)          (d)

**Fig. 4.14**

Any force **F** acting at a point $A$ of a rigid body may be replaced by a *force-couple system* at the origin $O$, consisting of the force **F** applied at $O$ and a couple **M**. The components $M_x$, $M_y$, and $M_z$ of the couple vector **M** are equal, respectively, to the moments of **F** (in its original position) about the $x$, $y$, and $z$ axes, and it was shown that the force **F** and the couple vector **M** are perpendicular to each other [Sec. 4.5].

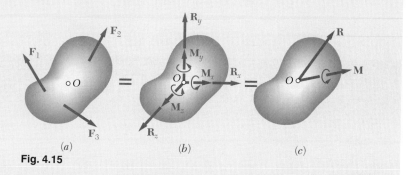

(a)                    (b)                    (c)

**Fig. 4.15**

*Any system of three-dimensional forces may be reduced to a force-couple system at the origin $O$* by replacing each of the forces of the system by an equivalent force-couple system at $O$ and adding all the forces and all the couples obtained in this manner (Fig. 4.15). The computations involved are facilitated if projections on the three coordinate planes are used and if the results are expressed in a tabular form [Sample Prob. 4.3]. Since, in general, the resulting force **R** and couple vector **M** are not perpendicular to each other, the force-couple system at $O$ *cannot* be reduced further to a single resultant force or couple; it may, however, be reduced to a special force-couple system consisting of the resultant **R** and a couple vector $\mathbf{M}_1$ directed along **R**; such a system is called a *wrench* [Sec. 4.6]. We should note that three different types of forces may be reduced to a single resultant force or couple: concurrent forces [Chap. 2], coplanar forces [Chap. 3], and parallel forces [Sample Prob. 4.4].

Reduction of a system of forces to a force-couple system

In the second part of the chapter we considered the *equilibrium of a three-dimensional rigid body*, i.e., the situation when the external forces exerted on the body form a system equivalent to zero. We saw that *six equilibrium equations* may be written [Sec. 4.7], namely,

Equilibrium equations

$$\Sigma F_x = 0 \qquad \Sigma F_y = 0 \qquad \Sigma F_z = 0 \qquad (4.4)$$
$$\Sigma M_x = 0 \qquad \Sigma M_y = 0 \qquad \Sigma M_z = 0 \qquad (4.5)$$

These equations may be solved for *six unknowns*, and while they cannot be augmented by additional equations, any of them may be replaced by another equation, such as a moment equation with respect to a line parallel to a coordinate axis.

Before applying the equilibrium equations (4.4) and (4.5) to the solution of a given problem, it is necessary to identify correctly all the forces acting on the rigid body and draw the corresponding *free-body diagram*. In addition to the forces *applied* to the body, we had to consider the *reactions* exerted on the body

by its supports [Sec. 4.8]. Depending upon the type of support, a reaction could
involve from one to six unknowns, as shown in Fig. 4.12.

If the reactions involve more than six unknowns, some of the reactions are
*statically indeterminate;* if they involve less than six unknowns, the rigid body is
only *partially constrained.* Even with six or more unknowns, the rigid body will
be *improperly constrained* if the supports are such that the reactions must either
be parallel or intersect the same line.

# REVIEW PROBLEMS

**4.76**  A crane is oriented so that the end of the 15-m boom *AO* lies in the *yz*
plane. At the instant shown the tension in cable *AB* is 3.5 kN. Determine the
moment about each of the coordinate axes of the force exerted on *A* by the
cable *AB*.

**4.77**  The 15-m crane boom *AO* lies in the *yz* plane. Determine the maximum permissible tension in the cable *AB* if the absolute value of the moments
about the coordinate axes of the force exerted on *A* must be as follows:
$|M_x| \leq 40$ kN · m, $|M_y| \leq 7.5$ kN · m, $|M_z| \leq 8$ kN · m.

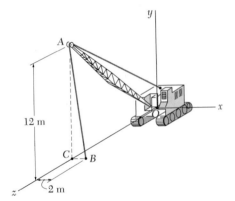

**Fig. P4.76 and P4.77**

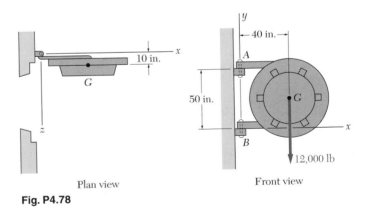

Plan view

Front view

**Fig. P4.78**

**4.78**  The door of a bank vault weighs 12,000 lb and is supported by two
hinges as shown. Determine the components of the reaction at each hinge.

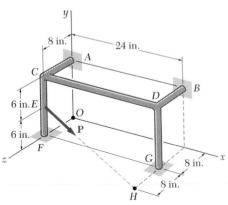

**Fig. P4.79**

**4.79**  A force **P** of magnitude 455 lb acts on the frame shown at point *E*.
Replace this force by an equivalent force-couple system at the origin *O*.

**4.80** Three rods are welded together to form the "corner" shown. The corner is supported by three frictionless eyebolts. Determine the reactions at $A$, $B$, and $C$ when $P = 750$ N, $a = 250$ mm, $b = 150$ mm, and $c = 200$ mm.

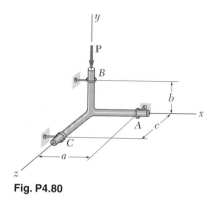

**Fig. P4.80**

**4.81** A concrete foundation mat of 6-m radius supports four equally spaced columns, each of which is located 5 m from the center of the mat. Determine the magnitude and the point of application of the resultant of the four loads.

**4.82** Determine the magnitude and point of application of the smallest additional load which must be applied to the foundation mat of Prob. 4.81 if the resultant of the five loads is to pass through the center of the mat.

**4.83** The table shown weighs 30 lb and has a diameter of 4 ft. It is supported by three legs equally spaced around the edge. A vertical load **P** of magnitude 75 lb is applied to the top of the table at $D$. Determine the maximum value of $a$ if the table is not to tip over. Show, on a sketch, the area of the table over which **P** can act without tipping the table.

**4.84** The 30-ft pole $AC$ is acted upon by a 1680-lb force as shown. It is held by a ball and socket at $A$ and by the two cables $BD$ and $BE$. Neglecting the weight of the pole, determine the tension in each cable and the reaction at $A$.

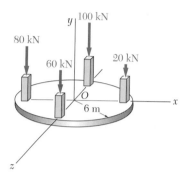

**Fig. P4.81**

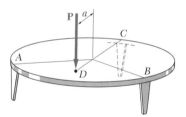

**Fig. P4.83**

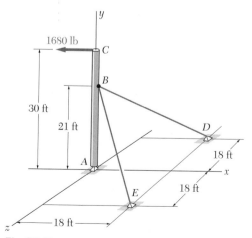

**Fig. P4.84**

**4.85** A 6 × 9 ft plate weighing 1200 lb is lifted by three cables which are joined
at point $D$ directly above the center of the plate. Determine the tension in each cable.

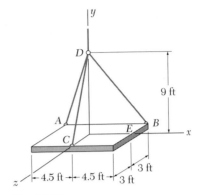

**Fig. P4.85**

**4.86** A 2.5-m boom is held by a ball and socket at $A$ and by two cables $EBF$
and $DC$; cable $EBF$ passes around a frictionless pulley at $B$. Determine the tension
in each cable.

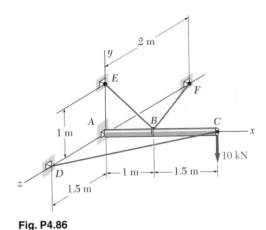

**Fig. P4.86**

**4.87** Replace the three forces shown by ($a$) a force-couple system at the
origin, ($b$) a wrench. (Specify the pitch and axis of the wrench.)

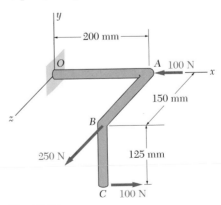

**Fig. P4.87**

# COMPUTER PROBLEMS

**4.C1**  The line of action of a force **P** of given magnitude $P$ is defined as shown by the points $A(x_A, y_A, z_A)$ and $B(x_B, y_B, z_B)$. Write a computer program which can be used to calculate the components of an equivalent force-couple system at the origin $O$. Use this program to solve (*a*) Prob. 4.20, (*b*) Prob. 4.21, (*c*) Prob. 4.79.

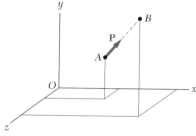

**Fig. P4.C1**

**4.C2**  Write a computer program which can be used to determine the magnitude and point of application of the resultant of the vertical forces $\mathbf{P}_1, \mathbf{P}_2, \ldots,$ $\mathbf{P}_n$ which act at points $A_1, A_2, \ldots, A_n$ located in the $xz$ plane. Use this program to solve (*a*) Sample Prob. 4.4, (*b*) Prob. 4.31, (*c*) Prob. 4.32.

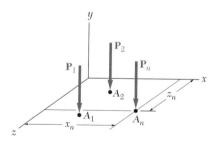

**Fig. P4.C2**

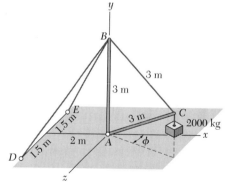

**Fig. P4.C3**

**4.C3**  The derrick shown supports a 2000-kg crate. It is held by a ball and socket at $A$ and by two cables attached at $D$ and $E$. Knowing that the derrick stands in a vertical plane forming an angle $\phi$ with the $xy$ plane, write a computer program and use it to calculate the tension in each cable for values of $\phi$ from 0 to 40° at 4° intervals.

**4.C4**  The lid of a roof scuttle weighs 90 lb. It is hinged at corners $A$ and $B$ and maintained in the desired position by a rod $CD$ pivoted at $C$; a pin at end $D$ of the rod fits into one of several holes drilled in the edge of the lid. Write a computer program and use it to calculate the magnitude of the force exerted by rod $CD$ and the components of the reactions at $A$ and $B$ for values of $\alpha$ from 0 to 90° at 5° intervals. Assume that the hinge at $B$ does not exert any thrust.

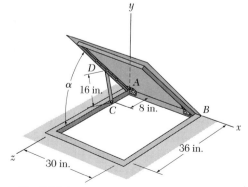

**Fig. P4.C4**

# CHAPTER 5

# Distributed Forces: Centroids and Centers of Gravity

## 5.1. INTRODUCTION

We have assumed so far that the attraction exerted by the earth on a rigid body could be represented by a single force $\mathbf{W}$. This force, called the force of gravity, or the weight of the body, was to be applied at the *center of gravity* of the body (Sec. 3.2). Actually, the earth exerts a force on each of the particles forming the body. The action of the earth on a rigid body should thus be represented by a large number of small forces distributed over the entire body. We shall see in this chapter, however, that all these small forces may be replaced by a single equivalent force $\mathbf{W}$. We shall also learn to determine the center of gravity, i.e., the point of application of the resultant $\mathbf{W}$, for various shapes of bodies.

In the first part of the chapter, we shall consider two-dimensional bodies, such as flat plates and wires contained in a given plane. We shall introduce two concepts closely associated with the determination of the center of gravity of a plate or wire, namely, the concept of *centroid* of an area or line, and the concept of *first moment* of an area or line with respect to a given axis.

We shall also find that the computation of the area of a surface of revolution or of the volume of a body of revolution is directly related to the determination of the centroid of the line or area used to generate that surface or body of revolution (Theorems of Pappus-Guldinus). And, as we shall see in Secs. 5.8 and 5.9, the determination of the centroid of an area simplifies the analysis of beams subjected to distributed loads and the computation of the forces exerted on submerged rectangular surfaces such as hydraulic gates and portions of dams.

In the last part of the chapter, we shall learn to determine the center of gravity of a three-dimensional body, as well as the centroid of a volume and the first moments of that volume with respect to the coordinate planes.

## 5.2. CENTER OF GRAVITY OF A TWO-DIMENSIONAL BODY

Let us first consider a flat horizontal plate (Fig. 5.1). We may divide the plate into $n$ small elements. The coordinates of the first element are denoted by $x_1$ and $y_1$, those of the second element by $x_2$ and $y_2$, etc. The forces exerted by the earth on the elements of plate will be denoted, respectively, by $\Delta\mathbf{W}_1, \Delta\mathbf{W}_2, \ldots, \Delta\mathbf{W}_n$. These forces or weights are directed toward the center of the earth; however, for all practical purposes they may be assumed parallel. Their resultant is therefore a single force in the same direction. The magnitude $W$ of this force is obtained by adding the magnitudes of the elementary weights,

$$\Sigma F_z: \qquad W = \Delta W_1 + \Delta W_2 + \cdots + \Delta W_n$$

To obtain the coordinates $\bar{x}$ and $\bar{y}$ of the point $G$ where the resultant $\mathbf{W}$ should be applied, we write that the moments of $\mathbf{W}$ about the $y$ and $x$ axes are equal to the sum of the corresponding moments of the elementary weights,

$$\Sigma M_y: \qquad \bar{x}W = x_1\,\Delta W_1 + x_2\,\Delta W_2 + \cdots + x_n\,\Delta W_n$$
$$\Sigma M_x: \qquad \bar{y}W = y_1\,\Delta W_1 + y_2\,\Delta W_2 + \cdots + y_n\,\Delta W_n \qquad (5.1)$$

If we now increase the number of elements into which the plate is divided and simultaneously decrease the size of each element, we obtain at the limit the following expressions:

$$W = \int dW \qquad \bar{x}W = \int x\,dW \qquad \bar{y}W = \int y\,dW \qquad (5.2)$$

These equations define the weight $\mathbf{W}$ and the coordinates $\bar{x}$ and $\bar{y}$ of the center of gravity $G$ of a flat plate. The same equations may be derived for a wire lying in the $xy$ plane (Fig. 5.2). We shall observe, in the latter case, that the center of gravity $G$ will generally not be located on the wire.

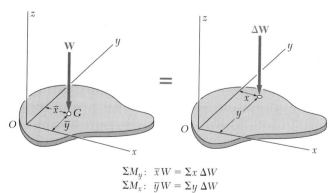

$$\Sigma M_y: \quad \bar{x}W = \Sigma x\,\Delta W$$
$$\Sigma M_x: \quad \bar{y}W = \Sigma y\,\Delta W$$

**Fig. 5.1**    Center of gravity of a plate.

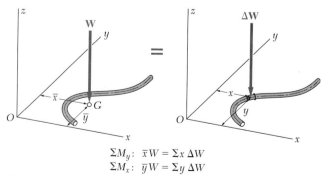

$$\Sigma M_y: \quad \bar{x}W = \Sigma x\,\Delta W$$
$$\Sigma M_x: \quad \bar{y}W = \Sigma y\,\Delta W$$

**Fig. 5.2**    Center of gravity of a wire.

## 5.3. CENTROIDS OF AREAS AND LINES

In the case of a homogeneous plate of uniform thickness, the magnitude $\Delta W$ of the weight of an element of plate may be expressed as

$$\Delta W = \gamma t\, \Delta A$$

where $\gamma$ = specific weight (weight per unit volume) of material
$t$ = thickness of plate
$\Delta A$ = area of element

Similarly, we may express the magnitude $W$ of the weight of the entire plate in the form

$$W = \gamma t A$$

where $A$ is the total area of the plate.

If U.S. customary units are used, the specific weight $\gamma$ should be expressed in $\text{lb/ft}^3$, the thickness $t$ in feet, and the areas $\Delta A$ and $A$ in square feet. We check that $\Delta W$ and $W$ will then be expressed in pounds. If SI units are used, $\gamma$ should be expressed in $\text{N/m}^3$, $t$ in meters, and the areas $\Delta A$ and $A$ in square meters; the weights $\Delta W$ and $W$ will then be expressed in newtons.[†]

Substituting for $\Delta W$ and $W$ in the moment equations (5.1) and dividing throughout by $\gamma t$, we write

$\Sigma M_y$:    $\bar{x}A = x_1\,\Delta A_1 + x_2\,\Delta A_2 + \cdots + x_n\,\Delta A_n$
$\Sigma M_x$:    $\bar{y}A = y_1\,\Delta A_1 + y_2\,\Delta A_2 + \cdots + y_n\,\Delta A_n$

If we increase the number of elements into which the area $A$ is divided and simultaneously decrease the size of each element, we obtain at the limit

$$\bar{x}A = \int x\, dA \qquad \bar{y}A = \int y\, dA \tag{5.3}$$

These equations define the coordinates $\bar{x}$ and $\bar{y}$ of the center of gravity of a homogeneous plate. The point of coordinates $\bar{x}$ and $\bar{y}$ is also known as the *centroid C of the area A* of the plate (Fig. 5.3). If the plate is not homogeneous, the equations cannot be used to determine the center of gravity of the plate; they still define, however, the centroid of the area.

In the case of a homogeneous wire of uniform cross section, the magnitude $\Delta W$ of the weight of an element of wire may be expressed as

$$\Delta W = \gamma a\, \Delta L$$

where $\gamma$ = specific weight of material
$a$ = cross-sectional area of wire
$\Delta L$ = length of element

[†] It should be noted that in the SI system of units a given material is generally characterized by its density $\rho$ (mass per unit volume) rather than by its specific weight $\gamma$. The specific weight of the material can then be obtained by writing

$$\gamma = \rho g$$

where g = $9.81 \text{ m/s}^2$. Since $\rho$ is expressed in $\text{kg/m}^3$, we check that $\gamma$ will be expressed in $(\text{kg/m}^3)(\text{m/s}^2)$, that is, in $\text{N/m}^3$.

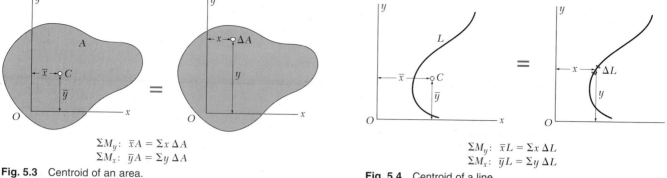

$$\Sigma M_y: \quad \bar{x}A = \Sigma x\,\Delta A$$
$$\Sigma M_x: \quad \bar{y}A = \Sigma y\,\Delta A$$

**Fig. 5.3**   Centroid of an area.

$$\Sigma M_y: \quad \bar{x}L = \Sigma x\,\Delta L$$
$$\Sigma M_x: \quad \bar{y}L = \Sigma y\,\Delta L$$

**Fig. 5.4**   Centroid of a line.

The center of gravity of the wire then coincides with the *centroid C of the line L* defining the shape of the wire (Fig. 5.4). The coordinates $\bar{x}$ and $\bar{y}$ of the centroid of the line $L$ are obtained from the equations

$$\bar{x}L = \int x\,dL \qquad \bar{y}L = \int y\,dL \tag{5.4}$$

## 5.4. FIRST MOMENTS OF AREAS AND LINES

The integral $\int x\,dA$ in Eqs. (5.3) of the preceding section is known as the *first moment of the area A with respect to the y axis* and is denoted by $Q_y$. Similarly, the integral $\int y\,dA$ defines the *first moment of A with respect to the x axis* and is denoted by $Q_x$. We write

$$Q_y = \int x\,dA \qquad Q_x = \int y\,dA \tag{5.5}$$

Comparing Eqs. (5.3) with Eqs. (5.5), we note that the first moments of the area $A$ may be expressed as the products of the area and of the coordinates of its centroid:

$$Q_y = \bar{x}A \qquad Q_x = \bar{y}A \tag{5.6}$$

It follows from Eqs. (5.6) that the coordinates of the centroid of an area may be obtained by dividing the first moments of that area by the area itself. The computation of the first moments of an area is also useful in the determination of shearing stresses under transverse loadings in mechanics of materials. Finally, we observe from Eqs. (5.6) that if the centroid of an area is located on a coordinate axis, the first moment of the area with respect to that axis is zero, and conversely.

Relations similar to Eqs. (5.5) and (5.6) may be used to define the first moments of a line with respect to the coordinate axes and to express these moments as the products of the length $L$ of the line and of the coordinates $\bar{x}$ and $\bar{y}$ of its centroid.

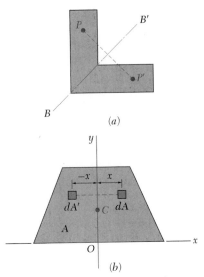

**Fig. 5.5**

An area $A$ is said to be *symmetric with respect to an axis BB'* if to every point $P$ of the area corresponds a point $P'$ of the same area such that the line $PP'$ is perpendicular to $BB'$ and is divided into two equal parts by that axis (Fig. 5.5a). A line $L$ is said to be symmetric with respect to $BB'$ if it satisfies similar conditions. When an area $A$ or a line $L$ possesses an axis of symmetry $BB'$, its first moment with respect to $BB'$ is zero and its centroid is located on that axis. Indeed, considering the area $A$ of Fig. 5.5b, which is symmetric with respect to the $y$ axis, we observe that to every element of area $dA$ of abscissa $x$ corresponds an element of area $dA'$ of abscissa $-x$. It follows that the integral in the first of Eqs. (5.5) is zero and, thus, that $Q_y = 0$. It also follows from the first of the relations (5.3) that $\bar{x} = 0$. Thus, if an area $A$ or a line $L$ possesses an axis of symmetry, its centroid $C$ is located on that axis.

We further note that if an area or line possesses two axes of symmetry, its centroid $C$ must be located at the intersection of the two axes of symmetry (Fig. 5.6). This property enables us to determine immediately the centroid of areas such as circles, ellipses, squares, rectangles, equilateral triangles, or other symmetric figures, as well as the centroid of lines in the shape of the circumference of a circle, the perimeter of a square, etc.

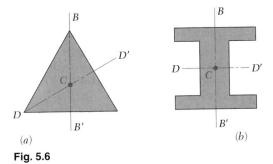

**Fig. 5.6**

An area $A$ is said to be *symmetric with respect to a center O* if to every element of area $dA$ of coordinates $x$ and $y$ corresponds an element of area $dA'$ of coordinates $-x$ and $-y$ (Fig. 5.7). It then follows that the integrals in Eqs. (5.5) are both zero, and that $Q_x = Q_y = 0$. It also follows from Eqs. (5.3) that $\bar{x} = \bar{y} = 0$, that is, the centroid of the area coincides with its center of symmetry $O$. It may be shown in a similar way that the centroid of a line symmetric with respect to a center $O$ coincides with $O$.

It should be noted that a figure possessing *a center of symmetry* does not necessarily possess *an axis of symmetry* (Fig. 5.7), while a figure possessing two axes of symmetry does not necessarily possess a center of symmetry (Fig. 5.6a). However, if a figure possesses two axes of symmetry at a right angle to each other, the point of intersection of these axes will be a center of symmetry (Fig. 5.6b).

Centroids of unsymmetrical areas and lines and of areas and lines possessing only one axis of symmetry will be determined by the methods of Secs. 5.6 and 5.7. Centroids of common shapes of areas and lines are shown in Fig. 5.8A and B. The formulas defining these centroids will be derived in the Sample Problems and Problems following Secs. 5.6 and 5.7.

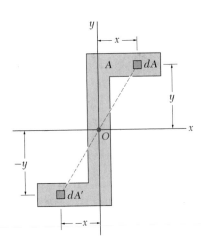

**Fig. 5.7**

| Shape | | $\bar{x}$ | $\bar{y}$ | Area |
|-------|---|-----------|-----------|------|
| Triangular area | 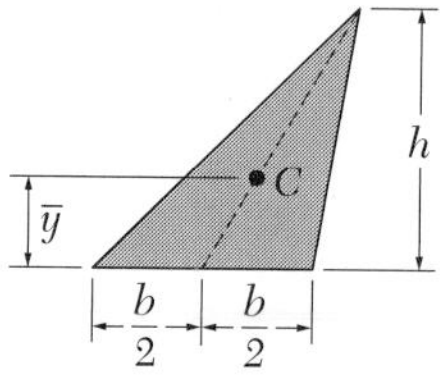 | | $\dfrac{h}{3}$ | $\dfrac{bh}{2}$ |
| Quarter-circular area | 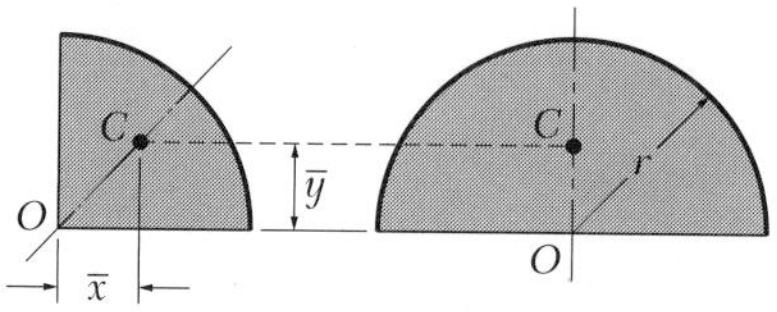 | $\dfrac{4r}{3\pi}$ | $\dfrac{4r}{3\pi}$ | $\dfrac{\pi r^2}{4}$ |
| Semicircular area | | $0$ | $\dfrac{4r}{3\pi}$ | $\dfrac{\pi r^2}{2}$ |
| Quarter-elliptical area | 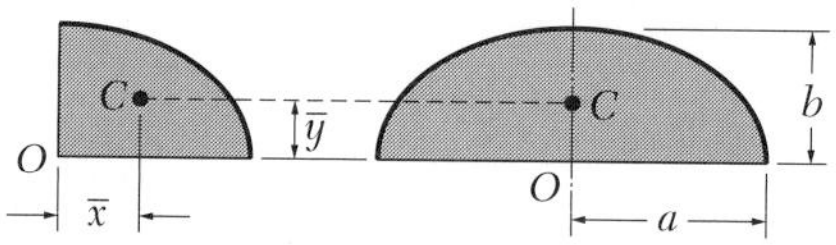 | $\dfrac{4a}{3\pi}$ | $\dfrac{4b}{3\pi}$ | $\dfrac{\pi ab}{4}$ |
| Semielliptical area | | $0$ | $\dfrac{4b}{3\pi}$ | $\dfrac{\pi ab}{2}$ |
| Semiparabolic area | 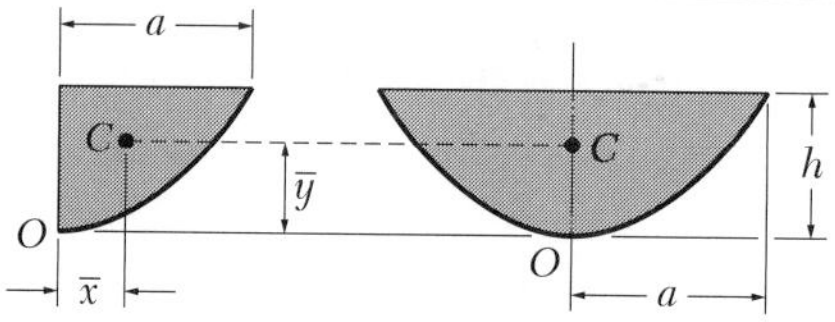 | $\dfrac{3a}{8}$ | $\dfrac{3h}{5}$ | $\dfrac{2ah}{3}$ |
| Parabolic area | | $0$ | $\dfrac{3h}{5}$ | $\dfrac{4ah}{3}$ |
| Parabolic spandrel | 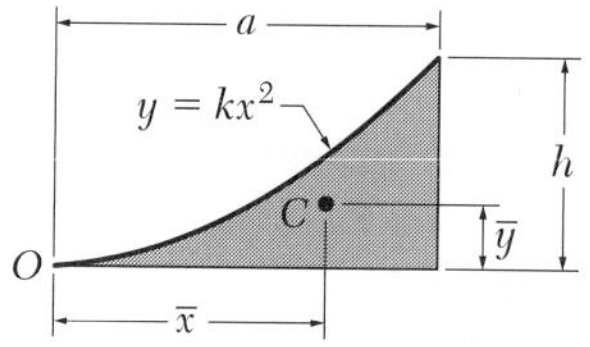 | $\dfrac{3a}{4}$ | $\dfrac{3h}{10}$ | $\dfrac{ah}{3}$ |
| General spandrel | 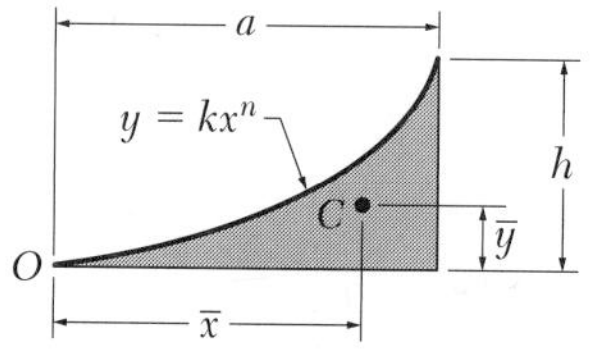 | $\dfrac{n+1}{n+2}a$ | $\dfrac{n+1}{4n+2}h$ | $\dfrac{ah}{n+1}$ |
| Circular sector | 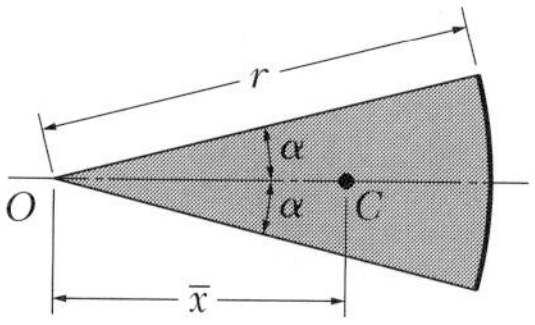 | $\dfrac{2r\sin\alpha}{3\alpha}$ | $0$ | $\alpha r^2$ |

**Fig. 5.8A** Centroids of common shapes of areas.

| Shape | | $\bar{x}$ | $\bar{y}$ | Length |
|-------|--|-----------|-----------|--------|
| Quarter-circular arc | 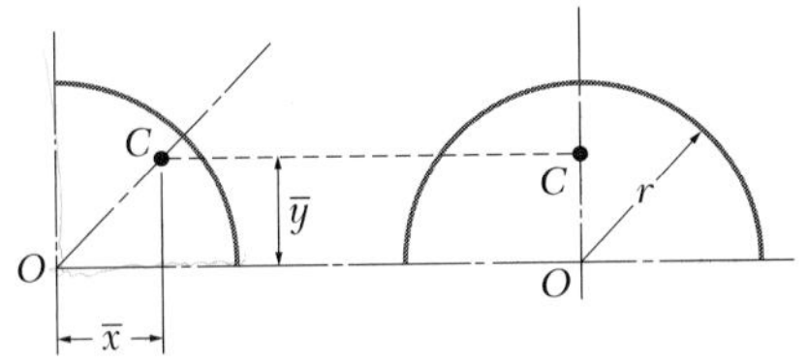 | $\dfrac{2r}{\pi}$ | $\dfrac{2r}{\pi}$ | $\dfrac{\pi r}{2}$ |
| Semicircular arc | | $0$ | $\dfrac{2r}{\pi}$ | $\pi r$ |
| Arc of circle | 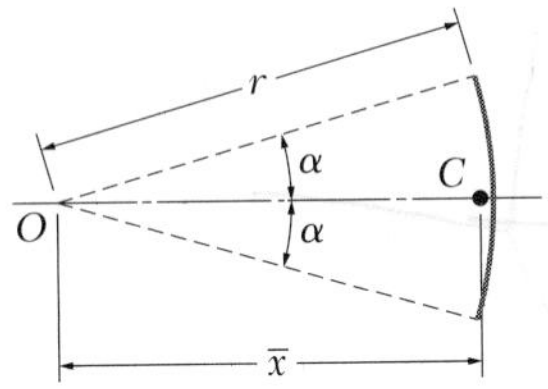 | $\dfrac{r \sin \alpha}{\alpha}$ | $0$ | $2\alpha r$ |

**Fig. 5.8B**  Centroids of common shapes of lines.

## 5.5. COMPOSITE PLATES AND WIRES

In many instances, a flat plate may be divided into rectangles, triangles, or other common shapes shown in Fig. 5.8A. The abscissa $\overline{X}$ of its center of gravity $G$ may be determined from the abscissas $\bar{x}_1, \bar{x}_2, \ldots$ of the centers of gravity of the various parts by expressing that the moment of the weight of the whole plate about the $y$ axis is equal to the sum of the moments of the weights of the various parts about the same axis (Fig. 5.9). The ordinate $\overline{Y}$ of the center of gravity of the plate is found in a similar way by equating moments about the $x$ axis. We write

$$\Sigma M_y: \quad \overline{X}(W_1 + W_2 + \cdots + W_n) = \bar{x}_1 W_1 + \bar{x}_2 W_2 + \cdots + \bar{x}_n W_n$$
$$\Sigma M_x: \quad \overline{Y}(W_1 + W_2 + \cdots + W_n) = \bar{y}_1 W_1 + \bar{y}_2 W_2 + \cdots + \bar{y}_n W_n$$

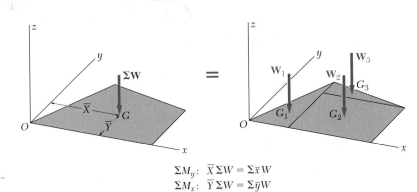

$$\Sigma M_y: \quad \overline{X}\,\Sigma W = \Sigma \bar{x}\,W$$
$$\Sigma M_x: \quad \overline{Y}\,\Sigma W = \Sigma \bar{y}\,W$$

**Fig. 5.9**  Center of gravity of a composite plate.

or, for short,

$$\overline{X}\Sigma W = \Sigma \overline{x}W \qquad \overline{Y}\Sigma W = \Sigma \overline{y}W \tag{5.7}$$

These equations may be solved for the coordinates $\overline{X}$ and $\overline{Y}$ of the center of gravity of the plate.

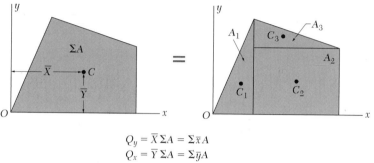

$$Q_y = \overline{X}\Sigma A = \Sigma \overline{x}A$$
$$Q_x = \overline{Y}\Sigma A = \Sigma \overline{y}A$$

**Fig. 5.10**   Centroid of a composite area.

If the plate is homogeneous and of uniform thickness, the center of gravity coincides with the centroid $C$ of its area. The abscissa $\overline{X}$ of the centroid of the area may be determined by noting that the first moment $Q_y$ of the composite area with respect to the $y$ axis may be expressed both as the product of $\overline{X}$ and the total area and as the sum of the first moments of the elementary areas with respect to the $y$ axis (Fig. 5.10). The ordinate $\overline{Y}$ of the centroid is found in a similar way by considering the first moment $Q_x$ of the composite area. We have

$$Q_y = \overline{X}(A_1 + A_2 + \cdots + A_n) = \overline{x}_1A_1 + \overline{x}_2A_2 + \cdots + \overline{x}_nA_n$$
$$Q_x = \overline{Y}(A_1 + A_2 + \cdots + A_n) = \overline{y}_1A_1 + \overline{y}_2A_2 + \cdots + \overline{y}_nA_n$$

or, for short,

$$Q_y = \overline{X}\Sigma A = \Sigma \overline{x}A \qquad Q_x = \overline{Y}\Sigma A = \Sigma \overline{y}A \tag{5.8}$$

These equations yield the first moments of the composite area, or they may be solved for the coordinates $\overline{X}$ and $\overline{Y}$ of its centroid.

Care should be taken to record the moment of each area with the appropriate sign. First moments of areas, just like moments of forces, may be positive or negative. For example, an area whose centroid is located to the left of the $y$ axis will have a negative first moment with respect to that axis. Also, the area of a hole should be recorded with a negative sign (Fig. 5.11).

Similarly, it is possible in many cases to determine the center of gravity of a composite wire or the centroid of a composite line by dividing the wire or line into simpler elements (see Sample Prob. 5.2).

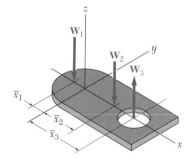

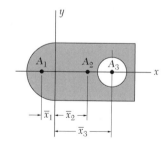

|  | $\overline{x}$ | $A$ | $\overline{x}A$ |
|---|---|---|---|
| $A_1$ Semicircle | $-$ | $+$ | $-$ |
| $A_2$ Full rectangle | $+$ | $+$ | $+$ |
| $A_3$ Circular hole | $+$ | $-$ | $-$ |

**Fig. 5.11**

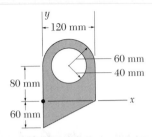

# SAMPLE PROBLEM 5.1

For the plane area shown, determine (a) the first moments with respect to the $x$ and $y$ axes, (b) the location of the centroid.

**Solution.** The area is obtained by adding a rectangle, a triangle, and a semi-circle and subtracting a circle. Using the coordinate axes shown, the area and the coordinates of the centroid of each of the component areas are determined and entered in the table below. The area of the circle is indicated as negative, since it is to be subtracted from the other areas. We note that the coordinate $\bar{y}$ of the centroid of the triangle is negative for the axes shown. The first moments of the component areas with respect to the coordinate axes are computed and entered in the table.

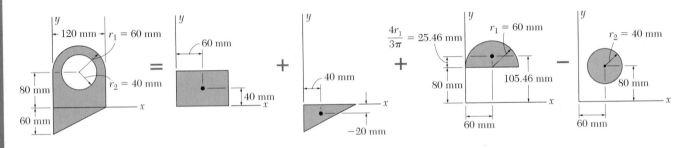

| Component | $A$, mm² | $\bar{x}$, mm | $\bar{y}$, mm | $\bar{x}A$, mm³ | $\bar{y}A$, mm³ |
|---|---|---|---|---|---|
| Rectangle | $(120)(80) = 9.6 \times 10^3$ | 60 | 40 | $+576 \times 10^3$ | $+384 \times 10^3$ |
| Triangle | $\frac{1}{2}(120)(60) = 3.6 \times 10^3$ | 40 | $-20$ | $+144 \times 10^3$ | $-72 \times 10^3$ |
| Semicircle | $\frac{1}{2}\pi(60)^2 = 5.655 \times 10^3$ | 60 | 105.46 | $+339.3 \times 10^3$ | $+596.4 \times 10^3$ |
| Circle | $-\pi(40)^2 = -5.027 \times 10^3$ | 60 | 80 | $-301.6 \times 10^3$ | $-402.2 \times 10^3$ |
| | $\Sigma A = 13.828 \times 10^3$ | | | $\Sigma\bar{x}A = +757.7 \times 10^3$ | $\Sigma\bar{y}A = +506.2 \times 10^3$ |

**a. First Moments of Area.** Using Eqs. (5.8), we write

$$Q_x = \Sigma\bar{y}A = 506.2 \times 10^3 \text{ mm}^3$$
$$Q_x = 506 \times 10^3 \text{ mm}^3 \blacktriangleleft$$
$$Q_y = \Sigma\bar{x}A = 757.7 \times 10^3 \text{ mm}^3$$
$$Q_y = 758 \times 10^3 \text{ mm}^3 \blacktriangleleft$$

**b. Location of Centroid.** Substituting the values obtained from the table into the equations defining the centroid of a composite area, we obtain

$$\bar{X}\Sigma A = \Sigma\bar{x}A: \quad \bar{X}(13.828 \times 10^3 \text{ mm}^2) = 757.7 \times 10^3 \text{ mm}^3$$
$$\bar{X} = 54.8 \text{ mm} \blacktriangleleft$$
$$\bar{Y}\Sigma A = \Sigma\bar{y}A: \quad \bar{Y}(13.828 \times 10^3 \text{ mm}^2) = 506.2 \times 10^3 \text{ mm}^3$$
$$\bar{Y} = 36.6 \text{ mm} \blacktriangleleft$$

# SAMPLE PROBLEM 5.2

The figure shown is made of a thin homogeneous wire. Determine its center of gravity.

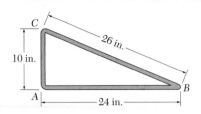

**Solution.** Since the figure is formed of homogeneous wire, its center of gravity may be located by determining the centroid of the corresponding line. Choosing the coordinate axes shown, with origin at $A$, we determine the coordinates of the centroid of each segment of line and compute its first moments with respect to the coordinate axes.

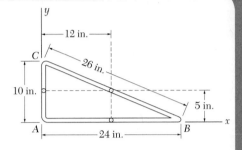

| Segment | $L$, in. | $\bar{x}$, in. | $\bar{y}$, in. | $\bar{x}L$, in.$^2$ | $\bar{x}L$, in.$^2$ |
|---------|----------|----------------|----------------|---------------------|---------------------|
| $AB$ | 24 | 12 | 0 | 288 | 0 |
| $BC$ | 26 | 12 | 5 | 312 | 130 |
| $CA$ | 10 | 0 | 5 | 0 | 50 |
| | $\Sigma L = 60$ | ... | ... | $\Sigma \bar{x}L = 600$ | $\Sigma \bar{y}L = 180$ |

Substituting the values obtained from the table into the equations defining the centroid of a composite line, we obtain

$$\bar{X}\Sigma L = \Sigma \bar{x}L: \qquad \bar{X}(60 \text{ in.}) = 600 \text{ in.}^2 \qquad\qquad \bar{X} = 10 \text{ in.} \blacktriangleleft$$
$$\bar{Y}\Sigma L = \Sigma \bar{y}L: \qquad \bar{Y}(60 \text{ in.}) = 180 \text{ in.}^2 \qquad\qquad \bar{Y} = \phantom{0}3 \text{ in.} \blacktriangleleft$$

## SAMPLE PROBLEM 5.3

A uniform semicircular rod of weight $W$ and radius $r$ is attached to a pin at $A$ and bears against a frictionless surface at $B$. Determine the reactions at $A$ and $B$.

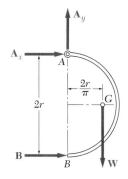

**Solution.** A free-body diagram of the rod is drawn. The forces acting on the rod consist of its weight $\mathbf{W}$ applied at the center of gravity $G$, whose position is obtained from Fig. 5.8B, of a reaction at $A$ represented by its components $\mathbf{A}_x$ and $\mathbf{A}_y$, and of a horizontal reaction at $B$.

$$+\,\uparrow\Sigma M_A = 0: \qquad B(2r) - W\left(\frac{2r}{\pi}\right) = 0$$

$$B = +\frac{W}{\pi} \qquad\qquad \mathbf{B} = \frac{W}{\pi} \rightarrow \quad \blacktriangleleft$$

$$\xrightarrow{+}\Sigma F_x = 0: \qquad A_x + B = 0$$

$$A_x = -B = -\frac{W}{\pi} \qquad \mathbf{A}_x = \frac{W}{\pi} \leftarrow$$

$$+\uparrow\Sigma F_y = 0: \qquad A_y - W = 0 \qquad A_y = W\uparrow$$

Adding the two components of the reaction at $A$:

$$A = \left[W^2 + \left(\frac{W}{\pi}\right)^2\right]^{1/2} \qquad\qquad A = W\left(1 + \frac{1}{\pi^2}\right)^{1/2} \quad \blacktriangleleft$$

$$\tan\alpha = \frac{W}{W/\pi} = \pi \qquad\qquad \alpha = \tan^{-1}\pi \quad \blacktriangleleft$$

The answers may also be expressed as follows:

$$\mathbf{A} = 1.049W \;\diagdown\; 72.3° \qquad \mathbf{B} = 0.318W \rightarrow \quad \blacktriangleleft$$

# PROBLEMS

**5.1 through 5.10** Locate the centroid of the plane area shown.

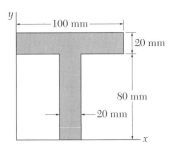

**Fig. P5.1**

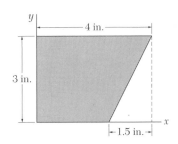

**Fig. P5.2**

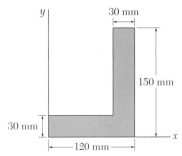

**Fig. P5.3**

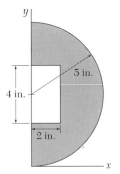

**Fig. P5.4**

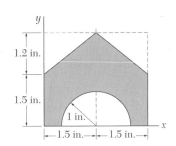

**Fig. P5.5**

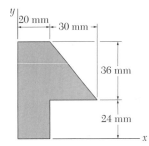

**Fig. P5.6**

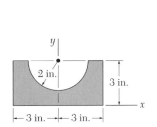

**Fig. P5.7**

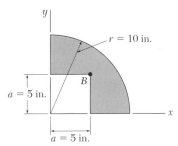

**Fig. P5.8**

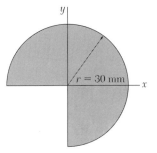

**Fig. P5.9**

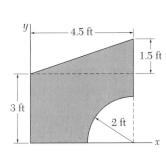

**Fig. P5.10**

**5.11 through 5.14**  Locate the centroid of the plane area shown.

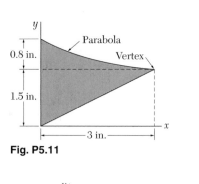

**Fig. P5.11**

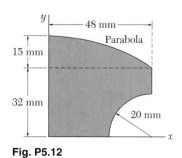

**Fig. P5.12**

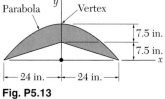

**Fig. P5.13**

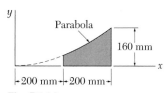

**Fig. P5.14**

**5.15**  Determine the abscissa of the centroid of the circular segment in terms of $r$ and $\alpha$.

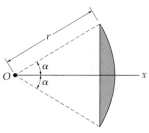

**Fig. P5.15**

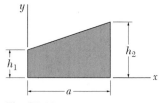

**Fig. P5.16**

**5.16**  Determine the abscissa of the centroid of the trapezoid shown in terms of $h_1$, $h_2$, and $a$.

**5.17 through 5.19**  Locate the centroid of the plane area shown.

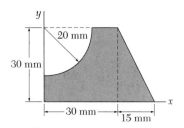

**Fig. P5.17**

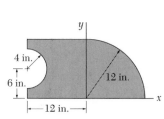

**Fig. P5.18**

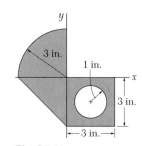

**Fig. P5.19**

**5.20** Shown is an area that is the composite of two subareas $A$ and $B$. The centroids of $A$ and $B$ are marked. Determine the centroid of the composite area and show that it plots on line $AB$. One quick check on a two-area composite centroid problem is to visually inspect that the computed centroid lies on this line. Where it plots on that line is a weighted average of the areas of $A$ and $B$. Since $A$ is twice the size of area $B$, the centroid should plot one-third of $AB$ from $A$ and two-thirds of $AB$ from $B$. A centroid is essentially a weighted average.

**5.21** A built-up beam has been constructed by nailing together seven planks as shown. The nails are equally spaced along the beam and the beam supports a vertical load. As proved in mechanics of materials, the shearing forces exerted on the nails at $A$ and $B$ are proportional to the first moments with respect to the centroidal $x$ axis of the shaded areas shown, respectively, in parts $a$ and $b$ of the figure. Knowing that the force exerted on the nail at $A$ is 120 N, determine the force exerted on the nail at $B$.

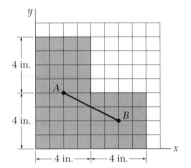

**Fig. P5.20**

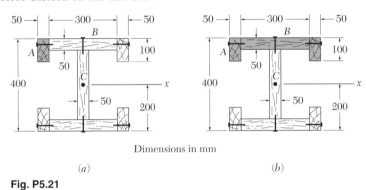

Dimensions in mm

(a)            (b)

**Fig. P5.21**

**5.22 and 5.23** The horizontal $x$ axis is drawn through the centroid $C$ of the area shown and divides it into two component areas $A_1$ and $A_2$. Determine the first moment of each component area with respect to the $x$ axis and explain the result obtained.

**5.24** Shown is an area that is the composite of three subareas $A$, $B$, and $C$, whose individual centroids are marked. Determine the centroid of the composite area and show that it plots within the $ABC$ triangle. One can intuit the location more precisely than that by estimating the centroid of areas $A$ and $B$ to be at $D$. Then, the overall centroid will lie on $CD$. Since the combined area of $A$ and $B$ is much larger than $C$, the overall centroid is expected to be approximately $E$. Is your calculated centroid located near $E$?

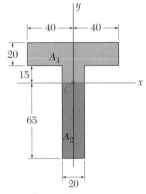

Dimensions in mm

**Fig. P5.22**

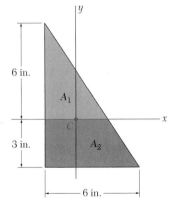

**Fig. P5.23**

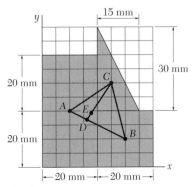

**Fig. P5.24**

**5.25 through 5.27** A thin homogeneous rod is bent to form the shape indicated in the figure. Locate the center of gravity of the shape.

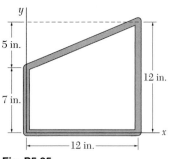

Fig. P5.25

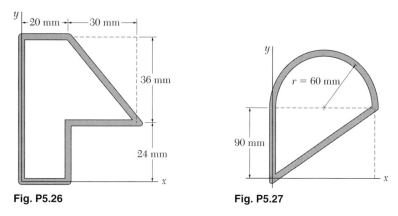

Fig. P5.26

Fig. P5.27

**5.28** A 0.75-in. diameter homogeneous steel rod (weighing 1.5 lb per foot) is bent to form the shape indicated in the figure. Determine the support reactions at $A$ and $B$ required for equilibrium.

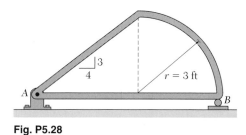

Fig. P5.28

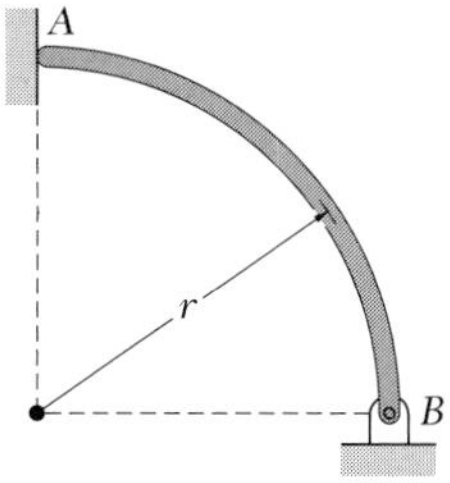

**5.29** A uniform circular rod of weight 20 lb and radius 18 in. is attached to a pin at $B$ and bears against a frictionless surface at $A$. Determine the reactions at $A$ and $B$.

**5.30** The homogeneous wire $ABCD$ is bent as shown and is attached to a hinge at $C$. Determine the length $L$ for which portion $BCD$ of the wire is horizontal.

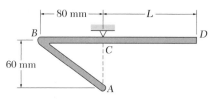

Fig. P5.30 and P5.31

Fig. P5.29

**5.31** The homogeneous wire $ABCD$ is bent as shown and is attached to a hinge at $C$. Determine the length $L$ for which portion $AB$ of the wire is horizontal.

**5.32** Knowing that the figure shown is formed of a thin homogeneous wire, determine the length $L$ of portion $CE$ of the wire for which the center of gravity of the figure is located at point $C$, when (a) $\theta = 30°$, (b) $\theta = 90°$.

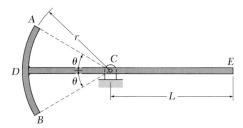

Fig. P5.32

## 5.6. DETERMINATION OF CENTROIDS BY INTEGRATION

The centroid of an area bounded by analytical curves (i.e., curves defined by algebraic equations) is usually determined by computing the integrals in Eqs. (5.3) of Sec. 5.3:

$$\bar{x}A = \int x\, dA \qquad \bar{y}A = \int y\, dA \qquad (5.3)$$

If the element of area $dA$ is chosen equal to a small square of sides $dx$ and $dy$, the determination of each of these integrals requires a *double integration* in $x$ and $y$. A double integration is also necessary if polar coordinates are used and if $dA$ is chosen equal to a small square of sides $dr$ and $r\, d\theta$.

In most cases, however, it is possible to determine the coordinates of the centroid of an area by performing a single integration. This is achieved by choosing for $dA$ a thin rectangle or strip, or a thin sector or pie-shaped element (Fig. 5.12); the centroid of the thin rectangle is located at its center, and the centroid of the thin sector at a distance $\frac{2}{3}r$ from its vertex (as for a triangle). The coordinates of the centroid of the area under consideration are then obtained by expressing that the first moment of the entire area with respect to each of the coordinate axes is equal to the sum (or integral) of the corresponding moments of the elements of area. Denoting by $\bar{x}_{el}$ and $\bar{y}_{el}$ the coordinates of the centroid of the element $dA$, we write

$$Q_y = \bar{x}A = \int \bar{x}_{el}\, dA$$
$$Q_x = \bar{y}A = \int \bar{y}_{el}\, dA \qquad (5.9)$$

If the area itself is not already known, it may also be computed from these elements.

The coordinates $\bar{x}_{el}$ and $\bar{y}_{el}$ of the centroid of the element of area should be expressed in terms of the coordinates of a point located on the curve bounding the area under consideration. Also, the element of area $dA$ should be expressed in terms of the coordinates of the point and their differentials. This has been done in Fig. 5.12 for three common types of elements; the pie-shaped element of part $c$ should be used when the equation of the curve bounding the area is given in polar coordinates. The appropriate expressions should be substituted in formulas (5.9), and the equation of the curve should be used to express one of the coordinates in terms of the other. The integration is thus reduced to a single integration which may be performed according to the usual rules of calculus. Once the area and the integrals in Eqs. (5.9) have been determined, these equations may be solved for the coordinates $\bar{x}$ and $\bar{y}$ of the centroid of the area.

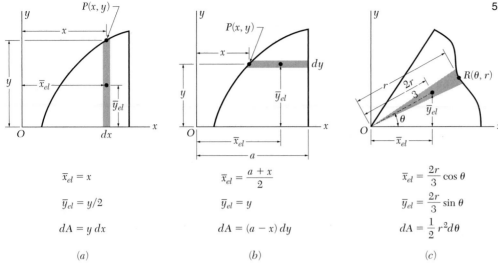

$$\bar{x}_{el} = x$$

$$\bar{y}_{el} = y/2$$

$$dA = y\,dx$$

(a)

$$\bar{x}_{el} = \frac{a+x}{2}$$

$$\bar{y}_{el} = y$$

$$dA = (a - x)\,dy$$

(b)

$$\bar{x}_{el} = \frac{2r}{3}\cos\theta$$

$$\bar{y}_{el} = \frac{2r}{3}\sin\theta$$

$$dA = \frac{1}{2}r^2 d\theta$$

(c)

**Fig. 5.12** Centroids and areas of
differential elements.

The centroid of a line defined by an algebraic equation may be determined by computing the integrals in Eqs. (5.4) of Sec. 5.3:

$$\bar{x}L = \int x\,dL \qquad \bar{y}L = \int y\,dL \qquad (5.4)$$

The element $dL$ should be replaced by one of the following expressions, depending upon the type of equation used to define the line (these expressions may be derived by using the Pythagorean theorem):

$$dL = \sqrt{1 + \left(\frac{dy}{dx}\right)^2}\,dx$$

$$dL = \sqrt{1 + \left(\frac{dx}{dy}\right)^2}\,dy$$

$$dL = \sqrt{r^2 + \left(\frac{dr}{d\theta}\right)^2}\,d\theta$$

The equation of the line is then used to express one of the coordinates in terms of the other, the integration may be performed by the methods of calculus, and Eqs. (5.4) may be solved for the coordinates $\bar{x}$ and $\bar{y}$ of the centroid of the line.

## 5.7. THEOREMS OF PAPPUS-GULDINUS

These theorems, which were first formulated by the Greek geometer Pappus during the third century A.D. and later restated by the Swiss mathematician Guldinus, or Guldin (1577–1643), deal with surfaces and bodies of revolution.

A *surface of revolution* is a surface which may be generated by rotating a plane curve about a fixed axis. For example (Fig. 5.13), the surface of

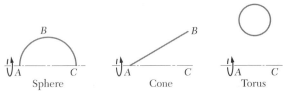

**Fig. 5.13** Generating a surface of revolution.

a sphere may be obtained by rotating a semicircular arc *ABC* about the diameter *AC*; the surface of a cone by rotating a straight line *AB* about an axis *AC*; the surface of a torus or ring by rotating the circumference of a circle about a nonintersecting axis. A *body of revolution* is a body which may be generated by rotating a plane area about a fixed axis. A solid sphere may be obtained by rotating a semicircular area, a cone by rotating a triangular area, and a solid torus by rotating a full circular area (Fig. 5.14).

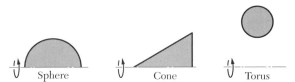

**Fig. 5.14** Generating a body of revolution.

**THEOREM I.**   *The area of a surface of revolution is equal to the length of the generating curve times the distance traveled by the centroid of the curve while the surface is being generated.*

*Proof.*   Consider an element $dL$ of the line $L$ (Fig. 5.15) which is revolved about the $x$ axis. The area $dA$ generated by the element $dL$ is equal to $2\pi y\, dL$. Thus, the entire area generated by $L$ is $A = \int 2\pi y\, dL$. But we saw in Sec. 5.3 that the integral $\int y\, dL$ is equal to $\bar{y}L$. We have therefore

$$A = 2\pi\bar{y}L \tag{5.10}$$

where $2\pi\bar{y}$ is the distance traveled by the centroid of $L$. It should be noted that the generating curve should not cross the axis about which it is rotated;

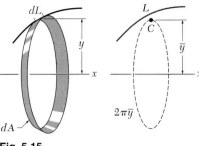

**Fig. 5.15**

if it did, the two sections on either side of the axis would generate areas of opposite signs and the theorem would not apply.

**THEOREM II.** *The volume of a body of revolution is equal to the generating area times the distance traveled by the centroid of the area while the body is being generated.*

*Proof.* Consider an element $dA$ of the area $A$ which is revolved about the $x$ axis (Fig. 5.16). The volume $dV$ generated by the element $dA$ is equal

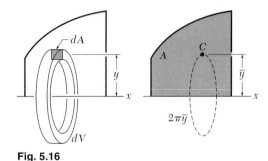

**Fig. 5.16**

to $2\pi y\, dA$. Thus, the entire volume generated by $A$ is $V = \int 2\pi y\, dA$. But since the integral $\int y\, dA$ is equal to $\bar{y}A$ (Sec. 5.3), we have

$$V = 2\pi\bar{y}A \tag{5.11}$$

where $2\pi\bar{y}$ is the distance traveled by the centroid of $A$. Again, it should be noted that the theorem does not apply if the axis of rotation intersects the generating area.

The theorems of Pappus-Guldinus offer a simple way for computing the area of surfaces of revolution and the volume of bodies of revolution. They may also be used conversely to determine the centroid of a plane curve when the area of the surface generated by the curve is known or to determine the centroid of a plane area when the volume of the body generated by the area is known (see Sample Prob. 5.8).

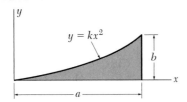

$y = kx^2$

# SAMPLE PROBLEM 5.4

Determine by direct integration the centroid of a parabolic spandrel.

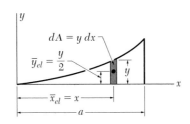

$dA = y\,dx$

$\bar{y}_{el} = \dfrac{y}{2}$

$\bar{x}_{el} = x$

**Solution.** The value of $k$ is determined by substituting $x = a$ and $y = b$ in the given equation. We have $b = ka^2$ and $k = b/a^2$. The equation of the curve is thus

$$y = \frac{b}{a^2}x^2 \qquad \text{or} \qquad x = \frac{a}{b^{1/2}}y^{1/2}$$

**Vertical Differential Element.** We choose the differential element shown and find the total area of the figure.

$$A = \int dA = \int y\,dx = \int_0^a \frac{b}{a^2}x^2\,dx = \left[\frac{b}{a^2}\frac{x^3}{3}\right]_0^a = \frac{ab}{3}$$

The first moment of the differential element with respect to the $y$ axis is $\bar{x}_{el}\,dA$; hence, the first moment of the entire area with respect to this axis is

$$Q_y = \int \bar{x}_{el}\,dA = \int xy\,dx = \int_0^a x\left(\frac{b}{a^2}x^2\right)dx = \left[\frac{b}{a^2}\frac{x^4}{4}\right]_0^a = \frac{a^2 b}{4}$$

Since $Q_y = \bar{x}A$, we have

$$\bar{x}A = \int \bar{x}_{el}\,dA \qquad \bar{x}\frac{ab}{3} = \frac{a^2 b}{4} \qquad\qquad \bar{x} = \tfrac{3}{4}a \quad \blacktriangleleft$$

Likewise, the first moment of the differential element with respect to the $x$ axis is $\bar{y}_{el}\,dA$, and the first moment of the entire area is

$$Q_x = \int \bar{y}_{el}\,dA = \int \frac{y}{2}y\,dx = \int_0^a \frac{1}{2}\left(\frac{b}{a^2}x^2\right)^2 dx = \left[\frac{b^2}{2a^4}\frac{x^5}{5}\right]_0^a = \frac{ab^2}{10}$$

Since $Q_x = \bar{y}A$, we have

$$\bar{y}A = \int \bar{y}_{el}\,dA \qquad \bar{y}\frac{ab}{3} = \frac{ab^2}{10} \qquad\qquad \bar{y} = \tfrac{3}{10}b \quad \blacktriangleleft$$

**Horizontal Differential Element.** The same result may be obtained by considering a horizontal element. The first moments of the area are

$$Q_y = \int \bar{x}_{el}\,dA = \int \frac{a+x}{2}(a-x)\,dy = \int_0^b \frac{a^2 - x^2}{2}\,dy$$

$$= \frac{1}{2}\int_0^b \left(a^2 - \frac{a^2}{b}y\right)dy = \frac{a^2 b}{4}$$

$$Q_x = \int \bar{y}_{el}\,dA = \int y(a-x)\,dy = \int y\left(a - \frac{a}{b^{1/2}}y^{1/2}\right)dy$$

$$= \int_0^b \left(ay - \frac{a}{b^{1/2}}y^{3/2}\right)dy = \frac{ab^2}{10}$$

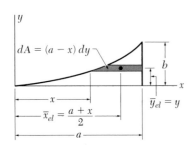

$dA = (a - x)\,dy$

$\bar{x}_{el} = \dfrac{a+x}{2}$

$\bar{y}_{el} = y$

The expressions obtained are again substituted in the equations defining the centroid of the area to obtain $\bar{x}$ and $\bar{y}$.

# SAMPLE PROBLEM 5.5

Determine the centroid of the arc of circle shown.

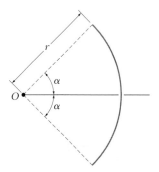

**Solution.** Since the arc is symmetrical with respect to the $x$ axis, we have $\bar{y} = 0$. A differential element is chosen as shown, and the length of the arc is determined by integration.

$$L = \int dL = \int_{-\alpha}^{\alpha} r\, d\theta = r \int_{-\alpha}^{\alpha} d\theta = 2r\alpha$$

The first moment of the arc with respect to the $y$ axis is

$$Q_y = \int x\, dL = \int_{-\alpha}^{\alpha} (r \cos \theta)(r\, d\theta) = r^2 \int_{-\alpha}^{\alpha} \cos \theta\, d\theta$$

$$= r^2 [\sin \theta]_{-\alpha}^{\alpha} = 2r^2 \sin \alpha$$

Since $Q_y = \bar{x}L$, we write

$$\bar{x}(2r\alpha) = 2r^2 \sin \alpha \qquad\qquad \bar{x} = \frac{r \sin \alpha}{\alpha} \quad \blacktriangleleft$$

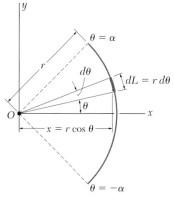

# SAMPLE PROBLEM 5.6

Determine the area of the surface of revolution shown, obtained by rotating a quarter-circular arc about a vertical axis.

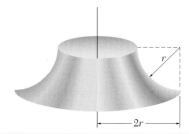

**Solution.** According to Theorem I of Pappus-Guldinus, the area generated is equal to the product of the length of the arc and the distance traveled by its centroid. Referring to Fig. 5.8B, we have

$$\bar{x} = 2r - \frac{2r}{\pi} = 2r\left(1 - \frac{1}{\pi}\right)$$

$$A = 2\pi\bar{x}L = 2\pi 2r\left(1 - \frac{1}{\pi}\right)\frac{\pi r}{2}$$

$$A = 2\pi r^2(\pi - 1) \quad \blacktriangleleft$$

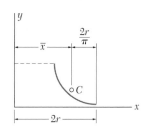

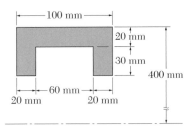

# SAMPLE PROBLEM 5.7

The outside diameter of a pulley is 0.8 m, and the cross section of its rim is as shown. Determine the mass and weight of the rim, knowing that the pulley is made of steel and that the density of steel is $\rho = 7.85 \times 10^3$ kg/m$^3$.

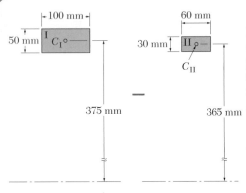

**Solution.** The volume of the rim may be found by applying Theorem II of Pappus-Guldinus, which states that the volume equals the product of the given cross-sectional area and of the distance traveled by its centroid in one complete revolution. However, the volume will be more easily obtained if we observe that the cross section consists of rectangle (I) with a positive area and rectangle (II) with a negative area.

| | Area, mm² | $\bar{y}$, mm | Distance Traveled by $C$, mm | Volume, mm³ |
|---|---|---|---|---|
| I | +5000 | 375 | $2\pi(375) = 2356$ | $(5000)(2356) = 11.78 \times 10^6$ |
| II | −1800 | 365 | $2\pi(365) = 2293$ | $(-1800)(2293) = -4.13 \times 10^6$ |
| | | | | Volume of rim $= 7.65 \times 10^6$ |

Since 1 mm $= 10^{-3}$ m, we have 1 mm$^3 = (10^{-3}$ m$)^3 = 10^{-9}$ m$^3$, and we obtain $V = 7.65 \times 10^6$ mm$^3 = (7.65 \times 10^6) (10^{-9}$ m$^3) = 7.65 \times 10^{-3}$ m$^3$.

$$m = \rho V = (7.85 \times 10^3 \text{ kg/m}^3)(7.65 \times 10^{-3} \text{ m}^3) \qquad m = 60.0 \text{ kg} \blacktriangleleft$$
$$W = mg = (60.0 \text{ kg})(9.81 \text{ m/s}^2) = 589 \text{ kg} \cdot \text{m/s}^2 \qquad W = 589 \text{ N} \blacktriangleleft$$

# SAMPLE PROBLEM 5.8

Using the theorems of Pappus-Guldinus, determine (a) the centroid of a semicircular area, (b) the centroid of a semicircular arc. We recall that the volume of a sphere is $\frac{4}{3}\pi r^3$ and that its surface area is $4\pi r^2$.

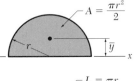

**Solution.** The volume of a sphere is equal to the product of the area of a semicircle and of the distance traveled by the centroid of the semicircle in one revolution about the $x$ axis.

$$V = 2\pi\bar{y}A \qquad \tfrac{4}{3}\pi r^3 = 2\pi\bar{y}(\tfrac{1}{2}\pi r^2) \qquad \bar{y} = \frac{4r}{3\pi} \blacktriangleleft$$

Likewise, the area of a sphere is equal to the product of the length of the generating semicircle and of the distance traveled by its centroid in one revolution.

$$A = 2\pi\bar{y}L \qquad 4\pi r^2 = 2\pi\bar{y}(\pi r) \qquad \bar{y} = \frac{2r}{\pi} \blacktriangleleft$$

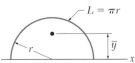

# PROBLEMS

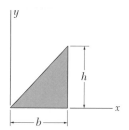

**Fig. P5.39**

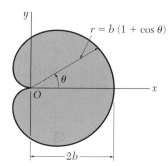

**Fig. P5.40**

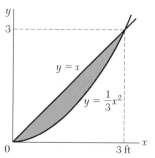

**Fig. P5.41**

**5.39 through 5.42**  Determine by direct integration the centroid of the area shown.

**5.43 through 5.48**  Derive by direct integration the expressions for $\bar{x}$ and $\bar{y}$ given in Fig. 5.8 for

**5.43**  A general spandrel $(y = kx^n)$.
**5.44**  A quarter-elliptical area.
**5.45**  A semicircular area.
**5.46**  A semiparabolic area.
**5.47**  A circular sector.
**5.48**  A quarter-circular arc.

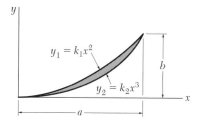

**Fig. P5.42**

**5.49**  Determine by direct integration the $x$ coordinate of the centroid of the area shown.

**5.50**  Determine by direct integration the $y$ coordinate of the centroid of the area shown.

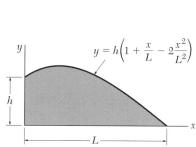

**Fig. P5.49 and P5.50**

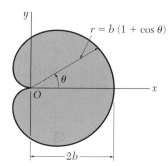

**Fig. P5.51**

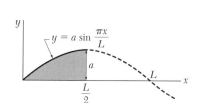

**Fig. P5.52**

**\*5.51 and \*5.52**  Determine by direct integration the centroid of the area shown.

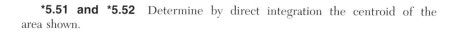

**5.53** Determine the centroid of the area shown when $b = h$.

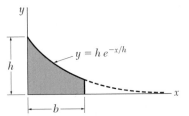

$y = h\,e^{-x/h}$

**Fig. P5.53 and P5.54**

**5.54** Determine the centroid of the area shown when $b = \infty$.

**5.55** Determine the volume of the solid obtained by rotating the area of Prob. 5.6 about (a) the x axis, (b) the y axis.

**5.56** Determine the volume of the solid obtained by rotating the trapezoid of Prob. 5.2 about (a) the x axis, (b) the y axis.

**5.57** Determine the volume of the solid generated by rotating the parabolic spandrel shown about (a) the x axis, (b) the axis $AA'$.

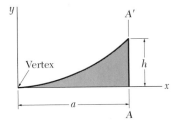

**Fig. P5.57**

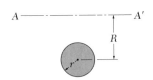

**Fig. P5.58**

**5.58** Determine the surface area and the volume of the torus obtained by rotating the circular area shown about the axis $AA'$.

**5.59** Knowing that two equal caps have been cut from a 10-in.-diameter wooden sphere, determine the total surface area of the remaining portion.

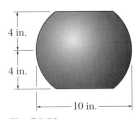

**Fig. P5.59**

**5.60** In Prob. 5.59, determine the volume and weight of the portion remaining after the two equal caps have been removed from the 10-in.-diameter sphere. (Specific weight of wood $= 40$ lb/ft$^3$.)

**5.61** A spherical pressure vessel has an inside diameter of 1.6 m. Determine (a) the volume of liquefied propane required to fill the vessel to a depth of 1.2 m, (b) the corresponding mass of the liquefied propane. (Density of liquefied propane $= 580$ kg/m$^3$.)

**5.62** For the pressure vessel of Prob. 5.61, determine the area of the surface in contact with the liquefied propane.

**5.63** Determine the volume and total surface area of the body shown.

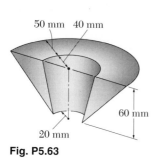

50 mm   40 mm

60 mm

20 mm

**Fig. P5.63**

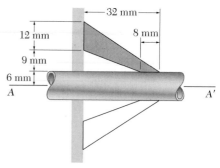

32 mm

12 mm

8 mm

9 mm

6 mm

A                                                                    A'

**Fig. P5.64**

**5.64** Determine the volume and mass of the cast-iron pipe collar obtained by rotating the shaded area about the centerline of the 12-mm-diameter pipe. (Density of cast iron = 7200 kg/m³.)

**5.65** Determine the volume and weight of the solid brass knob shown. (Specific weight of brass = 0.306 lb/in³.)

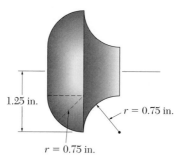

1.25 in.

r = 0.75 in.

r = 0.75 in.

**Fig. P5.65 and P5.66**

**5.66** Determine the total surface area of the solid brass knob shown.

**5.67** The rim of a double-groove V-belt sheave of 12-in. outside diameter is made of cast iron and weighs 7.50 lb. Knowing that the cross-sectional area of the rim is 0.790 in², determine the distance $\bar{y}$ from the axis of rotation AA' to the centroid of the cross section of the rim. (Specific weight of cast iron = 0.264 lb/in³.)

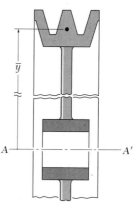

$\bar{y}$

A                          A'

**Fig. P5.67**

**5.68** The rotating mixing drum of a transit-mix concrete truck has the cross section shown. Determine by approximate means (*a*) the volume of the drum, (*b*) the surface area of the curved portions of the drum.

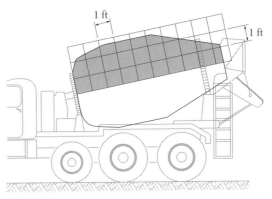

1 ft

1 ft

**Fig. P5.68**

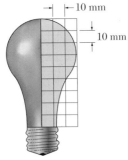

10 mm

10 mm

**Fig. P5.69**

**5.69** A 60-W standard lightbulb has the cross section shown. For the portion of the bulb above the base and neglecting the thickness of the glass, determine (*a*) the volume of inert gas inside the bulb, (*b*) the outside surface area of the bulb.

**5.70** Determine the volume of the solid of revolution formed by revolving each of the plane areas shown about its vertical edge *AB*. Shown that the volumes of the solids formed are in the ratio 6:4:3:2:1.

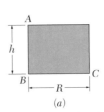
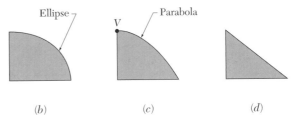

*A*

*h*

*B*

*C*

*R*

(*a*)

Ellipse

(*b*)

*V*

Parabola

(*c*)

(*d*)

Parabola

*V*

(*e*)

**Fig. P5.70 and P5.71**

**5.71** Determine the volume of the solid formed by revolving each of the plane areas shown about its base *BC*. Show that the volumes of the solids formed are in the ratio 15:10:8:5:3.

# *5.8. DISTRIBUTED LOADS ON BEAMS

The concept of centroid of an area may be used to solve other problems besides those dealing with the weight of flat plates. Consider, for example, a beam supporting a *distributed load;* this load may consist of the weight of materials supported directly or indirectly by the beam, or it may be caused by wind or hydrostatic pressure. The distributed load may be represented by plotting the load $w$ supported per unit length (Fig. 5.17); this load will be expressed in N/m or in lb/ft. The magnitude of the force exerted on an element of beam of length $dx$ is $dW = w \, dx$, and the total load supported by the beam is

$$W = \int_0^L w \, dx$$

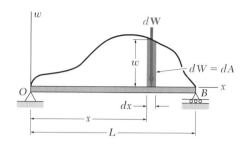

(a)

But the product $w \, dx$ is equal in magnitude to the element of area $dA$ shown in Fig. 5.17a, and $W$ is thus equal in magnitude to the total area $A$ under the load curve,

$$W = \int dA = A$$

$=$

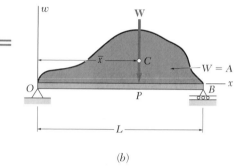

(b)

**Fig. 5.17**

We shall now determine where a *single concentrated load* **W**, of the same magnitude $W$ as the total distributed load, should be applied on the beam if it is to produce the same reactions at the supports (Fig. 5.17b). This concentrated load **W**, which represents the resultant of the given distributed loading, should be equivalent to this loading as far as the free-body diagram of the entire beam is concerned. The point of application $P$ of the equivalent concentrated load **W** will therefore be obtained by expressing that the moment of **W** about point $O$ is equal to the sum of the moments of the elementary loads $d$**W** about $O$:

$$(OP)W = \int x \, dW$$

or, since $dW = w \, dx = dA$ and $W = A$,

$$(OP)A = \int_0^L x \, dA \qquad (5.12)$$

Since the integral represents the first moment with respect to the $w$ axis of the area under the load curve, it may be replaced by the product $\bar{x}A$. We have therefore $OP = \bar{x}$, where $\bar{x}$ is the distance from the $w$ axis to the centroid $C$ of the area $A$ (this is *not* the centroid of the beam).

A *distributed load on a beam may thus be replaced by a concentrated load; the magnitude of this single load is equal to the area under the load curve, and its line of action passes through the centroid of that area.* It should be noted, however, that the concentrated load is equivalent to the given loading only as far as external forces are concerned. It may be used to determine reactions but should not be used to compute internal forces and deflections.

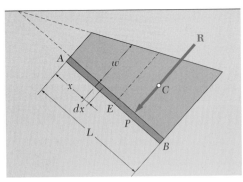

**Fig. 5.18**

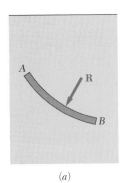

(a)

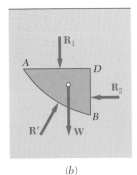

(b)

**Fig. 5.19**

## *5.9. FORCES ON SUBMERGED SURFACES

The approach used in the preceding section may be used to determine the resultant of the hydrostatic pressure forces exerted on a *rectangular surface* submerged in a liquid. Consider the rectangular plate shown in Fig. 5.18, of length $L$ and width $b$ perpendicular to the plane of the figure. As noted in Sec. 5.8, the load exerted on an element of plate of length $dx$ is $w\,dx$, where $w$ is the load per unit length. But this load may also be expressed as $p\,dA = pb\,dx$, where $p$ is the gage pressure in the liquid[†] and $b$ is the width of the plate; thus, $w = bp$. Since, on the other hand, the gage pressure in a liquid is $p = \gamma h$, where $\gamma$ is the specific weight of the liquid and $h$ is the vertical distance from the free surface, it follows that

$$w = bp = b\gamma h \qquad (5.13)$$

which shows that the load per unit length $w$ is proportional to $h$ and, thus, varies linearly with $x$.

Recalling the results of Sec. 5.8, we observe that the resultant **R** of the hydrostatic forces exerted on one side of the plate is equal in magnitude to the trapezoidal area under the load curve and that its line of action passes through the centroid $C$ of that area. The point $P$ of the plate where **R** is applied is known as the *center of pressure*.[‡]

We shall consider next the forces exerted by a liquid on a curved surface of constant width (Fig. 5.19*a*). Since the determination of the resultant **R** of these forces by direct integration would not be easy, we shall consider the free body obtained by detaching the volume of liquid $ABD$ bounded by the curved surface $AB$ and by the two plane surfaces $AD$ and $DB$ shown in Fig. 5.19*b*. The forces acting on the free body $ABD$ consist of the weight **W** of the volume of liquid detached, the resultant $\mathbf{R_1}$ of the forces exerted on $AD$, the resultant $\mathbf{R_2}$ of the forces exerted on $BD$, and the resultant **R′** of the forces exerted *by the curved surface on the liquid*. The resultant **R′** is equal and opposite to, and has the same line of action as the resultant **R** of the forces exerted *by the liquid on the curved surface*. The forces **W**, $\mathbf{R_1}$, and $\mathbf{R_2}$ may be determined by standard methods; after their values have been found, the force **R′** will be obtained by solving the equations of equilibrium for the free body of Fig. 5.19*b*. The resultant **R** of the hydrostatic forces exerted on the curved surface will then be obtained by reversing the sense of **R′**.

The methods outlined in this section may be used to determine the resultant of the hydrostatic forces exerted on the surface of dams or on rectangular gates and vanes. Resultants of forces on submerged surfaces of variable width should be determined by the methods of Chap. 9.

---

[†] The pressure $p$, which represents a load per unit area, is expressed in N/m² or in lb/ft². The derived SI unit N/m² is called a *pascal* (Pa).

[‡] Noting that the area under the load curve is equal to $w_E L$ and recalling Eq. (5.13), we may write

$$R = w_E L = (bp_E)L = p_E(bL) = p_E A$$

where $A$ denotes the area of the *plate*. Thus, the magnitude of **R** may be obtained by multiplying the area of the plate by the pressure at its center $E$. The resultant **R**, however, *should be applied at P, not at E*.

# SAMPLE PROBLEM 5.9

A beam supports a distributed load as shown. (*a*) Determine the equivalent concentrated load. (*b*) Determine the reactions at the supports.

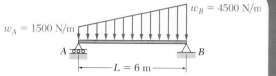

## a. Equivalent Concentrated Load.

The magnitude of the resultant of the load is equal to the area under the load curve, and the line of action of the resultant passes through the centroid of the same area. We divide the area under the load curve into two triangles and construct the table below. To simplify the computations and tabulation, the given loads per unit length have been converted into kN/m.

| Component | A, kN | $\bar{x}$, m | $\bar{x}A$, kN · m |
|-----------|-------|--------------|---------------------|
| Triangle I | 4.5 | 2 | 9 |
| Triangle II | 13.5 | 4 | 54 |
| | $\Sigma A = 18.0$ | ... | $\Sigma \bar{x}A = 63$ |

Thus, $\bar{X}\Sigma A = \Sigma \bar{x}A$:    $\bar{X}(18 \text{ kN}) = 63 \text{ kN·m}$    $\bar{X} = 3.5 \text{ m}$

The equivalent concentrated load is

$$\mathbf{W} = 18 \text{ kN} \downarrow \quad \blacktriangleleft$$

and its line of action is located at a distance

$$\bar{X} = 3.5 \text{ m to the right of } A \quad \blacktriangleleft$$

## b. Reactions.

The reaction at $A$ is vertical and is denoted by $\mathbf{A}$; the reaction at $B$ is represented by its components $\mathbf{B}_x$ and $\mathbf{B}_y$. The given load may be considered as the sum of two triangular loads as shown. The resultant of each triangular load is equal to the area of the triangle and acts at its centroid. We write the following equilibrium equations for the free body shown:

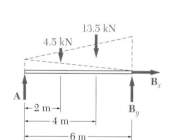

$\xrightarrow{+} \Sigma F_x = 0$:                                        $\mathbf{B}_x = 0$  ◀

$+\uparrow \Sigma M_A = 0$:    $-(4.5 \text{ kN})(2 \text{ m}) - (13.5 \text{ kN})(4 \text{ m}) + B_y(6 \text{ m}) = 0$

$\mathbf{B}_y = 10.5 \text{ kN} \uparrow$  ◀

$+\uparrow \Sigma M_B = 0$:    $+(4.5 \text{ kN})(4 \text{ m}) + (13.5 \text{ kN})(2 \text{ m}) - A(6 \text{ m}) = 0$

$\mathbf{A} = 7.5 \text{ kN} \uparrow$  ◀

**Alternate Solution.**    The given distributed load may be replaced by its resultant, which was found in part *a*. The reactions may be determined by writing the equilibrium equations $\Sigma F_x = 0$, $\Sigma M_A = 0$, and $\Sigma M_B = 0$. We again obtain

$$\mathbf{B}_x = 0 \qquad \mathbf{B}_y = 10.5 \text{ kN} \uparrow \qquad \mathbf{A} = 7.5 \text{ kN} \uparrow \quad \blacktriangleleft$$

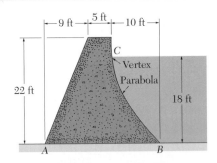

# SAMPLE PROBLEM 5.10

The cross section of a concrete dam is as shown. Consider a section of the dam 1 ft thick, and determine (a) the resultant of the reaction forces exerted by the ground on the base of the dam AB, (b) the resultant of the pressure forces exerted by the water on the face BC of the dam. Specific weight of concrete = 150 lb/ft$^3$; of water = 62.4 lb/ft$^3$.

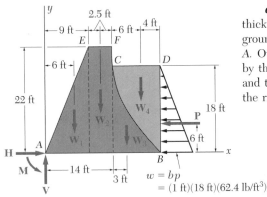

**a. Ground Reaction.** As a free body we choose a section AEFCDB, 1 ft thick, of the dam and water as shown. The reaction forces exerted by the ground on the base AB are represented by an equivalent force-couple system at A. Other forces acting on the free body are the weight of the dam, represented by the weights of its components $W_1$, $W_2$, and $W_3$, the weight of the water $W_4$, and the resultant **P** of the pressure forces exerted on section BD by the water to the right of section BD. We have

$$W_1 = \tfrac{1}{2}(9 \text{ ft})(22 \text{ ft})(1 \text{ ft})(150 \text{ lb/ft}^3) = 14{,}850 \text{ lb}$$
$$W_2 = (5 \text{ ft})(22 \text{ ft})(1 \text{ ft})(150 \text{ lb/ft}^3) = 16{,}500 \text{ lb}$$
$$W_3 = \tfrac{1}{3}(10 \text{ ft})(18 \text{ ft})(1 \text{ ft})(150 \text{ lb/ft}^3) = 9000 \text{ lb}$$
$$W_4 = \tfrac{2}{3}(10 \text{ ft})(18 \text{ ft})(1 \text{ ft})(62.4 \text{ lb/ft}^3) = 7488 \text{ lb}$$
$$P = \tfrac{1}{2}(18 \text{ ft})(1 \text{ ft})(18 \text{ ft})(62.4 \text{ lb/ft}^3) = 10{,}109 \text{ lb}$$

**Equilibrium Equations**

$\xrightarrow{+} \Sigma F_x = 0$:    $H - 10{,}109 \text{ lb} = 0$      **H** = 10,110 lb → ◀

$+\uparrow \Sigma F_y = 0$:    $V - 14{,}850 \text{ lb} - 16{,}500 \text{ lb} - 9000 \text{ lb} - 7488 \text{ lb} = 0$
               **V** = 47,840 lb ↑ ◀

$+\uparrow \Sigma M_A = 0$:    $-(14{,}850 \text{ lb})(6 \text{ ft}) - (16{,}500 \text{ lb})(11.5 \text{ ft})$
      $-(9000 \text{ lb})(17 \text{ ft}) - (7488 \text{ lb})(20 \text{ ft}) + (10{,}109 \text{ lb})(6 \text{ ft}) + M = 0$
               **M** = 520,960 lb · ft ↑ ◀

We may replace the force-couple system obtained by a single force acting at a distance d to the right of A, where

$$d = \frac{520{,}960 \text{ lb} \cdot \text{ft}}{47{,}840 \text{ lb}} = 10.89 \text{ ft}$$

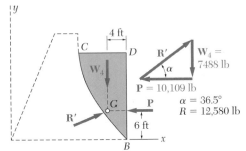

**b. Resultant R of Water Forces.** The parabolic section of water BCD is chosen as a free body. The forces involved are the resultant **R′** of the forces exerted by the dam on the water, the weight $W_4$, and the force **P**. Since these forces must be concurrent, **R′** passes through the point of intersection G of $W_4$ and **P**. A force triangle is drawn from which the magnitude and direction of **R′** are determined. The resultant **R** of the forces exerted by the water on the face BC is equal and opposite:

**R** = 12,580 lb ⬈ 36.5° ◀

**5.33**  The shape shown is formed from a 0.75-in. thick steel plate (weighing 30.6 lb/ft$^2$) and a 0.75-in. diameter homogeneous steel rod (weighing 1.5 lb/ft). Determine the support reactions at $A$ and $B$ required for equilibrium.

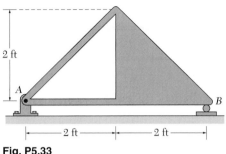

**Fig. P5.33**

**5.34**  The shape shown is formed from a 0.75-in. thick steel plate (weighing 30.6 lb/ft$^2$) and a 0.75-in. diameter homogeneous steel rod (weighing 1.5 lb/ft). Determine the support reactions at $A$ and $B$ required for equilibrium.

**5.35**  Determine by approximate means the $x$ coordinate of the centroid of the area shown.

**5.36**  Determine by approximate means the $y$ coordinate of the centroid of the area shown.

**5.37**  Divide the parabolic spandrel shown into five vertical sections and determine by approximate means the $x$ coordinate of its centroid; approximate the spandrel by rectangles of the form $bcc'b'$. What is the percentage error in the answer obtained? (See Fig. 5.8$A$ for exact answer.)

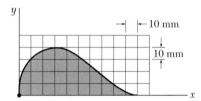

**Fig. P5.34**

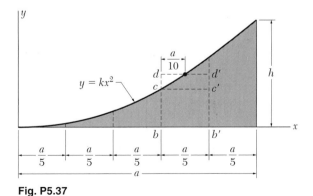

**Fig. P5.35 and P5.36**

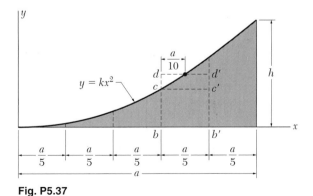

**Fig. P5.37**

**5.38**  Solve Prob. 5.37 using rectangles of the form $bdd'b'$.

# PROBLEMS

**5.72 and 5.73** Determine the magnitude and location of the resultant of the distributed load shown. Also calculate the reactions at $A$ and $B$.

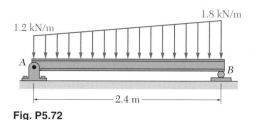

**Fig. P5.72**

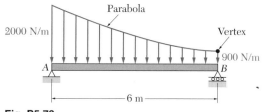

**Fig. P5.73**

**5.74 through 5.79** Determine the reactions at the beam supports for the given loading condition.

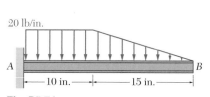

**Fig. P5.74**

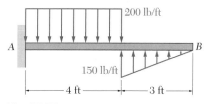

**Fig. P5.75**

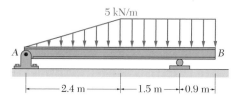

**Fig. P5.76**

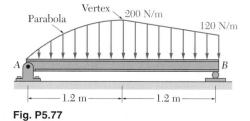

**Fig. P5.77**

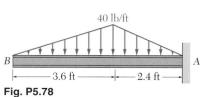

**Fig. P5.78**

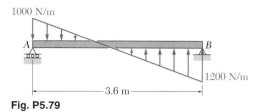

**Fig. P5.79**

**5.80** Determine the reactions at the beam supports for the given loading condition when $w_0 = 400$ lb/ft.

**5.81** Determine (*a*) the distributed load $w_0$ at the end $A$ of the beam $ABC$ for which the reaction at $C$ is zero, (*b*) the corresponding reaction at $B$.

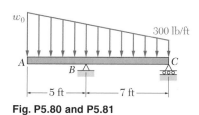

**Fig. P5.80 and P5.81**

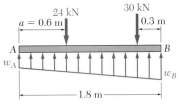

**Fig. P5.82**

**5.82** The beam $AB$ supports two concentrated loads and rests on soil which exerts a linearly distributed upward load as shown. Determine the values of $w_A$ and $w_B$ corresponding to equilibrium.

**5.83** For the beam and loading of Prob. 5.82, determine (a) the distance $a$ for which $w_A = 20$ kN/m, (b) the corresponding value of $w_B$.

In the following problems, use $\gamma = 62.4$ lb/ft³ for the specific weight of fresh water and $\gamma_c = 150$ lb/ft³ for the specific weight of concrete if U.S. customary units are used. With SI units, use $\rho = 10^3$ kg/m³ for the density of fresh water and $\rho_c = 2.40 \times 10^3$ kg/m³ for the density of concrete. (See footnote page 160 for the determination of the specific weight of a material from its density.)

**5.84** The cross section of a concrete dam is shown. Consider a section of the dam 1 m thick and determine (a) the resultant of the pressure forces exerted by the water on the face $BC$ of the dam, (b) the resultant of the reaction forces exerted by the ground on the base $AB$ of the dam. (This is affected by the resultant of the pressure forces found in [a]).

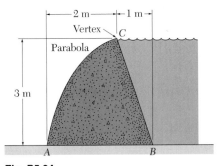

**Fig. P5.84**

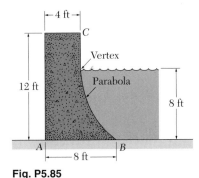

**Fig. P5.85**

**5.85** The cross section of a concrete dam is shown. Consider a section of the dam 1 ft thick and determine (a) the resultant of the pressure forces exerted by the water on the face $BC$ of the dam, (b) the resultant of the reaction forces exerted by the ground on the base $AB$ of the dam. (This is affected by the resultant of the pressure forces found in [a]).

**5.86** A 1.5 × 1.5 m gate is located in a wall below water level. (a) Determine the magnitude and location of the resultant of the forces exerted by the water on the gate. (b) If the gate is hinged along its top edge $A$, determine the force exerted on the gate by the brace $BC$.

**5.87** In Prob. 5.86, determine the depth of water $d$ for which the force exerted on the gate by the brace $BC$ is (a) 20 kN, (b) 9 kN.

**5.88** An automatic valve consists of a 15 × 15 in. square plate which pivots about a horizontal axis through $A$ located at a distance $h = 6$ in. above the lower edge. Determine the depth of water at which the valve will open.

**5.89** An automatic valve consists of a 15 × 15 in. square plate which pivots about a horizontal axis through $A$. If the valve is to open when the depth of water is $d = 24$ in., determine the distance $h$ from the bottom of the valve to the pivot $A$.

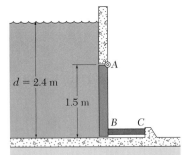

**Fig. P5.86**

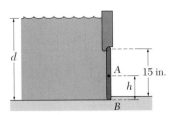

**Fig. P5.88 and P5.89**

**5.90** Determine the resultant force, as a vector, of the pressure forces acting on the gate AB. Consider a 1-ft-wide section of the gate.

**5.91** Determine the resultant force, as a vector, of the pressure forces acting on the quarter-circular gate CD. Consider a 1-ft-wide section of the gate.

**5.92** Determine the resultant force, as a vector, of the pressure forces acting on/ the gate AB. Consider a 1-ft-wide section of the gate.

**5.93** Determine the resultant force, as a vector, of the pressure forces acting on the quarter-circular section BC of the gate. Consider a 1-ft-wide section of the gate.

**5.94** At B and D are two 10-m-diameter circular areas at the bottom of a 50-m-deep lake. Determine the resultant of the forces exerted by the water on the two circles by two methods: (a) For column AB, use the total weight of water in the cylindrical column. (b) For column CD, determine the pressure P and multiply this times the area of the circle at D. (c) Explain your two answers.

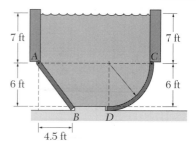

**Fig. P5.90 and P5.91**

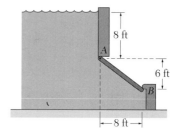

**Fig. P5.92**

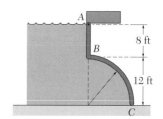

**Fig. P5.93**

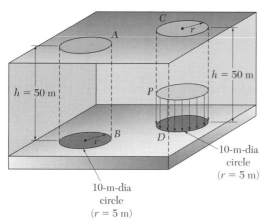

**Fig. P5.94**

**5.95** Determine the minimum allowable value of the width a of the rectangular concrete dam if the dam is not to overturn about point A when d = 10 ft. Assume that a seal exists at B and that no water pressure is present under the dam.

**5.96** Solve Prob. 5.95, assuming that a seal exists only at A and that the full hydrostatic pressure at B is present under the dam from A to B.

**5.97** Knowing that the width of the rectangular concrete dam is a = 3 ft, determine the maximum allowable value of the depth d of water if the dam is not to overturn about A. Assume that a seal exists at B and that no water pressure is present under the dam.

**5.98** The end of a freshwater channel consists of a plate ABCD which is hinged at B and is 0.4 m wide. Knowing that a = 0.5 m, determine the reactions at A and B.

**5.99** The end of a freshwater channel consists of a plate ABCD which is hinged at B and is 0.4 m wide. Determine the length a for which the reaction at A is zero.

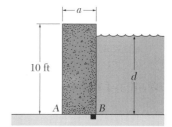

**Fig. P5.95 and P5.97**

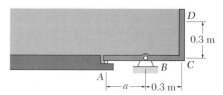

**Fig. P5.98 and P5.99**

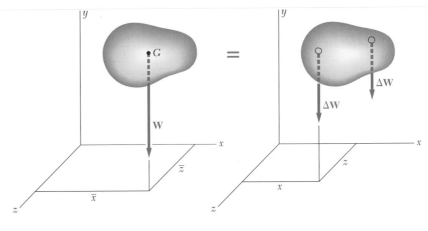

$$\Sigma M_z: \quad \overline{x}W = \Sigma x\, \Delta W$$
$$\Sigma M_x: \quad \overline{z}W = \Sigma z\, \Delta W$$

**Fig. 5.20**

## VOLUMES

### 5.10. CENTER OF GRAVITY OF A THREE-DIMENSIONAL BODY. CENTROID OF A VOLUME

The *center of gravity* G of a three-dimensional body is obtained by dividing the body into small elements and expressing that the weight **W** of the body attached at G is equivalent to the system of distributed forces **ΔW** representing the weights of the small elements. This has been done in Figs. 5.20 and 5.21 for two positions of the body. In Fig. 5.20, the $y$ axis is vertical; in Fig. 5.21, the body and the axes have been rotated so that the $z$ axis is vertical. Three independent equations are obtained, which may be used to determine the coordinates $\overline{x}$, $\overline{y}$, and $\overline{z}$ of the center of gravity G. Written in terms of infinitesimal elements, these equations are

$$\overline{x}W = \int x\, dW \qquad \overline{y}W = \int y\, dW \qquad \overline{z}W = \int z\, dW \qquad (5.14)$$

If the body is made of a homogeneous material of specific weight $\gamma$, the magnitude $dW$ of the weight of an infinitesimal element may be expressed in terms of the volume $dV$ of the element, and the magnitude $W$ of the total weight in terms of the total volume $V$. We write

$$dW = \gamma\, dV \qquad W = \gamma V$$

Substituting for $dW$ and $W$ in Eqs. (5.14), we obtain

$$\overline{x}V = \int x\, dV \qquad \overline{y}V = \int y\, dV \qquad \overline{z}V = \int z\, dV \qquad (5.15)$$

5.10. Center of Gravity of a **193**
Three-Dimensional Body.
Centroid of a Volume

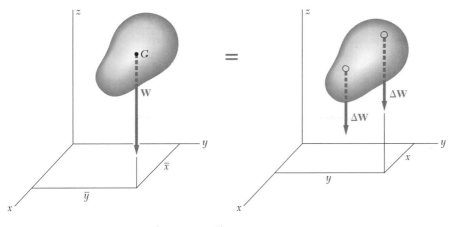

$$\Sigma M_x: \quad \bar{y}W = \Sigma y\,\Delta W$$
$$\Sigma M_y: \quad \bar{x}W = \Sigma x\,\Delta W$$

**Fig. 5.21**

The point of coordinates $\bar{x}$, $\bar{y}$, $\bar{z}$ is also known as the *centroid C of the volume V* of the body. If the body is not homogeneous, Eqs. (5.15) cannot be used to determine the center of gravity of the body; they still define, however, the centroid of the volume.

The integral $\int x\,dV$ is known as the *first moment of the volume with respect to the yz plane.* Similarly, the integrals $\int y\,dV$ and $\int z\,dV$ define the first moments of the volume with respect to the *zx* plane and the *xy* plane, respectively. It is seen from Eqs. (5.15) that if the centroid of a volume is located in a coordinate plane, the first moment of the volume with respect to that plane is zero.

A volume is said to be symmetrical with respect to a given plane if to every point P of the volume corresponds a point P′ of the same volume, such that the line PP′ is perpendicular to the given plane and divided into two equal parts by that plane. The plane is said to be a *plane of symmetry* for the given volume. When a volume V possesses a plane of symmetry, the first moment of V with respect to that plane is zero and the centroid of the volume must be located in the plane of symmetry. When a volume possesses two planes of symmetry, the centroid of the volume must be located on the line of intersection of the two planes. Finally, when a volume possesses three planes of symmetry which intersect in a well-defined point (i.e., not along a common line), the point of intersection of the three planes must coincide with the centroid of the volume. This property enables us to determine immediately the centroid of the volume of spheres, ellipsoids, cubes, rectangular parallelepipeds, etc.

Centroids of unsymmetrical volumes or of volumes possessing only one or two planes of symmetry should be determined by integration (Sec. 5.12). Centroids of common shapes of volumes are shown in Fig. 5.22. It should be observed that the centroid of a volume of revolution in general *does not coincide* with the centroid of its cross section. Thus, the centroid of a hemisphere is different from that of a semicircular area, and the centroid of a cone is different from that of a triangle.

| Shape | | $\bar{x}$ | Volume |
|---|---|---|---|
| Hemisphere | 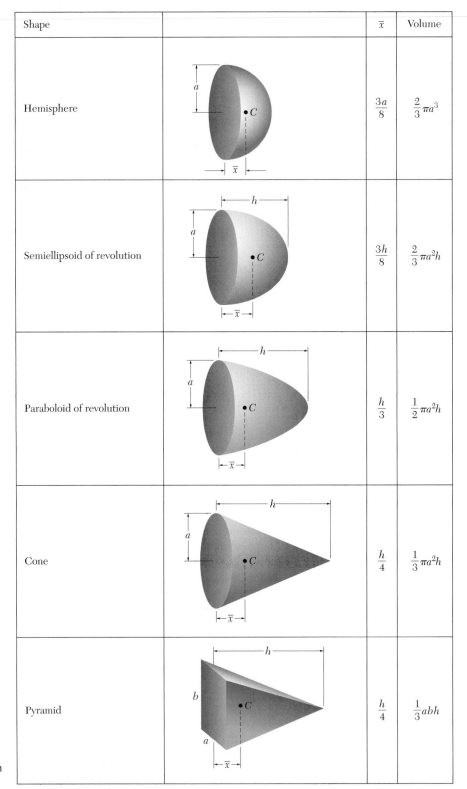 | $\dfrac{3a}{8}$ | $\dfrac{2}{3}\pi a^3$ |
| Semiellipsoid of revolution | | $\dfrac{3h}{8}$ | $\dfrac{2}{3}\pi a^2 h$ |
| Paraboloid of revolution | | $\dfrac{h}{3}$ | $\dfrac{1}{2}\pi a^2 h$ |
| Cone | | $\dfrac{h}{4}$ | $\dfrac{1}{3}\pi a^2 h$ |
| Pyramid | | $\dfrac{h}{4}$ | $\dfrac{1}{3}abh$ |

**Fig. 5.22**   Centroids of common
shapes of volumes.

## 5.11. COMPOSITE BODIES

If a body can be divided into several of the common shapes shown in Fig. 5.22, the coordinates $\bar{X}$, $\bar{Y}$, $\bar{Z}$ of its center of gravity $G$ may be determined by equating the moments of its total weight and the sums of the corresponding moments of the weights of its various component parts. Proceeding as in Sec. 5.10, we obtain

$$\bar{X}\Sigma W = \Sigma \bar{x}\,W \qquad \bar{Y}\Sigma W = \Sigma \bar{y}\,W \qquad \bar{Z}\Sigma W = \Sigma \bar{z}\,W \qquad (5.16)$$

If the body is made of a homogeneous material, its center of gravity coincides with the centroid of its volume and the following equations may be used:

$$\bar{X}\Sigma V = \Sigma \bar{x}\,V \qquad \bar{Y}\Sigma V = \Sigma \bar{y}\,V \qquad \bar{Z}\Sigma V = \Sigma \bar{z}\,V \qquad (5.17)$$

## 5.12. DETERMINATION OF CENTROIDS OF VOLUMES BY INTEGRATION

The centroid of a volume bounded by analytical surfaces may be determined by computing the integrals given in Sec. 5.10:

$$\bar{x}V = \int x\,dV \qquad \bar{y}V = \int y\,dV \qquad \bar{z}V = \int z\,dV \qquad (5.18)$$

If the element of volume $dV$ is chosen equal to a small cube of sides $dx$, $dy$, and $dz$, the determination of each of these integrals requires a *triple integration* in $x$, $y$, and $z$. However, it is possible to determine the coordinates of the centroid of most volumes by *double integration* if $dV$ is chosen equal to the volume of a thin filament as shown in Fig. 5.23. The coordinates of the centroid of the volume are then obtained by writing

$$\bar{x}V = \int \bar{x}_{el}\,dV \qquad \bar{y}V = \int \bar{y}_{el}\,dV \qquad \bar{z}V = \int \bar{z}_{el}\,dV \qquad (5.19)$$

and substituting for the volume $dV$ and the coordinates $\bar{x}_{el}$, $\bar{y}_{el}$, $\bar{z}_{el}$ the expressions given in Fig. 5.23. Using the equation of the surface to express $z$ in terms of $x$ and $y$, the integration is reduced to a double integration in $x$ and $y$.

If the volume under consideration possesses *two planes of symmetry*, its centroid must be located on their line of intersection. Choosing the $x$ axis along this line, we have

$$\bar{y} = \bar{z} = 0$$

and the only coordinate to determine is $\bar{x}$. This will be done most conveniently by dividing the given volume into thin slabs parallel to the $yz$ plane. In the particular case of a body of revolution, these slabs are circular; their volume $dV$ is given in Fig. 5.24. Substituting for $\bar{x}_{el}$ and $dV$ into the equation

$$\bar{x}V = \int \bar{x}_{el}\,dV \qquad (5.20)$$

and expressing the radius $r$ of the slab in terms of $x$, we may determine $\bar{x}$ by a single integration.

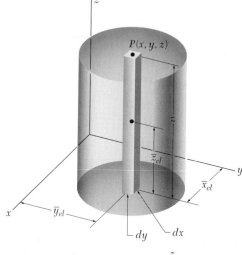

$$\bar{x}_{el} = x, \ \bar{y}_{el} = y, \ \bar{z}_{el} = \frac{z}{2}$$

$$dV = z\,dx\,dy$$

**Fig. 5.23** Determination of the centroid of a volume by double integration.

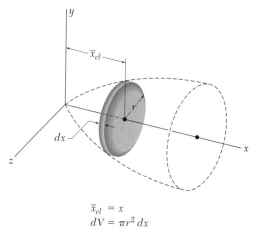

$$\bar{x}_{el} = x$$

$$dV = \pi r^2\,dx$$

**Fig. 5.24** Determination of the centroid of a body of revolution.

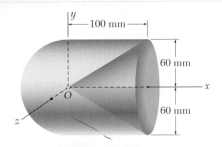

## SAMPLE PROBLEM 5.11

Determine the center of gravity of the homogeneous body of revolution shown.

**Solution.** Because of symmetry, the center of gravity lies on the $x$ axis. The body is seen to consist of a hemisphere, plus a cylinder, minus a cone, as shown. The volume and the abscissa of the centroid of each of these components are obtained from Fig. 5.22 and entered in the table below. The total volume of the body and the first moment of its volume with respect to the $yz$ plane are then determined.

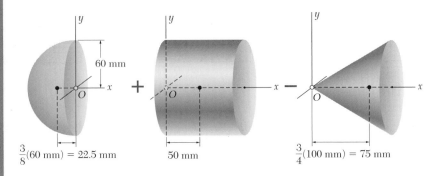

| Component | Volume, mm³ | | $\bar{x}$, mm | $\bar{x}V$, mm⁴ |
|---|---|---|---|---|
| Hemisphere | $\dfrac{1}{2}\dfrac{4\pi}{3}(60)^3$ | $= \quad 0.452 \times 10^6$ | $-22.5$ | $-10.17 \times 10^6$ |
| Cylinder | $\pi(60)^2(100)$ | $= \quad 1.131 \times 10^6$ | $+50$ | $+56.55 \times 10^6$ |
| Cone | $-\dfrac{\pi}{3}(60)^2(100)$ | $= -0.377 \times 10^6$ | $+75$ | $-28.28 \times 10^6$ |
| | | $\Sigma V = \quad 1.206 \times 10^6$ | $\cdots$ | $\Sigma \bar{x}V = +18.10 \times 10^6$ |

Thus,

$$\bar{X}\Sigma V = \Sigma \bar{x}V: \quad \bar{X}(1.206 \times 10^6 \text{ mm}^3) = 18.10 \times 10^6 \text{ mm}^4$$

$$\bar{X} = 15 \text{ mm} \quad \blacktriangleleft$$

## SAMPLE PROBLEM 5.12

Locate the center of gravity of the steel machine element shown. Both holes are of 1-in. diameter.

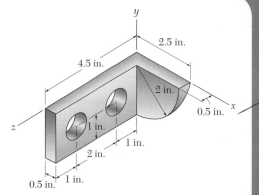

**Solution.** The machine element is seen to consist of a rectangular paral-lelepiped (I) plus a quarter cylinder (II) minus two 1-in.-diameter cylinders (III and IV). The volume and the coordinates of the centroid of each component are determined and entered in the table below. Using the data accumulated in the table, we then determine the total volume and the moments of the volume about each of the coordinate planes.

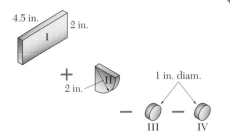

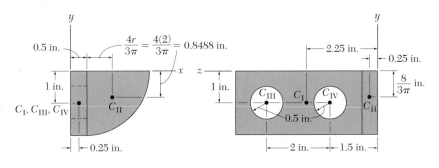

$$\frac{4r}{3\pi} = \frac{4(2)}{3\pi} = 0.8488 \text{ in.}$$

|  | $V$, in$^3$ | $\bar{x}$, in. | $\bar{y}$, in. | $\bar{z}$, in. | $\bar{x}V$, in$^4$ | $\bar{y}V$, in$^4$ | $\bar{z}V$, in$^4$ |
|---|---|---|---|---|---|---|---|
| I | $(4.5)(2)(0.5) = 4.5$ | 0.25 | $-1$ | 2.25 | 1.125 | $-4.5$ | 10.125 |
| II | $\frac{1}{4}\pi(2)^2(0.5) = 1.571$ | 1.3488 | $-0.8488$ | 0.25 | 2.119 | $-1.333$ | 0.393 |
| III | $-\pi(0.5)^2(0.5) = -0.3927$ | 0.25 | $-1$ | 3.5 | $-0.098$ | 0.393 | $-1.374$ |
| IV | $-\pi(0.5)^2(0.5) = -0.3927$ | 0.25 | $-1$ | 1.5 | $-0.098$ | 0.393 | $-0.589$ |
|  | $\Sigma V = 5.286$ |  |  |  | $\Sigma \bar{x}V = 3.048$ | $\Sigma \bar{y}V = -5.047$ | $\Sigma \bar{z}V = 8.555$ |

Thus,

$$\bar{X}\Sigma V = \Sigma \bar{x}V: \qquad \bar{X}(5.286 \text{ in}^3) = 3.048 \text{ in}^4 \qquad \bar{X} = \phantom{-}0.577 \text{ in.} \blacktriangleleft$$
$$\bar{Y}\Sigma V = \Sigma \bar{y}V: \qquad \bar{Y}(5.286 \text{ in}^3) = -5.047 \text{ in}^4 \qquad \bar{Y} = -0.955 \text{ in.} \blacktriangleleft$$
$$\bar{Z}\Sigma V = \Sigma \bar{z}V: \qquad \bar{Z}(5.286 \text{ in}^3) = 8.555 \text{ in}^4 \qquad \bar{Z} = \phantom{-}1.618 \text{ in.} \blacktriangleleft$$

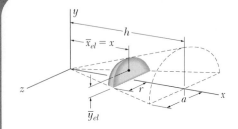

## SAMPLE PROBLEM 5.13

Determine the location of the centroid of the half right circular cone shown.

**Solution.** Since the $xy$ plane is a plane of symmetry, the centroid lies in this plane and $\bar{z} = 0$. A slab of thickness $dx$ is chosen as a differential element. The volume of this element is

$$dV = \tfrac{1}{2}\pi r^2\, dx$$

The coordinates $\bar{x}_{el}$ and $\bar{y}_{el}$ of the centroid of the element are obtained from Fig. 5.8 (semicircular area).

$$\bar{x}_{el} = x \qquad \bar{y}_{el} = \frac{4r}{3\pi}$$

We observe that $r$ is proportional to $x$ and write

$$\frac{r}{x} = \frac{a}{h} \qquad r = \frac{a}{h}x$$

The volume of the body is

$$V = \int dV = \int_0^h \tfrac{1}{2}\pi r^2\, dx = \int_0^h \tfrac{1}{2}\pi\left(\frac{a}{h}x\right)^2 dx = \frac{\pi a^2 h}{6}$$

The moment of the differential element with respect to the $yz$ plane is $\bar{x}_{el}\, dV$; and the total moment of the body with respect to this plane is

$$\int \bar{x}_{el}\, dV = \int_0^h x(\tfrac{1}{2}\pi r^2)\, dx = \int_0^h x(\tfrac{1}{2}\pi)\left(\frac{a}{h}x\right)^2 dx = \frac{\pi a^2 h^2}{8}$$

Thus,

$$\bar{x}V = \int \bar{x}_{el}\, dV \qquad \bar{x}\frac{\pi a^2 h}{6} = \frac{\pi a^2 h^2}{8} \qquad \bar{x} = \tfrac{3}{4}h \quad \blacktriangleleft$$

Likewise, the moment of the differential element with respect to the $zx$ plane is $\bar{y}_{el}\, dV$; and the total moment is

$$\int \bar{y}_{el}\, dV = \int_0^h \frac{4r}{3\pi}(\tfrac{1}{2}\pi r^2)\, dx = \frac{2}{3}\int_0^h \left(\frac{a}{h}x\right)^3 dx = \frac{a^3 h}{6}$$

Thus,

$$\bar{y}V = \int \bar{y}_{el}\, dV \qquad \bar{y}\frac{\pi a^2 h}{6} = \frac{a^3 h}{6} \qquad \bar{y} = \frac{a}{\pi} \quad \blacktriangleleft$$

# PROBLEMS

**5.100** Determine the location of the centroid of the composite body shown when $h = 3b$.

**5.101** Determine the ratio $h/b$ for which the centroid of the composite body is located in the plane between the cone and the cylinder.

**5.102** Locate the centroid of the frustum of a right circular cone when $r_1 = 40$ mm, $r_2 = 50$ mm, and $h = 60$ mm.

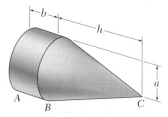

**Fig. P5.100 and P5.101**

**Fig. P5.102**

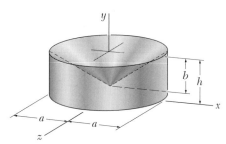

**Fig. P5.103**

**5.103** Determine the $y$ coordinate of the centroid of the body shown when (a) $b = \frac{1}{2}h$, (b) $b = h$.

**5.104** Determine the location of the center of gravity of the parabolic reflector shown, which is formed by machining a rectangular block so that the curved surface is a paraboloid of revolution of base radius $a$ and height $h$.

**5.105** For the machine element shown, locate the $x$ coordinate of the center of gravity.

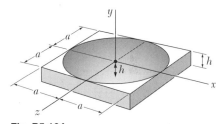

**Fig. P5.104**

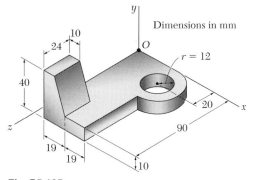

**Fig. P5.105**

**5.106 and 5.107** For the machine element shown, locate the $y$ coordinate of the center of gravity.

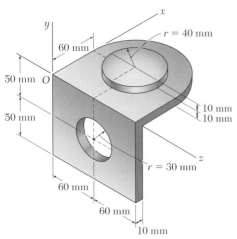

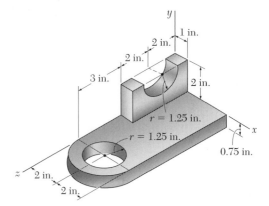

**Fig. P5.107 and P.5.108**

**5.108** For the machine element shown, locate the $z$ coordinate of the center of gravity.

**5.109** For the machine element shown, locate the $x$ coordinate of the center of gravity.

**5.110** For the machine element shown, locate the $z$ coordinate of the center of gravity.

**Fig. P5.106 and P5.109**

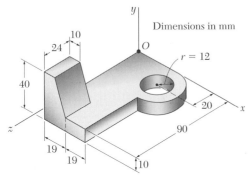

**Fig. P5.110**

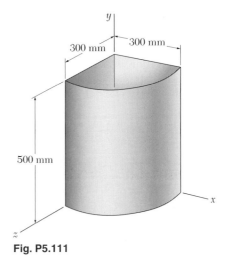

**Fig. P5.111**

**5.111** A wastebasket, designed to fit in the corner of a room, is 500 mm high and has a base in the shape of a quarter circle of radius 300 mm. Locate the center of gravity of the wastebasket, knowing that it is made of sheet metal of uniform thickness.

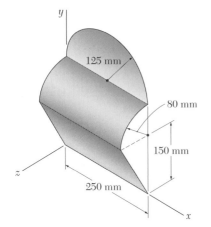

**Fig. P5.112**

**5.112** Locate the center of gravity of the sheet-metal form shown.

**5.113** Locate the center of gravity of the sheet-metal form shown.

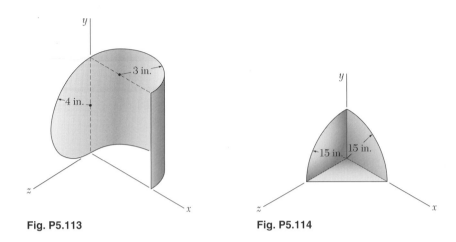

**Fig. P5.113**

**Fig. P5.114**

**5.114** A corner reflector for tracking by radar has two sides in the shape of a quarter circle of radius 15 in. and one side in the shape of a triangle. Locate the center of gravity of the reflector, knowing that it is made of sheet metal of uniform thickness.

**5.115 and 5.116** Locate the center of gravity of the figure shown, knowing that it is made of thin brass rods of uniform diameter.

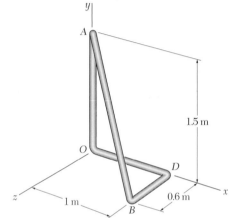

**Fig. P5.115**

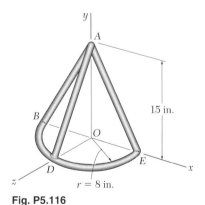

**Fig. P5.116**

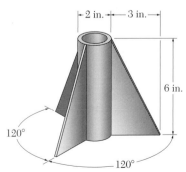

**5.117** Three brass plates are brazed to a steel pipe to form the flagpole base shown. Knowing that the pipe has a wall thickness of 0.25 in. and that each plate is 0.1875 in. thick, determine the location of the center of gravity of the base. (Specific weights: brass = 0.306 lb/in³; steel = 0.284 lb/in³.)

**Fig. P5.117**

**5.118** A brass collar, of length $h = 60$ mm, is mounted on an aluminum rod of length 100 mm. Locate the center of gravity of the composite body. (Densities: brass $= 8470$ kg/m$^3$; aluminum $= 2800$ kg/m$^3$.)

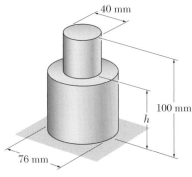

40 mm

100 mm

$h$

76 mm

**Fig. P5.118**

**5.119** In Prob. 5.118, determine (*a*) the length $h$ for which the center of gravity of the composite body is as low as possible, (*b*) the corresponding location of the center of gravity.

**5.120 through 5.122** Derive by direct integration the expression given for $\bar{x}$ in Fig. 5.22 for:

**5.120** A hemisphere.
**5.121** A semiellipsoid of revolution.
**5.122** A paraboloid of revolution.

**5.123** Locate by direct integration the centroid of the frustum of the right circular cone of Prob. 5.102, expressing the result in terms of $r_1$, $r_2$, and $h$.

**5.124** Locate the centroid of the volume obtained by rotating the shaded area about the $x$ axis.

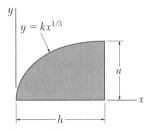

$y = kx^{1/3}$

$a$

$h$

**Fig. P5.124**

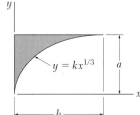

$y = kx^{1/3}$

$a$

$h$

**Fig. P5.125**

**5.125** Locate the centroid of the volume obtained by rotating the shaded area about the $y$ axis.

**\*5.126** Locate the centroid of the volume generated by revolving the portion shown of the cosine curve about the $x$ axis.

**\*5.127** Locate the centroid of the volume generated by revolving the portion shown of the cosine curve about the $y$ axis. (*Hint.* Use as an element of volume a thin cylindrical shell of radius $r$ and thickness $dr$.)

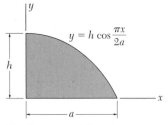

$y = h \cos \dfrac{\pi x}{2a}$

$h$

$a$

**Fig. P5.126 and P5.127**

**5.128**  Locate the centroid of the volume of the irregular pyramid shown.

Problems **203**

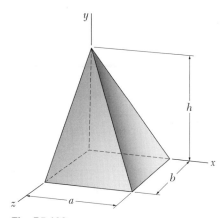

**Fig. P5.128**

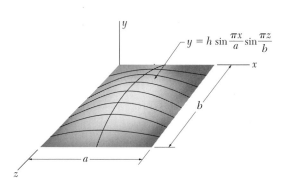

$$y = y_0 - y_1 \frac{x}{a} - y_2 \frac{z}{b}$$

**Fig. P5.129**

**5.129**  Determine by direct integration the $z$ coordinate of the centroid of the volume shown, which was cut from a rectangular prism by an oblique plane.

**5.130**  Solve Prob. 5.129, when $a = 120$ mm, $b = 90$ mm, $y_0 = 90$ mm, $y_1 = 40$ mm, and $y_2 = 20$ mm.

**5.131**  Locate the centroid of the section shown, which was cut from a thin circular pipe by an oblique plane.

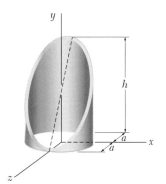

**Fig. P5.131**

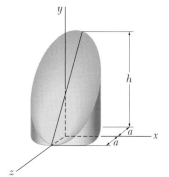

**Fig. P5.132**

$$y = h \sin\frac{\pi x}{a}\sin\frac{\pi z}{b}$$

**Fig. P5.133**

**5.132**  Locate the centroid of the section shown, which was cut from a circular cylinder by an oblique plane.

**5.133**  Determine by direct integration the location of the centroid of the volume between the $xz$ plane and the portion shown of the surface $y = h \sin(\pi x/a) \sin(\pi z/b)$.

# REVIEW AND SUMMARY
# FOR CHAPTER 5

This chapter was devoted chiefly to the determination of the *center of gravity* of a rigid body, i.e., to the determination of the point $G$ where a single force **W**, called the *weight* of the body, may be applied to represent the effect of the earth's attraction on the body.

**Center of gravity of a two-dimensional body**

In the first part of the chapter, we considered *two-dimensional bodies*, such as flat plates and wires contained in the $xy$ plane. By adding components in the vertical $z$ direction and moments about the horizontal $y$ and $x$ axes [Sec. 5.2], we derived the relations

$$W = \int dW \qquad \bar{x}W = \int x\, dW \qquad \bar{y}W = \int y\, dW \qquad (5.2)$$

which define the weight of the body and the coordinates $\bar{x}$ and $\bar{y}$ of its center of gravity.

**Centroid of an area or line**

In the case of a *homogeneous flat plate of uniform thickness* [Sec. 5.3], the center of gravity $G$ of the plate coincides with the *centroid $C$ of the area $A$* of the plate, the coordinates of which are defined by the relations

$$\bar{x}A = \int x\, dA \qquad \bar{y}A = \int y\, dA \qquad (5.3)$$

Similarly, the determination of the center of gravity of a *homogeneous flat wire of uniform cross section* reduces to that of the *centroid $C$ of the line $L$* representing the wire; we have

$$\bar{x}L = \int x\, dL \qquad \bar{y}L = \int y\, dL \qquad (5.4)$$

**First moments**

The integrals in Eqs. (5.3) are referred to as the *first moments* of the area $A$ with respect to the $y$ and $x$ axes and denoted by $Q_y$ and $Q_x$, respectively [Sec. 5.4]. We have

$$Q_y = \bar{x}A \qquad Q_x = \bar{y}A \qquad (5.6)$$

The first moments of a line may be defined in a similar way.

**Properties of symmetry**

The determination of the centroid $C$ of an area or line is simplified when the area or line possesses certain *properties of symmetry*. If the area or line is symmetric with respect to an axis, its centroid $C$ will lie on that axis; if it is symmetric with respect to two axes, $C$ will be located at the intersection of the two axes; if it is symmetric with respect to a center $O$, $C$ will coincide with $O$.

**Center of gravity of a composite body**

The *areas and centroids of various common shapes* have been tabulated in Fig. 5.8. When a flat plate may be divided into several of these shapes, the coordinates $\bar{X}$ and $\bar{Y}$ of its center of gravity $G$ may be determined from the coordinates $\bar{x}_1, \bar{x}_2, \ldots$ and $\bar{y}_1, \bar{y}_2, \ldots$ of the centers of gravity $G_1, G_2, \ldots$ of the various

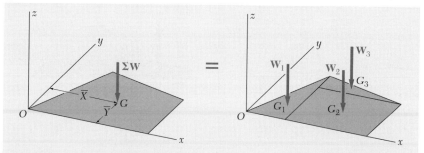

**Fig. 5.25**

parts [Sec. 5.5]. Equating moments about the $y$ and $x$ axes, respectively (Fig. 5.25), we have

$$\bar{X}\Sigma W = \Sigma \bar{x}W \qquad \bar{Y}\Sigma W = \Sigma \bar{y}W \qquad (5.7)$$

If the plate is homogeneous and of uniform thickness, the center of gravity coincides with the centroid $C$ of the area of the plate and Eqs. (5.7) reduce to

$$Q_y = \bar{X}\Sigma A = \Sigma \bar{x}A \qquad Q_x = \bar{Y}\Sigma A = \Sigma \bar{y}A \qquad (5.8)$$

These equations yield the first moments of the composite area, or they may be solved for the coordinates $\bar{X}$ and $\bar{Y}$ of its centroid [Sample Prob. 5.1]. The determination of the center of gravity of a composite wire is carried out in a similar fashion [Sample Prob. 5.2].

**Determination of centroid by integration**

When an area is bounded by analytical curves, the coordinates of its centroid may be determined *by integration* [Sec. 5.6]. This may be done by computing the double integrals in Eqs. (5.3), or through a *single integration*, using one of the thin rectangular or pie-shaped elements of area shown in Fig. 5.12. Denoting by $\bar{x}_{el}$ and $\bar{y}_{el}$ the coordinates of the centroid of the element $dA$, we have

$$Q_y = \bar{x}A = \int \bar{x}_{el}\, dA \qquad Q_x = \bar{y}A = \int \bar{y}_{el}\, dA \qquad (5.9)$$

It is advantageous to use the same element of area to compute both of the first moments $Q_y$ and $Q_x$; the same element may also be used to determine the area $A$ [Sample Prob. 5.4].

**Theorems of Pappus-Guldinus**

The *theorems of Pappus-Guldinus* relate the determination of the area of a surface of revolution or the volume of a body of revolution to the determination of the centroid of the generating curve or area [Sec. 5.7]. The area $A$ of the surface generated by rotating a curve of length $L$ about a fixed axis (Fig. 5.26$a$) is

$$A = 2\pi\bar{y}L \qquad (5.10)$$

where $\bar{y}$ represents the distance from the centroid $C$ of the curve to the fixed axis. Similarly, the volume $V$ of the body generated by rotating an area $A$ about a fixed axis (Fig. 5.26$b$) is

$$V = 2\pi\bar{y}A \qquad (5.11)$$

where $\bar{y}$ represents the distance from the centroid $C$ of the area to the fixed axis.

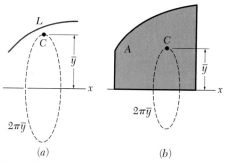

**Fig. 5.26**

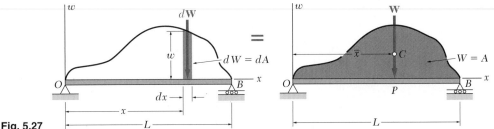

**Fig. 5.27**

**Distributed loads**

The concept of centroid of an area may be used to solve problems other than those dealing with the weight of flat plates. For example, to determine the reactions at the supports of a beam [Sec. 5.8], we may replace a *distributed load w* by a concentrated load $W$ equal in magnitude to the area $A$ under the load curve and passing through the centroid $C$ of that area (Fig. 5.27). The same approach may be used to determine the resultant of the hydrostatic forces exerted on a *rectangular plate submerged in a liquid* [Sec. 5.9].

**Center of gravity of a three-dimensional body**

The last part of the chapter was devoted to the determination of the *center of gravity G of a three-dimensional body*. The coordinates $\bar{x}$, $\bar{y}$, $\bar{z}$ of $G$ were defined by the relations

$$\bar{x}W = \int x\,dW \qquad \bar{y}W = \int y\,dW \qquad \bar{z}W = \int z\,dW \qquad (5.14)$$

**Centroid of a volume**

In the case of a *homogeneous body*, the center of gravity $G$ coincides with the *centroid C of the volume V* of the body, the coordinates of which are defined by the relations

$$\bar{x}V = \int x\,dV \qquad \bar{y}V = \int y\,dV \qquad \bar{z}V = \int z\,dV \qquad (5.15)$$

If the volume possesses a *plane of symmetry*, its centroid $C$ will lie in that plane; if it possesses two planes of symmetry, $C$ will be located on the line of intersection of the two planes; if it possesses three planes of symmetry intersecting at one point only, $C$ will coincide with that point [Sec. 5.10].

**Center of gravity of a composite body**

The *volumes and centroids of various common three-dimensional shapes* have been tabulated in Fig. 5.22. When a body may be divided into several of these shapes, the coordinates $\bar{X}$, $\bar{Y}$, $\bar{Z}$ of its center of gravity $G$ may be determined from the corresponding coordinates of the centers of gravity of its various parts [Sec. 5.11]. We have

$$\bar{X}\Sigma W = \Sigma\bar{x}W \qquad \bar{Y}\Sigma W = \Sigma\bar{y}W \qquad \bar{Z}\Sigma W = \Sigma\bar{z}W \qquad (5.16)$$

If the body is made of a homogeneous material, its center of gravity coincides with the centroid $C$ of its volume and we write [Sample Probs. 5.11 and 5.12]

$$\bar{X}\Sigma V = \Sigma\bar{x}V \qquad \bar{Y}\Sigma V = \Sigma\bar{y}V \qquad \bar{Z}\Sigma V = \Sigma\bar{z}V \qquad (5.17)$$

When a volume is bounded by analytical surfaces, the coordinates of its centroid may be determined *by integration* [Sec. 5.12]. To avoid the computation of the triple integrals in Eqs. (5.15), we may use elements of volume in the shape of thin filaments, as shown in Fig. 5.23. Denoting by $\bar{x}_{el}$, $\bar{y}_{el}$, and $\bar{z}_{el}$ the coordinates of the centroid of the element $dV$, we write the relations

$$\bar{x}V = \int \bar{x}_{el}\,dV \qquad \bar{y}V = \int \bar{y}_{el}\,dV \qquad \bar{z}V = \int \bar{z}_{el}\,dV \qquad (5.19)$$

which involve only double integrals. If the volume possesses *two planes of symmetry*, its centroid $C$ is located on their line of intersection. Choosing the $x$ axis along that line and dividing the volume into thin slabs parallel to the $yz$ plane, we determine $C$ from the relation

$$\bar{x}V = \int \bar{x}_{el}\,dV \qquad (5.20)$$

through a *single integration* [Sample Prob. 5.13].

Determination of centroid by integration

# REVIEW PROBLEMS

**5.134**  Locate the centroid of the plane area shown.

**5.135**  Determine the volume and the total surface area of the solid generated by rotating the area shown about the $y$ axis.

**5.136**  The bin shown is made of sheet metal of uniform thickness. Determine (*a*) the center of gravity of the bin when $a = 8$ in. and $b = 24$ in., (*b*) the ratio $a/b$ for which the $y$ coordinate of the center of gravity is $\bar{y} = -\frac{1}{2}a$.

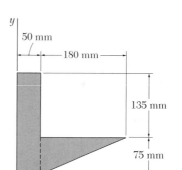

**Fig. P5.134 and P5.135**

**Fig. P5.136**

**5.137**  The bent plate $ABCD$ is 5 ft wide and is hinged at $A$. Determine the reactions at $A$ and $D$ for the water level shown.

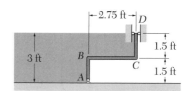

**Fig. P5.137**

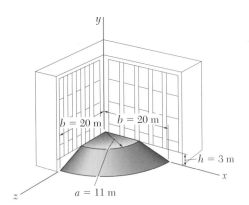

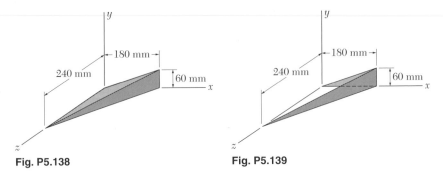

**Fig. P5.138**

**Fig. P5.139**

**5.138** Locate the centroid of the irregular pyramid shown. (*Hint.* See Fig. 5.22.)

**5.139** Locate the center of gravity of the sheet-metal form shown.

**5.140** The original design geometry of a plaza proposed for the corner entrance to a building is shown. Determine the change in the volume of the plaza if the dimensions used in the final design are (*a*) $a = 13$ m, $b = 22$ m, $h = 3$ m; (*b*) $a = 11$ m, $b = 22$ m, $h = 3$ m.

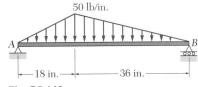

**Fig. P5.140**

**5.141** Solve Prob. 5.140, assuming that the final dimensions of the plaza are (*a*) $a = 9$ m, $b = 18$ m, $h = 3$ m; (*b*) $a = 9$ m, $b = 20$ m, $h = 3$ m.

**5.142** Locate the center of gravity of the machine element shown.

**5.143** Determine the reactions at the beam supports for the given loading condition.

**5.144** The production-line balancing of an automobile speedometer cup is done as follows. The unbalanced cup is placed on a frictionless shaft at $O$ and is allowed to come to rest. A hole is then punched at $A$; the cup rotates through an angle $\theta$ and again comes to rest. The balancing is completed by punching additional holes at $B$ and $C$. Knowing that the three holes are equal in size and are at the same distance from the shaft $O$, determine the required angle $\alpha$ in terms of the angle $\theta$.

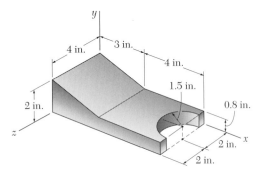

**Fig. P5.142**

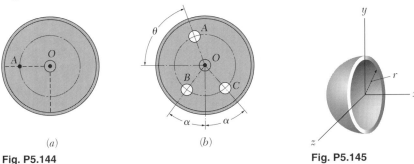

**Fig. P5.144**

**Fig. P5.145**

**5.145** Locate the center of gravity of a thin hemispherical shell of radius $r$ and thickness $t$. (*Hint.* Consider the shell as formed by removing a hemisphere of radius $r$ from a hemisphere of radius $r + t$; then neglect the terms containing $t^2$ and $t^3$, and keep those terms containing $t$.)

**Fig. P5.143**

# COMPUTER PROBLEMS

**5.C1** Approximate the plane area shown by a series of rectangles of the form $bb'd'd$ and write a computer program which can be used to calculate the coordinates $\overline{X}$ and $\overline{Y}$ of the centroid of the area shown in part $a$ of the figure. Use this program to calculate $\overline{X}$ and $\overline{Y}$ for the areas shown in parts $b$ and $c$ of the figure.

**5.C2** The area of the parabolic reflector of Prob. 5.104 may be obtained by rotating the parabola shown about the $y$ axis. Approximate the parabola by 10 straight-line segments and write a computer program which can be used to calculate the area of the parabolic reflector. Use this program to determine the area of the reflector when ($a$) $a = 10$ in., $h = 3$ in.; ($b$) $a = 12$ in., $h = 8$ in.; ($c$) $a = 15$ in., $h = 18$ in.

**5.C3** Write a computer program and use it to solve Prob. 5.90 for values of $d$ from 0 to 2.4 m at 0.3-m intervals.

**5.C4** Sheet metal of uniform thickness is used to form the bin shown which has a base in the shape of a quarter circle and three sides. Write a computer program which can be used to calculate the coordinates of the center of gravity of the bin. Use this program to locate the center of gravity for ($a$) $a = 400$ mm, $b = c = d = 200$ mm; ($b$) $a = 400$ mm, $b = d = 600$ mm, $c = 300$ mm; ($c$) $a = 400$ mm, $b = 500$ mm, $c = 200$ mm, $d = 300$ mm.

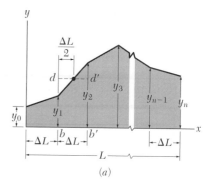

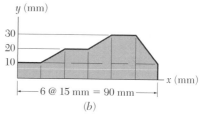
(a)

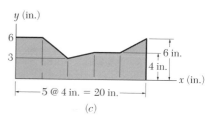
(b)

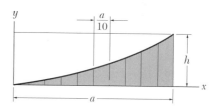
(c)

**Fig. P5.C1**

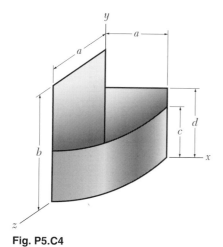

**Fig. P5.C4**

**Fig. P5.C2**

# CHAPTER 6

# Analysis of Structures

## 6.1. INTRODUCTION

The problems considered in the preceding chapters concerned the equilibrium of a single rigid body, and all forces involved were external to the rigid body. We shall now consider problems dealing with the equilibrium of structures made of several connected parts. These problems call for the determination not only of the external forces acting on the structure but also of the forces which hold together the various parts of the structure. From the point of view of the structure as a whole, these forces are *internal forces*.

Consider, for example, the crane shown in Fig. 6.1a, which carries a load W. The crane consists of three beams AD, CF, and BE connected by frictionless pins; it is supported by a pin at A and by a cable DG. The free-body diagram of the crane has been drawn in Fig. 6.1b. The external

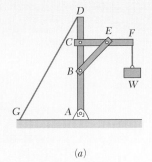
(a)

Fig. 6.1

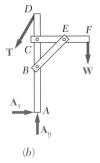
(b)

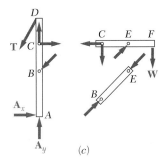
(c)

forces are shown in the diagram and include the weight **W**, the two components **A**$_x$ and **A**$_y$ of the reaction at $A$, and the force **T** exerted by the cable at $D$. The internal forces holding the various parts of the crane together do not appear in the diagram. If, however, the crane is dismembered and if a free-body diagram is drawn for each of its component parts, the forces holding the three beams together must also be represented, since these forces are external forces from the point of view of each component part (Fig. 6.1c).

It will be noted that the force exerted at $B$ by member $BE$ on member $AD$ has been represented as equal and opposite to the force exerted at the same point by member $AD$ on member $BE$; similarly, the force exerted at $E$ by $BE$ on $CF$ is shown equal and opposite to the force exerted by $CF$ on $BE$; and the components of the force exerted at $C$ by $CF$ on $AD$ are shown equal and opposite to the components of the force exerted by $AD$ on $CF$. This is in conformity with Newton's third law, which states that *the forces of action and reaction between bodies in contact have the same magnitude, same line of action, and opposite sense.* As pointed out in Chap. 1, this law is one of the six fundamental principles of elementary mechanics and is based on experimental evidence. Its application is essential to the solution of problems involving connected bodies.

In this chapter, we shall consider three broad categories of engineering structures:

1. *Trusses,* which are designed to support loads and are usually stationary, fully constrained structures. Trusses consist exclusively of straight members connected at joints located at the ends of each member. Members of a truss, therefore, are *two-force members,* i.e., members acted upon by two equal and opposite forces directed along the member.

2. *Frames,* which are also designed to support loads and are also usually stationary, fully constrained structures. However, like the crane of Fig. 6.1, frames always contain at least one *multiforce member,* i.e., a member acted upon by three or more forces which, in general, are not directed along the member.

3. *Machines,* which are designed to transmit and modify forces and are structures containing moving parts. Machines, like frames, always contain at least one multiforce member.

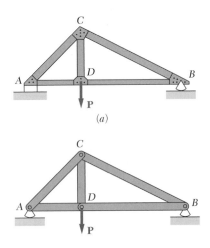

*(a)*

*(b)*

**Fig. 6.2**

## TRUSSES

### 6.2. DEFINITION OF A TRUSS

The truss is one of the major types of engineering structures. It provides both a practical and an economical solution to many engineering situations, especially in the design of bridges and buildings. A truss consists of straight members connected at joints; a typical truss is shown in Fig. 6.2*a*. Truss members are connected at their extremities only; thus no member is continuous through a joint. In Fig. 6.2*a*, for example, there is no member *AB*; there are instead two distinct members *AD* and *DB*. Most actual structures are made of several trusses joined together to form a space framework. Each truss is designed to carry those loads which act in its plane and thus may be treated as a two-dimensional structure.

In general, the members of a truss are slender and can support little lateral load; all loads, therefore, must be applied to the various joints, and not to the members themselves. When a concentrated load is to be applied between two joints, or when a distributed load is to be supported by the truss, as in the case of a bridge truss, a floor system must be provided which, through the use of stringers and floor beams, transmits the load to the joints (Fig. 6.3).

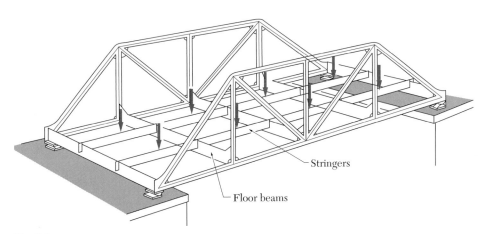

Stringers

Floor beams

**Fig. 6.3**

The weights of the members of the truss are also assumed to be applied to the joints, half of the weight of each member being applied to each of the two joints the member connects. Although the members are actually joined together by means of welded, bolted, or riveted connections, it is customary to assume that the members are pinned together; therefore, the forces acting at each end of a member reduce to a single force and no couple. Thus, the only forces assumed to be applied to a truss member are a single force at each end of the member. Each member may then be treated as a two-force member, and the entire truss may be considered as a group of pins and two-force members (Fig. 6.2*b*). An individual member may be acted upon as shown in either of the two sketches of Fig. 6.4. In the first sketch, the forces tend to pull the member apart, and the member is in tension, while in the second sketch, the forces tend to compress the member, and the member is in compression. Several typical trusses are shown in Fig. 6.5.

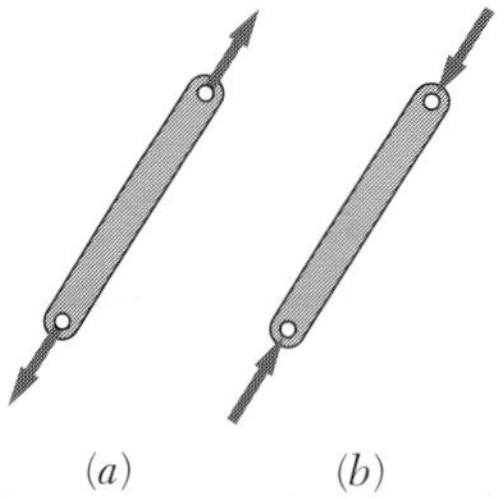

(*a*)          (*b*)

**Fig. 6.4**

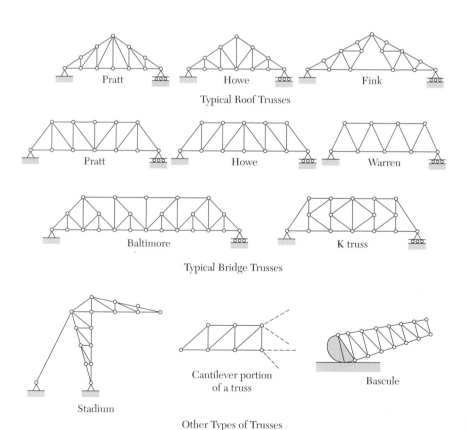

Pratt          Howe          Fink

Typical Roof Trusses

Pratt          Howe          Warren

Baltimore          K truss

Typical Bridge Trusses

Stadium

Cantilever portion
of a truss

Bascule

Other Types of Trusses

**Fig. 6.5**

## 6.3. SIMPLE TRUSSES

Consider the truss of Fig. 6.6*a*, which is made of four members connected by pins at *A*, *B*, *C*, and *D*. If a load is applied at *B*, the truss will greatly deform and lose completely its original shape. On the other hand, the truss of Fig. 6.6*b*, which is made of three members connected by pins at *A*, *B*, and *C*, will deform only slightly under a load applied at *B*. The only possible deformation for this truss is one involving small changes in the length of its members. The truss of Fig. 6.6*b* is said to be a *rigid truss*, the term rigid being used here to indicate that the truss *will not collapse*.

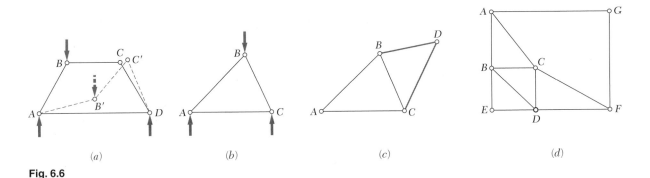

**Fig. 6.6**

As shown in Fig. 6.6*c*, a larger rigid truss may be obtained by adding two members *BD* and *CD* to the basic triangular truss of Fig. 6.6*b*. This procedure may be repeated as many times as desired, and the resulting truss will be rigid if each time we add two new members, we attach them to separate existing joints and connect them at a new joint.† A truss which may be constructed in this manner is called a *simple truss*.

It should be noted that a simple truss is not necessarily made only of triangles. The truss of Fig. 6.6*d*, for example, is a simple truss which was constructed from triangle *ABC* by adding successively the joints *D*, *E*, *F*, and *G*. On the other hand, rigid trusses are not always simple trusses, even when they appear to be made of triangles. The Fink and Baltimore trusses shown in Fig. 6.5, for instance, are not simple trusses, since they cannot be constructed from a single triangle in the manner described above. All the other trusses shown in Fig. 6.5 are simple trusses, as may be easily checked. (For the K truss, start with one of the central triangles.)

Returning to the basic triangular truss of Fig. 6.6*b*, we note that this truss has three members and three joints. The truss of Fig. 6.6*c* has two more members and one more joint, i.e., altogether five members and four joints. Observing that every time two new members are added, the number of joints is increased by one, we find that in a simple truss the total number of members is $m = 2n - 3$, where $n$ is the total number of joints.

---

† The three joints must not be in a straight line.

## 6.4. ANALYSIS OF TRUSSES BY THE METHOD OF JOINTS

We saw in Sec. 6.2 that a truss may be considered as a group of pins and two-force members. The truss of Fig. 6.2, whose free-body diagram is shown in Fig. 6.7a, may thus be dismembered, and a free-body diagram can be drawn for each pin and each member (Fig. 6.7b). Each member is acted upon by two forces, one at each end; these forces have the same magnitude, same line of action, and opposite sense (Sec. 3.17). Besides, Newton's third law indicates that the forces of action and reaction between a member and a pin are equal and opposite. Therefore, the forces exerted by a member on the two pins it connects must be directed along that member and be equal and opposite. The common magnitude of the forces exerted by a member on the two pins it connects is commonly referred to as the *force in the member* considered, even though this quantity is actually a scalar. Since the lines of action of all the internal forces in a truss are known, the analysis of a truss reduces to computing the forces in its various members and to determining whether each of its members is in tension or in compression.

Since the entire truss is in equilibrium, each pin must be in equilibrium. The fact that a pin is in equilibrium may be expressed by drawing its free-body diagram and writing two equilibrium equations (Sec. 2.9). If the truss contains $n$ pins, there will be therefore $2n$ equations available, which may be solved for $2n$ unknowns. In the case of a simple truss, we have $m = 2n - 3$, that is, $2n = m + 3$, and the number of unknowns which may be determined from the free-body diagrams of the pins is thus $m + 3$. This means that the forces in all the members, as well as the two components of the reaction $\mathbf{R}_A$, and the reaction $\mathbf{R}_B$ may be found by considering the free-body diagrams of the pins.

The fact that the entire truss is a rigid body in equilibrium may be used to write three more equations involving the forces shown in the free-body diagram of Fig. 6.7a. Since they do not contain any new information, these equations are not independent from the equations associated with the free-body diagrams of the pins. Nevertheless, they may be used to determine immediately the components of the reactions at the supports. The arrangement of pins and members in a simple truss is such that it will then always be possible to find a joint involving only two unknown forces. These forces may be determined by the methods of Sec. 2.11 and their values transferred to the adjacent joints and treated as known quantities at these joints. This procedure may be repeated until all unknown forces have been determined.

As an example, we shall analyze the truss of Fig. 6.7 by considering successively the equilibrium of each pin, starting with a joint at which only two forces are unknown. In the truss considered, all pins are subjected to at least three unknown forces. Therefore, the reactions at the supports must first be determined by considering the entire truss as a free body and using the equations of equilibrium of a rigid body. We find in this way that $\mathbf{R}_A$ is vertical and determine the magnitudes of $\mathbf{R}_A$ and $\mathbf{R}_B$.

The number of unknown forces at joint $A$ is thus reduced to two, and these forces may be determined by considering the equilibrium of pin $A$. The reaction $\mathbf{R}_A$ and the forces $\mathbf{F}_{AC}$ and $\mathbf{F}_{AD}$ exerted on pin $A$ by members

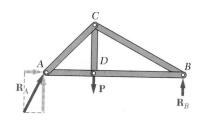

(a)

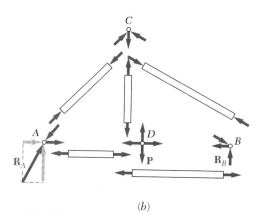

(b)

**Fig. 6.7**

$AC$ and $AD$, respectively, must form a force triangle. First we draw $\mathbf{R}_A$ (Fig. 6.8); noting that $\mathbf{F}_{AC}$ and $\boldsymbol{F}_{AD}$ are directed along $AC$ and $AD$, respectively, we complete the triangle and determine the magnitude and sense of $\mathbf{F}_{AC}$ and $\mathbf{F}_{AD}$. The magnitudes $\mathrm{F}_{AC}$ and $\mathrm{F}_{AD}$ represent the forces in members $AC$ and $AD$. Since $\mathbf{F}_{AC}$ is directed down and to the left, that is, toward joint $A$, member $AC$ pushes on pin $A$ and is in compression. On the other hand, since $\mathbf{F}_{AD}$ is directed away from the joint, member $AD$ pulls on pin $A$ and is in tension.

We may now proceed to joint $D$, where only two forces, $\mathbf{F}_{DC}$ and $\mathbf{F}_{DB}$, are still unknown. The other forces are the load $\mathbf{P}$, which is given, and the force $\mathbf{F}_{DA}$ exerted on the pin by member $AD$. As indicated above, this force is equal and opposite to the force $\mathbf{F}_{AD}$ exerted by the same member on pin $A$. We may draw the force polygon corresponding to joint $D$, as shown in Fig. 6.8, and determine the forces $\mathbf{F}_{DC}$ and $\mathbf{F}_{DB}$ from that polygon.

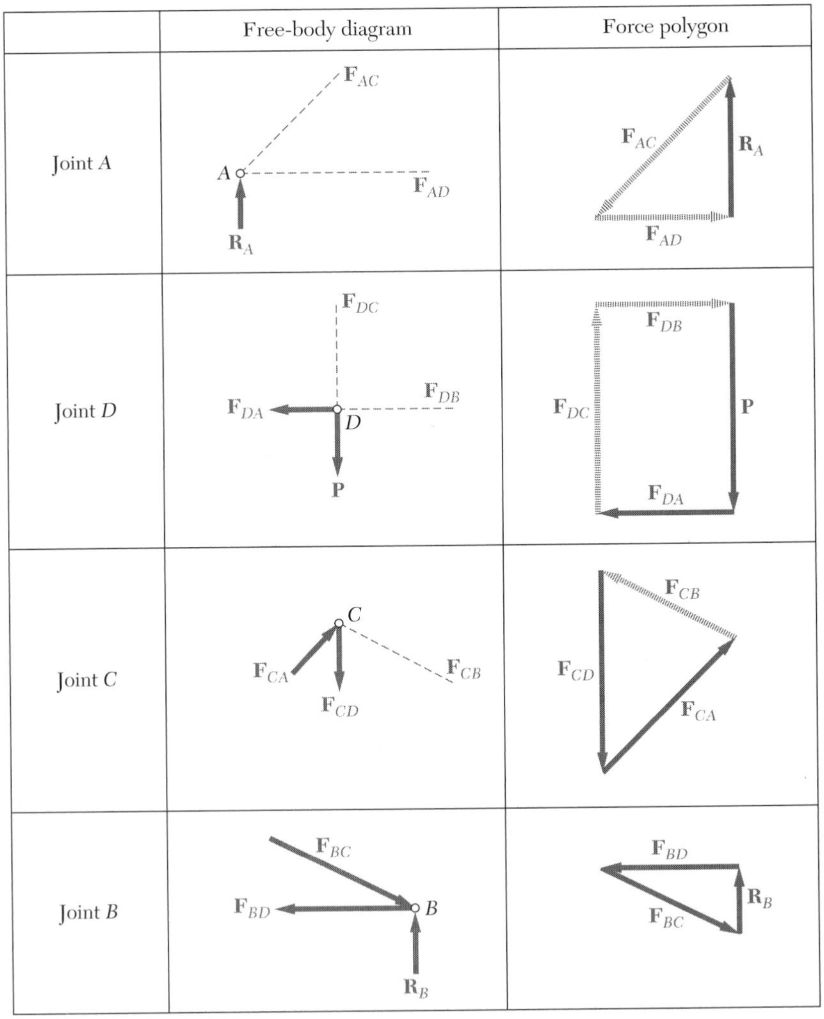

| | Free-body diagram | Force polygon |
|---|---|---|
| Joint $A$ | | |
| Joint $D$ | | |
| Joint $C$ | | |
| Joint $B$ | | |

**Fig. 6.8**

However, when more than three forces are involved, it is usually more convenient to write the equations of equilibrium $\Sigma F_x = 0$, $\Sigma F_y = 0$ and solve these equations for the two unknown forces. Since both of these forces are found to be directed away form joint $D$, members $DC$ and $DB$ pull on the pin and are in tension.

Next, joint $C$ is considered; its free-body diagram is shown in Fig. 6.8. It is noted that both $\mathbf{F}_{CD}$ and $\mathbf{F}_{CA}$ are known from the analysis of the preceding joints and that only $\mathbf{F}_{CB}$ is unknown. Since the equilibrium of each pin provides sufficient information to determine two unknowns, a check of our analysis is obtained at this joint. The force triangle is drawn, and the magnitude and sense of $\mathbf{F}_{CB}$ are determined. Since $\mathbf{F}_{CB}$ is directed toward joint $C$, member $CB$ pushes on pin $C$ and is in compression. The check is obtained by verifying that the force $\mathbf{F}_{CB}$ and member $CB$ are parallel.

At joint $B$, all the forces are known. Since the corresponding pin is in equilibrium, the force triangle must close and an additional check of the analysis is obtained.

It should be noted that the force polygons shown in Fig. 6.8 are not unique. Each of them could be replaced by an alternate configuration. For example, the force triangle corresponding to joint $A$ could be drawn as shown in Fig. 6.9. The triangle actually shown in Fig. 6.8 was obtained by drawing the three forces $\mathbf{R}_A$, $\mathbf{F}_{AC}$, and $\mathbf{F}_{AD}$ in tip-to-tail fashion in the order in which their lines of action are encountered when moving clockwise around joint $A$. The other force polygons in Fig. 6.8, having been drawn in the same way, can be made to fit into a single diagram, as shown in Fig. 6.10. Such a diagram, known as *Maxwell's diagram*, greatly facilitates the *graphical analysis* of truss problems.

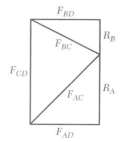

**Fig. 6.9**

**Fig. 6.10**

## *6.5. JOINTS UNDER SPECIAL LOADING CONDITIONS

Consider the joint shown in Fig. 6.11*a*, which connects four members lying in two intersecting straight lines. The free-body diagram of Fig. 6.11*b* shows that pin $A$ is subjected to two pairs of directly opposite forces. The corresponding force polygon, therefore, must be a parallelogram (Fig. 6.11*c*), and *the forces in opposite members must be equal.*

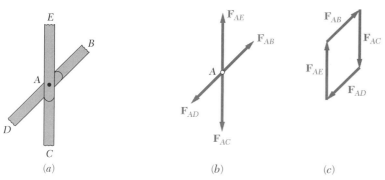

(a)          (b)          (c)

**Fig. 6.11**

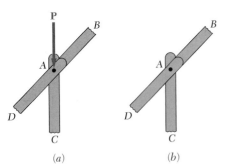

**Fig. 6.12**

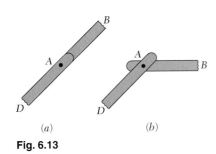

**Fig. 6.13**

Consider next the joint shown in Fig. 6.12*a*, which connects three members and supports a load **P**. Two of the members lie in the same line, and the load **P** acts along the third member. The free-body diagram of pin *A* and the corresponding force polygon will be as shown in Fig. 6.11*b* and *c* with $\mathbf{F}_{AE}$ replaced by the load **P**. Thus, *the forces in the two opposite members must be equal, and the force in the other member must equal P.* A particular case of special interest is shown in Fig. 6.12*b*. Since, in this case, no external load is applied to the joint, we have *P* = 0, and the force in member *AC* is zero. Member *AC* is said to be a *zero-force member*.

Consider now a joint connecting two members only. From Sec. 2.9, we know that a particle which is acted upon by two forces will be in equilibrium if the two forces have the same magnitude, same line of action, and opposite sense. In the case of the joint of Fig. 6.13*a*, which connects two members *AB* and *AD* lying in the same line, the equilibrium of pin *A* requires therefore that *the forces in the two members be equal.* In the case of the joint of Fig. 6.13*b*, the equilibrium of pin *A* is impossible unless the forces in both members are zero. Members connected as shown in Fig. 6.13*b*, therefore, must be *zero-force members*.

Spotting the joints which are under the special loading conditions listed above will expedite the analysis of a truss. Consider, for example, a Howe truss loaded as shown in Fig. 6.14. All the members represented by colored lines will be recognized as zero-force members. Joint *C* connects three members, two of which lie in the same line, and is not subjected to any external load; member *BC* is thus a zero-force member. Applying the same reasoning to joint *K*, we find that member *JK* is also a zero-force member. But joint *J* is now in the same situation as joints *C* and *K*, and member *IJ* must be a zero-force member. The examination of joints *C*, *J*, and *K* also shows that the forces in members *AC* and *CE* are equal, that the forces in members *HJ* and *JL* are equal, and that the forces in members *IK* and *KL* are equal. Furthermore, now turning our attention to joint *I*, where the 20-kN load and member *HI* are collinear, we note that the force in member *HI* is 20 kN (tension) and that the forces in members *GI* and *IK* are equal. Hence, the forces in members *GI*, *IK*, and *KL* are equal.

Students, however, should be warned against misusing the rules established in this section. For example, it would be wrong to assume that the force in member *DE* is 25 kN or that the forces in members *AB* and *BD* are equal. The conditions discussed above do not apply to joints *B* and *D*. The forces in these members and in all remaining members should be found by carrying out the analysis of joints *A*, *B*, *D*, *E*, *F*, *G*, *H*, and *L* in the usual manner. Until they have become thoroughly familiar with the conditions of application of the rules established in this section, students would be well advised to draw the free-body diagrams of all pins and to write the corresponding equilibrium equations (or draw the corresponding force polygons), whether or not the joints considered fall into the categories listed above.

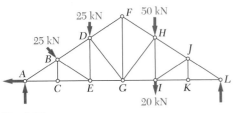

**Fig. 6.14**

A final remark concerning zero-force members: These members are not useless. While they do not carry any load under the particular loading conditions shown, the zero-force members of Fig. 6.14 will probably carry loads if the loading conditions are changed. Besides, even in the case considered, these members are needed to support the weight of the truss and to maintain the truss in the desired shape.

## *6.6. SPACE TRUSSES

When several straight members are joined together at their extremities to form a three-dimensional configuration, the structure obtained is called a *space truss*.

We recall from Sec. 6.3 that the most elementary two-dimensional rigid truss consisted of three members joined at their extremities to form the sides of a triangle; by adding two members at a time to this basic configuration, and connecting them at a new joint, it was possible to obtain a larger rigid structure which was defined as a simple truss. Similarly, the most elementary rigid space truss consists of six members joined at their extremities to form the edges of a tetrahedron $ABCD$ (Fig. 6.15a). By adding three members at a time to this basic configuration, such as $AE$, $BE$, and $CE$, attaching them at separate existing joints, and connecting them at a new joint, we can obtain a larger rigid structure which is defined as a *simple space truss* (Fig. 6.15b).† Observing that the basic tetrahedron has six members and four joints and that every time three members are added, the number of joints is increased by one, we conclude that in a simple space truss the total number of members is $m = 3n - 6$, where $n$ is the total number of joints.

If a space truss is to be completely constrained and if the reactions at its supports are to be statically determinate, the supports should consist of a combination of balls, rollers, and balls and sockets which provides six unknown reactions (see Sec. 4.8). These unknown reactions may be readily determined by solving the six equations expressing that the three-dimensional truss is in equilibrium.

Although the members of a space truss are actually joined together by means of riveted or welded connections, it is assumed that each joint consists of a ball-and-socket connection. Thus, no couple will be applied to the members of the truss, and each member may be treated as a two-force member. The conditions of equilibrium for each joint will be expressed by the three equations $\Sigma F_x = 0$, $\Sigma F_y = 0$, and $\Sigma F_z = 0$. In the case of a simple space truss containing $n$ joints, writing the conditions of equilibrium for each joint will thus yield $3n$ equations. Since $m = 3n - 6$, these equations suffice to determine all unknown forces (forces in $m$ members and six reactions at the supports). However, to avoid solving many simultaneous equations, care should be taken to select joints in such an order that no selected joint will involve more than three unknown forces.

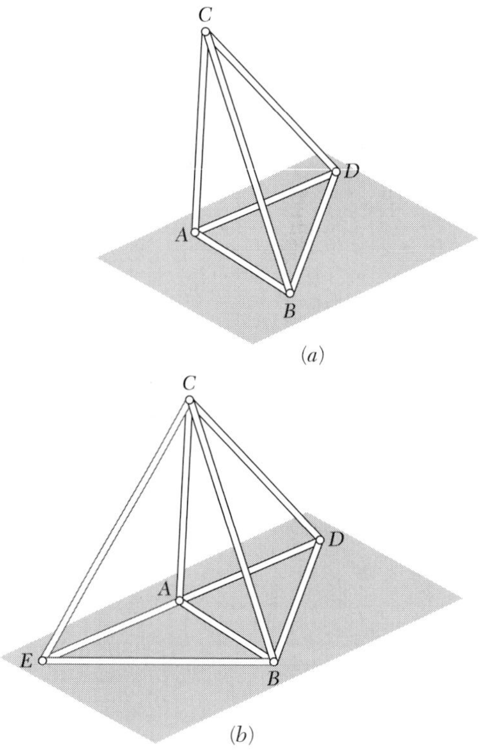

† The four joints must not lie in a plane.

(a)

(b)

**Fig. 6.15**

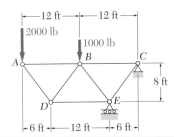

# SAMPLE PROBLEM 6.1

Using the method of joints, determine the force in each member of the truss shown.

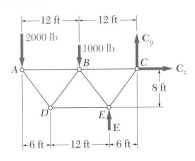

**Solution.** A free-body diagram of the entire truss is drawn; external forces acting on this free body consist of the applied loads and the reactions at $C$ and $E$.

### Equilibrium of Entire Truss

$+\uparrow\Sigma M_C = 0$: 

$$(2000 \text{ lb})(24 \text{ ft}) + (1000 \text{ lb})(12 \text{ ft}) - E(6 \text{ ft}) = 0$$

$$E = +10,000 \text{ lb} \qquad\qquad \mathbf{E} = 10,000 \text{ lb}\uparrow$$

$\xrightarrow{+}\Sigma F_x = 0$: 

$$\mathbf{C}_x = 0$$

$+\uparrow\Sigma F_y = 0$: 

$$-2000 \text{ lb} - 1000 \text{ lb} + 10,000 \text{ lb} + C_y = 0$$

$$C_y = -7000 \text{ lb} \qquad\qquad \mathbf{C}_y = 7000 \text{ lb}\downarrow$$

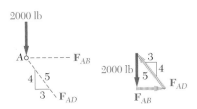

***Joint A.*** This joint is subjected to only two unknown forces, namely, the forces exerted by members $AB$ and $AD$. A force triangle is used to determine $\mathbf{F}_{AB}$ and $\mathbf{F}_{AD}$. We note that member $AB$ pulls on the joint and thus is in tension and that member $AD$ pushes on the joint and thus is in compression. The magnitudes of the two forces are obtained from the proportion

$$\frac{2000 \text{ lb}}{4} = \frac{F_{AB}}{3} = \frac{F_{AD}}{5}$$

$$F_{AB} = 1500 \text{ lb } T \blacktriangleleft$$
$$F_{AD} = 2500 \text{ lb } C \blacktriangleleft$$

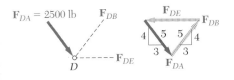

***Joint D.*** Since the force exerted by member $AD$ has been determined, only two unknown forces are now involved at this joint. Again, a force triangle is used to determine the unknown forces in members $DB$ and $DE$.

$$F_{DB} = F_{DA} \qquad\qquad F_{DB} = 2500 \text{ lb } T \blacktriangleleft$$
$$F_{DE} = 2(\tfrac{3}{5})F_{DA} \qquad\qquad F_{DE} = 3000 \text{ lb } C \blacktriangleleft$$

**Joint B.** Since more than three forces act at this joint, we determine the two unknown forces $\mathbf{F}_{BC}$ and $\mathbf{F}_{BE}$ by solving the equilibrium equations $\Sigma F_x = 0$ and $\Sigma F_y = 0$. We arbitrarily assume that both unknown forces act away from the joint, i.e., that the members are in tension. The positive value obtained for $F_{BC}$ indicates that our assumption was correct; member $BC$ is in tension. The negative value of $F_{BE}$ indicates that our assumption was wrong; member $BE$ is in compression.

$+\uparrow \Sigma F_y = 0:\qquad -1000 - \tfrac{4}{5}(2500) - \tfrac{4}{5}F_{BE} = 0$

$$F_{BE} = -3750\text{ lb} \qquad F_{BE} = 3750\text{ lb } C \quad \blacktriangleleft$$

$\xrightarrow{+} \Sigma F_x = 0:\qquad F_{BC} - 1500 - \tfrac{3}{5}(2500) - \tfrac{3}{5}(3750) = 0$

$$F_{BC} = +5250\text{ lb} \qquad F_{BC} = 5250\text{ lb } T \quad \blacktriangleleft$$

**Joint E.** The unknown force $\mathbf{F}_{EC}$ is assumed to act away from the joint. Summing $x$ components, we write

$\xrightarrow{+} \Sigma F_x = 0:\qquad \tfrac{3}{5}F_{EC} + 3000 + \tfrac{3}{5}(3750) = 0$

$$F_{EC} = -8750\text{ lb} \qquad F_{EC} = 8750\text{ lb } C \quad \blacktriangleleft$$

Summing $y$ components, we obtain a check of our computations:

$$+\uparrow \Sigma F_y = 10{,}000 - \tfrac{4}{5}(3750) - \tfrac{4}{5}(8750)$$
$$= 10{,}000 - 3000 - 7000 = 0 \qquad \text{(checks)}$$

**Joint C.** Using the computed values of $\mathbf{F}_{CB}$ and $\mathbf{F}_{CE}$, we may determine the reactions $\mathbf{C}_x$ and $\mathbf{C}_y$ by considering the equilibrium of this joint. Since these reactions have already been determined from the equilibrium of the entire truss, we will obtain two checks of our computations. We may also merely use the computed values of all forces acting on the joint (forces in members and reactions) and check that the joint is in equilibrium:

$$\xrightarrow{+} \Sigma F_n = -5250 + \tfrac{3}{5}(8750) = -5250 + 5250 = 0 \qquad \text{(checks)}$$
$$+\uparrow \Sigma F_y = -7000 + \tfrac{4}{5}(8750) = -7000 + 7000 = 0 \qquad \text{(checks)}$$

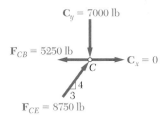

# PROBLEMS

**6.1 through 6.12** Using the method of joints, determine the force in each member of the truss shown. State whether each member is in tension or compression.

**Fig. P6.1**

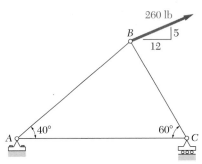

**Fig. P6.2**

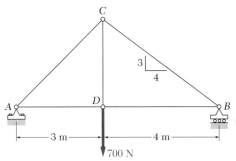

**Fig. P6.3**

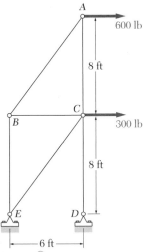

**Fig. P6.4**

**Fig. P6.5**

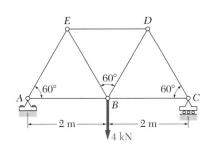

**Fig. P6.6**

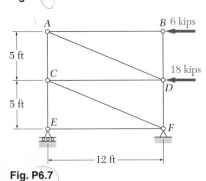

**Fig. P6.7**

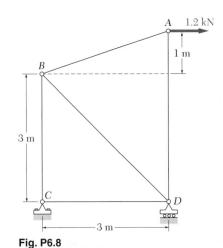

**Fig. P6.8**

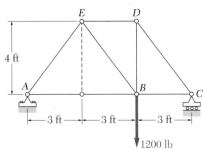

**Fig. P6.9**

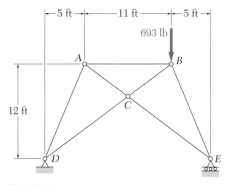

Fig. P6.10

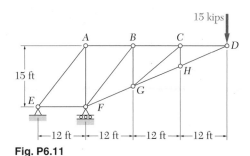

Fig. P6.11

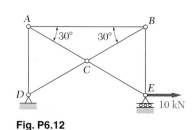

Fig. P6.12

**6.13** Determine whether the trusses given in Probs. 6.10, 6.14, 6.15, and 6.16 are simple trusses.

**6.14** Determine the zero-force members in the truss shown for the given loading. Re-draw the truss with the zero-force members removed. Determine the force in members UV, UB and UC.

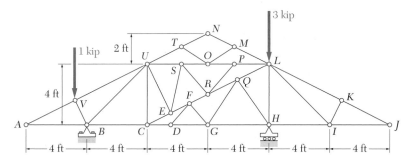

Fig. P6.14

**6.15** Determine the zero-force members in the truss shown for the given loading. Re-draw the truss with the zero-force members removed. Determine the force in members AC, AN, CF and CN.

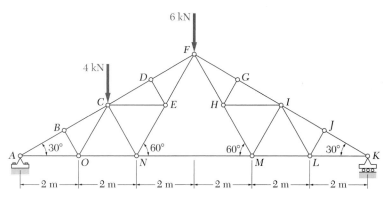

Fig. P6.15

**6.16** Determine the zero-force members in the truss shown for the given loading. Re-draw the truss with the zero-force members removed. Determine the force in members BK, JK and KD.

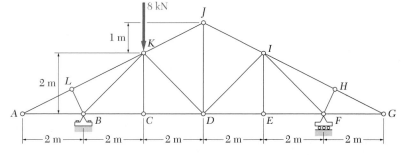

Fig. P6.16

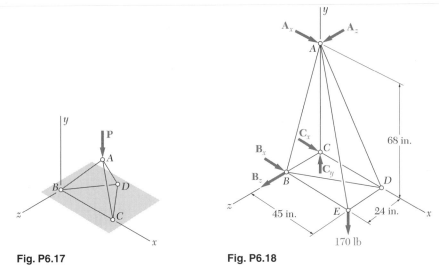

**Fig. P6.17**

**Fig. P6.18**

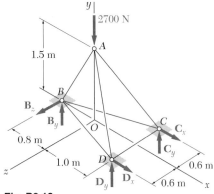

**Fig. P6.19**

**\*6.17**   Six bars, each of length $L$, are connected to form a regular tetrahedron which rests on a frictionless horizontal surface. Determine the force in each of the six members when a vertical force **P** is applied at $A$.

**\*6.18**   The three-dimensional truss is supported by the six reactions shown. If a vertical 170-lb load is applied at $E$, determine ($a$) the reactions, ($b$) the force in each member.

**\*6.19**   The three-dimensional truss is supported by the six reactions shown. If a vertical 2700-N load is applied at $A$, determine ($a$) the reactions, ($b$) the force in each member.

**\*6.20**   Solve Prob. 6.19, assuming that the 2700-N load is applied at $A$ in a direction parallel to the $z$ axis (with positive sense).

## 6.7. ANALYSIS OF TRUSSES BY THE METHOD OF SECTIONS

The method of joints is most effective when the forces in all the members of a truss are to be determined. If, however, the force in only one member or the forces in a very few members are desired, another method, the method of sections, will prove more efficient.

Assume, for example, that we want to determine the force in member $BD$ of the truss shown in Fig. 6.16$a$. To do this, we must determine the force with which member $BD$ acts on either joint $B$ or joint $D$. If we were to use the method of joints, we would choose either joint $B$ or joint $D$ as a free body. However, we may also choose as a free body a larger portion of the truss, composed of several joints and members, provided that the desired force is one of the external forces acting on that portion. If, in addition, the portion of the truss is chosen so that there is a total of only three unknown forces acting upon it, the desired force may be obtained by solving the equations of equilibrium for this portion of the truss. In practice, the portion of the truss to be utilized is obtained by *passing a section* through three members of the truss, one of which is the desired member, i.e., by drawing a line which divides the truss into two completely separate parts but does not intersect more than three members. Either of the two portions of the truss obtained after the intersected members have been removed may then be used as a free body.†

In Fig. 6.16a, the section $nn$ has been passed through members $BD$, $BE$, and $CE$, and the portion $ABC$ of the truss is chosen as the free body (Fig. 6.16b). The forces acting on the free body are the loads $\mathbf{P}_1$ and $\mathbf{P}_2$ at points $A$ and $B$ and the three unknown forces $\mathbf{F}_{BD}$, $\mathbf{F}_{BE}$, and $\mathbf{F}_{CE}$. Since it is not known whether the members removed were in tension or compression, the three forces have been arbitrarily drawn away from the free body as if the members were in tension.

The fact that rigid body $ABC$ is in equilibrium can be expressed by writing three equations which may be solved for the three unknown forces. If only the force $\mathbf{F}_{BD}$ is desired, we need write only one equation, provided that the equation does not contain the other unknowns. Thus the equation $\Sigma M_E = 0$ yields the value of the magnitude $F_{BD}$ of the force $\mathbf{F}_{BD}$. A positive sign in the answer will indicate that our original assumption regarding the sense of $\mathbf{F}_{BD}$ was correct and that member $BD$ is in tension; a negative sign will indicate that our assumption was incorrect and that $BD$ is in compression.

On the other hand, if only the force $\mathbf{F}_{CE}$ is desired, an equation which does not involve $\mathbf{F}_{BD}$ or $\mathbf{F}_{BE}$ should be written; the appropriate equation is $\Sigma M_B = 0$. Again a positive sign for the magnitude $F_{CE}$ of the desired force indicates a correct assumption, hence tension; and a negative sign indicates an incorrect assumption, hence compression.

If only the force $\mathbf{F}_{BE}$ is desired, the appropriate equation is $\Sigma F_y = 0$. Whether the member is in tension or compression is again determined from the sign of the answer.

When the force in only one member is determined, no independent check of the computation is available. However, when all the unknown forces acting on the free body are determined, the computations can be checked by writing an additional equation. For instance, if $\mathbf{F}_{BD}$, $\mathbf{F}_{BE}$, and $\mathbf{F}_{CE}$ are determined as indicated above, the computation can be checked by verifying that $\Sigma F_x = 0$.

## *6.8. TRUSSES MADE OF SEVERAL SIMPLE TRUSSES

Consider two simple trusses $ABC$ and $DEF$. If they are connected by three bars $BD$, $BE$, and $CE$ as shown in Fig. 6.17a, they will form together a rigid truss $ABDF$. The trusses $ABC$ and $DEF$ can also be combined into a single rigid truss by joining joints $B$ and $D$ into a single joint $B$ and by connecting joints $C$ and $E$ by a bar $CE$ (Fig. 6.17b). The truss thus obtained is known as a Fink truss. It should be noted that the trusses of Fig. 6.17a and $b$ are *not* simple trusses; they cannot be constructed from a triangular truss by adding successive pairs of members as prescribed in Sec. 6.3. They are rigid trusses, however, as we may check by comparing the systems of

†In the analysis of certain trusses, sections are passed which intersect more than three members; the forces in one, or possibly two, of the intersected members may be obtained if equilibrium equations can be found, each of which involves only one unknown (see Probs. 6.37 through 6.39).

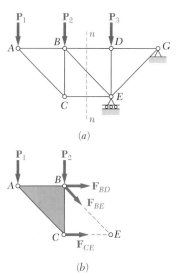

(a)

(b)

**Fig. 6.16**

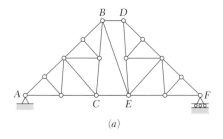

(a)

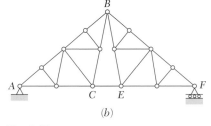

(b)

**Fig. 6.17**

connections used to hold the simple trusses *ABC* and *DEF* together (three bars in Fig. 6.17*a*, one pin and one bar in Fig. 6.17*b*) with the systems of supports discussed in Secs. 4.4 and 4.5. Trusses made of several simple trusses rigidly connected are known as *compound trusses*.

It may be checked that in a compound truss the number of members *m* and the number of joints *n* are still related by the formula $m = 2n - 3$. If a compound truss is supported by a frictionless pin and a roller (involving three unknown reactions), the total number of unknowns is $m + 3$ and this number is therefore equal to the number $2n$ of equations obtained by expressing that the *n* pins are in equilibrium. Compound trusses supported by a pin and a roller, or by an equivalent system of supports, are *statically determinate, rigid,* and *completely constrained.* This means that all unknown reactions and forces in members can be determined by the methods of statics and that all equilibrium equations being satisfied, the truss will neither collapse nor move. The forces in the members, however, cannot all be determined by the method of joints, except by solving a large number of simultaneous equations. In the case of the compound truss of Fig. 6.17*a*, for example, it will be found more expeditious to pass a section through members *BD*, *BE*, and *CE* to determine their forces.

Suppose, now, that the simple trusses *ABC* and *DEF* are connected by *four* bars *BD*, *BE*, *CD*, and *CE* (Fig. 6.18). The number of members *m* is now larger than $2n - 3$; the truss obtained is *overrigid,* and one of the four members *BD*, *BE*, *CD*, or *CE* is said to be *redundant.* If the truss is supported by a pin at *A* and a roller at *F*, the total number of unknowns is $m + 3$. This number is now larger than the number $2n$ of available independent equations; the truss is *statically indeterminate.*

Finally, we shall assume that the two simple trusses *ABC* and *DEF* are joined by a pin as shown in Fig. 6.19*a*. The number of members *m* is smaller than $2n - 3$. If the truss is supported by a pin at *A* and a roller at *F*, the total number of unknowns is $m + 3$. This number is now smaller than the number $2n$ of equilibrium equations which should be satisfied; the truss is *nonrigid* and will collapse under its own weight. However, if two pins are used to support it, the truss becomes *rigid* and will not collapse (Fig. 6.19*b*). We note that the total number of unknowns is now $m + 4$ and is equal to the number $2n$ of equations. More generally, if the reactions at the supports involve *r* unknowns, the condition for a compound truss to be statically determinate, rigid, and completely constrained is $m + r = 2n$. While necessary, this condition, however, is not sufficient for the equilibrium of a structure which ceases to be rigid when detached from its supports (see Sec. 6.11).

**Fig. 6.18**

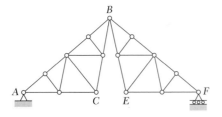

**Fig. 6.19**                    (*a*)

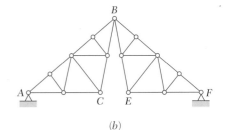

(*b*)

# SAMPLE PROBLEM 6.2

Determine the force in members $EF$ and $GI$ of the truss shown.

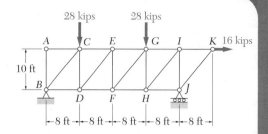

**Solution.** A free-body diagram of the entire truss is drawn; external forces acting on this free body consist of the applied loads and the reactions at $B$ and $J$.

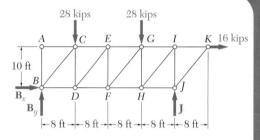

### Equilibrium of the Entire Truss

$+\circlearrowleft\Sigma M_B = 0$:

$-(28 \text{ kips})(8 \text{ ft}) - (28 \text{ kips})(24 \text{ ft}) - (16 \text{ kips})(10 \text{ ft}) + J(32 \text{ ft}) = 0$

$$J = +33 \text{ kips} \qquad \mathbf{J} = 33 \text{ kips} \uparrow$$

$\xrightarrow{+} \Sigma F_x = 0$: $\qquad B_x + 16 \text{ kips} = 0$

$$B_x = -16 \text{ kips} \qquad \mathbf{B}_x = 16 \text{ kips} \leftarrow$$

$+\circlearrowleft\Sigma M_J = 0$:

$(28 \text{ kips})(24 \text{ ft}) + (28 \text{ kips})(8 \text{ ft}) - (16 \text{ kips})(10 \text{ ft}) + B_y(32 \text{ ft}) = 0$

$$B_y = +23 \text{ kips} \qquad \mathbf{B}_y = 23 \text{ kips} \uparrow$$

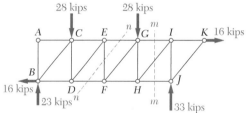

**Force in Member EF.** Section $nn$ is passed through the truss so that it intersects member $EF$ and only two additional members. After the intersected members have been removed, the left-hand portion of the truss is chosen as a free body. Three unknowns are involved; to eliminate the two horizontal forces, we write

$+\uparrow\Sigma F_y = 0$: $\qquad +23 \text{ kips} - 28 \text{ kips} - F_{EF} = 0$

$$F_{EF} = -5 \text{ kips}$$

The sense of $\mathbf{F}_{EF}$ was chosen assuming member $EF$ to be in tension; the negative sign obtained indicates that the member is in compression.

$$F_{EF} = 5 \text{ kips } C \qquad \blacktriangleleft$$

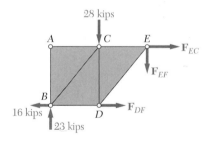

**Force in Member GI.** Section $mm$ is passed through the truss so that it intersects member $GI$ and only two additional members. After the intersected members have been removed, we choose the right-hand portion of the truss as a free body. Three unknown forces are again involved; to eliminate the two forces passing through point $H$, we write

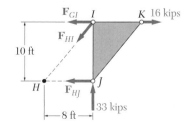

$+\circlearrowleft\Sigma M_H = 0$: $\qquad (33 \text{ kips})(8 \text{ ft}) - (16 \text{ kips})(10 \text{ ft}) + F_{GI}(10 \text{ ft}) = 0$

$$F_{GI} = -10.4 \text{ kips} \qquad F_{GI} = 10.4 \text{ kips } C \qquad \blacktriangleleft$$

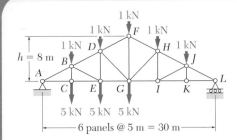

1 kN
1 kN  F  1 kN
1 kN  D  H  1 kN
h = 8 m  B  J
A
C  E  G  I  K  L

5 kN  5 kN  5 kN

6 panels @ 5 m = 30 m

# SAMPLE PROBLEM 6.3

Determine the force in members *FH*, *GH*, and *GI* of the roof truss shown.

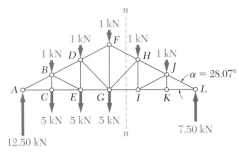

1 kN
1 kN  F  1 kN
1 kN  D  H  1 kN
B  J  α = 28.07°
A
C  E  G  I  K  L

5 kN  5 kN  5 kN

12.50 kN  7.50 kN

**Solution.**  Section *nn* is passed through the truss as shown. The right-hand portion of the truss will be taken as a free body. Since the reaction at *L* acts on this free body, the value of **L** must be calculated separately, using the entire truss as a free body; the equation $\Sigma M_A = 0$ yields **L** = 7.50 kN↑.

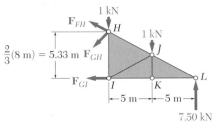

1 kN
$\mathbf{F}_{FH}$  H
$\frac{2}{3}$(8 m) = 5.33 m  $\mathbf{F}_{GH}$  1 kN
J
$\mathbf{F}_{GI}$  I  K  L
5 m  5 m
7.50 kN

***Force in Member GI.***  Using the portion *HLI* of the truss as a free body, the value of $F_{GI}$ is obtained by writing

$$+\uparrow\Sigma M_H = 0: \qquad (7.50 \text{ kN})(10 \text{ m}) - (1 \text{ kN})(5 \text{ m}) - F_{GI}(5.33 \text{ m}) = 0$$
$$F_{GI} = +13.13 \text{ kN} \qquad F_{GI} = 13.13 \text{ kN } T \blacktriangleleft$$

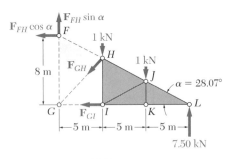

$\mathbf{F}_{FH} \sin \alpha$
$\mathbf{F}_{FH} \cos \alpha$  F
1 kN
8 m  $\mathbf{F}_{GH}$  H  1 kN
J
G  $\mathbf{F}_{GI}$  I  K  α = 28.07°  L
5 m  5 m  5 m
7.50 kN

***Force in Member FH.***  The value of $F_{FH}$ is obtained from the equation $\Sigma M_G = 0$. We move $\mathbf{F}_{FH}$ along its line of action until it acts at point *F*, where it is resolved into its *x* and *y* components. The moment of $\mathbf{F}_{FH}$ with respect to point *G* is now equal to $(F_{FH} \cos \alpha)(8 \text{ m})$.

$$+\uparrow\Sigma M_G = 0:$$
$$(7.50 \text{ kN})(15 \text{ m}) - (1 \text{ kN})(10 \text{ m}) - (1 \text{ kN})(5 \text{ m}) + (F_{FH} \cos \alpha)(8 \text{ m}) = 0$$
$$F_{FH} = -13.81 \text{ kN} \qquad F_{FH} = 13.81 \text{ kN } C \blacktriangleleft$$

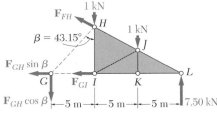

1 kN
$\mathbf{F}_{FH}$  H
$\beta = 43.15°$  1 kN
J
$\mathbf{F}_{GH} \sin \beta$
G  $\mathbf{F}_{GI}$  I  K  L
$\mathbf{F}_{GH} \cos \beta$  5 m  5 m  5 m  7.50 kN

***Force in Member GH.***  The value of $F_{GH}$ is determined by first resolving the force $\mathbf{F}_{GH}$ into *x* and *y* components at point *G* and then solving the equation $\Sigma M_L = 0$.

$$+\uparrow\Sigma M_L = 0: \qquad (1 \text{ kN})(10 \text{ m}) + (1 \text{ kN})(5 \text{ m}) + (F_{GH} \cos \beta)(15 \text{ m}) = 0$$
$$F_{GH} = -1.371 \text{ kN} \qquad F_{GH} = 1.371 \text{ kN } C \blacktriangleleft$$

# PROBLEMS

**6.21** Determine the force in members $CE$ and $CF$ of the truss shown.

**6.22** Determine the force in members $FG$ and $FH$ of the truss shown.

**6.23** Determine the force in members $HC$ and $HG$ of the truss shown.

**6.24** Determine the force in members $CG$ and $CD$ of the truss shown.

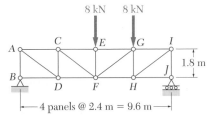

**Fig. P6.21 and P6.22**

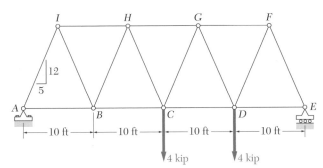

**Fig. P6.23 and P6.24**

**6.25** Determine the force in members $DE$ and $EF$ of the truss shown.

**6.26** Determine the force in members $FG$ and $FH$ of the truss shown.

**6.27** Determine the force in members $FG$ and $FC$ of the truss shown.

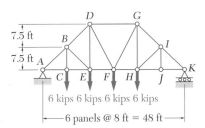

**Fig. P6.25 and P6.26**

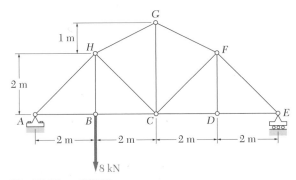

**Fig. P6.27 and P6.28**

**6.28** Determine the force in members $FG$ and $FC$ of the truss shown.

**6.29** Determine the force in members $BD$ and $DE$ of the truss shown.

**6.30** Determine the force in members $DG$ and $EG$ of the truss shown.

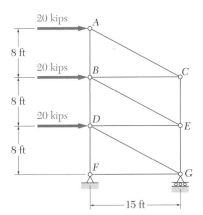

**Fig. P6.29 and P6.30**

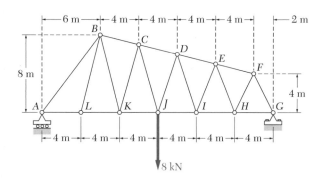

Fig. P6.31 and P6.32

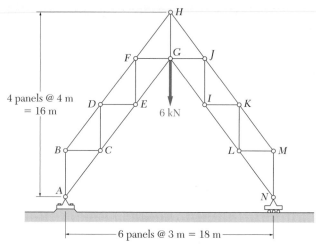

Fig. P6.33 and P6.34

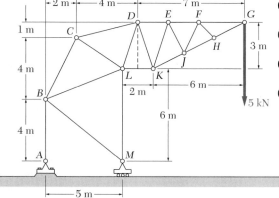

Fig. P6.35 and P6.36

**6.31** Determine the force in members $CD$, $CJ$ and $KJ$ of the truss shown.

**6.32** Determine the force in members $DE$, $DI$, and $JI$ of the truss shown.

**6.33** Determine the force in members $DF$, $DE$, and $CE$ of the truss shown.

**6.34** Determine the force in members $JK$, $IJ$, and $GI$ of the truss shown.

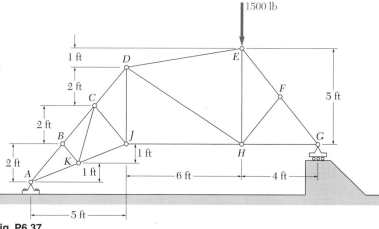

Fig. P6.37

**6.35** Determine the force in members $CD$, $DL$, and $LK$ of the truss shown. Also, identify all zero-force members.

**6.36** Determine the force in members $BC$, $BL$, and $LM$ of the truss shown. Also, identify all zero-force members.

**6.37** Determine the force in members $DE$, $DH$ and $JH$ of the truss shown. Also, identify all zero-force members.

**6.38** Determine the force in members $AC$ and $BE$ of the truss shown. (*Hint.* Use section *a-a*.)

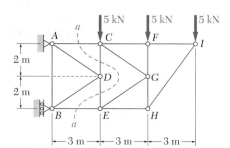

Fig. P6.38

**6.39**  Determine the force in member *GJ* of the truss shown. (*Hint.* Use section *a-a*.)

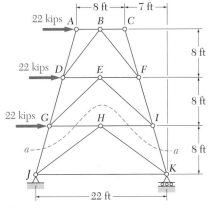

Fig. P6.39

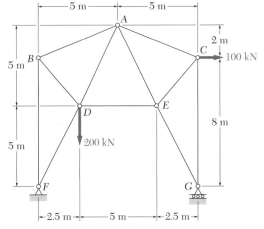

Fig. P6.40 and P6.41

**6.40**  Determine the force in members *DE* and *AE* of the truss shown.

**6.41**  Determine the force in members *AB* and *AD* of the truss shown.

**6.42**  The diagonal members in the center panel of the truss shown are very slender and can act only in tension; such members are known as *counters*. Determine the force in member *DE* and in the counters which are acting under the given loading.

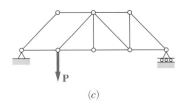

Fig. P6.42

**6.43**  Solve Prob. 6.42, assuming that the 6-kip load has been removed.

**6.44**  Solve Prob. 6.42, assuming that the 9-kip load has been removed.

**6.45 and 6.46**  Classify each of the given structures as completely, partially, or improperly constrained; if completely constrained, further classify as determinate or indeterminate. (All members can act both in tension and in compression.)

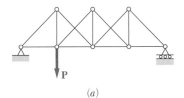
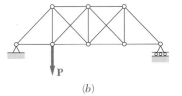
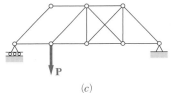

(a)   (b)   (c)

Fig. P6.45

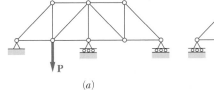
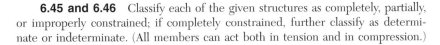

(a)   (b)   (c)

Fig. P6.46

**6.47 and 6.48** Classify each of the given structures as completely, partially, or improperly constrained; if completely constrained, further classify as determinate or indeterminate. (All members can act both in tension and in compression.)

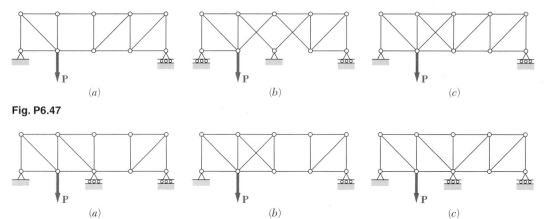

(a)                    (b)                    (c)

**Fig. P6.47**

(a)                    (b)                    (c)

**Fig. P6.48**

## FRAMES AND MACHINES

### 6.9. STRUCTURES CONTAINING MULTIFORCE MEMBERS

Under Trusses, we have considered structures consisting entirely of pins and of straight two-force members. The forces acting on the two-force members were known to be directed along the members themselves. We shall now consider structures in which at least one of the members is a *multiforce* member, i.e., a member acted upon by three or more forces. These forces will generally not be directed along the members on which they act; their direction is unknown, and they should be represented therefore by two unknown components.

Frames and machines are structures containing multiforce members. *Frames* are designed to support loads and are usually stationary, fully constrained structures. *Machines* are designed to transmit and modify forces; they may or may not be stationary and will always contain moving parts.

### 6.10. ANALYSIS OF A FRAME

As a first example of analysis of a frame, we shall consider again the crane described in Sec. 6.1, which carries a given load W (Fig. 6.20a). The free-body diagram of the entire frame is shown in Fig. 6.20b. This diagram may be used to determine the external forces acting on the frame. Summing moments about A, we first determine the force **T** exerted by the cable; summing x and y components, we then determine the components $A_x$ and $A_y$ of the reaction at the pin A.

In order to determine the internal forces holding the various parts of a frame together, we must dismember the frame and draw a free-body diagram for each of its component parts (Fig. 6.20c). First, the two-force members should be considered. In this frame, member BE is the only two-force member. The forces acting at each end of this member must have the same magnitude, same line of action, and opposite sense (Sec. 3.18). They are therefore directed along $BE$ and will be denoted, respectively, by $\mathbf{F}_{BE}$ and $\mathbf{F}'_{BE}$. Their sense will be arbitrarily assumed as shown in Fig. 6.20c, and the correctness of this assumption will be checked later by the sign obtained for the common magnitude $F_{BE}$ of the two forces.

Next, we consider the multiforce members, i.e., the members which are acted upon by three or more forces. According to Newton's third law, the force exerted at $B$ by member $BE$ on member $AD$ must be equal and opposite to the force $\mathbf{F}_{BE}$ exerted by $AD$ on $BE$. Similarly, the force exerted at $E$ by member $BE$ on member $CF$ must be equal and opposite to the force $\mathbf{F}'_{BE}$ exerted by $CF$ on $BE$. The forces that the two-force member $BE$ exerts on $AD$ and $CF$ are therefore respectively equal to $\mathbf{F}'_{BE}$ and $\mathbf{F}_{BE}$; they have the same magnitude $F_{BE}$ and opposite sense, and should be directed as shown in Fig. 6.20c.

At $C$ two multiforce members are connected. Since neither the direction nor the magnitude of the forces acting at $C$ is known, these forces will be represented by their $x$ and $y$ components. The components $\mathbf{C}_x$ and $\mathbf{C}_y$ of the force acting on member $AD$ will be arbitrarily directed to the right and upward. Since, according to Newton's third law, the forces exerted by member $CF$ on $AD$ and by member $AD$ on $CF$ are equal and opposite, the components $\mathbf{C}'_x$ and $\mathbf{C}'_y$ of the force acting on member $CF$ *must* be directed to the left and downward. Whether the force $\mathbf{C}_x$ is actually directed to the right and the force $\mathbf{C}'_x$ is actually directed to the left will be determined later from the sign of their common magnitude $C_x$, a plus sign indicating that the assumption made was correct, and a minus sign that it was wrong. The free-body diagrams of the multiforce members are completed by showing the external forces acting at $A$, $D$, and $F$.†

The internal forces may now be determined by considering the free-body diagram of either of the two multiforce members. Choosing the free-body diagram of $CF$, for example, we write the equations $\Sigma M_C = 0$, $\Sigma M_E = 0$, and $\Sigma F_x = 0$, which yield the values of the magnitudes $F_{BE}$, $C_y$, and $C_x$, respectively. These values may be checked by verifying that member $AD$ is also in equilibrium.

It should be noted that the free-body diagrams of the pins were not shown in Fig. 6.20c. This was because the pins were assumed to form an integral part of one of the two members they connected. This assumption can always be used to simplify the analysis of frames and machines. When a pin connects three or more members, however, or when a pin connects a support and two or more members, or when a load is applied to a pin, a clear decision must be made in choosing the member to which the pin will be assumed to belong. (If multiforce members are involved, the pin should be attached to one of these members.) The various forces exerted on the pin should then be clearly identified. This is illustrated in Sample Prob. 6.6.

† The use of primes (′) to distinguish the force exerted by one member on another from the equal and opposite force exerted by the second member on the first is not strictly necessary, since the two forces belong to different free-body diagrams and thus cannot easily be confused. In the Sample Problems, we shall represent by the same symbol equal and opposite forces which are applied to different free bodies.

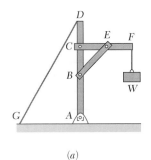

(a)

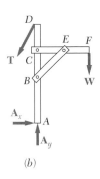

(b)

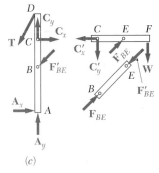

(c)

**Fig. 6.20**

## 6.11. FRAMES WHICH CEASE TO BE RIGID WHEN DETACHED FROM THEIR SUPPORTS

The crane analyzed in Sec. 6.10 was so constructed that it could keep the same shape without the help of its supports; it was therefore considered as a rigid body. Many frames, however, will collapse if detached from their supports; such frames cannot be considered as rigid bodies. Consider, for example, the frame shown in Fig. 6.21$a$, which consists of two members $AC$ and $CB$ carrying loads $\mathbf{P}$ and $\mathbf{Q}$ at their midpoints; the members are supported by pins at $A$ and $B$ and are connected by a pin at $C$. If detached from its supports, this frame will not maintain its shape; it should therefore be considered as made of *two distinct rigid parts* $AC$ and $CB$.

The equations $\Sigma F_x = 0$, $\Sigma F_y = 0$, $\Sigma M = 0$ (about any given point) express the conditions for the *equilibrium of a rigid body* (Chap. 3); we should use them, therefore, in connection with the free-body diagrams of rigid bodies, namely, the free-body diagrams of members $AC$ and $CB$ (Fig. 6.21$b$). Since these members are multiforce members, and since pins are used at the supports and at the connection, the reactions at $A$ and $B$ and the

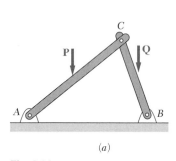

(a)

**Fig. 6.21**

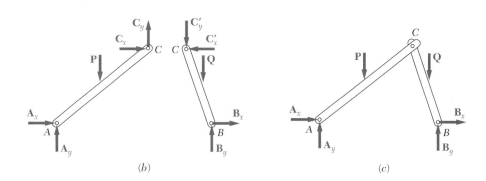

(b)                    (c)

forces at $C$ will each be represented by two components. In accordance with Newton's third law, the components $\mathbf{C}_x$ and $\mathbf{C}_y$ of the force exerted by $CB$ on $AC$ and the components $\mathbf{C}'_x$ and $\mathbf{C}'_y$ of the force exerted by $AC$ on $CB$ will be represented by vectors of the same magnitude and opposite sense. We note that four unknown force components act on free body $AC$, while only three independent equations may be used to express that the body is in equilibrium; similarly, four unknowns, but only three equations, are associated with $CB$. However, only six different unknowns are involved in the analysis of the two members, and altogether six equations are available to express that the members are in equilibrium. Writing $\Sigma M_A = 0$ for free body $AC$ and $\Sigma M_B = 0$ for $CB$, we obtain two simultaneous equations which may be solved for the common magnitude $C_x$ of the components $\mathbf{C}_x$ and $\mathbf{C}'_x$, and for the common magnitude $C_y$ of the components $\mathbf{C}_y$ and $\mathbf{C}'_y$. Writing, then, $\Sigma F_x = 0$ and $\Sigma F_y = 0$ for each of the two free bodies, we obtain successively the magnitudes $A_x$, $A_y$, $B_x$, and $B_y$.

We shall observe now that since the equations of equilibrium $\Sigma F_x = 0$, $\Sigma F_y = 0$, $\Sigma M = 0$ (about any given point) are satisfied by the forces acting on free body $AC$, and since they are also satisfied by the forces acting on free body $CB$, they must be satisfied when the forces acting on the two free bodies are considered simultaneously. Since the internal forces at $C$ cancel each other, we find that the equations of equilibrium must be satisfied by the external forces shown on the free-body diagram of the frame $ACB$ itself (Fig. 6.21$c$), although the frame is not a rigid body. These equations may be used to determine some of the components of the reactions at $A$ and $B$. We shall note, however, that *the reactions cannot be completely determined from the free-body diagram of the whole frame.* It is thus necessary to dismember the frame and to consider the free-body diagrams of its component parts (Fig.6.21$b$), even when we are interested only in finding external reactions. This may be explained by the fact that the equilibrium equations obtained for free body $ACB$ are *necessary conditions* for the equilibrium of a nonrigid structure, *but not sufficient conditions.*

The method of solution outlined in the second paragraph of this section involved simultaneous equations. We shall now discuss a more expeditious method, which utilizes the free body $ACB$ as well as the free bodies $AC$ and $CB$. Writing $\Sigma M_A = 0$ and $\Sigma M_B = 0$ for free body $ABC$, we obtain $B_y$ and $A_y$. Writing $\Sigma M_C = 0$, $\Sigma F_x = 0$ and $\Sigma F_y = 0$ for free body $AC$, we obtain successively $A_x$, $C_x$, and $C_y$. Finally, writing $\Sigma F_x = 0$ for $ACB$, we obtain $B_x$.

We noted above that the analysis of the frame of Fig. 6.21 involves six unknown force components and six independent equilibrium equations (the equilibrium equations for the whole frame were obtained from the original six equations and, therefore, are not independent). Moreover, we checked that all unknowns could be actually determined and that all equations could be satisfied. The frame considered is *statically determinate and rigid.*† In general, to determine whether a structure is statically determinate and rigid, we should draw a free-body diagram for each of its component parts and count the reactions and internal forces involved. We should also determine the number of independent equilibrium equations (excluding equations expressing the equilibrium of the whole structure or of groups of component parts already analyzed). If there are more unknowns than equations, the structure is *statically indeterminate.* If there are fewer unknowns than equations, the structure is *nonrigid.* If there are as many unknowns as equations, *and if all unknowns may be determined and all equations satisfied* under general loading conditions, the structure is *statically determinate and rigid;* if, however, due to an *improper arrangement* of members and supports, all unknowns cannot be determined and all equations cannot be satisfied, the structure is *statically indeterminate and nonrigid.*

---

† The word "rigid" is used to indicate that the frame will maintain its shape as long as it remains attached to its supports.

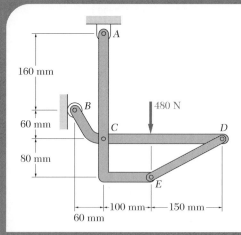

# SAMPLE PROBLEM 6.4

In the frame shown, members $ACE$ and $BCD$ are connected by a pin at $C$ and by the link $DE$. For the loading shown, determine the force in link $DE$ and the components of the force exerted at $C$ on member $BCD$.

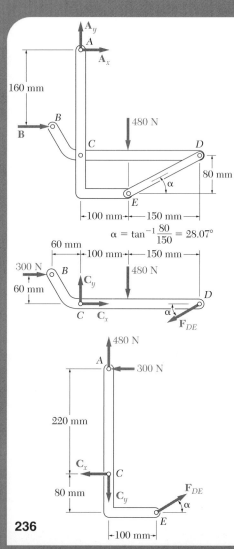

**Entire Frame.** Since the external reactions involve only three unknowns, we compute the reactions by considering the free-body diagram of the entire frame.

$+\uparrow\Sigma F_y = 0:$ $\qquad A_y - 480\ N = 0$ $\qquad A_y = +480\ N = 0$ $\qquad \mathbf{A}_y = 480\ N\uparrow$

$+\curvearrowleft\Sigma M_A = 0:$ $\qquad -(480\ N)(100\ mm) + B(160\ mm) = 0$

$\qquad\qquad\qquad\qquad\qquad B = +300\ N$ $\qquad\qquad \mathbf{B} = 300\ N \rightarrow$

$\xrightarrow{+}\Sigma F_x = 0:$ $\qquad B + A_x = 0$

$\qquad\qquad\qquad 300\ N + A_x = 0$ $\qquad A_x = -300\ N$ $\qquad \mathbf{A}_x = 300\ N \leftarrow$

**Members.** We now dismember the frame. Since only two members are connected at $C$, the components of the unknown forces acting on $ACE$ and $BCD$ are, respectively, equal and opposite and are assumed directed as shown. We assume that link $DE$ is in tension and exerts at $D$ and $E$ equal and opposite forces directed as shown.

**Member $BCD$.** Using the free body $BCD$, we write

$+\curvearrowright\Sigma M_C = 0:$ $\qquad (F_{DE} \sin \alpha)(250\ mm) + (300\ N)(60\ mm) + (480\ N)(100\ mm) = 0$

$\qquad\qquad\qquad F_{DE} = -561\ N$ $\qquad\qquad\qquad\qquad F_{DE} = 561\ N\ C$ ◄

$\xrightarrow{+}\Sigma F_x = 0:$ $\qquad C_x - F_{DE} \cos \alpha + 300\ N = 0$

$\qquad\qquad\qquad C_x - (-561\ N) \cos 28.07° + 300\ N = 0$ $\qquad C_x = -795\ N$

$+\uparrow\Sigma F_y = 0:$ $\qquad C_y - F_{DE} \sin \alpha - 480\ N = 0$

$\qquad\qquad\qquad C_y - (-561\ N) \sin 28.07° - 480\ N = 0$ $\qquad C_y = +216\ N$

From the signs obtained for $C_x$ and $C_y$ we conclude that the force components $\mathbf{C}_x$ and $\mathbf{C}_y$ exerted on member $BCD$ are directed respectively to the left and up. We have

$$\mathbf{C}_x = 795\ N \leftarrow, \mathbf{C}_y = 216\ N\uparrow \blacktriangleleft$$

**Member $ACE$ (Check).** The computations are checked by considering the free body $ACE$. For example,

$+\curvearrowleft\Sigma M_A = (F_{DE} \cos \alpha)(300\ mm) + (F_{DE} \sin \alpha)(100\ mm) - C_x(220\ mm)$

$\qquad\qquad = (-561 \cos \alpha)(300) + (-561 \sin \alpha)(100) - (-795)(220) = 0$

# SAMPLE PROBLEM 6.5

Determine the components of the forces acting on each member of the frame shown.

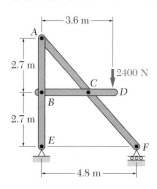

**Entire Frame.** Since the external reactions involve only three unknowns, we compute the reactions by considering the free-body diagram of the entire frame.

$+\uparrow\Sigma M_E = 0$:   $-(2400\text{ N})(3.6\text{ m}) + F(4.8\text{ m}) = 0$
$\qquad\qquad F = +1800\text{ N}$   $\mathbf{F} = 1800\text{ N}\uparrow$ ◀

$+\uparrow\Sigma F_y = 0$:   $-2400\text{ N} + 1800\text{ N} + E_y = 0$
$\qquad\qquad E_y = +600\text{ N}$   $\mathbf{E}_y = 600\text{ N}\uparrow$ ◀

$\xrightarrow{+}\Sigma F_x = 0$:   $\mathbf{E}_x = 0$ ◀

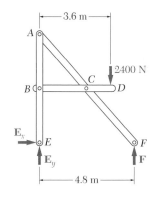

The frame is now dismembered; since only two members are connected at each joint, equal and opposite components are shown on each member at each joint.

### Member BCD

$+\uparrow\Sigma M_B = 0$:   $-(2400\text{ N})(3.6\text{ m}) + C_y(2.4\text{ m}) = 0$   $C_y = +3600\text{ N}$ ◀
$+\uparrow\Sigma M_C = 0$:   $-(2400\text{ N})(1.2\text{ m}) + B_y(2.4\text{ m}) = 0$   $B_y = +1200\text{ N}$ ◀

$\xrightarrow{+}\Sigma F_x = 0$:   $-B_x + C_x = 0$

We note that neither $B_x$ nor $C_x$ can be obtained by considering only member BCD. The positive values obtained for $B_y$ and $C_y$ indicate that the force components $\mathbf{B}_y$ and $\mathbf{C}_y$ are directed as assumed.

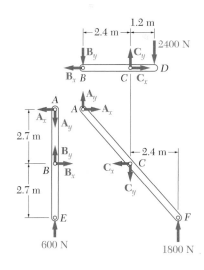

### Member ABE

$+\uparrow\Sigma M_A = 0$:   $B_x(2.7\text{ m}) = 0$   $B_x = 0$ ◀
$\xrightarrow{+}\Sigma F_x = 0$:   $+B_x - A_x = 0$   $A_x = 0$ ◀
$+\uparrow\Sigma F_y = 0$:   $-A_y + B_y + 600\text{ N} = 0$
$\qquad\qquad -A_y + 1200\text{ N} + 600\text{ N} = 0$   $A_y = +1800\text{ N}$ ◀

**Member BCD.**   Returning now to member BCD, we write

$\xrightarrow{+}\Sigma F_x = 0$:   $-B_x + C_x = 0$   $0 + C_x = 0$   $C_x = 0$ ◀

**Member ACF (Check).**   All unknown components have now been found; to check the results, we verify that member ACF is in equilibrium.

$$+\uparrow\Sigma M_C = (1800\text{ N})(2.4\text{ m}) - A_y(2.4\text{ m}) - A_x(2.7\text{ m})$$
$$= (1800\text{ N})(2.4\text{ m}) - (1800\text{ N})(2.4\text{ m}) - 0 = 0 \quad\text{(checks)}$$

**237**

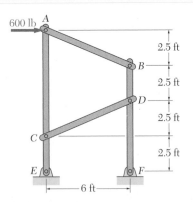

## SAMPLE PROBLEM 6.6

A 600-lb horizontal force is applied to pin $A$ of the frame shown. Determine the forces acting on the two vertical members of the frame.

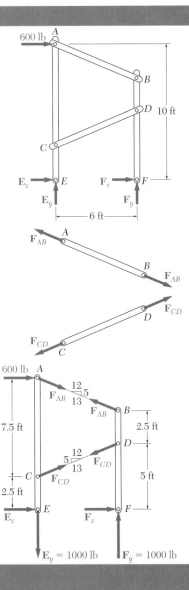

**Entire Frame.** The entire frame is chosen as a free body; although the reactions involve four unknowns, $\mathbf{E}_y$ and $\mathbf{F}_y$ may be determined by writing

$+\circlearrowleft \Sigma M_E = 0$:  $\quad -(600 \text{ lb})(10 \text{ ft}) + F_y(6 \text{ ft}) = 0$
$\qquad\qquad\qquad F_y = +1000 \text{ lb}$  $\qquad\qquad \mathbf{F}_y = 1000 \text{ lb} \uparrow$ ◄

$+\uparrow \Sigma F_y = 0$:  $\quad E_y + F_y = 0$
$\qquad\qquad\qquad E_y = -1000 \text{ lb}$  $\qquad\qquad \mathbf{E}_y = 1000 \text{ lb} \downarrow$ ◄

The equations of equilibrium of the entire frame are not sufficient to determine $\mathbf{E}_x$ and $\mathbf{F}_x$. The equilibrium of the various members must now be considered in order to proceed with the solution. In dismembering the frame we shall assume that pin $A$ is attached to the multiforce member $ACE$ and, thus, that the 600-lb force is applied to that member. We also note that $AB$ and $CD$ are two-force members.

### Member *ACE*

$+\uparrow \Sigma F_y = 0$:  $\quad -\frac{5}{13}F_{AB} + \frac{5}{13}F_{CD} - 1000 \text{ lb} = 0$
$+\circlearrowleft \Sigma M_E = 0$:  $\quad -(600 \text{ lb})(10 \text{ ft}) - (\frac{12}{13}F_{AB})(10 \text{ ft}) - (\frac{12}{13}F_{CD})(2.5 \text{ ft}) = 0$

Solving these equations simultaneously, we find

$$F_{AB} = -1040 \text{ lb} \qquad F_{CD} = +1560 \text{ lb} \quad ◄$$

The signs obtained indicate that the sense assumed for $F_{CD}$ was correct and the sense for $F_{AB}$ incorrect. Summing now $x$ components,

$\xrightarrow{+} \Sigma F_x = 0$:  $\quad 600 \text{ lb} + \frac{12}{13}(-1040 \text{ lb}) + \frac{12}{13}(+1560 \text{ lb}) + E_x = 0$
$\qquad\qquad\qquad E_x = -1080 \text{ lb}$  $\qquad\qquad \mathbf{E}_x = 1080 \text{ lb} \leftarrow$ ◄

**Entire Frame.** Since $\mathbf{E}_x$ has been determined, we may return to the free-body diagram of the entire frame and write

$\xrightarrow{+} \Sigma F_x = 0$:  $\quad 600 \text{ lb} - 1080 \text{ lb} + F_x = 0$
$\qquad\qquad\qquad F_x = +480 \text{ lb}$  $\qquad\qquad \mathbf{F}_x = 480 \text{ lb} \rightarrow$ ◄

**Member *BDF* (Check).** We may check our computations by verifying that the equation $\Sigma M_B = 0$ is satisfied by the forces acting on member *BDF*.

$\begin{aligned} +\circlearrowleft \Sigma M_B &= -(\frac{12}{13}F_{CD})(2.5 \text{ ft}) + (F_x)(7.5 \text{ ft}) \\ &= -\frac{12}{13}(1560 \text{ lb})(2.5 \text{ ft}) + (480 \text{ lb})(7.5 \text{ ft}) \\ &= -3600 \text{ lb} \cdot \text{ft} + 3600 \text{ lb} \cdot \text{ft} = 0 \qquad \text{(checks)} \end{aligned}$

# PROBLEMS

**6.49 and 6.50**   Determine the force in member $BD$ and the components of the reaction at $C$.

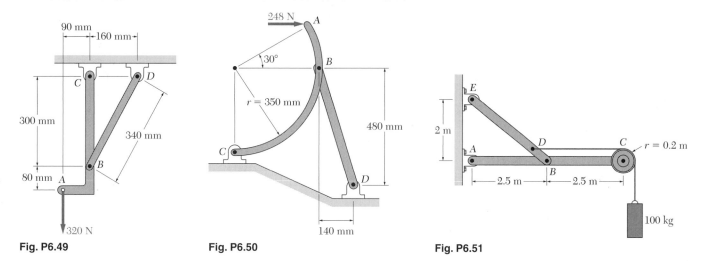

**Fig. P6.49**          **Fig. P6.50**          **Fig. P6.51**

**6.51**   Determine the components of all forces acting on member $ABC$ of the frame shown.

**6.52**   Determine the components of all forces acting on member $ABD$ of the frame shown.

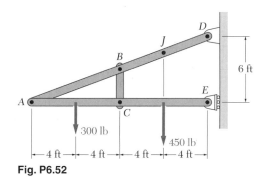

**Fig. P6.52**

**6.53**   Determine the components of all forces acting on member $ABCD$ when $\theta = 0$.

**6.54**   Determine the components of all forces acting on member $ABCD$ of the frame shown when $\theta = 90°$.

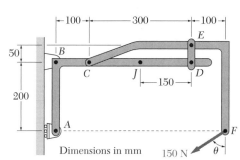

**Fig. P6.53 and P6.54**

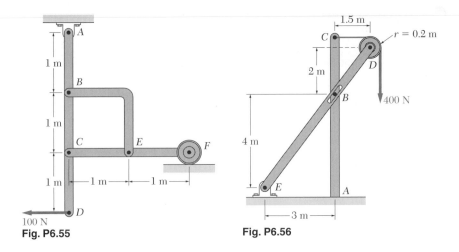

Fig. P6.55       Fig. P6.56

**6.55** Determine the components of all forces acting on member *ABCD* of the frame shown.

**6.56** Determine the support reactions at *A* for the frame shown.

**6.57** Knowing that *P* = 75 lb and *Q* = 45 lb, determine the components of all forces acting on member *BCDE* of the assembly shown.

**6.58** Knowing that *P* = 45 lb and *Q* = 75 lb, determine the components of all forces acting on member *BCDE* of the assembly shown.

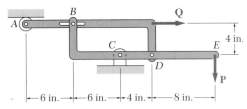

**Fig. P6.57 and P6.58**

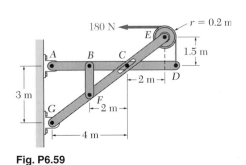

**Fig. P6.59**

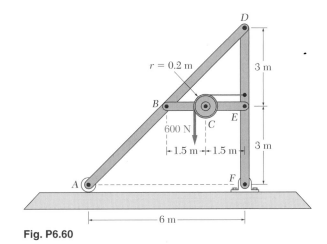

**Fig. P6.60**

**6.59** Determine all forces acting on member *ABCD* of the frame shown.

**6.60** Determine all forces acting on member *DEF* of the frame shown.

**6.61**  Determine all the forces exerted on member *AI* if the frame is loaded by a clockwise couple of moment 1200 lb·in. applied (*a*) at point *D*, (*b*) at point *E*.

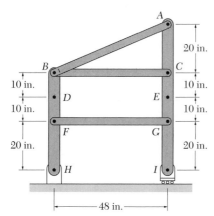

**Fig. P6.61 and P6.62**

**6.62**  Determine all the forces exerted on member *AI* if the frame is loaded by a 40-lb force directed horizontally to the right and applied (*a*) at point *D*, (*b*) at point *E*.

**6.63**  Determine the components of the reactions at *A* and *E*, (*a*) if the 800-N load is applied as shown, (*b*) if the 800-N load is moved along its line of action and is applied at point *D*.

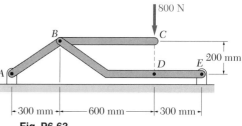

**Fig. P6.63**

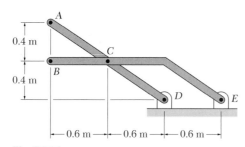

**Fig. P6.64**

**6.64**  Determine the components of the reactions at *D* and *E* if the frame is loaded by a clockwise couple of moment 150 N·m applied (*a*) at point *A*, (*b*) at point *B*.

**6.65**  A 70-kg worker stands at *C* on the step stool shown. Half the worker's weight is carried by the legs shown. Determine the components of the force exerted at *E* on leg *BE*, assuming that the bottom of the legs is not quite parallel to the floor and that bearing occurs only at points *A* and *B*. Neglect the weight of the stool, and assume the floor to be frictionless.

**6.66**  In Prob. 6.65, bearing between the legs of the step stool and the floor can occur in four ways: at *A* and *B*, *A* and *B'*, *A'* and *B*, or *A'* and *B'*. Determine (*a*) for which combination of bearing points the force in member *FG* is maximum, (*b*) the corresponding value of the force in *FG*.

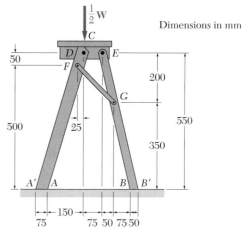

**Fig. P6.65**

**6.67** The hydraulic cylinder *CF*, which partially controls the position of rod *DE*, has been locked in the position shown. Knowing that $\theta = 70°$, determine (*a*) the force **P** for which the tension in link *AB* is 125 lb, (*b*) the corresponding force exerted on member *BCD* at point *C*.

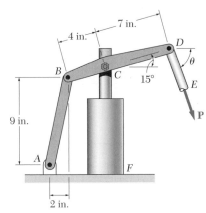

**Fig. P6.67 and P6.68**

**6.68** The hydraulic cylinder *CF*, which partially controls the position of rod *DE*, has been locked in the position shown. Knowing that $P = 90$ lb and $\theta = 80°$, determine (*a*) the force in link *AB*, (*b*) the force exerted on member *BCD* at point *C*.

**6.69** Knowing that the pulley has a radius of 0.5 m, determine the components of the reactions at *A* and *E*.

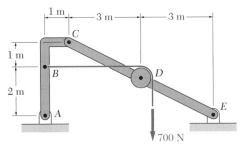

**Fig. P6.69**

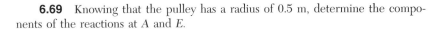

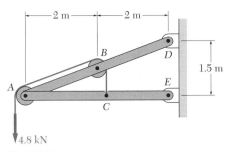

**Fig. P6.70**

**6.70** Knowing that each pulley has a radius of 250 mm, determine the components of the reactions at *D* and *E*.

**6.71** A 3-ft-diameter pipe is supported every 16 ft by a small frame; a typical frame is shown. Knowing that the combined weight of the pipe and its contents is 500 lb/ft and neglecting the effect of friction, determine the components (*a*) of the reaction at *E*, (*b*) of the force exerted at *C* on member *CDE*.

**6.72** Solve Prob. 6.71, for a frame where $h = 6$ ft.

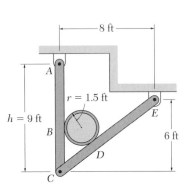

**Fig. P6.71**

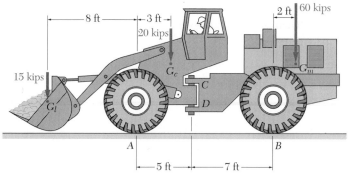

**Fig. P6.73**

**6.73**  The cab and motor units of the front-end loader shown are connected by a vertical pin located 5 ft behind the cab wheels. The distance from $C$ to $D$ is 30 in. The center of gravity of the 60-kip motor unit is located at $G_m$, while the centers of gravity of the 20-kip cab and 15-kip load are located, respectively at $G_c$ and $G_l$. Knowing that the machine is at rest with its brakes released, determine (*a*) the reactions at each of the four wheels, (*b*) the forces exerted on the motor unit at $C$ and $D$.

**6.74**  Solve Prob. 6.73, assuming that the 15-kip load has been removed.

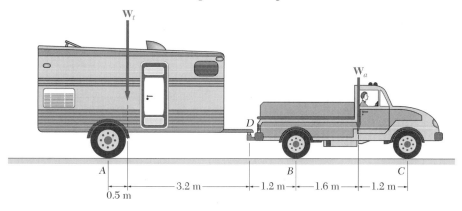

**Fig. P6.75**

**6.75**  A 1050-kg trailer is attached to a 1200-kg automobile by a ball-and-socket trailer hitch at $D$. Determine (*a*) the reactions at each of the six wheels when the automobile and trailer are at rest, (*b*) the additional load on each of the automobile wheels due to the trailer.

**6.76**  In order to obtain a better weight distribution over the four wheels of the automobile of Prob. 6.75, a compensating hitch of the type shown is used to attach the trailer to the automobile. This hitch consists of two bar springs (only one is shown in the figure) which fit into bearings inside a support rigidly attached to the automobile. The springs are also connected by chains to the trailer frame, and specially designed hooks make it possible to place both chains under tension. (*a*) Determine the tension $T$ required in each of the two chains if the additional load due to the trailer is to be evenly distributed over the four wheels of the automobile. (*b*) What are the corresponding reactions at each of the six wheels of the trailer-automobile combination?

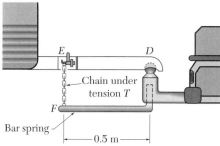

**Fig. P6.76**

**6.77 and 6.78** For the frame and loading shown, determine the components of all forces acting on member *ABD*.

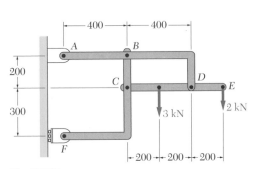

**Fig. P6.77**

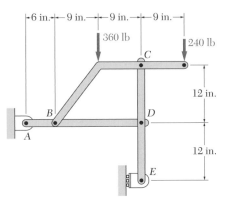

**Fig. P6.78**

**6.79** Solve Prob. 6.78, assuming that the 360-lb load has been removed.

**6.80** Solve Prob. 6.77, assuming that the 3-kN load has been removed.

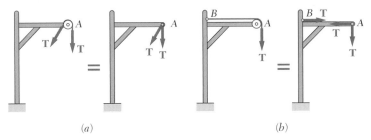

**Fig. P6.81**

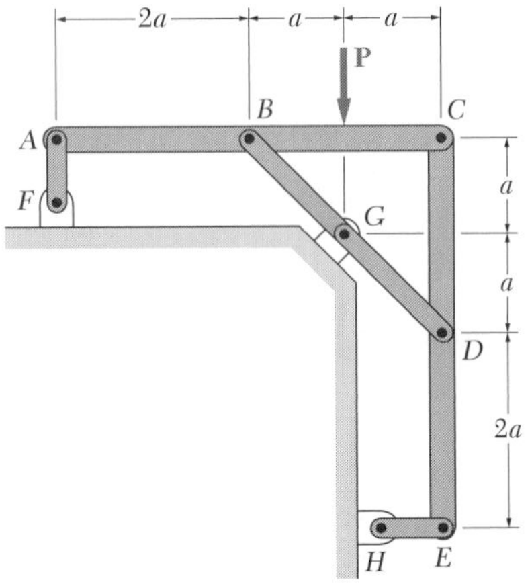

**6.81** (*a*) Show that when a frame supports a pulley at *A*, an equivalent loading of the frame and of each of its component parts may be obtained by removing the pulley and applying at *A* two forces equal and parallel to the forces of tension in the cable. (*b*) Further, show that if one end of the cable is attached to the frame at a point *B*, a force of magnitude equal to the tension should also be applied at *B*.

**6.82** Members *ABC* and *CDE* are pin-connected at *C* and supported by the four links *AF*, *BG*, *GD*, and *EH*. For the loading shown, determine the force in each link.

**6.83** Solve Prob. 6.82, assuming that the force **P** has been replaced by a clockwise couple of moment $M_o$ applied at the same point.

**Fig. P6.82**

**6.84 through 6.86** The frame shown consists of members AD and EH, which are connected by two links. Determine the force in each link for the given loading.

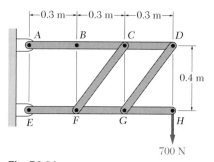

Fig. P6.84

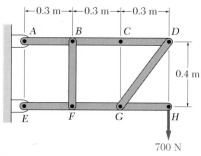

Fig. P6.85

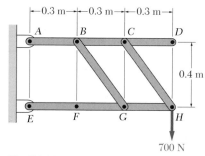

Fig. P6.86

**6.87** Knowing that the surfaces at A and D are frictionless, determine the components of all forces exerted on member ACD.

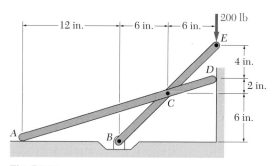

Fig. P6.87

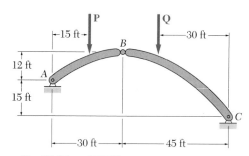

Fig. P6.89 and P6.90

**6.88** Solve Prob. 6.87, assuming that the 200-lb force applied at E is directed horizontally to the right.

**6.89** The axis of the three-hinged arch ABC is a parabola with vertex at B. Knowing that P = 90 kips and Q = 60 kips, determine (a) the components of the reaction at C, (b) the components of the force exerted at B on segment AB.

**6.90** The axis of the three-hinged arch ABC is a parabola with vertex at B. Knowing that P = 60 kips and Q = 90 kips, determine (a) the components of the reaction at C, (b) the components of the force exerted at B on segment AB.

**6.91** Three beams are nailed together at their midpoints to form the support system shown. Assuming that only vertical forces are exerted at the connections, determine the vertical reactions at A, D, and F.

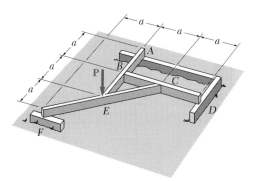

Fig. P6.91

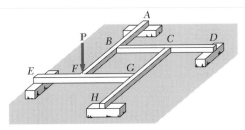

**Fig. P6.92**

**6.92**   Four beams, each of length $2a$, are nailed together at their midpoints to form the support system shown. Assuming that only vertical forces are exerted at the connections, determine the vertical reactions at $A$, $D$, $E$, and $H$.

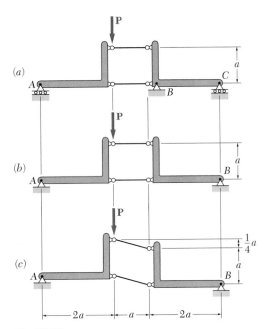

**Fig. P6.93**

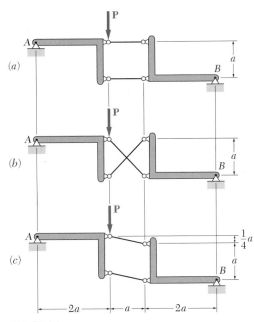

**Fig. P6.94**

**6.93 and 6.94**   Each of the frames shown consists of two L-shaped members connected by two rigid links. For each frame, determine the reactions at the supports and indicate whether the frame is rigid.

**6.95**   A vertical load **P** of magnitude 900 N is applied to member $ABC$. Members $ABC$ and $DEF$ are parallel and are placed between two frictionless walls and connected by the link $BE$. Knowing that $a = 0.8$ m, determine the reactions at points $A$, $C$, $D$, and $F$.

**6.96**   The two parallel members $ABC$ and $DEF$ are placed between two frictionless walls and connected by a link $BE$. Determine the range of values of the distance $a$ for which the load **P** can be supported.

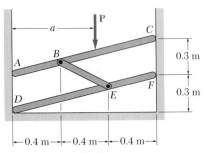

**Fig. P6.95 and P6.96**

## 6.12. MACHINES

Machines are structures designed to transmit and modify forces. Whether they are simple tools or include complicated mechanisms, their main purpose is to transform *input forces* into *output forces*. Consider, for example, a pair of cutting pliers used to cut a wire (Fig. 6.22a). If we apply two equal and opposite forces **P** and **P′** on their handles, they will exert two equal and opposite forces **Q** and **Q′** on the wire (Fig. 6.22b).

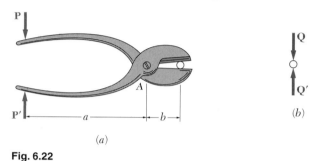

(a)

(b)

**Fig. 6.22**

To determine the magnitude $Q$ of the output forces when the magnitude $P$ of the input forces is known (or, conversely, to determine $P$ when $Q$ is known), we draw a free-body diagram of the pliers *alone*, showing the input forces **P** and **P′** and the *reactions* **Q′** and **Q** that the wire exerts on the pliers (Fig. 6.23). However, since a pair of pliers forms a nonrigid

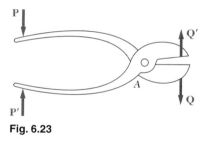

**Fig. 6.23**

structure, we must use one of the component parts as a free body in order to determine the unknown forces. Considering Fig. 6.24a, for example, and taking moments about $A$, we obtain the relation $Pa = Qb$, which defines the magnitude $Q$ in terms of $P$ or $P$ in terms of $Q$. The same free-body diagram may be used to determine the components of the internal force at $A$; we find $A_x = 0$ and $A_y = P + Q$.

In the case of more complicated machines, it generally will be necessary to use several free-body diagrams and, possibly, to solve simultaneous equations involving various internal forces. The free bodies should be chosen to include the input forces and the reactions to the output forces, and the total number of unknown force components involved should not exceed the number of available independent equations. While it is advisable to check whether the problem is determinate before attempting to solve it, there is no point in discussing the rigidity of a machine. A machine includes moving parts and thus must be nonrigid.

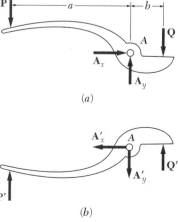

(a)

(b)

**Fig. 6.24**

# SAMPLE PROBLEM 6.7

A hydraulic-lift table is used to raise a 1000-kg crate. It consists of a platform and two identical linkages on which hydraulic cylinders exert equal forces. (Only one linkage and one cylinder are shown.) Members $EDB$ and $CG$ are each of length $2a$, and member $AD$ is pinned to the midpoint of $EDB$. If the crate is placed on the table, so that half of its weight is supported by the system shown, determine the force exerted by each cylinder in raising the crate for $\theta = 60°$, $a = 0.70$ m, and $L = 3.20$ m. Show that the result obtained is independent of the distance $d$.

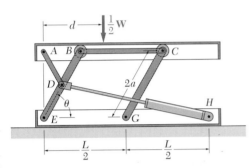

**Solution.** The machine considered consists of the platform and of the linkage, with an input force $\mathbf{F}_{DH}$ exerted by the cylinder and an output force equal and opposite to $\frac{1}{2}\mathbf{W}$. Since more than three unknowns would be involved, the entire mechanism is not used as a free body. The mechanism is dismembered and free-body diagrams and drawn for platform $ABC$, for roller $C$, and for member $EDB$. Since $AD$, $BC$, and $CG$ are two-force members, their free-body diagrams have been omitted, and the forces they exert on the other parts of the mechanism have been drawn parallel to these members.

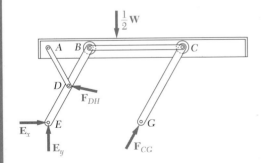

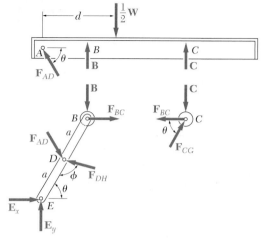

## Platform ABC

$\xrightarrow{+}\Sigma F_x = 0$:  $\quad -F_{AD}\cos\theta = 0$  $\qquad F_{AD} = 0$
$+\uparrow\Sigma F_y = 0$:  $\quad B + C - \frac{1}{2}W = 0$  $\qquad B + C = \frac{1}{2}W$  $\qquad$ (1)

**Roller C.**  We draw a force triangle and obtain $F_{BC} = C\cot\theta$.

**Member EDB.**  Recalling that $F_{AD} = 0$,

$+\uparrow\Sigma M_E = 0$:  $\quad F_{DH}\cos(\phi - 90°)a - B(2a\cos\theta) - F_{BC}(2a\sin\theta) = 0$
$\qquad\qquad\qquad F_{DH}a\sin\phi - B(2a\cos\theta) - (C\cot\theta)(2a\sin\theta) = 0$
$\qquad\qquad\qquad F_{DH}\sin\phi - 2(B + C)\cos\theta = 0$

Recalling Eq. (1), we have

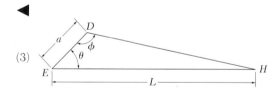

$$F_{DH} = W\frac{\cos\theta}{\sin\phi} \qquad (2)$$

and we observe that the result obtained is independent of $d$.
Applying first the law of sines to triangle $EDH$, we write

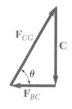

$$\frac{\sin\phi}{EH} = \frac{\sin\theta}{DH} \qquad \sin\phi = \frac{EH}{DH}\sin\theta \qquad (3)$$

Using now the law of cosines, we have

$$(DH)^2 = a^2 + L^2 - 2aL\cos\theta$$
$$= (0.70)^2 + (3.20)^2 - 2(0.70)(3.20)\cos 60°$$
$$(DH)^2 = 8.49 \qquad DH = 2.91\text{ m}$$

We also note that

$$W = mg = (1000\text{ kg})(9.81\text{ m/s}^2) = 9810\text{ N} = 9.81\text{ kN}$$

Substituting for $\sin\phi$ from (3) into (2) and using the numerical data, we write

$$F_{DH} = W\frac{DH}{EH}\cot\theta = (9.81\text{ kN})\frac{2.91\text{ m}}{3.20\text{ m}}\cot 60°$$

$$F_{DH} = 5.15\text{ kN} \qquad \blacktriangleleft$$

# PROBLEMS

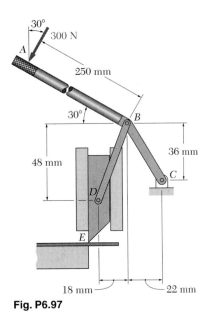

**Fig. P6.97**

**6.97**   The shear shown is used to cut and trim electronic-circuit-board laminates. For the position shown, determine (*a*) the vertical component of the force exerted on the shearing blade at *D*, (*b*) the reaction at *C*.

**6.98**   An 84-lb force is applied to the toggle vise at *C*. Knowing that $\theta = 90°$, determine (*a*) the vertical force exerted on the block at *D*, (*b*) the force exerted on member *ABC* at *B*.

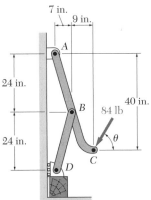

**Fig. P6.98**

**6.99**   Solve Prob. 6.98 when $\theta = 0$.

**6.100**   Solve Prob. 6.97, assuming that the 300-N force is applied vertically downward at *A*.

**6.101**   The jaw crusher shown consists of a fixed jaw and a moving jaw *AB* which is attached to a large pin at *A*. Power is delivered to shaft *C* to which a circular disk *D* is attached eccentrically. As shaft *C* rotates, the connecting rod *EF* moves toggle *GFH*. For the position shown, the magnitude of the resultant force **F** on jaw *AB* is 250 kN and the arms *GF* and *FH* of the toggle each form an angle of 6° with the horizontal. Neglecting the effect of friction, determine the force in rod *EF*. (*Note:* At the end of each crushing cycle springs, which are not shown, return the jaw and toggle to their initial positions.)

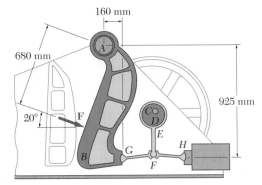

**Fig. P6.101**

**6.102**  A 9-m length of railroad track of mass 40 kg/m is lifted by the rail tongs shown. Determine the forces exerted at $D$ and $F$ on tong $BDF$.

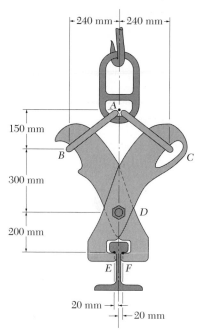

**Fig. P6.102**

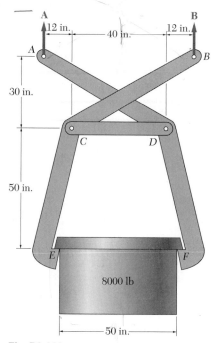

**Fig. P6.103**

**6.103**  A steel ingot weighing 8000 lb is lifted by a pair of tongs as shown. Determine the forces exerted at $C$ and $E$ on tong $BCE$.

**6.104**  If the toggle shown is added to the tongs of Prob. 6.103 and the load is lifted by applying a single force at $G$, determine the forces exerted at $C$ and $E$ on the tong $BCE$.

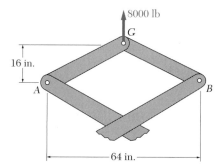

**Fig. P6.104**

**6.105**  The action of the roll clamp is controlled by the two hydraulic cylinders shown. In order to hold firmly the paper roll, a vertical 1000-lb force is applied at the top of the roll by arm $CAF$. Knowing that the weight of the paper roll is 4500 lb, determine (a) the force exerted by each cylinder, (b) the force exerted at $C$ on arm $BCEH$.

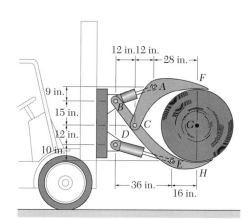

**Fig. P6.105**

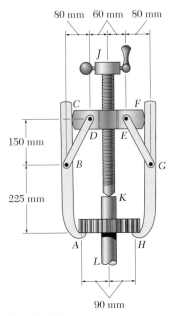

80 mm   60 mm   80 mm

150 mm

225 mm

90 mm

**Fig. P6.106**

**6.106**   The gear-pulling assembly shown consists of a crosshead *CF*, two grip arms *ABC* and *FGH*, two links *BD* and *EG*, and a threaded center rod *JK*. Knowing that the center rod *JK* must exert a 4500-N force on the vertical shaft *KL* in order to start the removal of the gear, determine all forces acting on grip arm *ABC*. Assume that the rounded ends of the crosshead are smooth and exert horizontal forces on the grip arms.

**6.107**   A force **P** of magnitude 1.8 kN is applied to the piston of the engine system shown. For each of the two positions shown, determine the couple **M** required to hold the system in equilibrium.

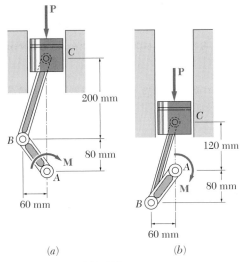

200 mm

80 mm

**M**

60 mm

120 mm

80 mm

**M**

60 mm

(*a*)                    (*b*)

**Fig. P6.107 and P6.108**

**6.108**   A couple **M** of moment 180 N·m is applied to the crank of the engine system shown. For each of the two positions shown, determine the force **P** required to hold the system in equilibrium.

**6.109 and 6.110**   Two rods are connected by a frictionless collar *B*. Knowing that the moment of the couple **M**$_A$ is 500 lb·in., determine (*a*) the couple **M**$_C$ required for equilibrium, (*b*) the corresponding components of the reaction at *C*.

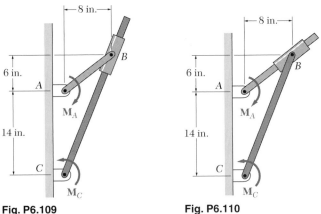

8 in.

6 in.

*B*

*A*

**M**$_A$

14 in.

*C*

**M**$_C$

8 in.

6 in.

*B*

*A*

**M**$_A$

14 in.

*C*

**M**$_C$

**Fig. P6.109**                    **Fig. P6.110**

**6.111**  A 48-mm-diameter pipe is gripped by the Stillson wrench shown. Portions $AB$ and $DE$ of the wrench are rigidly attached to each other and portion $CF$ is connected by a pin at $D$. Assuming that no slipping occurs between the pipe and the wrench, determine the components of the forces exerted on the pipe at $A$ and at $C$.

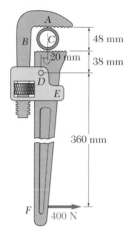

**Fig. P6.111**

**6.112**  Two 400-N forces are applied to the handles of the pliers as shown. Determine (*a*) the magnitude of the forces exerted on the rod, (*b*) the force exerted by the pin at $A$ on portion $AB$ of the pliers.

**6.113**  The tool shown is used to crimp terminals onto electric wires. Knowing that a worker will apply forces of magnitude $P = 120$ N to the handles, determine the magnitude of the crimping forces which will be exerted on the terminal.

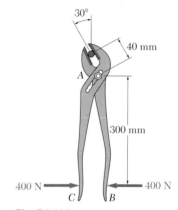

**Fig. P6.112**

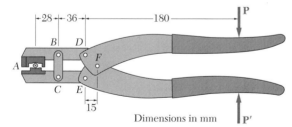

Dimensions in mm

**Fig. P6.113**

**6.114**  The pliers shown are used to attach electric connectors to flat cables which carry many separate wires. The clamping jaws remain parallel as the connectors are attached. Knowing that 300-N forces directed along line $a$-$a$ are required to complete the attachment, determine the magnitude $P$ of the forces which must be applied to the handles. Assume that pins $A$ and $D$ slide freely in slots cut in the jaws.

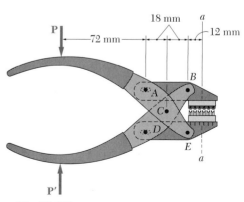

**Fig. P6.114**

**6.115** A hand-operated hydraulic cylinder has been designed for use where space is severely limited. Determine the magnitude of the force exerted on the piston at *D* when two 80-lb forces are applied as shown.

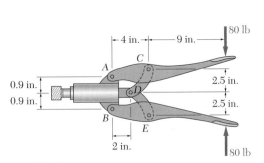

**Fig. P6.115**

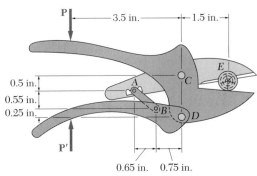

**Fig. P6.116**

**6.116** The compound-lever pruning shears shown can be adjusted by placing pin *A* at various ratchet positions on blade *ACE*. Knowing that 300-lb vertical forces are required to complete the pruning of a twig, determine the magnitude *P* of the forces which must be applied to the handles when the shears are adjusted as shown.

**6.117 and 6.118** Determine the force **P** which must be applied to the toggle *CDE* to maintain bracket *ABC* in the position shown.

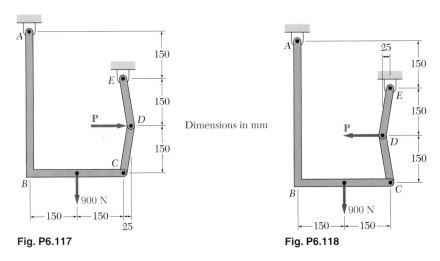

Dimensions in mm

**Fig. P6.117**

**Fig. P6.118**

**6.119** The shelf *ABC* is held horizontally by a self-locking brace which consists of two parts *BDE* and *EDF* hinged at *E* and bearing against each other at *D*. Knowing that the shelf is 16 in. wide and weighs 36 lb, determine the force **P** required to release the brace. (*Hint.* To release the brace, the forces of contact at *D* must be zero.)

**6.120** Solve Prob. 6.119, assuming that the force **P** is applied vertically downward at point *E*.

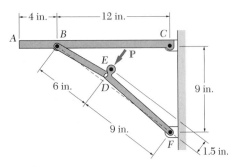

**Fig. P6.119**

**6.121**   The telescoping arm *ABC* is used to raise a worker to the elevation of overhead electric and telephone wires. For the extension shown, the center of gravity of the 1500-lb arm is located at point *G*. The worker, the bucket, and equipment attached to the bucket together weigh 500 lb and have a combined center of gravity at *C*. For the position shown, determine the force exerted at *B* by the single hydraulic cylinder used.

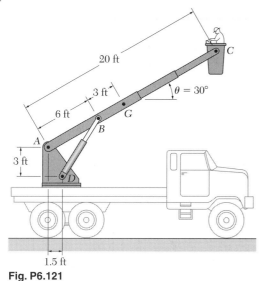

**Fig. P6.121**

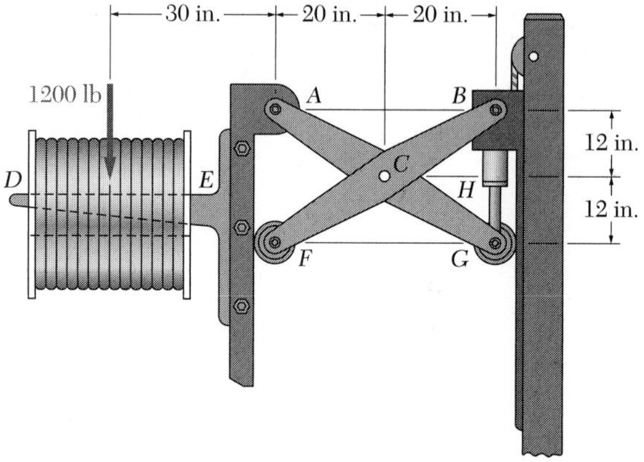

**6.122**   The position of the spindle *DE* of a lift truck is partially controlled by two identical linkage-and-hydraulic-cylinder systems, only one of which is shown. A 2400-lb spool of electric cable is held by the spindle in the position shown. Knowing that the load supported by the one system shown is 1200 lb, determine (*a*) the force exerted by the hydraulic cylinder on point *G*, (*b*) the components of the force exerted on member *BCF* at point *C*.

**Fig. P6.122**

**6.123**   A 400-kg concrete slab is supported by a chain and sling attached to the bucket of the front-end loader shown. The action of the bucket is controlled by two identical mechanisms, only one of which is shown. Knowing that the mechanism shown supports one-half of the 400-kg slab, determine the force exerted (*a*) by cylinder *CD*, (*b*) by cylinder *FH*.

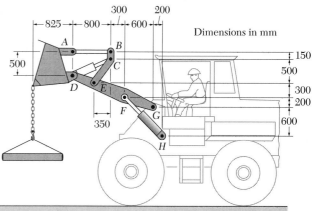

**Fig. P6.123**

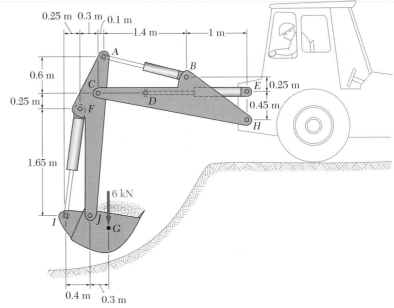

**Fig. P6.124**

**6.124**  The motion of the backhoe bucket is controlled by the hydraulic cylinders *AB*, *DE*, and *FI*. Determine the force exerted by each cylinder in supporting the 6-kN load shown.

**6.125**  The gears *D* and *G* are rigidly attached to shafts which are held by frictionless bearings. If $r_D = 4$ in. and $r_G = 2$ in., determine (*a*) the couple $\mathbf{M}_0$ which must be applied for equilibrium, (*b*) the reactions at *A* and *B*.

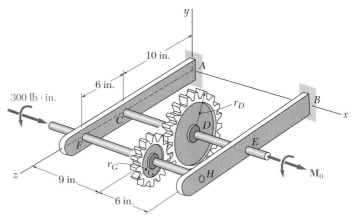

**Fig. P6.125**

**6.126**  Solve Prob. 6.125, if $r_D = 3.5$ in. and $r_G = 2.5$ in.

**6.127** In the planetary-gear system shown, the radius of the central gear $A$ is $a$, the radius of each planetary gear is $b$, and the radius of the outer gear $E$ is $(a + 2b)$. In a particular gear system where $a = b = 40$ mm, a clockwise couple $\mathbf{M}_A$ is applied to gear $A$. If the system is to be in equilibrium, determine (a) the couple $\mathbf{M}_S$ which must be applied to the spider $BCD$, (b) the couple $\mathbf{M}_E$ which must be applied to the outer gear $E$.

**6.128** In the planetary-gear system shown, the radius of the central gear $A$ is $a$, the radius of each of the planetary gears is $b$, and the radius of the outer gear $E$ is $(a + 2b)$. A clockwise couple of moment $M_A$ is applied to the central gear $A$ and a counterclockwise couple of moment $3.6M_A$ is applied to the spider $BCD$. If the system is to be in equilibrium, determine (a) the required ratio $b/a$, (b) the couple $\mathbf{M}_E$ which must be applied to the outer gear $E$.

**\* 6.129** Two shafts $AC$ and $EG$, which lie in the vertical $yz$ plane, are connected by a universal joint at $D$. The bearings at $B$ and $E$ do not exert any axial force. A couple of moment 25 N·m (clockwise when viewed from the positive $z$ axis) is applied to shaft $AC$ at $A$. At a time when the arm of the crosspiece attached to shaft $AC$ is vertical, determine (a) the moment of the couple $\mathbf{M}_G$ which must be applied to shaft $EG$ to maintain equilibrium, (b) the reactions at $B$, $C$, and $E$. (*Hint*. The sum of the couples exerted on the crosspiece must be zero.)

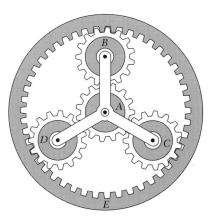

**Fig. P6.127 and P6.128**

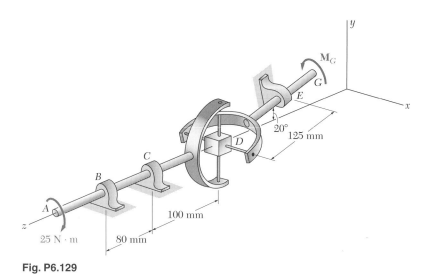

**Fig. P6.129**

**\* 6.130** Solve Prob. 6.129, assuming that the arm of the crosspiece attached to shaft $ABC$ is horizontal.

**\* 6.131** The large mechanical tongs shown are used to grab and lift a thick 7500-kg steel slab $HJ$. Knowing that slipping does not occur between the tong grips and the slab at $H$ and $J$, determine the components of all forces acting on member $EFH$. (*Hint*. Consider the symmetry of the tongs to establish relationships between the components of the force acting at $E$ on $EFH$ and the components of the force acting at $D$ on $CDF$.)

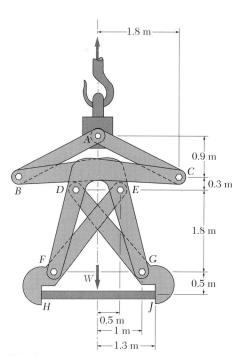

**Fig. P6.131**

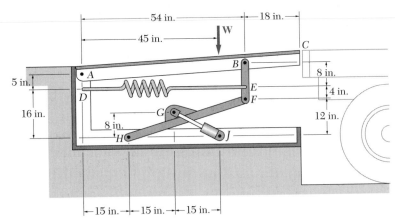

**Fig. P6.132**

**\*6.132** The mechanism shown is installed at some loading docks so that the dock edge $C$ can be brought to the same elevation as the beds of various trucks. The weight of the platform and mechanism is roughly balanced by the force exerted by spring $DE$, while the final adjustment in elevation is controlled by the hydraulic cylinder $GJ$. Knowing that the weight of all moving parts is equivalent to the single force $\mathbf{W}$ of magnitude 1250 lb, determine the required tension in spring $DE$ if the force exerted by the hydraulic cylinder is to be zero in the position shown.

**\*6.133** The weight of the platform and mechanism of Prob. 6.132 is exactly balanced by the force exerted by spring $DE$ in the position shown and the force exerted by the hydraulic cylinder $GJ$ is zero. A 180-lb force directed vertically downward is then applied at edge $C$. Determine the force which must be exerted by the hydraulic cylinder if the mechanism is to remain in the same position.

# REVIEW AND SUMMARY
# FOR CHAPTER 6

In this chapter we learned to determine the *internal forces* holding together the various parts of a structure.

**Analysis of trusses**

The first half of the chapter was devoted to the analysis of *trusses,* i.e., to the analysis of structures consisting of *straight members connected at their extremities only.* The members being slender and unable to support lateral loads, all the loads must be applied at the joints; a truss may thus be assumed to consist of *pins and two-force members* [Sec. 6.2].

**Simple trusses**

A truss is said to be *rigid* if it is designed in such a way that it will not greatly deform and collapse under a small load. A triangular truss consisting of three members connected at three joints is clearly a rigid truss (Fig. 6.25*a*), and as will be the truss obtained by adding two new members to the first one and connecting

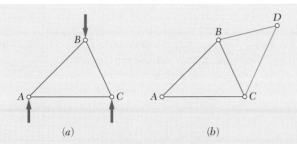

(a)                    (b)                    **Fig. 6.25**

them at a new joint (Fig. 6.25*b*). Trusses obtained by repeating this procedure are called *simple trusses*. We may check that in a simple truss the total number of members is $m = 2n - 3$, where $n$ is the total number of joints [Sec. 6.3].

The forces in the various members of a simple truss may be determined by the *method of joints* [Sec. 6.4]. First, the reactions at the supports may be obtained by considering the entire truss as a free body. The free-body diagram of each pin is then drawn, showing the forces exerted on the pin by the members or supports it connects. Since the members are straight two-force members, the force exerted by a member on the pin is directed along that member, and only the magnitude of the force is unknown. It is always possible in the case of a simple truss to draw the free-body diagrams of the pins in such an order that only two unknown forces are included in each diagram. These forces may be obtained from the corresponding two equilibrium equations or—if only three forces are involved—from the corresponding force triangle. If the force exerted by a member on a pin is directed toward that pin, the member is in *compression;* if it is directed away from the pin, the member is in *tension* [Sample Prob. 6.1]. The analysis of a truss may sometimes be expedited by recognizing *joints under special loading conditions* [Sec. 6.5]. The method of joints may also be extended to the analysis of three-dimensional or *space trusses* [Sec. 6.6].

Method of joints

The *method of sections* is usually preferred to the method of joints when the force in only one member—or very few members—of a truss is desired [Sec. 6.7]. To determine the force in member *BD* of the truss of Fig. 6.26*a*, for example, we *pass a section* through members *BD*, *BE*, and *CE*, remove these members, and use the portion *ABC* of the truss as a free body (Fig. 6.26*b*). Writing $\Sigma M_E = 0$, we determine the magnitude of the force $\mathbf{F}_{BD}$, which represents the force in member *BD*. A positive sign indicates that the member is in *tension*, a negative sign that it is in *compression* [Sample Probs. 6.2 and 6.3].

Method of sections

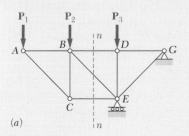

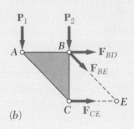

(a)                    (b)                    **Fig. 6.26**

The method of sections is particularly useful in the analysis of *compound trusses,* i.e., trusses which cannot be constructed from the basic triangular truss of Fig. 6.25*a.* These trusses may be obtained by rigidly connecting several simple trusses [Sec. 6.8]. If the component trusses have been properly connected (e.g., one pin and one link, or three nonconcurrent and nonparallel links) and if the resulting structure is properly supported (e.g., one pin and one roller), the compound truss is *statically determinate, rigid,* and *completely constrained.* The following necessary—but not sufficient—condition is then satisfied: $m + r = 2n$, where $m$ is the number of members, $r$ is the number of unknowns representing the reactions at the supports, and $n$ is the number of joints.

**Frames and machines**

The second part of the chapter was devoted to the analysis of *frames and machines.* Frames and machines are structures which contain *multiforce members,* i.e., members acted upon by three or more forces. Frames are designed to support loads and are usually stationary, fully constrained structures. Machines are designed to transmit or modify forces and always contain moving parts [Sec. 6.9].

**Analysis of a frame**

To *analyze a frame,* we first consider the *entire frame as a free body* and write three equilibrium equations [Sec. 6.10]. If the frame remains rigid when detached from its supports, the reactions involve only three unknowns and may be determined from these equations [Sample Probs. 6.4 and 6.5]. On the other hand, if the frame ceases to be rigid when detached from its supports, the reactions involve more than three unknowns and cannot be completely determined [Sec. 6.11; Sample Prob. 6.6].

**Multiforce members**

We then *dismember the frame* and identify the various members as either two-force members or multiforce members; pins are assumed to form an integral part of one of the members they connect. We draw the free-body diagram of each of the multiforce members, noting that when two multiforce members are connected to the same two-force member, they are acted upon by that member with *equal and opposite forces of unknown magnitude but known direction.* When two multiforce members are connected by a pin, they exert on each other *equal and opposite forces of unknown direction,* which should be represented by *two unknown components.* The equilibrium equations obtained from the free-body diagrams of the multiforce members may then be solved for the various internal forces [Sample Probs. 6.4 and 6.5]. They may also be used to complete the determination of the reactions at the supports [Sample Prob. 6.6]. Actually, if the frame is *statically determinate and rigid,* the free-body diagrams of the multiforce members could provide as many equations as there are unknown forces (including the reactions) [Sec. 6.11]. However, as suggested above, it is advisable to first consider the free-body diagram of the entire frame to minimize the number of equations to be solved simultaneously.

**Analysis of a machine**

To *analyze a machine,* we dismember it and, following the same procedure as for a frame, draw the free-body diagram of each of the multiforce members. The corresponding equilibrium equations yield the *output forces* exerted by the machine in terms of the *input forces* applied to it, as well as the *internal forces* at the various connections [Sec. 6.12; Sample Prob. 6.7].

# Review Problems

**6.134** The Whitworth mechanism shown is used to produce a quick-return motion of point $D$. The block at $B$ is pinned to the crank $AB$ and is free to slide in a slot cut in member $CD$. Determine the couple **M** which must be applied to the crank $AB$ to hold the mechanism in equilibrium when (a) $\alpha = 0$, (b) $\alpha = 30°$.

**6.135** Solve Prob. 6.134, when (a) $\alpha = 60°$, (b) $\alpha = 90°$.

**6.136** The elevation of the platform is controlled by two identical mechanisms, only one of which is shown. A load of 1500 lb is applied to the mechanism shown. Knowing that the pin at $C$ can transmit only a horizontal force, determine (a) the force in link $BE$, (b) the components of the force exerted by the hydraulic cylinder on pin $H$.

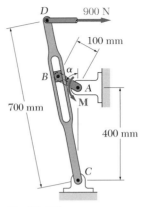

**Fig. P6.134**

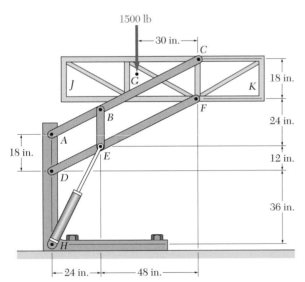

**Fig. P6.136**

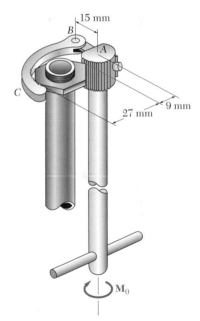

**Fig. P6.137**

**6.137** The specialized plumbing wrench shown is used in confined areas (e.g., under a basin or sink). It consists essentially of a jaw $BC$ pinned at $B$ to a long rod. Knowing that the forces exerted on the nut are equivalent to a clockwise couple (when viewed from above) of moment 13.5 N·m, determine (a) the magnitude of the force exerted by pin $B$ on jaw $BC$, (b) the moment of the couple $\mathbf{M}_0$ which is applied to the wrench.

**6.138** Determine the magnitude of the gripping forces produced when two 300-N forces are applied as shown.

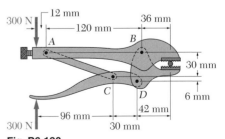

**Fig. P6.138**

**6.139** Knowing that the pulley has a radius of 75 mm, determine the components of the reactions at $A$ and $B$.

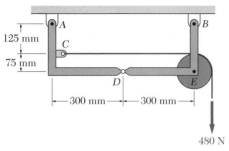

125 mm

75 mm

300 mm — 300 mm

480 N

**Fig. P6.139**

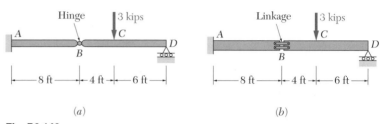

Hinge   3 kips

8 ft — 4 ft — 6 ft

(a)

Linkage   3 kips

8 ft — 4 ft — 6 ft

(b)

**Fig. P6.140**

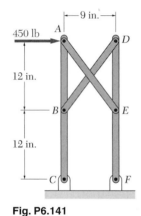

9 in.

450 lb   A

12 in.

B   E

12 in.

C   F

**Fig. P6.141**

**6.140** Determine the reactions at the supports of the beams shown.

**6.141** Determine the reaction at $F$ and the force in members $AE$ and $BD$.

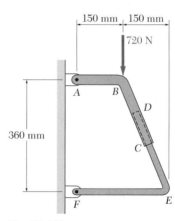

150 mm   150 mm

720 N

A   B

D

360 mm

C

F   E

**Fig. P6.142**

**6.142** The bent rod $DEF$ fits into the bent pipe $ABC$ as shown. Neglecting the effect of friction, determine the reactions at $A$ and $F$ due to the 720-N force applied at $B$.

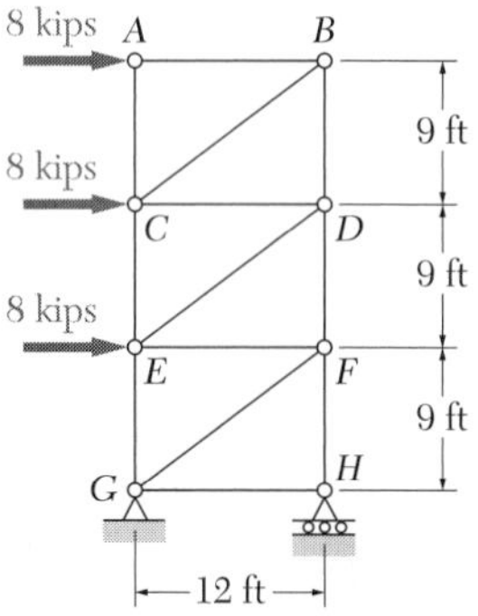

**6.143** Determine the force in members $EG$, $EF$, and $DF$ of the truss shown.

**6.144** Determine the force in members $CE$, $DE$, and $DF$ of the truss shown.

8 kips  A      B

9 ft

8 kips

C      D

9 ft

8 kips

E      F

9 ft

G      H

12 ft

**Fig. P6.143 and P6.144**

**6.145**  A 300-kg block may be supported by a small frame in each of the four ways shown. The diameter of the pulley is 250 mm. For each case, determine (*a*) the force components and the couple representing the reaction at *A*, (*b*) the force exerted at *D* on the vertical member.

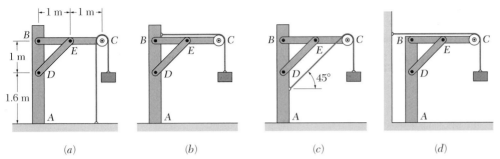

(*a*)          (*b*)          (*c*)          (*d*)

**Fig. P6.145**

# COMPUTER PROBLEMS

**6.C1**  For the truss and loading shown, write a computer program and calculate the force in each member for values of $\theta$ from 0 to 90° at 5° intervals.

**6.C2**  Several comparative designs of the truss of Sample Prob. 6.3 are to be made. Write a computer program and calculate the force in members *FH*, *GH*, and *GI* for values of *h* from 4 to 12m at 1-m intervals.

**6.C3**  Various configurations of the frame shown are being investigated. (*a*) Write a computer program and calculate the force in links *CF* and *DG* for values of *a* from 0 to 0.3 m at 0.05-m intervals. (*b*) Explain what happens when *a* = 0.15 m.

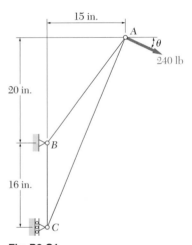

**Fig. P6.C1**

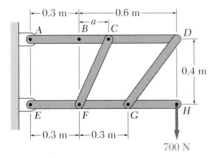

**Fig. P6.C3**

**6.C4**  Write a computer program which can be used to calculate the moment of the couple **M** required to hold the system shown in equilibrium for values of $\theta$ from 0 to 180° at 15° intervals. For the case when *b* = 100 mm, *l* = 350 mm, and *P* = 1.5 kN, calculate (*a*) the moment *M* at 15° intervals, (*b*) the value of $\theta$ for which *M* is maximum and the corresponding moment $M_{\max}$.

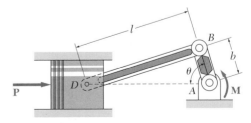

**Fig. P6.C4**

CHAPTER

# 7

# Forces in Beams and Cables

## *7.1. INTRODUCTION

In preceding chapters, two basic problems involving structures were considered: (1) the determination of the external forces acting on a structure (Chaps. 3 and 4) and (2) the determination of the forces which hold together the various members forming a structure (Chap. 6). We shall now consider the problem of determining the internal forces which hold together the various parts of a given member.

We shall first analyze the internal forces in the members of a frame, such as the crane considered earlier in Chap. 6 (Secs. 6.1 and 6.10) and note that whereas the internal forces in a straight two-force member can produce only *tension* or *compression* in that member, the internal forces in any other type of member will usually produce *shear* and *bending* as well.

Most of this chapter will be devoted to the analysis of the internal forces in two important types of engineering structures, namely,

1. *Beams*, which are usually long, straight prismatic members designed to support loads applied at various points along the member.
2. *Cables*, which are flexible members capable of withstanding only tension, designed to support either concentrated or distributed loads. Cables are used in many engineering applications, such as suspension bridges and transmission lines.

## *7.2. INTERNAL FORCES IN MEMBERS

We shall first consider a *straight two-force member AB* (Fig. 7.1a). From Sec. 3.18, we know that the forces **F** and **F'** acting at A and B, respectively, must be directed along AB in opposite sense and have the same magnitude F. Now, let us cut the member at C. To maintain the equilibrium of the free bodies AC and CB thus obtained, we must apply to AC a force **F'** equal and opposite to **F**, and to CB a force **F** equal and opposite to **F'** (Fig. 7.1b). These new forces are directed along AB in opposite sense and have the same magnitude F. Since the two parts AC and CB were in equilibrium before the member was cut, *internal forces* equivalent to these new

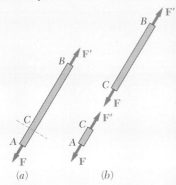

Fig. 7.1

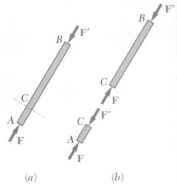

Fig. 7.2

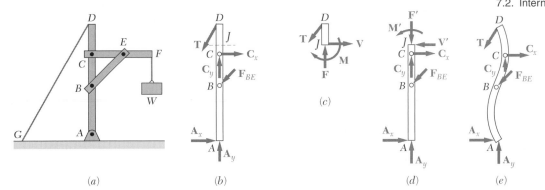

(a)        (b)        (c)        (d)        (e)

**Fig. 7.3**

forces must have existed in the member itself. We see that in the case of a straight two-force member, the internal forces acting on each part of the member are equivalent to an axial force. The magnitude $F$ of this force does not depend upon the location of the section $C$ and is referred to as the *force in member AB*. In the case considered, the member is in tension and will elongate under the action of the internal forces. In the case represented in Fig. 7.2, the member is in compression and will decrease in length under the action of the internal forces.

Next we shall consider a *multiforce member*. Take, for instance, member $AD$ of the crane analyzed in Sec. 6.10. This crane is shown again in Fig. 7.3$a$, and the free-body diagram of member $AD$ is drawn in Fig. 7.3$b$. We now cut member $AD$ at $J$ and draw a free-body diagram for each of the portions $JD$ and $AJ$ of the member (Fig. 7.3$c$ and $d$). Considering the free body $JD$, we find that its equilibrium will be maintained if we apply at $J$ a force $\mathbf{F}$ to balance the vertical component of $\mathbf{T}$, a force $\mathbf{V}$ to balance the horizontal component of $\mathbf{T}$, and a couple $\mathbf{M}$ to balance the moment of $\mathbf{T}$ about $J$. Again we conclude that internal forces must have existed at $J$ before member $AD$ was cut. The internal forces acting on the portion $JD$ of member $AD$ are equivalent to the force-couple system shown in Fig. 7.3$c$. According to Newton's third law, the internal forces acting on $AJ$ must be equivalent to an equal and opposite force-couple system, as shown in Fig. 7.3$d$. It clearly appears that the action of the internal forces in member $AD$ *is not limited to producing tension or compression* as in the case of straight two-force members; the internal forces *also produce shear and bending*. The force $\mathbf{F}$ is again in this case called an *axial force;* the force $\mathbf{V}$ is called a *shearing force;* and the moment $M$ of the couple is known as the *bending moment at J*. We note that when determining internal forces in a member, we should clearly indicate on which portion of the member the forces are supposed to act. The deformation which will occur in member $AD$ is sketched in Fig. 7.3$e$. The actual analysis of such a deformation is part of the study of mechanics of materials.

It should be noted that in a *two-force member which is not straight*, the internal forces are also equivalent to a force-couple system. This is shown in Fig. 7.4, where the two-force member $ABC$ has been cut at $D$.

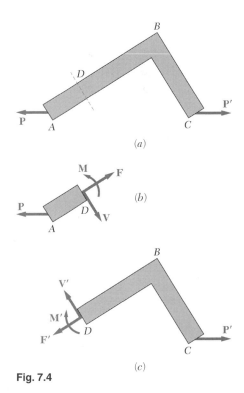

(a)

(b)

(c)

**Fig. 7.4**

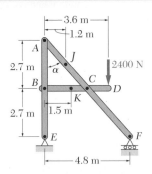

# SAMPLE PROBLEM 7.1

In the frame shown, determine the internal forces (*a*) in member *ACF* at point *J* and (*b*) in member *BCD* at point *K*. This frame has been previously considered in Sample Prob. 6.5.

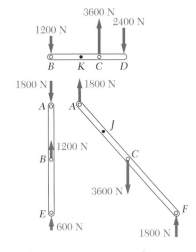

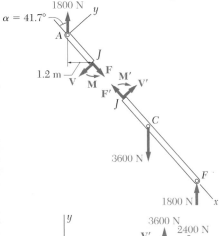

**Solution.** The reactions and the forces acting on each member of the frame are determined; this has been previously done in Sample Prob. 6.5, and the results are repeated here.

**a. Internal Forces at J.** Member *ACF* is cut at point *J*, and the two parts shown are obtained. The internal forces at *J* are represented by an equivalent force-couple system and may be determined by considering the equilibrium of either part. Considering the *free body AJ*, we write

$+\uparrow \Sigma M_J = 0:$      $-(1800 \text{ N})(1.2 \text{ m}) + M = 0$
$M = +2160 \text{ N} \cdot \text{m}$      $\mathbf{M} = 2160 \text{ N} \cdot \text{m} \uparrow$ ◀

$+\searrow \Sigma F_x = 0:$      $F - (1800 \text{ N}) \cos 41.7° = 0$
$F = +1344 \text{ N}$      $\mathbf{F} = 1344 \text{ N} \searrow$ ◀

$+\nearrow \Sigma F_y = 0:$      $-V + (1800 \text{ N}) \sin 41.7° = 0$
$V = +1197 \text{ N}$      $\mathbf{V} = 1197 \text{ N} \swarrow$ ◀

The internal forces at *J* are therefore equivalent to a couple **M**, an axial force **F**, and a shearing force **V**. The internal force-couple system acting on part *JCF* is equal and opposite.

**b. Internal Forces at K.** We cut member *BCD* at *K* and obtain the two parts shown. Considering the *free body BK*, we write

$+\uparrow \Sigma M_K = 0:$      $(1200 \text{ N})(1.5 \text{ m}) + M = 0$
$M = -1800 \text{ N} \cdot \text{m}$      $\mathbf{M} = 1800 \text{ N} \cdot \text{m} \downarrow$ ◀

$\xrightarrow{+} \Sigma F_x = 0:$      $F = 0$      $\mathbf{F} = 0$ ◀

$+\uparrow \Sigma F_y = 0:$      $-1200 \text{ N} - V = 0$
$V = -1200 \text{ N}$      $\mathbf{V} = 1200 \text{ N}\uparrow$ ◀

266

# PROBLEMS

**7.1 through 7.4** Determine the internal forces (axial force, shearing force, and bending moment) at point *J* of the structure indicated:

**7.1** Frame and loading of Prob. 6.51.
**7.2** Frame and loading of Prob. 6.52.
**7.3** Frame and loading of Prob. 6.53.
**7.4** Frame and loading of Prob. 6.54.

**7.5 and 7.6** Knowing that the turnbuckle has been tightened until the tension in wire *AD* is 850 N, determine the internal forces at the point indicated:

**7.5** Point *J*.
**7.6** Point *K*.

**7.7** It has been experimentally determined that the bending moment at point *K* of the frame shown is 300 N · m Determine (*a*) the tension in wire *AD*, (*b*) the corresponding internal forces at point *J*.

**7.8** The bracket *ABD* is supported by a pin at *A* and by the cable *DE*. Determine the internal forces just to the left of the load.

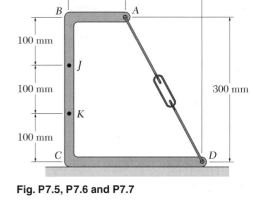

**Fig. P7.5, P7.6 and P7.7**

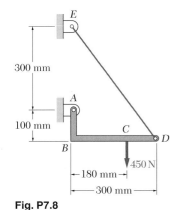

**Fig. P7.8**

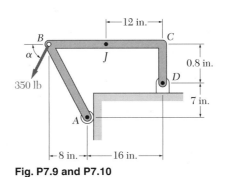

**Fig. P7.9 and P7.10**

**7.9** Determine the internal forces at point *J* when *α* = 90°.

**7.10** Determine the internal forces at point *J* when *α* = 0.

**7.11** Knowing that the radius of each pulley is 120 mm, determine the internal forces at (*a*) point *J*, (*b*) point *K*.

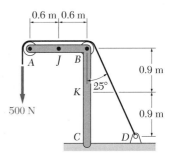

**Fig. P7.11**

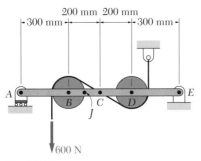

Fig. P7.12

**7.12**   Knowing that the radius of each pulley is 100 mm, determine the internal forces at (a) point $C$, (b) point $J$.

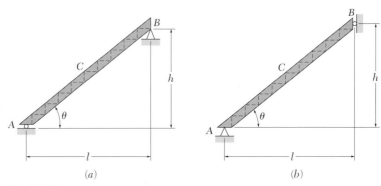

(a)                    (b)

Fig. P7.13

**7.13**   A steel channel forms one side of a flight of stairs. If one channel weighs $w$ lb/ft, determine the internal forces at the center of one channel due to its own weight when $w = 20$ lb/ft, $l = 12$ ft, and $h = 9$ ft. Consider each of the support conditions shown.

**7.14**   The axis of the curved member $AB$ is a parabola with vertex at $A$. If a vertical load **P** of magnitude 300 lb is applied at $A$, determine the internal forces at $J$ when $h = 9$ in., $L = 30$ in., and $a = 20$ in.

**7.15**   Knowing that the axis of the curved member $AB$ is a parabola with vertex at $A$, determine the magnitude and location of the maximum bending moment.

**7.16 and 7.17**   A half section of pipe rests on a frictionless horizontal surface as shown. If the half section of pipe has a mass of 12 kg and a diameter of 400 mm, determine the bending moment at point $J$ when $\theta = 90°$.

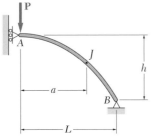

Fig. P7.14 and P7.15

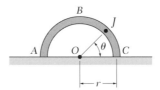

Fig. P7.16 and P7.18

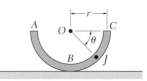

Fig. P7.17 and P7.19

**\*7.18 and 7.19**   A half section of pipe, of weight $W$ and of unit length, rests on a frictionless horizontal surface. Determine the internal force at point $J$ in terms of $W$, $r$, and $\theta$.

**\*7.20**   In Prob. 7.19, determine the magnitude and location of the maximum internal axial force.

## *7.3. VARIOUS TYPES OF LOADING AND SUPPORT

A structural member designed to support loads applied at various points along the member is known as a *beam*. In most cases, the loads are perpendicular to the axis of the beam and will cause only shear and bending in the beam. When the loads are not at a right angle to the beam, they will also produce axial forces in the beam.

Beams are usually long, straight prismatic bars. Designing a beam consists essentially in selecting the cross section which will provide the most effective resistance to the shear and bending produced by the applied loads. The design of the beam, therefore, includes two distinct parts. In the first part, the shearing forces and bending moments produced by the loads are determined. The second part is concerned with the selection of the cross section best suited to resist the shearing forces and bending moments determined in the first part. This portion of the chapter, Beams, deals with the first part of the problem of beam design, namely, the determination of the shearing forces and bending moments in beams subjected to various loading conditions and supported in various ways. The second part of the problem belongs to the study of mechanics of materials.

A beam may be subjected to *concentrated loads* $P_1$, $P_2$, . . . , expressed in newtons, pounds, or their multiples kilonewtons and kips (Fig. 7.5a), to a *distributed load* $w$, expressed in N/m, kN/m, lb/ft, or kips/ft (Fig. 7.5b), or to a combination of both. When the load $w$ per unit length has a constant value over part of the beam (as between $A$ and $B$ in Fig. 7.5b), the load is said to be *uniformly distributed* over that part of the beam. The determination of the reactions at the supports may be considerably simplified if distributed loads are replaced by equivalent concentrated loads, as explained in Sec. 5.8. This substitution, however, should not be performed, or at least should be performed with care, when internal forces are being computed (see Sample Prob. 7.3).

Beams are classified according to the way in which they are supported. Several types of beams frequently used are shown in Fig. 7.6. The distance $L$ between supports is called the *span*. It should be noted that the reactions

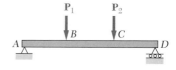

*(a)* Concentrated loads

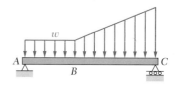

*(b)* Distributed load

**Fig. 7.5**  Types of loadings.

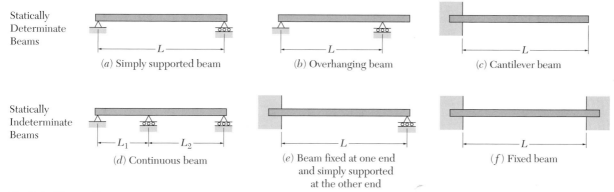

Statically
Determinate
Beams

*(a)* Simply supported beam     *(b)* Overhanging beam     *(c)* Cantilever beam

Statically
Indeterminate
Beams

*(d)* Continuous beam     *(e)* Beam fixed at one end and simply supported at the other end     *(f)* Fixed beam

**Fig. 7.6**  Types of beams.

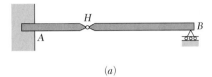

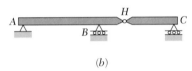

(a)

(b)

**Fig. 7.7** Combined beams.

will be determinate if the supports involve only three unknowns. The reactions will be statically indeterminate if more unknowns are involved; the methods of statics are not sufficient then to determine the reactions, and the properties of the beam with regard to its resistance to bending must be taken into consideration. Beams supported by two rollers are not shown here; such beams are only partially constrained and will move under certain loading conditions.

Sometimes two or more beams are connected by hinges to form a single continuous structure. Two examples of beams hinged at a point $H$ are shown in Fig. 7.7. It will be noted that the reactions at the supports involve four unknowns and cannot be determined from the free-body diagram of the two-beam system. They can be determined, however, by considering the free-body diagram of each beam separately; six unknowns are involved (including two force components at the hinge), and six equations are available.

## *7.4. SHEAR AND BENDING MOMENT IN A BEAM

Consider a beam $AB$ subjected to various concentrated and distributed loads (Fig. 7.8a). We propose to determine the shearing force and bending moment at any point of the beam. In the example considered here, the beam is simply supported, but the method used could be applied to any type of statically determinate beam.

First we determine the reactions at $A$ and $B$ by choosing the entire beam as a free body (Fig. 7.8b); writing $\Sigma M_A = 0$ and $\Sigma M_B = 0$, we obtain, respectively, $\mathbf{R}_B$ and $\mathbf{R}_A$.

To determine the internal forces at $C$, we cut the beam at $C$ and draw the free-body diagrams of the portions $AC$ and $CB$ of the beam (Fig. 7.8c).

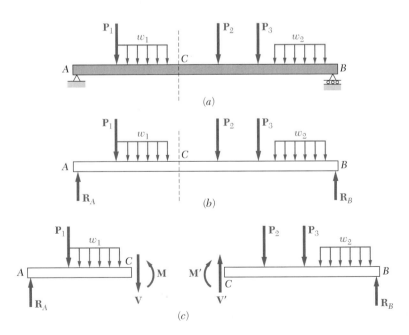

**Fig. 7.8**

Using the free-body diagram of *AC*, we may determine the shearing force **V** at *C* by equating to zero the sum of the vertical components of all forces acting on *AC*. Similarly, the bending moment *M* at *C* may be found by equating to zero the sum of the moments about *C* of all forces and couples acting on *AC*. We could have used just as well, however, the free-body diagram of *CB* and determined the shearing force **V′** and the bending moment *M′* by equating to zero the sum of the vertical components and the sum of the moments about *C* of all forces and couples acting on *CB*. While this possible choice of alternate free bodies may facilitate the computation of the numerical values of the shearing force and bending moment, it makes it necessary to indicate on which portion of the beam the internal forces considered are acting. If the shearing force and bending moment, however, are to be computed at every point of the beam and efficiently recorded, we should not have to specify every time which portion of the beam is used as a free body. We shall adopt, therefore, the following conventions:

To determine the shearing force in a beam, *we shall always assume that the internal forces* **V** *and* **V′** *are directed as shown in Fig. 7.8c.* A positive value obtained for their common magnitude *V* will indicate that this assumption was correct and that the shearing forces are actually directed as shown. A negative value obtained for *V* will indicate that the assumption was wrong and that the shearing forces are directed in the opposite way. Thus, only the magnitude *V*, together with a plus or minus sign, needs to be recorded to define completely the shearing forces at a given point of the beam. The scalar *V* is commonly referred to as the *shear* at the given point of the beam.

Similarly, *we shall always assume* that the internal couples **M** and **M′** are directed as shown in Fig. 7.8c. A positive value obtained for their magnitude *M*, commonly referred to as the bending moment, will indicate that this assumption was correct, and a negative value that it was wrong. Summarizing the sign conventions we have presented, we state:

*The shear V and the bending moment M at a given point of a beam are said to be positive when the internal forces and couples acting on each portion of the beam are directed as shown in Fig. 7.9a.*

These conventions may be more easily remembered if we note that:

1. *The shear at C is positive when the* **external** *forces (loads and reactions) acting on the beam tend to shear off the beam at C as indicate in Fig. 7.9b.*
2. *The bending moment at C is positive when the* **external** *forces acting on the beam tend to bend the beam at C as indicated in Fig. 7.9c.*

It may also help to note that the situation described in Fig. 7.9, and corresponding to positive values of the shear and of the bending moment, is precisely the situation which occurs in the left half of a simply supported beam carrying a single concentrated load at its midpoint. This particular example is fully discussed in the following section.

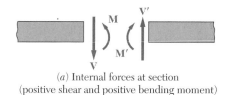

(*a*) Internal forces at section
(positive shear and positive bending moment)

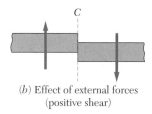

(*b*) Effect of external forces
(positive shear)

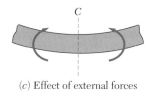

(*c*) Effect of external forces
(positive bending moment)

**Fig. 7.9**

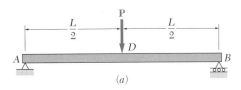

(a)

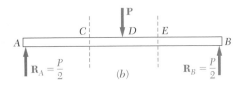

(b)

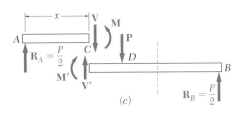

(c)

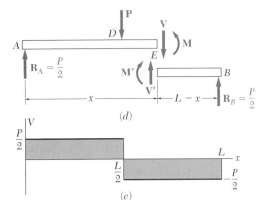

(d)

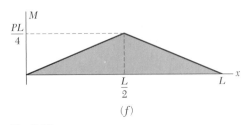

(e)

(f)

**Fig. 7.10**

## *7.5. SHEAR AND BENDING-MOMENT DIAGRAMS

Now that shear and bending moment have been clearly defined in sense as well as in magnitude, we may easily record their values at any point of a beam by plotting these values against the distance $x$ measured from one end of the beam. The graphs obtained in this way are called, respectively, the *shear diagram* and the *bending-moment diagram*. As an example, consider a simply supported beam $AB$ of span $L$ subjected to a single concentrated load **P** applied at this midpoint $D$ (Fig. 7.10a). We first determine the reactions at the supports from the free-body diagram of the entire beam (Fig. 7.10b); we find that the magnitude of each reaction is equal to $P/2$.

Next we cut the beam at a point $C$ between $A$ and $D$ and draw the free-body diagrams of $AC$ and $CB$ (Fig. 7.10c). *Assuming that shear and bending moment are positive*, we direct the internal forces **V** and **V′** and the internal couples **M** and **M′** as indicated in Fig. 7.9a. Considering the free body $AC$ and writing that the sum of the vertical components and the sum of the moments about $C$ of the forces acting on the free body are zero, we find $V = +P/2$ and $M = +Px/2$. Both shear and bending moment are therefore positive; this may be checked by observing that the reaction at $A$ tends to shear off and to bend the beam at $C$ as indicated in Fig. 7.9b and c. We may plot $V$ and $M$ between $A$ and $D$ (Fig. 7.10e and f); the shear has a constant value $V = P/2$, while the bending moment increases linearly from $M = 0$ at $x = 0$ to $M = PL/4$ at $x = L/2$.

Cutting, now, the beam at a point $E$ between $D$ and $B$ and considering the free body $EB$ (Fig. 7.10d), we write that the sum of the vertical components and the sum of the moments about $E$ of the forces acting on the free body are zero. We obtain $V = -P/2$ and $M = P(L - x)/2$. The shear is therefore negative and the bending moment positive; this may be checked by observing that the reaction at $B$ bends the beam at $E$ as indicated in Fig. 7.9c but tends to shear it off in a manner opposite to that shown in Fig. 7.9b. We can complete, now, the shear and bending-moment diagrams of Fig. 7.10e and f; the shear has a constant value $V = -P/2$ between $D$ and $B$, while the bending moment decreases linearly from $M = PL/4$ at $x = L/2$ to $M = 0$ and $x = L$.

We shall note that when a beam is subjected to concentrated loads only, the shear is of constant value between loads and the bending moment varies linearly between loads. On the other hand, when a beam is subjected to distributed loads, the shear and bending moment vary quite differently (see Sample Prob. 7.3).

# SAMPLE PROBLEM 7.2

Draw the shear and bending-moment diagram for the beam and loading shown.

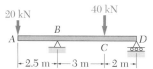

**Solution.** The reactions are determined by considering the entire beam as a free body; they are

$$\mathbf{R}_B = 46\ \text{kN}\uparrow \qquad \mathbf{R}_D = 14\ \text{kN}\uparrow$$

We first determine the internal forces just to the right of the 20-kN load at $A$. Considering the stub of beam to the left of section $1$ as a free body and assuming $V$ and $M$ to be positive (according to the standard convention), we write

$$+\uparrow\Sigma F_y = 0: \qquad -20\ \text{kN} - V_1 = 0 \qquad\qquad V_1 = -20\ \text{kN}$$

$$+\,\uparrow\Sigma M_1 = 0: \qquad (20\ \text{kN})(0\ \text{m}) + M_1 = 0 \qquad\qquad M_1 = 0$$

We next consider as a free body the portion of beam to the left of section $2$ and write

$$+\uparrow\Sigma F_y = 0: \qquad -20\ \text{kN} - V_2 = 0 \qquad\qquad V_2 = -20\ \text{kN}$$

$$+\,\uparrow\Sigma M_2 = 0: \qquad (20\ \text{kN})(2.5\ \text{m}) + M_2 = 0 \qquad\qquad M_2 = -50\ \text{kN} \cdot \text{m}$$

The shear and bending moment at sections $3$, $4$, $5$, and $6$ are determined in a similar way from the free-body diagrams shown. We obtain

$$V_3 = +26\ \text{kN} \qquad M_3 = -50\ \text{kN} \cdot \text{m}$$

$$V_4 = +26\ \text{kN} \qquad M_4 = +28\ \text{kN} \cdot \text{m}$$

$$V_5 = -14\ \text{kN} \qquad M_5 = +28\ \text{kN} \cdot \text{m}$$

$$V_6 = -14\ \text{kN} \qquad M_6 = 0$$

For several of the latter sections, the results may be more easily obtained by considering as a free body the portion of the beam to the right of the section. For example, considering the portion of the beam to the right of section $4$, we write

$$+\uparrow\Sigma F_y = 0: \qquad V_4 - 40\ \text{kN} + 14\ \text{kN} = 0 \qquad\qquad V_4 = +26\ \text{kN}$$

$$+\,\uparrow\Sigma M_4 = 0: \qquad -M_4 + (14\ \text{kN})(2\ \text{m}) = 0 \qquad\qquad M_4 = +28\ \text{kN} \cdot \text{m}$$

We may now plot the six points shown on the shear and bending-moment diagrams. As indicated in Sec. 7.5, the shear is of constant value between concentrated loads, and the bending moment varies linearly; we obtain therefore the shear and bending-moment diagrams shown.

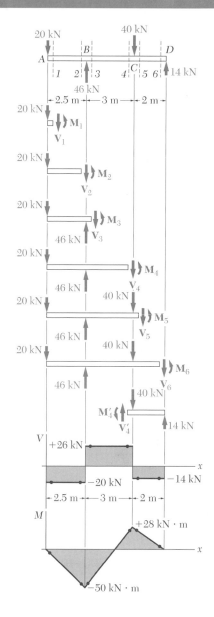

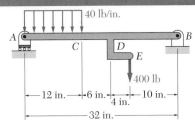

## SAMPLE PROBLEM 7.3

Draw the shear and bending-moment diagrams for the beam $AB$. The distributed load of 40 lb/in. extends over 12 in. of the beam, from $A$ to $C$, and the 400-lb load is applied at $E$.

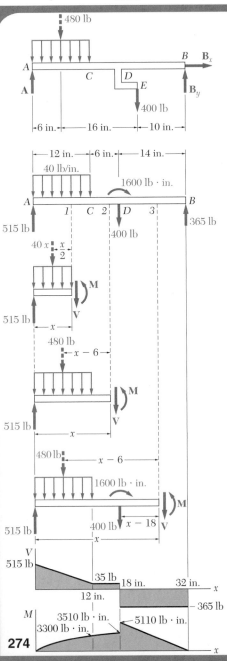

**Solution.** The reactions are determined by considering the entire beam as a free body.

$+\uparrow \Sigma M_A = 0$:    $B_y(32 \text{ in.}) - (480 \text{ lb})(6 \text{ in.}) - (400 \text{ lb})(22 \text{ in.}) = 0$
$B_y = +365 \text{ lb}$       $\mathbf{B}_y = 365 \text{ lb}\uparrow$

$+\uparrow \Sigma M_B = 0$:    $(480 \text{ lb})(26 \text{ in.}) + (400 \text{ lb})(10 \text{ in.}) - A(32 \text{ in.}) = 0$
$A = +515 \text{ lb}$       $\mathbf{A} = 515 \text{ lb}\uparrow$

$\xrightarrow{+} \Sigma F_x = 0$:    $B_x = 0$       $\mathbf{B}_x = 0$

The 400-lb load is now replaced by an equivalent force-couple system acting on the beam at point $D$.

***From A to C.*** We determine the internal forces at a distance $x$ from point $A$ by considering the portion of beam to the left of section $1$. That part of the distributed load acting on the free body is replaced by its resultant, and we write

$+\uparrow \Sigma F_y = 0$:    $515 - 40x - V = 0$       $V = 515 - 40x$

$+\uparrow \Sigma M_1 = 0$:    $-515x + 40x(\frac{1}{2}x) + M = 0$       $M = 515x - 20x^2$

Since the free-body diagram shown may be used for all values of $x$ smaller than 12 in., the expressions obtained for $V$ and $M$ and valid in the region $0 < x < 12$ in.

***From C to D.*** Considering the portion of beam to the left of section $2$ and again replacing the distributed load by its resultant, we obtain

$+\uparrow \Sigma F_y = 0$:    $515 - 480 - V = 0$       $V = 35 \text{ lb}$

$+\uparrow \Sigma M_2 = 0$:    $-515x + 480(x - 6) + M = 0$       $M = (2880 + 35x) \text{ lb} \cdot \text{in.}$

These expressions are valid in the region 12 in. $< x <$ 18 in.

***From D to B.*** Using the portion of beam to the left of section $3$, we obtain for the region 18 in. $< x <$ 32 in.

$+\uparrow \Sigma F_y = 0$:    $515 - 480 - 400 - V = 0$       $V = -365 \text{ lb}$

$+\uparrow \Sigma M_3 = 0$:    $-515x + 480(x - 6) - 1600 + 400(x - 18) + M = 0$
$M = (11,680 - 365x) \text{ lb} \cdot \text{in.}$

The shear and bending-moment diagrams for the entire beam may now be plotted. We note that the couple of moment 1600 lb · in. applied at point $D$ introduces a discontinuity into the bending-moment diagram.

**274**

# PROBLEMS

**7.21 through 7.26**  Draw the shear and bending-moment diagrams for the beams and loading shown.

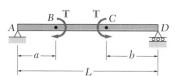

**Fig. P7.21**

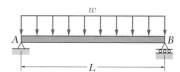

**Fig. P7.22**

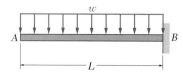

**Fig. P7.23**

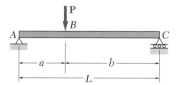

**Fig. P7.24**

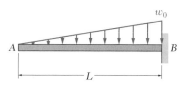

**Fig. P7.25**

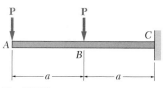

**Fig. P7.26**

**7.27 and 7.28**  Draw the shear and bending-moment diagrams for the beam *AB*.

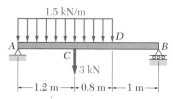

**Fig. P7.27**

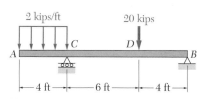

**Fig. P7.28**

**7.29**  Draw the shear and bending-moment diagrams for the beam *AB* if the magnitude of the force **P** is 75 lb.

**7.30**  Draw the shear and bending-moment diagrams for the beam *AB* if $a = 0.3$ m.

**\*7.31**  Determine the magnitude of the force **P** for which the maximum absolute value of the bending moment in the beam is as small as possible. (*Hint*. Draw the bending-moment diagram and then equate the absolute values of the largest positive and negative bending moments obtained.)

**\*7.32**  Determine the distance *a* for which the maximum absolute value of the bending moment in the beam is as small as possible. (See hint of Prob. 7.31.)

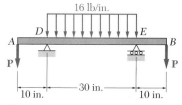

**Fig. P7.29 and P7.31**

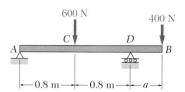

**Fig. P7.30 and P7.32**

**7.33 through 7.35** Assuming the upward reaction of the ground to be uniformly distributed, draw the shear and bending-moment diagrams for the beam AB.

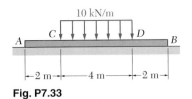

**Fig. P7.33**

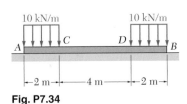

**Fig. P7.34**

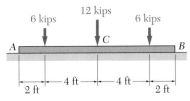

**Fig. P7.35**

**7.36** Solve Prob. 7.35, assuming that the 12-kip load has been removed.

**7.37 through 7.39** Draw the shear and bending-moment diagrams for the beam AB.

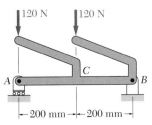

**Fig. P7.37**

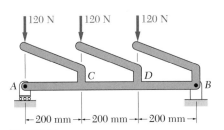

**Fig. P7.38**

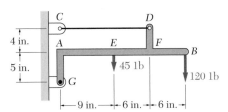

**Fig. P7.39**

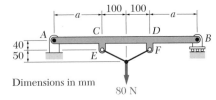

Dimensions in mm

**Fig. P7.41 and P7.42**

**7.40** Solve Prob. 7.39, assuming that the 45-lb load has been removed.

**7.41** For the beam and loading shown, (a) determine the dimension a for which the bending moment is zero from C to D, (b) draw the corresponding shear and bending-moment diagrams.

**7.42** Draw the shear and bending-moment diagrams for the beam AB if a = 200 mm.

**\*7.43** A uniform beam is to be picked up by crane cables attached at A and B. Determine the distance a from the ends of the beam to the points where the cables should be attached if the maximum absolute value of the bending moment in the beam is to be as small as possible. (*Hint.* Draw the bending-moment diagram in terms of a, L, and the weight w per unit length, and then equate the absolute values of the largest positive and negative bending moments obtained.)

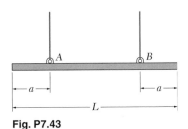

**Fig. P7.43**

**\*7.44**   In order to reduce the bending moment in the cantilever beam $AB$, a cable and counterweight are permanently attached at end $B$. Determine the magnitude of the counterweight for which the maximum absolute value of the bending moment in the beam is as small as possible. (*a*) Consider only the case when the force **P** is actually applied at $C$. (*b*) Consider the more general case when the force **P** may either be applied at $C$ or removed.

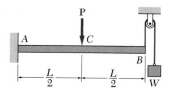

**Fig. P7.44**

## *7.6. RELATIONS AMONG LOAD, SHEAR, AND BENDING MOMENT

When a beam carries more than two or three concentrated loads, or when it carries distributed loads, the method outlined in Sec. 7.5 for plotting shear and bending moment may prove quite cumbersome. The construction of the shear diagram and, especially, of the bending-moment diagram will be greatly facilitated if certain relations existing among load, shear, and bending moment are taken into consideration.

Let us consider a simply supported beam $AB$ carrying a distributed load $w$ per unit length (Fig. 7.11*a*), and let $C$ and $C'$ be two points of the beam at a distance $\Delta x$ from each other. The shear and bending moment at $C$ will be denoted by $V$ and $M$, respectively, and will be assumed positive; the shear and bending moment at $C'$ will be denoted by $V + \Delta V$ and $M + \Delta M$.

We shall now detach the portion of beam $CC'$ and draw its free-body diagram (Fig. 7.11*b*). The forces exerted on the free body include a load of magnitude $w\,\Delta x$ and internal forces and couples at $C$ and $C'$. Since shear and bending moment have been assumed positive, the forces and couples will be directed as shown in figure.

**Relations between Load and Shear.**   We write that the sum of the vertical components of the forces acting on the free body $CC'$ is zero:

$$V - (V + \Delta V) - w\,\Delta x = 0$$
$$\Delta V = -w\,\Delta x$$

Dividing both members of the equation by $\Delta x$ and then letting $\Delta x$ approach zero, we obtain

$$\frac{dV}{dx} = -w \tag{7.1}$$

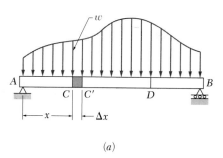

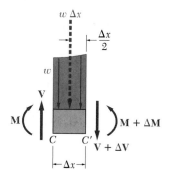

*(a)*

*(b)*

**Fig. 7.11**

Formula (7.1) indicates that for a beam loaded as shown in Fig. 7.11*a*, the slope $dV/dx$ of the shear curve is negative; the numerical value of the slope at any point is equal to the load per unit length at that point.

Integrating (7.1) between points $C$ and $D$, we obtain

$$V_D - V_C = -\int_{x_C}^{x_D} w\,dx \tag{7.2}$$

$$V_D - V_C = -(\text{area under load curve between } C \text{ and } D) \tag{7.2'}$$

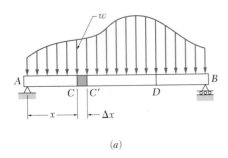

*(a)*

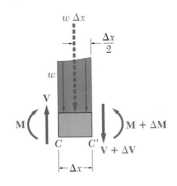

*(b)*

**Fig. 7.11 (*repeated*)**

Note that this result could also have been obtained by considering the equilibrium of the portion of beam $CD$, since the area under the load curve represents the total load applied between $C$ and $D$.

It should be observed that formula (7.1) *is not valid* at a point where a concentrated load is applied; the shear curve is discontinuous at such a point, as seen in Sec. 7.5. Similarly, Formulas (7.2) and (7.2′) cease to be valid when concentrated loads are applied between $C$ and $D$, since they do not take into account the sudden change in shear caused by a concentrated load. Formulas (7.2) and (7.2′), therefore, should be applied only between successive concentrated loads.

***Relations between Shear and Bending Moment.*** Returning to the free-body diagram of Fig. 7.11*b*, and writing now that the sum of the moments about $C'$ is zero, we obtain

$$(M + \Delta M) - M - V\Delta x + w\,\Delta x \frac{\Delta x}{2} = 0$$

$$\Delta M = V\Delta x - \tfrac{1}{2}w(\Delta x)^2$$

Dividing both members of the equation by $\Delta x$ and then letting $\Delta x$ approach zero, we obtain

$$\frac{dM}{dx} = V \tag{7.3}$$

Formula (7.3) indicates that the slope $dM/dx$ of the bending-moment curve is equal to the value of the shear. This is true at any point where the shear has a well-defined value, i.e., at any point where no concentrated load is applied. Formula (7.3) also shows that the shear is zero at points where the bending moment is maximum. The property facilitates the determination of the points where the beam is likely to fail under bending.

Integating (7.3) between points $C$ and $D$, we obtain

$$M_D - M_C = \int_{x_C}^{x_D} V\,dx \tag{7.4}$$

$$M_D - M_C = \text{area under shear curve between } C \text{ and } D \tag{7.4′}$$

Note that the area under the shear curve should be considered positive where the shear is positive and negative where the shear is negative. Formulas (7.4) and (7.4′) are valid even when concentrated loads are applied between $C$ and $D$, as long as the shear curve has been correctly drawn. The formulas cease to be valid, however, if a *couple* is applied at a point between $C$ and $D$, since they do not take into account the sudden change in bending moment caused by a couple (see Sample Prob. 7.7).

***Example.*** Let us consider a simply supported beam $AB$ of span $L$ carrying a uniformly distributed load $w$ (Fig. 7.12*a*). From the free-body diagram of the entire beam we determine the magnitude of the reactions at the supports: $R_A = R_B = wL/2$ (Fig. 7.12b). Next, we draw the shear diagram. Close to the end $A$ of the beam, the shear is equal to $R_A$, that is, to $wL/2$, as we may check by considering as a free body a very small portion of the beam. Using formula (7.2), we may then determine the shear $V$ at any distance $x$ from $A$; we write

$$V - V_A = -\int_0^x w\,dx = -wx$$

$$V = V_A - wx = \frac{wL}{2} - wx = w\left(\frac{L}{2} - x\right)$$

The shear curve is thus an oblique straight line which crosses the $x$ axis at $x = L/2$ (Fig. 7.12c). Considering, now, the bending moment, we first observe that $M_A = 0$. The value $M$ of the bending moment at any distance $x$ from $A$ may then be obtained from formula (7.4); we have

$$M - M_A = \int_0^x V\,dx$$

$$M = \int_0^x w\left(\frac{L}{2} - x\right)dx = \frac{w}{2}(Lx - x^2)$$

The bending-moment curve is a parabola. The maximum value of the bending moment occurs when $x = L/2$, since $V$ (and thus $dM/dx$) is zero for that value of $x$. Substituting $x = L/2$ in the last equation, we obtain $M_{max} = wL^2/8$.

In most engineering applications, the value of the bending moment needs to be known only at a few specific points. Once the shear diagram has been drawn, and after $M$ has been determined at one of the ends of the beam, the value of the bending moment may then be obtained at any given point by computing the area under the shear curve and using formula (7.4'). For instance, since $M_A = 0$ for the beam of Fig. 7.12, the maximum value of the bending moment for that beam may be obtained simply by measuring the area of the shaded triangle in the shear diagram:

$$M_{max} = \frac{1}{2}\frac{L}{2}\frac{wL}{2} = \frac{wL^2}{8}$$

We note that in this example, the load curve is a horizontal straight line, the shear curve an oblique straight line, and the bending-moment curve a parabola. If the load curve had been an oblique straight line (first degree), the shear curve would have been a parabola (second degree) and the bending-moment curve a cubic (third degree). The shear and bending-moment curves will always be, respectively, one and two degrees higher than the load curve. With this in mind, we should be able to sketch the shear and bending-moment diagrams without actually determining the functions $V(x)$ and $M(x)$, once a few values of the shear and bending moment have been computed. The sketches obtained will be more accurate if we make use of the fact that at any point where the curves are continuous, the slope of the shear curve is equal to $-w$ and the slope of the bending-moment curve is equal to $V$.

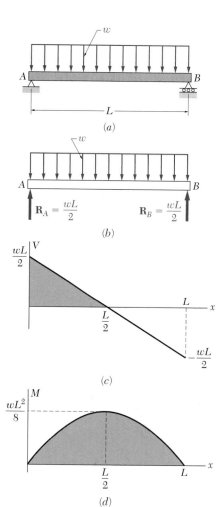

**Fig. 7.12**

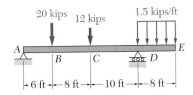

20 kips   12 kips

1.5 kips/ft

A

B   C   D

E

$\leftarrow$6 ft$\rightarrow$$\leftarrow$8 ft$\rightarrow$$\leftarrow$10 ft$\rightarrow$$\leftarrow$8 ft$\rightarrow$

# SAMPLE PROBLEM 7.4

Draw the shear and bending-moment diagrams for the beam and loading shown.

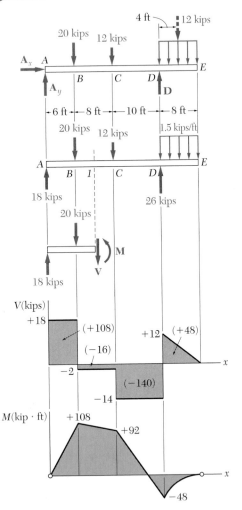

**Solution.** Considering the entire beam as a free body, we determine the reactions:

$+\Sigma\!\uparrow M_A = 0$:

$$D(24\text{ ft}) - (20\text{ kips})(6\text{ ft}) - (12\text{ kips})(14\text{ ft}) - (12\text{ kips})(28\text{ ft}) = 0$$
$$D = +26\text{ kips} \qquad\qquad \mathbf{D} = 26\text{ kips}\!\uparrow$$

$+\uparrow\Sigma F_y = 0$: $\quad A_y - 20\text{ kips} - 12\text{ kips} + 26\text{ kips} - 12\text{ kips} = 0$
$\qquad\qquad A_y = +18\text{ kips} \qquad\qquad\qquad \mathbf{A}_y = 18\text{ kips}\!\uparrow$

$\xrightarrow{+}\Sigma F_x = 0$: $\quad A_x = 0 \qquad\qquad\qquad\qquad\qquad \mathbf{A}_x = 0$

We also note that at both $A$ and $E$ the bending moment is zero; thus two points (indicated by dots) are obtained on the bending-moment diagram.

**Shear Diagram.** Since $dV/dx = -w$, we find that between concentrated loads and reactions the slope of the shear diagram is zero (i.e., the shear is constant). The shear at any point is determined by dividing the beam into two parts and considering either part as a free body. For example, using the portion of beam to the left of section *1*, we obtain the shear between $B$ and $C$:

$+\uparrow\Sigma F_y = 0$: $\quad +18\text{ kips} - 20\text{ kips} - V = 0 \qquad\qquad V = -2\text{ kips}$

We also find that the shear is $+12$ kips just to the right of $D$ and zero at end $E$. Since the slope $dV/dx = -w$ is constant between $D$ and $E$, the shear diagram between these two points is a straight line.

**Bending-Moment Diagram.** We recall that the area under the shear curve between two points is equal to the change in bending moment between the same two points. For convenience, the area of each portion of the shear diagram is computed and is indicated on the diagram. Since the bending moment $M_A$ at the left end is known to be zero, we write

$$
\begin{array}{ll}
M_B - M_A = +108 & M_B = +108\text{ kip}\cdot\text{ft} \\
M_C - M_B = -\ 16 & M_C = +\ 92\text{ kip}\cdot\text{ft} \\
M_D - M_C = -140 & M_D = -\ 48\text{ kip}\cdot\text{ft} \\
M_E - M_D = +\ 48 & M_E = 0
\end{array}
$$

Since $M_E$ is known to be zero, a check of the computations is obtained.

Between the concentrated loads and reactions the shear is constant, thus the slope $dM/dx$ is constant and the bending-moment diagram is drawn by connecting the known points with straight lines. Between $D$ and $E$ where the shear diagram is an oblique straight line, the bending-moment diagram is a parabola.

From the $V$ and $M$ diagrams we note that $V_{\text{max}} = 18$ kips and $M_{\text{max}} = 108$ kip $\cdot$ ft.

# SAMPLE PROBLEM 7.5

Draw the shear and bending-moment diagrams for the beam and loading shown.

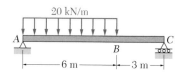

**Solution.** Considering the entire beam as a free body, we obtain the reactions

$$\mathbf{R}_A = 80 \text{ kN}\uparrow \qquad \mathbf{R}_C = 40 \text{ kN}\uparrow$$

**Shear Diagram.** The shear just to the right of $A$ is $V_A = +80$ kN. Since the change in shear between two points is equal to *minus* the area under the load curve between the same two points, we obtain $V_B$ by writing

$$V_B - V_A = -(20 \text{ kN/m})(6 \text{ m}) = -120 \text{ kN}$$
$$V_B = -120 + V_A = -120 + 80 = -40 \text{ kN}$$

The slope $dV/dx = -w$ being constant between $A$ and $B$, the shear diagram between these two points is represented by a straight line. Between $B$ and $C$, the area under the load curve is zero; therefore,

$$V_C - V_B = 0 \qquad V_C = V_B = -40 \text{ kN}$$

and the shear is constant between $B$ and $C$.

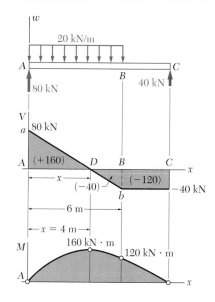

**Bending-Moment Diagram.** We note that the bending moment at each end of the beam is zero. In order to determine the maximum bending moment, we locate the section $D$ of the beam where $V = 0$. Considering the portion of the shear diagram between $A$ and $B$, we note that the triangles $DAa$ and $DBb$ are similar; thus,

$$\frac{x}{80 \text{ kN}} = \frac{6 - x}{40 \text{ kN}} \qquad x = 4 \text{ m}$$

The maximum bending moment occurs at point $D$, where we have $dM/dx = V = 0$. The areas of the various portions of the shear diagram are computed and are given (in parentheses) on the diagram. Since the area of the shear diagram between two points is equal to the change in bending moment between the same two points, we write

$$
\begin{aligned}
M_D - M_A &= +160 \text{ kN} \cdot \text{m} & M_D &= +160 \text{ kN} \cdot \text{m} \\
M_B - M_D &= -\ 40 \text{ kN} \cdot \text{m} & M_B &= +120 \text{ kN} \cdot \text{m} \\
M_C - M_B &= -120 \text{ kN} \cdot \text{m} & M_C &= 0
\end{aligned}
$$

The bending-moment diagram consists of an arc of parabola followed by a segment of straight line; the slope of the parabola at $A$ is equal to the value of $V$ at that point.

281

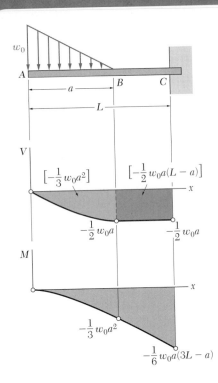

## SAMPLE PROBLEM 7.6

Sketch the shear and bending-moment diagrams for the cantilever beam shown.

**Solution.** *Shear Diagram.* At the free end of the beam, we find $V_A = 0$. Between $A$ and $B$, the area under the load curve is $\frac{1}{2}w_0a$; we find $V_B$ by writing

$$V_B - V_A = -\tfrac{1}{2}w_0a \qquad V_B = -\tfrac{1}{2}w_0a$$

Between $B$ and $C$, the beam is not loaded; thus $V_C = V_B$. At $A$, we have $w = w_0$ and, according to Eq. (7.1), the slope of the shear curve is $dV/dx = -w_0$, while at $B$ the slope is $dV/dx = 0$. Between $A$ and $B$, the loading decreases linearly, and the shear diagram is parabolic. Between $B$ and $C$, $w = 0$, and the shear diagram is a horizontal line.

*Bending-Moment Diagram.* The bending moment $M_A$ at the free end of the beam is zero. We compute the area under the shear curve and write

$$M_B - M_A = -\tfrac{1}{3}w_0a^2 \qquad M_B = -\tfrac{1}{3}w_0a^2$$
$$M_C - M_B = -\tfrac{1}{2}w_0a(L - a)$$
$$M_C = -\tfrac{1}{6}w_0a(3L - a)$$

The sketch of the bending-moment diagram is completed by recalling that $dM/dx = V$. We find that between $A$ and $B$ the diagram is represented by a cubic curve with zero slope at $A$, and between $B$ and $C$ by a straight line.

## SAMPLE PROBLEM 7.7

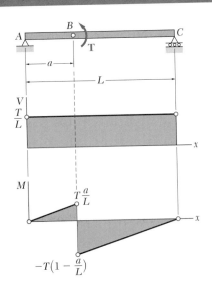

The simple beam $AC$ is loaded by a couple of moment $T$ applied at point $B$. Draw the shear and bending-moment diagrams of the beam.

**Solution.** The entire beam is taken as a free body, and we obtain

$$\mathbf{R}_A = \frac{T}{L}\!\uparrow \qquad \mathbf{R}_C = \frac{T}{L}\!\downarrow$$

The shear at any section is constant and equal to $T/L$. Since a couple is applied at $B$, the bending-moment diagram is discontinuous at $B$; the bending moment decreases suddenly by an amount equal to $T$.

# PROBLEMS

**7.45** Using the methods of Sec. 7.6, solve Prob. 7.21.
**7.46** Using the methods of Sec. 7.6, solve Prob. 7.22.
**7.47** Using the methods of Sec. 7.6, solve Prob. 7.25.
**7.48** Using the methods of Sec. 7.6, solve Prob. 7.26.
**7.49** Using the methods of Sec. 7.6, solve Prob. 7.27.
**7.50** Using the methods of Sec. 7.6, solve Prob. 7.28.
**7.51** Using the methods of Sec. 7.6, solve Prob. 7.29.
**7.52** Using the methods of Sec. 7.6, solve Prob. 7.30.
**7.53** Using the methods of Sec. 7.6, solve Prob. 7.33.
**7.54** Using the methods of Sec. 7.6, solve Prob. 7.34.
**7.55** Using the methods of Sec. 7.6, solve Prob. 7.35.
**7.56** Using the methods of Sec. 7.6, solve Prob. 7.38.

**7.57 through 7.60** Draw the shear and bending-moment diagrams for the beams and loading shown.

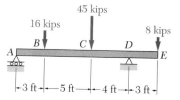

**Fig. P7.57**

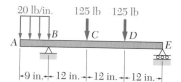

**Fig. P7.58**

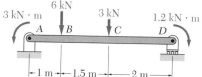

**Fig. P7.59**

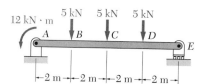

**Fig. P7.60**

**7.61 though 7.66** Draw the shear and bending-moment diagrams for the beams and loading shown, and determine the location and magnitude of the maximum bending moment.

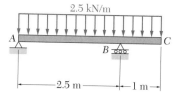

**Fig. P7.61**

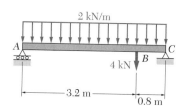

**Fig. P7.62**

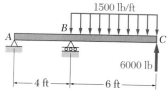

**Fig. P7.63**

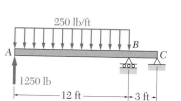

**Fig. P7.64**

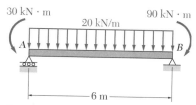

**Fig. P7.65**

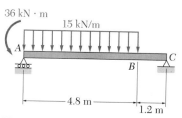

**Fig. P7.66**

**7.67** Determine the equations of the shear and bending-moment curves for the beam and loading of Prob. 7.25. (Place the origin at point $A$.)

**7.68 through 7.70** Determine the equations of the shear and bending moment curves for the given beams and loading. Also determine the magnitude and location of the maximum bending moment in the beam.

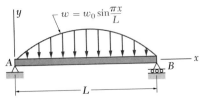

**Fig. P7.68**

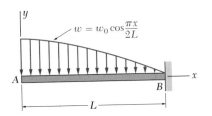

**Fig. P7.69**

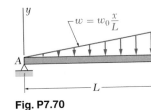

**Fig. P7.70**

**\*7.71** The beam $AB$ is acted upon by the uniformly distributed load of 1.5 kN/m and by two forces **P** and **Q**. It has been experimentally determined that the bending moment is $+360$ N $\cdot$ m at point $D$ and $+960$ N $\cdot$ m at point $E$. Determine **P** and **Q**, and draw the shear and bending-moment diagrams for the beam.

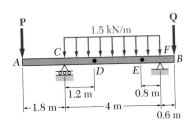

**Fig. P7.71**

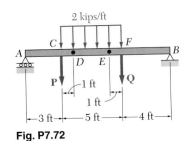

**Fig. P7.72**

**\*7.72** The beam $AB$ is acted upon by the uniformly distributed load of 2 kips/ft and by two forces **P** and **Q**. It has been experimentally determined that the bending moment is $+172$ kip $\cdot$ ft at point $D$ and $+235$ kip $\cdot$ ft at point $E$. Determine **P** and **Q**, and draw the shear and bending-moment diagrams for the beam.

**\*7.73** Solve Prob. 7.72, assuming that the uniformly distributed load of 2 kips/ft extends over the entire beam $AB$.

**\*7.74** Solve Prob. 7.71, assuming that the uniformly distributed load of 1.5 kN/m extends over the entire beam $AB$.

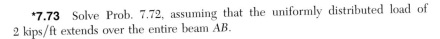
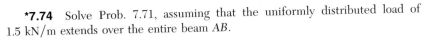

**Fig. P7.75**

**\*7.75** A uniform beam of weight $w$ per unit length is supported as shown. Determine the distance $a$ if the maximum absolute value of the bending moment is to be as small as possible. (See hint of Prob. 7.43.)

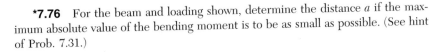

**Fig. P7.76**

**\*7.76** For the beam and loading shown, determine the distance $a$ if the maximum absolute value of the bending moment is to be as small as possible. (See hint of Prob. 7.31.)

## *7.7. CABLES WITH CONCENTRATED LOADS

Cables are used in many engineering applications, such as suspension bridges, transmission lines, aerial tramways, guy wires for high towers, etc. Cables may be divided into two categories, according to their loading: (1) cables supporting concentrated loads, (2) cables supporting distributed loads. In this section, we shall examine cables of the first category.

Consider a cable attached to two fixed points $A$ and $B$ and supporting $n$ given vertical concentrated loads $\mathbf{P}_1, \mathbf{P}_2, \ldots, \mathbf{P}_n$ (Fig. 7.13$a$). We assume that the cable is *flexible*, i.e., that its resistance to bending is small and may be neglected. We further assume that the *weight of the cable is negligible* compared with the loads supported by the cable. Any portion of cable between successive loads may therefore be considered as a two-force member, and the internal forces at any point in the cable reduce to a *force of tension directed along the cable.*

We assume that each of the loads lies in a given vertical line, i.e., that the horizontal distance from support $A$ to each of the loads is known; we also assume that the horizontal and vertical distances between the supports are known. We propose to determine the shape of the cable, i.e., the vertical distance from support $A$ to each of the points $C_1, C_2, \ldots, C_n$, and also the tension $T$ in each portion of the cable.

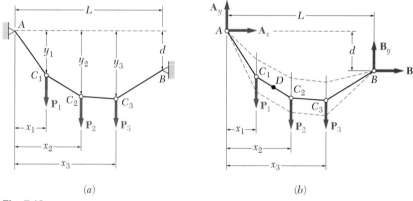

(a)                                         (b)

**Fig. 7.13**

We first draw the free-body diagram of the entire cable (Fig. 7.13$b$). Since the slope of the portions of cable attached at $A$ and $B$ is not known, the reactions at $A$ and $B$ must be represented by two components each. Thus, four unknowns are involved, and the three equations of equilibrium are not sufficient to determine the reactions at $A$ and $B$.[†] We must therefore obtain an additional equation by considering the equilibrium of a

---

[†] Clearly, the cable is not a rigid body; the equilibrium equations represent, therefore, *necessary but not sufficient conditions* (see Sec. 6.11).

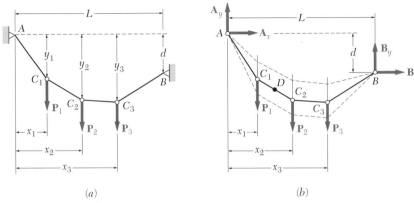

(a)                                    (b)

**Fig. 7.13 (*repeated*)**

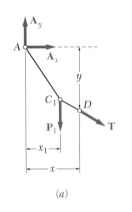

(a)

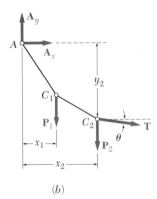

(b)

**Fig. 7.14**

portion of the cable. This is possible if we know the coordinates $x$ and $y$ of a point $D$ of the cable. Drawing the free-body diagram of the portion of cable $AD$ (Fig. 7.14$a$) and writing $\Sigma M_D = 0$, we obtain an additional relation between the scalar components $A_x$ and $A_y$ and may determine the reactions at $A$ and $B$. The problem would remain indeterminate, however, if we did not know the coordinates of $D$, unless some other relation between $A_x$ and $A_y$ (or between $B_x$ and $B_y$) were given. The cable might hang in any of various possible ways, as indicated by the dashed lines in Fig. 7.13$b$.

Once $A_x$ and $A_y$ have been determined, the vertical distance form $A$ to any point of the cable may be easily found. Considering point $C_2$, for example, we draw the free-body diagram of the portion of cable $AC_2$ (Fig. 7.14$b$). Writing $\Sigma M_{C_2} = 0$, we obtain an equation which may be solved for $y_2$. Writing $\Sigma F_x = 0$ and $\Sigma F_y = 0$, we obtain the components of the force $\mathbf{T}$ representing the tension in the portion of cable to the right of $C_2$. We observe that $T \cos \theta = -A_x$; *the horizontal component of the tension force is the same at any point of the cable.* It follows that the tension $T$ is maximum when $\cos \theta$ is minimum, i.e., in the portion of cable which has the largest angle of inclination $\theta$. Clearly, this portion of cable must be adjacent to one of the two supports of the cable.

## *7.8. CABLES WITH DISTRIBUTED LOADS

Consider a cable attached to two fixed points $A$ and $B$ and carrying a *distributed load* (Fig. 7.15$a$). We saw in the preceding section that for a cable supporting concentrated loads, the internal force at any point is a force of tension directed along the cable. In the case of a cable carrying a distributed load, the cable hangs in the shape of a curve, and the internal force at a point $D$ is a force of tension $\mathbf{T}$ *directed along the tangent to the curve.* Given a certain distributed load, we propose in this section to determine the tension at any point of the cable. We shall also see in the following sections how the shape of the cable may be determined for two particular types of distributed loads.

Considering the most general case of distributed load, we draw the free-body diagram of the portion of cable extending form the lowest point

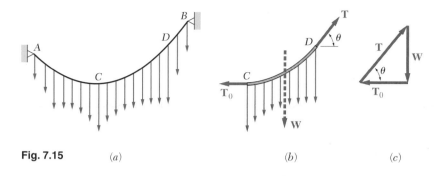

**Fig. 7.15**    (a)    (b)    (c)

$C$ to a given point $D$ of the cable (Fig. 7.15$b$). The forces acting on the free body are the tension force $\mathbf{T}_0$ at $C$, which is horizontal, the tension force $\mathbf{T}$ at $D$, directed along the tangent to the cable at $D$, and the resultant $\mathbf{W}$ of the distributed load supported by the portion of cable $CD$. Drawing the corresponding force triangle (Fig. 7.15$c$), we obtain the following relations:

$$T\cos\theta = T_0 \qquad T\sin\theta = W \tag{7.5}$$

$$T = \sqrt{T_0^2 + W^2} \qquad \tan\theta = \frac{W}{T_0} \tag{7.6}$$

From the relations (7.5), it appears that the horizontal component of the tension force $\mathbf{T}$ is the same at any point and that the vertical component of $\mathbf{T}$ is equal to the magnitude $W$ of the load measured from the lowest point. Relations (7.6) show that the tension $T$ is minimum at the lowest point and maximum at one of the two points of support.

## *7.9. PARABOLIC CABLE

Let us assume, now, that the cable $AB$ carries a load *uniformly distributed along the horizontal* (Fig. 7.16$a$). Cables of suspension bridges may be assumed loaded in this way, since the weight of the cables is small compared with the weight of the roadway. We denote by $w$ the load per unit length (*measured horizontally*) and express it in N/m or in lb/ft. Choosing coordinate axes with origin at the lowest point $C$ of the cable, we find that the magnitude $W$ of the total load carried by the portion of cable extending from $C$ to the point $D$ of coordinates $x$ and $y$ is $W = wx$. The relations (7.6) defining the magnitude and direction of the tension force at $D$ become

$$T = \sqrt{T_0^2 + w^2x^2} \qquad \tan\theta = \frac{wx}{T_0} \tag{7.7}$$

Moreover, the distance from $D$ to the line of action of the resultant $\mathbf{W}$ is equal to half the horizontal distance from $C$ to $D$ (Fig. 7.16$b$). Summing moments about $D$, we write

$$+\circlearrowleft \Sigma M_D = 0: \qquad wx\frac{x}{2} - T_0 y = 0$$

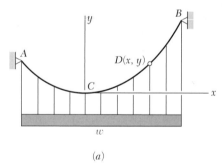

(a)

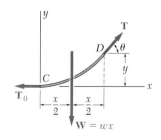

(b)

**Fig. 7.16**

and, solving for $y$,

$$y = \frac{wx^2}{2T_0} \quad (7.8)$$

This is the equation of a *parabola* with a vertical axis and its vertex at the origin of coordinates. The curve formed by cables loaded uniformly along the horizontal is thus a parabola.[†]

When the supports $A$ and $B$ of the cable have the same elevation, the distance $L$ between the supports is called the *span* of the cable and the vertical distance $h$ from the supports to the lowest point is called the *sag* of the cable (Fig. 7.17$a$). If the span and sag of a cable are known, and if the load $w$ per unit horizontal length is given, the minimum tension $T_0$ may be found by substituting $x = L/2$ and $y = h$ in formula (7.8). Formulas (7.7) and (7.8) will then define the tension at any point and the shape of the cable.

When the supports have different elevations, the position of the lowest point of the cable is not known and the coordinates $x_A$, $y_A$ and $x_B$, $y_B$ of the supports must be determined. To this effect, we express that the coordinates of $A$ and $B$ satisfy Eq. (7.8) and that $x_B - x_A = L$, $y_B - y_A = d$, where $L$ and $d$ denote, respectively, the horizontal and vertical distances between the two supports (Fig. 7.17$b$ and $c$).

The length of the cable from its lowest point $C$ to its support $B$ may be obtained from the formula

$$s_B = \int_0^{x_B} \sqrt{1 + \left(\frac{dy}{dx}\right)^2}\, dx \quad (7.9)$$

Differentiating (7.8), we obtain the derivative $dy/dx = wx/T_0$; substituting into (7.9) and using the binomial theorem to expand the radical in an infinite series, we have

$$s_B = \int_0^{x_B} \sqrt{1 + \frac{w^2 x^2}{T_0^2}}\, dx = \int_0^{x_B}\left(1 + \frac{w^2 x^2}{2T_0^2} - \frac{w^4 x^4}{8T_0^4} + \cdots\right) dx$$

$$s_B = x_B\left(1 + \frac{w^2 x_B^2}{6T_0^2} - \frac{w^4 x_B^4}{40T_0^4} + \cdots\right)$$

and, since $wx_B^2/2T_0 = y_B$,

$$s_B = x_B\left[1 + \frac{2}{3}\left(\frac{y_B}{x_B}\right)^2 - \frac{2}{5}\left(\frac{y_B}{x_B}\right)^4 + \cdots\right] \quad (7.10)$$

The series converges for values of the ratio $y_B/x_B$ less than 0.5; in most cases, this ratio is much smaller, and only the first two terms of the series need be computed.

[†] Cables hanging under their own weight are not loaded uniformly along the horizontal, and they do not form a parabola. The error introduced by assuming a parabolic shape for cables hanging under their own weight, however, is small when the cable is sufficiently taut. A complete discussion of cables hanging under their own weight is given in the next section.

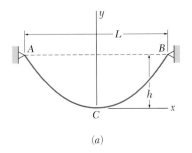

(a)

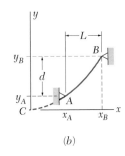

(b)

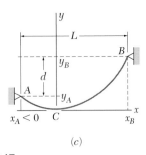

(c)

**Fig. 7.17**

# SAMPLE PROBLEM 7.8

The cable $AE$ supports three vertical loads from the points indicated. If point $C$ is 5 ft below the left support, determine (a) the elevations of points $B$ and $D$, (b) the maximum slope and the maximum tension in the cable.

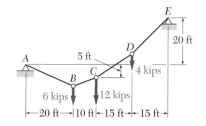

**Solution.** The reaction components $\mathbf{A}_x$ and $\mathbf{A}_y$ are determined as follows:

### Free Body: Entire Cable

$$+\uparrow\Sigma M_E = 0:$$
$$A_x(20\text{ ft}) - A_y(60\text{ ft}) + (6\text{ kips})(40\text{ ft}) + (12\text{ kips})(30\text{ ft}) + (4\text{ kips})(15\text{ ft}) = 0$$
$$20A_x - 60A_y + 660 = 0$$

### Free Body: ABC

$$+\uparrow\Sigma M_C = 0: \qquad -A_x(5\text{ ft}) - A_y(30\text{ ft}) + (6\text{ kips})(10\text{ ft}) = 0$$
$$-5A_x - 30A_y + 60 = 0$$

Solving the two equations simultaneously, we obtain

$$A_x = -18\text{ kips} \qquad \mathbf{A}_x = 18\text{ kips} \leftarrow$$
$$A_y = +5\text{ kips} \qquad \mathbf{A}_y = 5\text{ kips}\uparrow$$

**a. Elevation of Point B.** Considering the portion of cable $AB$ as a free body, we write

$$+\uparrow\Sigma M_B = 0: \qquad (18\text{ kips})y_B - (5\text{ kips})(20\text{ ft}) = 0$$
$$y_B = 5.56\text{ ft below } A \quad \blacktriangleleft$$

**Elevation of Point D.** Using the portion of cable $ABCD$ as a free body, we write

$$+\uparrow\Sigma M_D = 0:$$
$$-(18\text{ kips})y_D - (5\text{ kips})(45\text{ ft}) + (6\text{ kips})(25\text{ ft}) + (12\text{ kips})(15\text{ ft}) = 0$$
$$y_D = 5.83\text{ ft above } A \quad \blacktriangleleft$$

**b. Maximum Slope and Maximum Tension.** We observe that the maximum slope occurs in portion $DE$. Since the horizontal component of the tension is constant and equal to 18 kips, we write

$$\tan\theta = \frac{14.17}{15\text{ ft}} \qquad \theta = 43.4° \quad \blacktriangleleft$$

$$T_{max} = \frac{18\text{ kips}}{\cos\theta} \qquad T_{max} = 24.8\text{ kips} \quad \blacktriangleleft$$

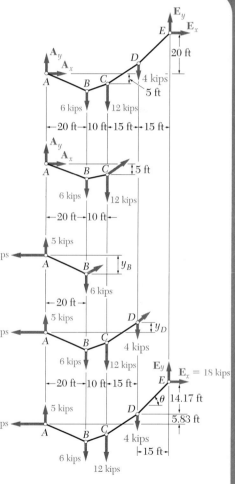

**289**

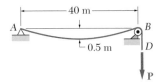

# SAMPLE PROBLEM 7.9

A light cable is attached to a support at $A$, passes over a small pulley at $B$, and supports a load $\mathbf{P}$. Knowing that the sag of the cable is 0.5 m and that the mass per unit length of the cable is 0.75 kg/m, determine (a) the magnitude of the load $\mathbf{P}$, (b) the slope of the cable at $B$, and (c) the total length of the cable from $A$ to $B$. Since the ratio of the sag to the span is small, assume the cable to be parabolic. Also, neglect the weight of the portion of cable from $B$ to $D$.

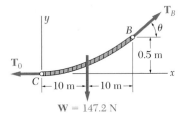

**a. Load P.** We denote by $C$ the lowest point of the cable and draw the free-body diagram of the portion $CB$ of cable. Assuming the load to be uniformly distributed along the horizontal, we write

$$w = (0.75 \text{ kg/m})(9.81 \text{ m/s}^2) = 7.36 \text{ N/m}$$

The total load for the portion $CB$ of the cable is

$$W = wx_B = (7.36 \text{ N/m})(20 \text{ m}) = 147.2 \text{ N}$$

and is applied halfway between $C$ and $B$. Summing moments about $B$, we write

$$+ \curvearrowleft \Sigma M_B = 0: \qquad (147.2 \text{ N})(10 \text{ m}) - T_0(0.5 \text{ m}) = 0 \qquad T_0 = 2944 \text{ N}$$

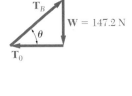

From the force triangle we obtain

$$T_B = \sqrt{T_0^2 + W^2}$$
$$= \sqrt{(2944 \text{ N})^2 + (147.2 \text{ N})^2} = 2948 \text{ N}$$

Since the tension on each side of the pulley is the same, we find

$$P = T_B = 2948 \text{ N} \blacktriangleleft$$

**b. Slope of Cable at B.** We also obtain from the force triangle

$$\tan \theta = \frac{W}{T_0} = \frac{147.2 \text{ N}}{2944 \text{ N}} = 0.05$$

$$\theta = 2.9° \blacktriangleleft$$

**c. Length of Cable.** Applying Eq. (7.10) between $C$ and $B$, we write

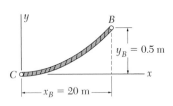

$$s_B = x_B \left[ 1 + \frac{2}{3}\left(\frac{y_B}{x_B}\right)^2 + \cdots \right]$$
$$= (20 \text{ m}) \left[ 1 + \frac{2}{3}\left(\frac{0.5 \text{ m}}{20 \text{ m}}\right)^2 + \cdots \right] = 20.00833 \text{ m}$$

The total length of the cable between $A$ and $B$ is twice this value,

$$\text{Length} = 2s_B = 40.0167 \text{ m} \blacktriangleleft$$

# PROBLEMS

**7.77** Three loads are suspended as shown from the cable. Knowing that $h_C = 3$ m, determine (a) the components of the reaction at $E$, (b) the maximum value of the tension in the cable.

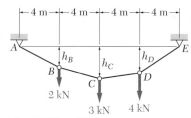

**Fig. P7.77 and P7.78**

**7.78** Determine the sag at point $C$ if the maximum tension in the cable is 13 kN.

**7.79** If $a = 16$ ft and $b = 18$ ft, determine the components of the reaction at $E$ for the cable and loading shown.

**7.80** If $a = b = 16$ ft, determine (a) the components of the reaction at $E$, (b) the maximum tension in the cable.

**7.81** If $h_C = 7$ m, determine (a) the components of the reaction at $E$, (b) the maximum tension in the cable.

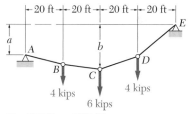

**Fig. P7.79 and P7.80**

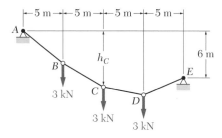

**Fig. P7.81 and P7.82**

**7.82** Determine (a) the distance $h_C$ for which the portion $CD$ of the cable is horizontal, (b) the corresponding components of the reaction at $E$.

**7.83** Cable $ABC$ supports two loads as shown. Knowing that the distance $b$ is to be 7 m, determine (a) the required magnitude of the force **P**, (b) the corresponding distance $a$.

**7.84** Cable $ABC$ supports two loads as shown. Determine the distances $a$ and $b$ when a horizontal force **P** of magnitude 1000 N is applied at $C$.

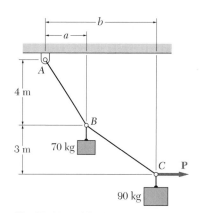

**Fig. P7.83 and P7.84**

**7.85**  A 400-lb load is attached to a small pulley which may roll on the cable *ACB*. The pulley and load are held in the position shown by a second cable *DE* which is parallel to the portion *CB* of the main cable. Determine (*a*) the reactions at *A* and *B*, (*b*) the tension in cable *ACB*, (*c*) the tension in cable *DE*. Neglect the radius of the pulleys and the weight of the cables.

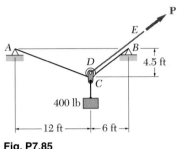

**Fig. P7.85**

**7.86**  Knowing that the 440-lb force applied at *C* and the block attached at *B* maintain the cable *ABCD* in the position shown, determine (*a*) the reaction at *D*, (*b*) the required weight *W* of the block, (*c*) the tension in each portion of the cable.

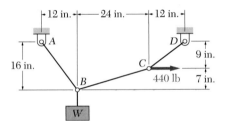

**Fig. P7.86**

**7.87**  An electric wire having a mass per unit length of 0.4 kg/m is strung between two insulators at the same elevation and 40 m apart. Knowing that the sag of the wire is 1.2 m, determine (*a*) the maximum tension in the wire, (*b*) the length of the wire.

**7.88**  The center span of the Verrazano-Narrows Bridge consists of two uniform roadways suspended from four cables. The uniform load supported by each cable is $w$ = 10.8 kips/ft along the horizontal. Knowing that the span $L$ is 4260 ft and the sag $h$ is 390 ft, determine (*a*) the maximum tension in each cable, (*b*) the length of each cable.

**7.89**  The design of the Verrazano-Narrows Bridge included the effect of extreme temperature changes which cause the sag of the center span to vary from $h_w$ = 386 ft in winter to $h_s$ = 394 ft in summer. Knowing that the span is $L$ = 4260 ft, determine the change in length of the cables due to extreme temperature variation.

**7.90**  A 61-m length of wire having a mass per unit length of 1.8 kg/m is used to span a horizontal distance of 60 m. Determine (*a*) the approximate sag of the wire, (*b*) the maximum tension in the wire. [*Hint.* Use only the first two terms of Eq. (7.10).]

**7.91**    The total mass of cable *ACB* is 10 kg. Assuming that the mass of the cable is distributed uniformly along the horizontal, determine (*a*) the sag *h*, (*b*) the slope of the cable at *A*.

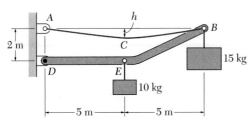

**Fig. P7.91**

**7.92**    Solve Prob. 7.91, assuming that the 10-kg load attached at *E* is removed.

**7.93**    Cable *AB* supports a load distributed uniformly along the horizontal as shown. The lowest point of the cable is located at a distance *a* = 12 ft below the support *A*. Determine the maximum and minimum values of the tension in the cable.

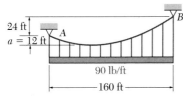

**Fig. P7.93**

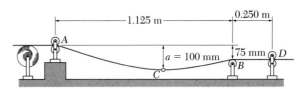

**Fig. P7.94**

**7.94**    Before being fed into a printing press located to the right of *D*, a continuous sheet of paper having a mass per unit length of 300 g/m passes over rollers at *A* and *B*. Assuming that the curve formed by the sheet is parabolic, determine (*a*) the location of the lowest point *C*, (*b*) the maximum tension in the sheet.

**7.95**    Solve Prob. 7.94, assuming that *a* = 90 mm.

**7.96**    Solve Prob. 7.93, assuming that *a* = 9 ft.

**\*7.97**    A cable *AB* of span *L* and a simple beam *A'B'* of the same span are subjected to identical vertical loadings as shown. Show that the magnitude of the bending moment at a point *C'* in the beam is equal to the product of $T_0 h$, where $T_0$ is the magnitude of the horizontal component of the tension force in the cable and *h* is the vertical distance between point *C* of the cable and the chord joining the points of support *A* and *B*.

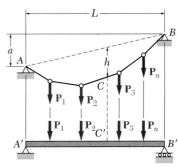

**Fig. P7.97**

**\*7.98**    Show that the curve assumed by a cable which carries a distributed load *w*(*x*) is defined by the differential equation $d^2y/dx^2 = w(x)/T_0$, where $T_0$ is the tension at the lowest point.

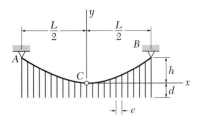

**Fig. P7.100**

**\*7.99** Using the property indicated in Prob. 7.98, determine the curve assumed by a cable of span $L$ and sag $h$ carrying a distributed load per unit length $w = w_0\cos(\pi x/L)$, where $x$ is measured from mid-span. Also determine the maximum and minimum values of the tension.

**\*7.100** A large number of ropes are tied to a light wire which is suspended from two points $A$ and $B$ at the same level. Show that, if the lower ends of the ropes have been cut so that they lie in the same horizontal line and if the ropes are kept uniformly spaced, the curve assumed by the wire $ACB$ is

$$y + d = d \cosh\left(\frac{w_r}{T_0 e}\right)^{1/2} x$$

where $w_r$ is the weight per unit length of the ropes, $d$ is the length of the shortest rope, and $e$ is the horizontal distance between adjacent ropes. (*Hint.* Use the property indicated in Prob. 7.98.)

**\*7.10. CATENARY**

We shall consider now a cable $AB$ carrying a load *uniformly distributed along the cable itself* (Fig. 7.18*a*). Cables hanging under their own weight are loaded in this way. We denote by $w$ the load per unit length (*measured along the cable*) and express it in N/m or in lb/ft. The magnitude $W$ of the total load carried by a portion of cable of length $s$ extending from the lowest point $C$ to a point $D$ is $W = ws$. Substituting this value for $W$ in formula (7.6), we obtain the tension at $D$:

$$T = \sqrt{T_0^2 + w^2 s^2}$$

In order to simplify the subsequent computations, we shall introduce the constant $c = T_0/w$. We thus write

$$T_0 = wc \qquad W = ws \qquad T = w\sqrt{c^2 + s^2} \qquad (7.11)$$

The free-body diagram of the portion of cable $CD$ is shown in Fig. 7.18*b*. This diagram, however, cannot be used to obtain directly the equation of the curve assumed by the cable, since we do not know the horizontal distance from $D$ to the line of action of the resultant $\mathbf{W}$ of the load. To obtain this equation, we shall write first that the horizontal projection

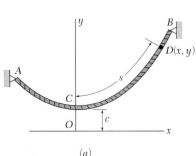

(*a*)

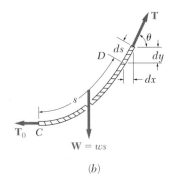

(*b*)

(*c*)

**Fig. 7.18**

of a small element of cable of length $ds$ is $dx = ds\cos\theta$. Observing from Fig. 7.18c that $\cos\theta = T_0/T$ and using (7.11), we write

$$dx = ds\cos\theta = \frac{T_0}{T}ds = \frac{wc\,ds}{w\sqrt{c^2 + s^2}} = \frac{ds}{\sqrt{1 + s^2/c^2}}$$

Selecting the origin $O$ of the coordinates at a distance $c$ directly below $C$ (Fig. 7.18a) and integrating from $C(0, c)$ to $D(x, y)$, we obtain[†]

$$x = \int_0^s \frac{ds}{\sqrt{1 + s^2/c^2}} = c\left[\sinh^{-1}\frac{s}{c}\right]_0^s = c\sinh^{-1}\frac{s}{c}$$

This equation, which relates the length $s$ of the portion of cable $CD$ and the horizontal distance $x$, may be written in the form

$$s = c\sinh\frac{x}{c} \tag{7.15}$$

The relation between the coordinates $x$ and $y$ may now be obtained by writing $dy = dx\tan\theta$. Observing from Fig. 7.18c that $\tan\theta = W/T_0$ and using (7.11) and (7.15), we write

$$dy = dx\tan\theta = \frac{W}{T_0}dx = \frac{s}{c}dx = \sinh\frac{x}{c}dx$$

Integrating from $C(0, c)$ to $D(x, y)$ and using (7.12) and (7.13), we obtain

$$y - c = \int_0^x \sinh\frac{x}{c}dx = c\left[\cosh\frac{x}{c}\right]_0^x = c\left(\cosh\frac{x}{c} - 1\right)$$

$$y - c = c\cosh\frac{x}{c} - c$$

---

[†]This integral may be found in all standard integral tables. The function

$$z = \sinh^{-1}u$$

(read "arc hyperbolic sine $u$") is the *inverse* of the function $u = \sinh z$ (read "hyperbolic sine $z$"). This function and the function $v = \cosh z$ (read "hyperbolic cosine $z$") are defined as follows:

$$u = \sinh z = \tfrac{1}{2}(e^z - e^{-z}) \qquad v = \cosh z = \tfrac{1}{2}(e^z + e^{-z})$$

Numerical values of the functions $\sinh z$ and $\cosh z$ are found in *tables of hyperbolic functions*. They may also be computed on most calculators either directly or from the above definitions. The student is referred to any calculus text for a complete description of the properties of these functions. In this section, we shall make use only of the following properties, which may be easily derived from the above definitions:

$$\frac{d\sinh z}{dz} = \cosh z \qquad \frac{d\cosh z}{dz} = \sinh z \tag{7.12}$$

$$\sinh 0 = 0 \qquad \cosh 0 = 1 \tag{7.13}$$

$$\cosh^2 z - \sinh^2 z = 1 \tag{7.14}$$

which reduces to

$$y = c \cosh \frac{x}{c} \qquad (7.16)$$

This is the equation of a *catenary* with vertical axis. The ordinate $c$ of the lowest point $C$ is called the *parameter* of the catenary. Squaring both sides of Eqs. (7.15) and (7.16), subtracting, and taking (7.14) into account, we obtain the following relation between $y$ and $s$:

$$y^2 - s^2 = c^2 \qquad (7.17)$$

Solving (7.17) for $s^2$ and carrying into the last of the relations (7.11), we write these relations as follows:

$$T_0 = wc \qquad W = ws \qquad T = wy \qquad (7.18)$$

The last relation indicates that the tension at any point $P$ of the cable is proportional to the vertical distance from $D$ to the horizontal line representing the $x$ axis.

When the supports $A$ and $B$ of the cable have the same elevation, the distance $L$ between the supports is called the *span* of the cable and the vertical distance $h$ from the supports to the lowest point $C$ is called the *sag* of the cable. These defnitions are the same that were given in the case of parabolic cables, but it should be noted that because of our choice of coordinate axes, the sag $h$ is now

$$h = y_A - c \qquad (7.19)$$

It should also be observed that certain catenary problems involve transcendental equations which must be solved by successive approximations (see Sample Prob. 7.10). When the cable is fairly taut, however, the load may be assumed uniformly distributed *along the horizontal* and the catenary may be replaced by a parabola. The solution of the problem is thus greatly simplified, while the error introduced is small.

When the supports $A$ and $B$ have different elevations, the position of the lowest point of the cable is not known. The problem may by solved then in a manner similar to that indicated for parabolic cables, by expressing that the cable must pass through the supports and that $x_B - x_A = L$, $y_B - y_A = d$, where $L$ and $d$ denote, respectively, the horizontal and vertical distances between the two supports.

# SAMPLE PROBLEM 7.10

A uniform cable weighing 3 lb/ft is suspended between two points $A$ and $B$ as shown. Determine (a) the maximum and minimum values of the tension in the cable, (b) the length of the cable.

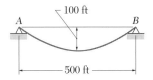

**Solution.** *Equation of Cable.* The origin of coordinates is placed at a distance $c$ below the lowest point of the cable. The equation of the cable is given by Eq. (7.16),

$$y = c \cosh \frac{x}{c}$$

The coordinates of point $B$ are

$$x_B = 250 \text{ ft} \qquad y_B = 100 + c$$

Substituting these coordinates into the equation of the cable, we obtain

$$100 + c = c \cosh \frac{250}{c}$$

$$\frac{100}{c} + 1 = \cosh \frac{250}{c}$$

The value of $c$ is determined by assuming successive trial values, as shown in the following table:

| C | $\dfrac{250}{c}$ | $\dfrac{100}{c}$ | $\dfrac{100}{c} + 1$ | $\cosh \dfrac{250}{c}$ |
|---|---|---|---|---|
| 300 | 0.833 | 0.333 | 1.333 | 1.367 |
| 350 | 0.714 | 0.286 | 1.286 | 1.266 |
| 330 | 0.758 | 0.303 | 1.303 | 1.301 |
| 328 | 0.762 | 0.305 | 1.305 | 1.305 |

Taking $c = 328$, we have

$$y_B = 100 + c = 428 \text{ ft}$$

*a. Maximum and Minimum Values of the Tension.* Using Eqs. (7.18), we obtain

$$T_{\min} = T_0 = wc = (3 \text{ lb/ft})(328 \text{ ft}) \qquad T_{\min} = 984 \text{ lb} \blacktriangleleft$$
$$T_{\max} = T_B = wy_B = (3 \text{ lb/ft})(428 \text{ ft}) \qquad T_{\max} = 1284 \text{ lb} \blacktriangleleft$$

*b. Length of Cable.* One-half the length of the cable is found by solving Eq. (7.17):

$$y_B^2 - s_{CB}^2 = c^2 \qquad s_{CB}^2 = y_B^2 - c^2 = (428)^2 - (328)^2 \qquad s_{CB} = 275 \text{ ft}$$

The total length of the cable is therefore

$$s_{AB} = 2s_{CB} = 2(275 \text{ ft}) \qquad s_{AB} = 550 \text{ ft} \blacktriangleleft$$

# PROBLEMS

**7.101** A 30-m rope is strung between the roofs of two buildings, each 9 m high. The maximum tension is found to be 250 N and the lowest point of the cable is observed to be 3 m above the ground. Determine (*a*) the horizontal distance between the buildings, (*b*) the total mass of the rope.

**7.102** A 50-m steel surveying tape has a mass of 1.6 kg. If the tape is stretched between two points at the same elevation and pulled until the tension at each end is 60 N, determine the horizontal distance between the ends of the tape. Neglect the elongation of the tape due to the tension.

**7.103** An aerial tramway cable of length 600 ft and weighing 3.0 lb/ft is suspended between two points at the same elevation. Knowing that the sag is 150 ft, find (*a*) the horizontal distance between the supports, (*b*) the maximum tension in the cable.

**7.104** A 20-ft chain of weight 30 lb is suspended between two points at the same elevation. Knowing that the sag is observed to be 8 ft, determine (*a*) the distance between the supports, (*b*) the maximum tension in the chain.

**7.105** A counterweight *D* of mass 50 kg is attached to a cable which passes over a small pulley at *A* and is attached to a support at *B*. Knowing that *L* = 12 m and *h* = 4 m, determine (*a*) the length of the cable from *A* to *B*, (*b*) the mass per unit length of the cable. Neglect the mass of the cable from *A* to *D*.

**7.106** Solve Prob. 7.105, assuming that the mass of the counterweight *D* is 40 kg.

**7.107** A wire of length 60 ft and total weight 20 lb is suspended between two points at the same elevation and 30 ft apart. Determine the sag and the maximum tension.

**7.108** A 240-ft wire is suspended between two points at the same elevation and 160 ft apart. Knowing that the maximum tension is 50 lb, determine the sag and the total weight of the wire.

**7.109** A uniform cord 750 mm long passes over a frictionless pulley at *B* and is attached to a rigid support at *A*. Knowing that *L* = 250 mm, determine the smaller of the two values of *h* for which the cord is in equilibrium.

**7.110** A uniform cord 50 in. long passes over a frictionless pulley at *B* and is attached to a rigid support at *A*. Knowing that *L* = 20 in., determine the smaller of the two values of *h* for which the cord is in equilibrium.

**7.111** To the left of point *B* the long cable *ABDE* rests on the rough horizontal surface shown. Knowing that the cable weighs 1.5 lb/ft, determine the force **F** required when *a* = 6 ft.

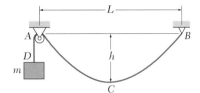
Fig. P7.105

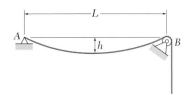
Fig. P7.109 and P7.110

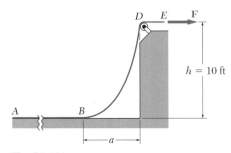
Fig. P7.111

**7.112**   Solve Prob. 7.111 when $a = 10$ ft.

**7.113**   Denoting by $\theta$ the angle formed by a uniform cable and the horizontal, show that at any point $y = c/\cos\theta$

**\*7.114**   (*a*) Determine the maximum allowable horizontal span for a uniform cable of weight $w$ per unit length if the tension in the cable is not to exceed the value $T_m$. (*b*) Using the result of part *a*, find the maximum span of a drawn-steel wire for which $w = 0.25$ lb/ft and $T_m = 8000$ lb.

**\*7.115**   A cable has a mass per unit length of 4 kg/m and is supported as shown. Knowing that the span $L$ is 6 m, determine the *two* values of the sag $h$ for which the maximum tension is 500 N.

**\*7.116**   Determine the sag-to-span ratio for which the maximum tension in the cable is equal to the total weight of the entire cable $AB$.

**\*7.117**   A cable, of weight $w$ per unit length, is suspended between two points at the same elevation and a distance $L$ apart. Determine the sag-to-span ratio for which the maximum tension is as small as possible. What are the corresponding values of $\theta_B$ and $T_m$?

**Fig. P7.115, P7.116, and P7.117**

# REVIEW AND SUMMARY
# FOR CHAPTER 7

In this chapter we learned to determine the internal forces which hold together the various parts of a given member in a structure.

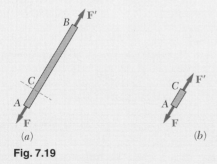

**Fig. 7.19**

Considering first a *straight two-force member AB* [Sec. 7.2], we recall that such a member is subjected at $A$ and $B$ to equal and opposite forces directed along $AB$ (Fig. 7.19*a*). Cutting member $AB$ at $C$ and drawing the free-body diagram of portion $AC$, we conclude that the internal forces which existed at $C$ in member $AB$ are equivalent to an *axial force* $\mathbf{F'}$ equal and opposite to $\mathbf{F}$ (Fig. 7.19*b*). We note that in the case of a two-force member which is not straight, the internal forces reduce to a force-couple system and not to a single force.

Forces in straight two-force members

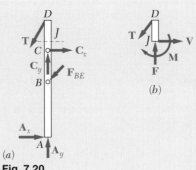

**Fig. 7.20**

**Forces in multiforce members**

Considering next a *multiforce member AD* (Fig. 7.20*a*), cutting it at *J*, and drawing the free-body diagram of portion *JD*, we conclude that the internal forces at *J* are equivalent to a force-couple system consisting of the *axial force* **F**, the *shearing force* **V**, and a couple **M** (Fig. 7.20*b*). The magnitude *V* of the shearing force measures the *shear* at point *J* and the moment *M* of the couple is referred to as the *bending moment* at *J*. Since an equal and opposite force-couple system would have been obtained by considering the free-body diagram of portion *AJ*, it is necessary to specify which portion of member *AD* was used when recording the answers [Sample Prob. 7.1].

**Forces in beams**

Most of the chapter was devoted to the analysis of the internal forces in two important types of engineering structures: *beams and cables. Beams* are usually long, straight prismatic members designed to support loads applied at various points along the member. In general the loads are perpendicular to the axis of the beam and produce only *shear and bending* in the beam. The loads may be either *concentrated* at specific points, or *distributed* along the entire length or a portion of the beam. The beam itself may be supported in various ways; since only statically determinate beams are considered in this text, we limited our analysis to that of *simply supported beams, overhanging beams*, and *cantilever beams* [Sec. 7.3].

**Shear and bending moment in a beam**

To obtain the *shear V* and *bending moment M* at a given point *C* of a beam, we first determine the reactions at the supports by considering the entire beam as a free body. We then cut the beam at *C* and use the free-body diagram of one of the two portions obtained in this fashion to determine *V* and *M*. In order to avoid any confusion regarding the sense of the shearing force **V** and couple **M** (which act in opposite directions on the two portions of the beam), the sign convention illustrated in Fig. 7.21 was adopted [Sec. 7.4]. Once the values of the shear and bending moment have been determined at a few selected points of the beam, it is usually possible to draw a *shear diagram* and a *bending-moment diagram* representing, respectively, the shear and the bending moment at any point of the beam [Sec. 7.5]. When a beam is subjected to concentrated loads only, the shear is of constant value between loads and the bending moment varies linearly between loads [Sample Prob. 7.2]. On the other hand, when a beam is subjected to distributed loads, the shear and bending moment vary quite differently [Sample Prob. 7.3].

Internal forces at section
(positive shear and positive bending moment)
**Fig. 7.21**

The construction of the shear and bending-moment diagrams is facilitated if the following relations are taken into account. Denoting by $w$ the distributed load per unit length (assumed positive if directed downward), we have [Sec.7.5]:

$$\frac{dV}{dx} = -w \tag{7.1}$$

$$\frac{dM}{dx} = V \tag{7.3}$$

or, in integrated form,

$$V_D - V_C = -(\text{area under load curve between } C \text{ and } D) \tag{7.2'}$$

$$M_D - M_C = \text{area under shear curve between } C \text{ and } D \tag{7.4'}$$

Equation (7.2′) makes it possible to draw the shear diagram of a beam from the curve representing the distributed load on that beam and the value of $V$ at one end of the beam. Similarly, Eq. (7.4′) makes it possible to draw the bending-moment diagram from the shear diagram and the value of $M$ at one end of the beam. However, concentrated loads introduce discontinuities in the shear diagram and concentrated couples in the bending-moment diagram, none of which is accounted for in these equations [Sample Probs. 7.4 and 7.7]. Finally, we note from Eq. (7.3) that the points of the beam where the bending moment is maximum or minimum are also the points where the shear is zero [Sample Prob. 7.5].

**Relations among load, shear, and bending moment**

The second half of the chapter was devoted to the analysis of *flexible cables*. We first considered a cable of negligible weight supporting *concentrated loads* [Sec. 7.7]. Using the entire cable $AB$ as a free body (Fig. 7.22), we noted that the three available equilibrium equations were not sufficient to determine the four unknowns representing the reactions at the supports $A$ and $B$. However, if the coordinates of a point $D$ of the cable are known, an additional equation may be obtained by considering the free-body diagram of the portion $AD$ or $DB$ of the cable. Once the reactions at the supports have been determined, the elevation of any point of the cable and the tension in any portion of the cable may be found from the appropriate free-body diagram [Sample Prob. 7.8]. It was noted that the horizontal component of the force $\mathbf{T}$ representing the tension is the same at any point of the cable.

**Cables with concentrated loads**

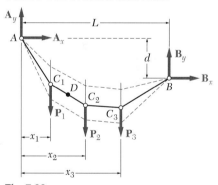

**Fig. 7.22**

We next considered cables carrying *distributed loads* [Sec. 7.8]. Using as a free body a portion of cable $CD$ extending from the lowest point $C$ to an arbitrary point $D$ of the cable (Fig. 7.23), we observed that the horizontal component of the tension force $\mathbf{T}$ at $D$ is constant and equal to the tension $T_0$ at $C$, while its vertical component is equal to the weight $W$ of the portion of cable $CD$. The magnitude and direction of $\mathbf{T}$ were obtained from the force triangle:

**Cables with distributed loads**

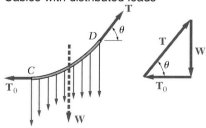

**Fig. 7.23**

$$T = \sqrt{T_0^2 + W^2} \qquad \tan \theta = \frac{W}{T_0} \tag{7.6}$$

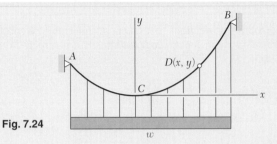

**Fig. 7.24**

**Parabolic cable**

In the case of a load *uniformly distributed along the horizontal*—as in a suspension bridge (Fig. 7.24)—the load supported by portion $CD$ is $W = wx$, where $w$ is the constant load per unit horizontal length [Sec. 7.9]. We also found that the curve formed by the cable is a parabola of equation

$$y = \frac{wx^2}{2T_0} \tag{7.8}$$

and that the length of the cable may be found by using the expansion in series given in Eq. (7.10) [Sample Prob. 7.9].

**Catenary**

In the case of a load *uniformly distributed along the cable itself*—e.g., a cable hanging under its own weight (Fig. 7.25)—the load supported by portion $CD$ is $W = ws$, where $s$ is the length measured along the cable and $w$ is the constant load per unit length [Sec. 7.10]. Choosing the origin $O$ of the coordinate axes at a distance $c = T_0/w$ below $C$, we derived the relations

$$s = c \sinh\frac{x}{c} \tag{7.15}$$

$$y = c \cosh\frac{x}{c} \tag{7.16}$$

$$y^2 - s^2 = c^2 \tag{7.17}$$

$$T_0 = wc \qquad W = ws \qquad T = wy \tag{7.18}$$

which may be used to solve problems involving cables hanging under their own weight [Sample Prob. 7.10]. Equation (7.16), which defines the shape of the cable, is the equation of a *catenary*.

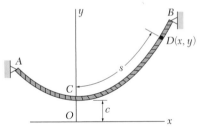

**Fig. 7.25**

# Review Problems

**7.118**  A 14-mm-diameter wire rope has a mass per unit length of 0.7 kg/m and is suspended from two supports at the same elevation and 120 m apart. If the sag is 30 m, determine (*a*) the total length of the cable, (*b*) the maximum tension.

**7.119**  Two cables of the same gage are attached to a transmission tower at *B*. Since the tower is slender, the horizontal component of the resultant of the forces exerted by the cables at *B* is to be zero. Assuming the cables to be parabolic, determine the required sag *h* of cable *AB*.

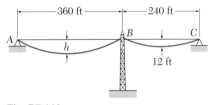

**Fig. P7.119**

**7.120 and 7.121** Draw the shear and bending-moment diagrams for the beam and loading shown.

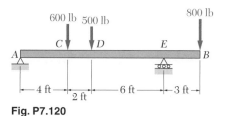

**Fig. P7.120**

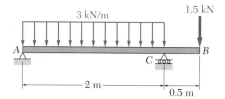

**Fig. P7.121**

**7.122** A 5-kN load is applied at point $F$ of plank $EFG$, which is suspended as shown from cable $ABCD$. If $h_B = 1.5$ m, determine (a) the distance $a$, (b) the corresponding maximum tension in cable $ABCD$.

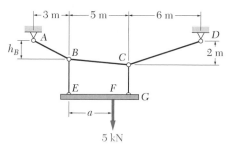

**Fig. P7.122 and P7.123**

**7.123** A 5-kN load is applied at point $F$ of plank $EFG$, which is suspended as shown from cable $ABCD$. If $a = 2.5$ m, determine (a) the sag $h_B$, (b) the corresponding maximum tension in cable $ABCD$.

**7.124** For each of the arrangements shown, determine the internal forces at point $F$ of beam $AB$.

**7.125** For arrangements $a$ and $b$, draw the shear and bending-moment diagrams for beam $AB$.

**7.126** For arrangements $c$ and $d$, draw the shear and bending-moment diagrams for beam $AB$.

**7.127** A cable of length $L + \Delta$ is suspended between two points which are at the same elevation and a distance $L$ apart. (a) Assuming that $\Delta$ is small compared with $L$ and that the cable is parabolic, determine the approximate sag in terms of $L$ and $\Delta$. (b) If $L = 30$ m and $\Delta = 0.8$ m and the mass of the wire is 0.35 kg/m, use the result of part $a$ to determine approximate values for the sag and the maximum tension in the wire. [*Hint.* Use only the first two terms of Eq. (7.10).]

**7.128** A uniform rod $ABCDE$ has been bent and is held by a wire sling as shown. Knowing that the rod weighs 3.6 lb, draw the shear and bending-moment diagrams for portion $BCD$ of the rod.

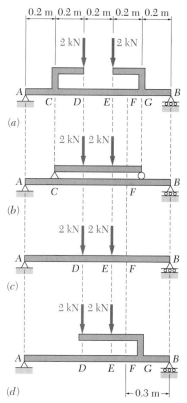

**Fig. P7.124, P7.125, and P7.126**

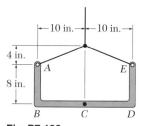

**Fig. P7.128**

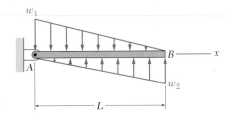

**Fig. P7.129**

**7.129**  The rod $AB$ is attached to a hinge at $A$ and is free at $B$. Determine (a) the ratio of $w_2$ to $w_1$ for which the rod is in equilibrium, (b) the equations of the shear and bending-moment curves for the rod, (c) the magnitude and location of the maximum bending moment.

# COMPUTER PROBLEMS

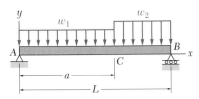

**Fig. P7.C1**

**7.C1**  Write a computer program which can be used to calculate the shearing force and the bending moment along the beam $AB$ from $x = 0$ to $x = L$ at intervals $\Delta L$. Use this program to calculate the values of $V$ and $M$ for each of the following cases:

|  | **L** | **$\Delta L$** | **a** | **$w_1$** | **$w_2$** |
|---|---|---|---|---|---|
| (a) | 9 m | 1 m | 6 m | 20 kN/m | 0 |
| (b) | 9 m | 1 m | 6 m | 0 | 20 kN/m |
| (c) | 6 m | 0.5 m | 2 m | 6 kN/m | 15 kN/m |

**7.C2**  Write a computer program which can be used to calculate the shearing force and the bending moment along the beam $AB$ from $x = 0$ to $x = L$ at intervals $\Delta L$. Use this program to calculate values of $V$ and $M$ for each of the following cases:

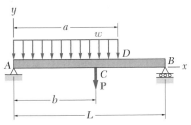

**Fig. P7.C2**

|  | **L** | **$\Delta L$** | **a** | **b** | **w** | **P** |  |
|---|---|---|---|---|---|---|---|
| (a) | 9 m | 1 m | 6 m | — | 20 kN/m | 0 | [Samp. Prob. 7.5] |
| (b) | 4 m | 0.4 m | 4 m | 3.2 m | 2 kN/m | 4 kN | [Prob. 7.62] |
| (c) | 3 m | 0.2 m | 2 m | 1.2 m | 1.5 kN/m | 3 kN | [Prob. 7.27] |

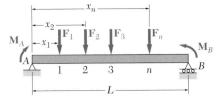

**Fig. P7.C3**

**7.C3**  For the beam and loading shown, write a computer program which can be used to calculate the bending moment in beam $AB$ at the location of each concentrated load. Use this program to solve (a) Prob. 7.59, (b) Prob. 7.60.

**7.C4**  Write a computer program which can be used to calculate the horizontal and vertical components of the reaction at the support $A_n$ from the values of the loads $P_1, P_2, \ldots, P_{n-1}$, the horizontal distances $d_1, d_2, \ldots, d_n$, and the two vertical distances $h_0$ and $h_k$. Use this program to solve part $a$ of Probs. 7.77, 7.79, 7.80, and 7.81.

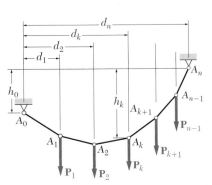

**Fig. P7.C4**

# CHAPTER 8

# Friction

## 8.1. INTRODUCTION

In the preceding chapters, it was assumed that surfaces in contact were either *frictionless* or *rough*. If they were frictionless, the force each surface exerted on the other was normal to the surfaces and the two surfaces could move freely with respect to each other. If they were rough, it was assumed that tangential forces could develop to prevent the motion of one surface with respect to the other.

The view was a simplified one. Actually, no perfectly frictionless surface exists. When two surfaces are in contact, tangential forces, called *friction forces*, will always develop if one attempts to move one surface with respect to the other. On the other hand, these friction forces are limited in magnitude and will not prevent motion if sufficiently large forces are applied. The distinction between frictionless and rough surfaces is thus a matter of degree. This will be seen more clearly in the present chapter, which is devoted to the study of friction and of its applications to common engineering situations.

There are two types of friction: *dry friction*, sometimes called *Coulomb friction*, and *fluid friction*. Fluid friction develops between layers of fluid moving at different velocities. Fluid friction is of great importance in problems involving the flow of fluids through pipes and orifices or dealing with bodies immersed in moving fluids. It is also basic in the analysis of the motion of *lubricated mechanisms*. Such problems are considered in texts on fluid mechanics. We shall limit our present study to dry friction, i.e., to problems involving rigid bodies which are in contact along *nonlubricated* surfaces.

In the first part of this chapter, we shall analyze the equilibrium of various rigid bodies and structures, assuming dry friction at the surfaces of contact. Later we shall consider a number of specific engineering applications where dry friction plays an important role. These are wedges, square-threaded screws, journal bearings, thrust bearings, rolling resistance, and belt friction.

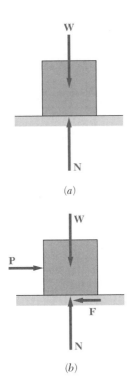

(a)

(b)

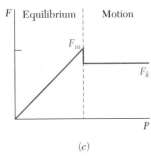

(c)

**Fig. 8.1**

## 8.2. THE LAWS OF DRY FRICTION. COEFFICIENTS OF FRICTION

The laws of dry friction are best understood by the following experiment. A block of weight **W** is placed on a horizontal plane surface (Fig. 8.1a). The forces acting on the block are its weight **W** and the reaction of the surface. Since the weight has no horizontal component, the reaction of the surface also has no horizontal component; the reaction is therefore *normal* to the surface and is represented by **N** in Fig. 8.1a. Suppose, now, that a horizontal force **P** is applied to the block (Fig. 8.1b). If **P** is small, the block will not move; some other horizontal force must therefore exist, which balances **P**. This other force is the *static-friction force* **F**, which is actually the resultant of a great number of forces acting over the entire surface of contact between the block and the plane. The nature of these forces is not known exactly, but it is generally assumed that these forces are due to the irregularities of the surfaces in contact and also, to a certain extent, to molecular attraction.

If the force **P** is increased, the friction force **F** also increases, continuing to oppose **P**, until its magnitude reaches a certain *maximum value* $F_m$ (Fig. 8.1c). If **P** is further increased, the friction force cannot balance it any more and the block starts sliding. As soon as the block has been set in motion, the magnitude of **F** drops from $F_m$ to a lower value $F_k$. This is because there is less interpenetration between the irregularities of the surfaces in contact when these surfaces move with respect to each other. From then on, the block keeps sliding with increasing velocity while the friction force, denoted by $\mathbf{F}_k$ and called the *kinetic-friction force,* remains approximately constant.

Experimental evidence shows that the maximum value $F_m$ of the static-friction force is proportional to the normal component $N$ of the reaction of the surface. We have

$$F_m = \mu_s N \tag{8.1}$$

where $\mu_s$ is a constant called the *coefficient of static friction.* Similarly, the magnitude $F_k$ of the kinetic-friction force may be put in the form

$$F_k = \mu_k N \tag{8.2}$$

where $\mu_k$ is a constant called the *coefficient of kinetic friction.* The coefficients of friction $\mu_s$ and $\mu_k$ do not depend upon the area of the surfaces in contact. Both coefficients, however, depend strongly on the *nature* of the surfaces in contact. Since they also depend upon the exact condition of the surfaces, their value is seldom known with an accuracy greater than 5 percent. Approximate values of coefficients of static friction are given in Table 8.1 for various dry surfaces. The corresponding values of the coefficient of kinetic friction would be about 25 percent smaller. Since coefficients of

## Table 8.1   Approximate Values of Coefficient of Static Friction for Dry Surfaces

| | |
|---|---|
| Metal on metal | 0.15–0.60 |
| Metal on wood | 0.20–0.60 |
| Metal on stone | 0.30–0.70 |
| Metal on leather | 0.30–0.60 |
| Wood on wood | 0.25–0.50 |
| Wood on leather | 0.25–0.50 |
| Stone on stone | 0.40–0.70 |
| Earth on earth | 0.20–1.00 |
| Rubber on concrete | 0.60–0.90 |

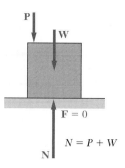

$F = 0$

$N = P + W$

(a) No friction ($P_x = 0$)

friction are dimensionless quantities, the values given in Table 8.1 may be used with both SI units and U.S. customary units.

From the description given above, it appears that four different situations may occur when a rigid body is in contact with a horizontal surface:

1.  The forces applied to the body do not tend to move it along the surface of contact; there is no friction force (Fig. 8.2*a*).

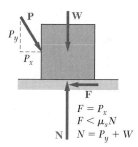

$F = P_x$
$F < \mu_s N$
$N = P_y + W$

(b) No motion ($P_x < F_m$)

2.  The applied forces tend to move the body along the surface of contact but are not large enough to set it in motion. The friction force **F** which has developed may be found by solving the equations of equilibrium for the body. Since there is no evidence that the maximum value of the static-friction force has been reached, the equation $F_m = \mu_s N$ *cannot be used* to determine the friction force (Fig. 8.2*b*).

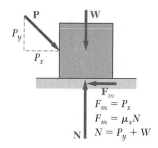

$F_m$
$F_m = P_x$
$F_m = \mu_s N$
$N = P_y + W$

(c) Motion impending ⟶ ($P_x = F_m$)

3.  The applied forces are such that the body is just about to slide. We say that *motion is impending*. The friction force **F** has reached its maximum value $F_m$ and, together with the normal force **N**, balances the applied forces. Both the equations of equilibrium and the equation $F_m = \mu_s N$ *may be used*. We also note that the friction force has a sense opposite to the sense of impending motion (Fig. 8.2*c*).

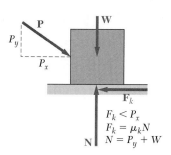

$F_k$
$F_k < P_x$
$F_k = \mu_k N$
$N = P_y + W$

(d) Motion ⟶ ($P_x > F_m$)

**Fig. 8.2**

4.  The body is sliding under the action of the applied forces, and the equations of equilibrium do not apply any more. However, **F** is now equal to the kinetic-friction force $\mathbf{F}_k$ and the equation $F_k = \mu_k N$ may be used. The sense of $\mathbf{F}_k$ is opposite to the sense of motion (Fig. 8.2*d*).

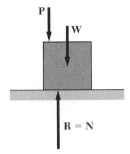

$R = N$

(a) No friction

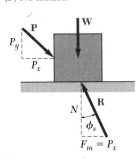

$< \phi_s$

$F = P_x$

(b) No motion

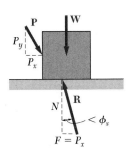

$R$

$\phi_s$

$F_m = P_x$

(c) Motion impending ⟶

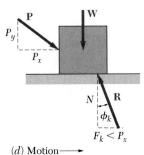

$R$

$\phi_k$

$F_k < P_x$

(d) Motion ⟶

**Fig. 8.3**

## 8.3. ANGLES OF FRICTION

It is sometimes found convenient to replace the normal force **N** and the friction force **F** by their resultant **R**. Let us consider again a block of weight **W** resting on a horizontal plane surface. If no horizontal force is applied to the block, the resultant **R** reduces to the normal force **N** (Fig. 8.3*a*). However, if the applied force **P** has a horizontal component $P_x$ which tends to move the block, the force **R** will have a horizontal component **F** and, thus, will form a certain angle with the vertical (Fig. 8.3*b*). If $P_x$ is increased until motion becomes impending, the angle between **R** and the vertical grows and reaches a maximum value (Fig. 8.3*c*). This value is called the *angle of static friction* and is denoted by $\phi_s$. From the geometry of Fig. 8.3*c*, we note that

$$\tan \phi_s = \frac{F_m}{N} = \frac{\mu_s N}{N}$$

$$\tan \phi_s = \mu_s \tag{8.3}$$

If motion actually takes place, the magnitude of the friction force drops to $F_k$; similarly, the angle between **R** and **N** drops to a lower value $\phi_k$, called the *angle of kinetic friction* (Fig. 8.3*d*). From the geometry of Fig. 8.3*d*, we write

$$\tan \phi_k = \frac{F_k}{N} = \frac{\mu_k N}{N}$$

$$\tan \phi_k = \mu_k \tag{8.4}$$

Another example will show how the angle of friction may be used to advantage in the analysis of certain types of problems. Consider a block resting on a board which may be given any desired inclination; the block is subjected to no other force than its weight **W** and the reaction **R** of the board. If the board is horizontal, the force **R** exerted by the board on the block is perpendicular to the board and balances the weight **W** (Fig. 8.4*a*). If the board is given a small angle of inclination $\theta$, the force **R** will deviate from the perpendicular to the board by the angle $\theta$ and will keep balancing **W** (Fig. 8.4*b*); it will then have a normal component **N** of magnitude $N = W \cos \theta$ and a tangential component **F** of magnitude $F = W \sin \theta$.

If we keep increasing the angle of inclination, motion will soon become impending. At that time, the angle between **R** and the normal will have reached its maximum value $\phi_s$ (Fig. 8.4*c*). The value of the angle of inclination corresponding to impending motion is called the *angle of repose*. Clearly, the angle of repose is equal to the angle of static friction $\phi_s$. If the angle of inclination $\theta$ is further increased, motion starts and the angle between **R** and the normal drops to the lower value $\phi_k$ (Fig. 8.4*d*). The reaction **R** is not vertical any more, and the forces acting on the block are unbalanced.

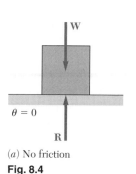

$\theta = 0$

**R**

(*a*) No friction

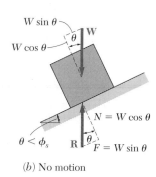

$W \sin \theta$

$W \cos \theta$

$\theta$ **W**

$N = W \cos \theta$

$\theta < \phi_s$  **R**  $\theta$  $F = W \sin \theta$

(*b*) No motion

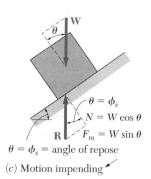

$\theta$ **W**

$N = W \cos \theta$

**R**  $F_m = W \sin \theta$

$\theta = \phi_s = $ angle of repose

(*c*) Motion impending

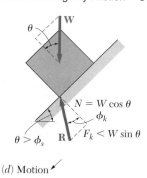

$\theta$ **W**

$N = W \cos \theta$

$\phi_k$

$\theta > \phi_s$  **R**  $F_k < W \sin \theta$

(*d*) Motion

**Fig. 8.4**

## 8.4. PROBLEMS INVOLVING DRY FRICTION

Problems involving dry friction are found in many engineering applications. Some deal with simple situations such as the block sliding on a plane described in the preceding sections. Others involve more complicated situations as in Sample Prob. 8.3; many deal with the stability of rigid bodies in accelerated motion and will be studied in dynamics. Also, a number of common machines and mechanisms may be analyzed by applying the laws of dry friction. These include wedges, screws, journal and thrust bearings, and belt transmissions. They will be studied in the following sections.

The *methods* which should be used to solve problems involving dry friction are the same that were used in the preceding chapters. If a problem involves only a motion of translation, with no possible rotation, the body under consideration may usually be treated as a particle, and the methods of Chap. 2 may be used. If the problem involves a possible rotation, the body must be considered as a rigid body and the methods of Chap. 3 should be used. If the structure considered is made of several parts, the principle of action and reaction must be used as was done in Chap. 6.

If the body considered is acted upon by more than three forces (including the reactions at the surfaces of contact), the reaction at each surface will be represented by its components **N** and **F** and the problem will be solved from the equations of equilibrium. If only three forces act on the body under consideration, it may be found more convenient to represent each reaction by the single force **R** and to solve the problem by drawing a force triangle.

Most problems involving friction fall into one of the following *three groups:* In the *first group* of problems, all applied forces are given, and the coefficients of friction are known; we are to determine whether the body considered will remain at rest or slide. The friction force **F** *required to maintain equilibrium* is unknown (its magnitude is *not* equal to $\mu_s N$) and should be determined, together with the normal force **N**, by drawing a free-body diagram and *solving the equations of equilibrium* (Fig. 8.5*a*). The

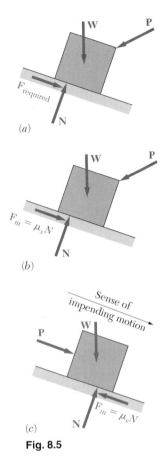

(a)

(b)

(c)

**Fig. 8.5**

value found for the magnitude $F$ of the friction force is then compared with the maximum value $F_m = \mu_s N$. If $F$ is smaller than or equal to $F_m$, the body remains at rest. If the value found for $F$ is larger than $F_m$, equilibrium cannot be maintained and motion takes place; the actual magnitude of the friction force is then $F_k = \mu_k N$.

In problems of the *second group*, all applied forces are given, and the motion is known to be impending; we are to determine the value of the coefficient of static friction. Here again, we determine the friction force and the normal force by drawing a free-body diagram and solving the equations of equilibrium (Fig. 8.5$b$). Since we know that the value found for $F$ is the maximum value $F_m$, the coefficient of friction may be found by writing and solving the equation $F_m = \mu_s N$.

In problems of the *third group*, the coefficient of static friction is given, and it is known that the motion is impending in a given direction; we are to determine the magnitude or the direction of one of the applied forces. The friction force should be shown in the free-body diagram with a *sense opposite to that of the impending motion* and with a magnitude $F_m = \mu_s N$ (Fig. 8.5$c$). The equations of equilibrium may then be written, and the desired force may be determined.

As noted above, it may be more convenient when only three forces are involved to represent the reaction of the surface by a single force $\mathbf{R}$ and to solve the problem by drawing a force triangle. Such a solution is used in Sample Prob. 8.2.

When two bodies $A$ and $B$ are in contact (Fig. 8.6$a$), the forces of friction exerted, respectively, by $A$ on $B$ and by $B$ on $A$ are equal and opposite (Newton's third law). It is important in drawing the free-body diagram of one of the bodies, to include the appropriate friction force with its correct sense. The following rule should then be observed: *The sense of the friction force acting on $A$ is opposite to that of the motion (or impending motion) of $A$ as observed from $B$* (Fig. 8.6$b$).† *The sense of the friction force acting on $B$ is determined in a similar way* (Fig. 8.6$c$). Note that the motion of $A$ as observed from $B$ is a *relative motion*. Body $A$ may be fixed; yet it will have a relative motion with respect to $B$ if $B$ itself moves. Also, $A$ may actually move down yet be observed from $B$ to move up if $B$ moves down faster than $A$.

---

†It is therefore *the same as that of the motion of $B$ as observed from $A$.*

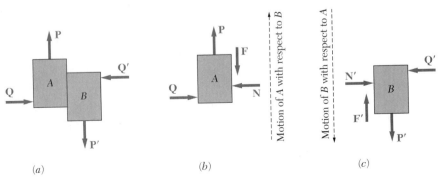

**Fig. 8.6**  (a)  (b)  (c)

# SAMPLE PROBLEM 8.1

A 100-lb force acts as shown on a 300-lb block placed on an inclined plane. The coefficients of friction between the block and the plane are $\mu_s = 0.25$ and $\mu_k = 0.20$. Determine whether the block is in equilibrium, and find the value of the friction force.

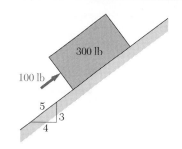

**Force Required for Equilibrium.** We first determine the value of the friction force *required to maintain equilibrium.* Assuming that **F** is directed down and to the left, we draw the free-body diagram of the block and write

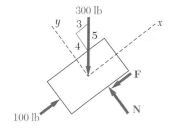

$$+\nearrow \Sigma F_x = 0: \qquad 100 \text{ lb} - \tfrac{3}{5}(300 \text{ lb}) - F = 0$$
$$F = -80 \text{ lb} \qquad \mathbf{F} = 80 \text{ lb} \nearrow$$

$$+\nwarrow \Sigma F_y = 0: \qquad N - \tfrac{4}{5}(300 \text{ lb}) = 0$$
$$N = +240 \text{ lb} \qquad \mathbf{N} = 240 \text{ lb} \nwarrow$$

The force **F** required to maintain equilibrium is an 80-lb force directed up and to the right; the tendency of the block is thus to move down the plane.

**Maximum Friction Force.** The magnitude of the maximum friction force which may be developed is

$$F_m = \mu_s N \qquad F_m = 0.25(240 \text{ lb}) = 60 \text{ lb}$$

Since the value of the force required to maintain equilibrium (80 lb) is larger than the maximum value which may be obtained (60 lb), equilibrium will not be maintained and *the block will slide down the plane.*

**Actual Value of Friction Force.** The magnitude of the actual friction force is obtained as follows:

$$F_{\text{actual}} = F_k = \mu_k N$$
$$= 0.20(240 \text{ lb}) = 48 \text{ lb}$$

The sense of this force is opposite to the sense of motion; the force is thus directed up and to the right:

$$\mathbf{F}_{\text{actual}} = 48 \text{ lb} \nearrow \quad \blacktriangleleft$$

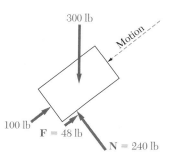

It should be noted that the forces acting on the block are not balanced; the resultant is

$$\tfrac{3}{5}(300 \text{ lb}) - 100 \text{ lb} - 48 \text{ lb} = 32 \text{ lb} \swarrow$$

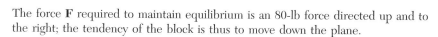

**311**

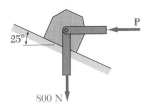

# SAMPLE PROBLEM 8.2

A support block is acted upon by two forces as shown. Knowing that the coefficients of friction between the block and the incline are $\mu_s = 0.35$ and $\mu_k = 0.25$, determine the force **P** required (*a*) to start the block moving up the incline, (*b*) to keep it moving up, (*c*) to prevent it from sliding down.

**Solution.** For each part of the problem we draw a free-body diagram of the block and a force triangle including the 800-N vertical force, the horizontal force **P**, and the force **R** exerted on the block by the incline. The direction of **R** must be determined in each separate case. We note that since **P** is perpendicular to the 800-N force, the force triangle is a right triangle, which may easily be solved for **P**. In most other problems, however, the force triangle will be an oblique triangle and should be solved by applying the law of sines.

### a. Force **P** to Start Block Moving Up

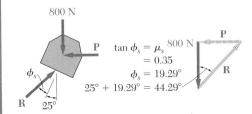

$$\tan \phi_s = \mu_s$$
$$= 0.35$$
$$\phi_s = 19.29°$$
$$25° + 19.29° = 44.29°$$

$$P = (800 \text{ N}) \tan 44.29°$$

$$\mathbf{P} = 780 \text{ N} \leftarrow \quad \blacktriangleleft$$

### b. Force **P** to Keep Block Moving Up

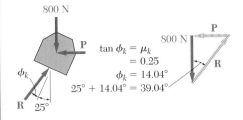

$$\tan \phi_k = \mu_k$$
$$= 0.25$$
$$\phi_k = 14.04°$$
$$25° + 14.04° = 39.04°$$

$$P = (800 \text{ N}) \tan 39.04°$$

$$\mathbf{P} = 649 \text{ N} \leftarrow \quad \blacktriangleleft$$

### c. Force **P** to Prevent Block from Sliding Down

$$\phi_s = 19.29°$$
$$25° - 19.29° = 5.71°$$

$$P = (800 \text{ N}) \tan 5.71°$$

$$\mathbf{P} = 80.0 \text{ N} \leftarrow \quad \blacktriangleleft$$

# SAMPLE PROBLEM 8.3

The movable bracket shown may be placed at any height on the 3-in.-diameter pipe. If the coefficient of static friction between the pipe and bracket is 0.25, determine the minimum distance $x$ at which the load $\mathbf{W}$ can be supported. Neglect the weight of the bracket.

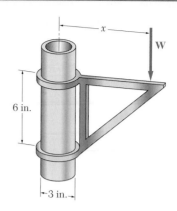

**Solution.** We draw the free-body diagram of the bracket. When $\mathbf{W}$ is placed at the minimum distance $x$ from the axis of the pipe, the bracket is just about to slip, and the forces of friction at $A$ and $B$ have reached their maximum values:

$$F_A = \mu_s N_A = 0.25\,N_A$$
$$F_B = \mu_s N_B = 0.25\,N_B$$

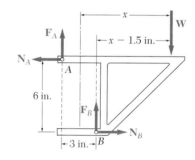

## Equilibrium Equations

$\xrightarrow{+} \Sigma F_x = 0:$  
$\qquad N_B - N_A = 0$  
$\qquad N_B = N_A$  

$+\uparrow \Sigma F_y = 0:$  
$\qquad F_A + F_B - W = 0$  
$\qquad 0.25N_A + 0.25N_B = W$

And, since $N_B$ has been found equal to $N_A$,

$$0.50N_A = W$$
$$N_A = 2\,W$$

$+\uparrow \Sigma M_B = 0:$  
$\qquad N_A(6 \text{ in.}) - F_A(3 \text{ in.}) - W(x - 1.5 \text{ in.}) = 0$  
$\qquad 6N_A - 3(0.25N_A) - Wx + 1.5W = 0$  
$\qquad 6(2W) - 0.75(2W) - Wx + 1.5W = 0$

Dividing through by $W$ and solving for $x$,

$$x = 12 \text{ in.} \quad \blacktriangleleft$$

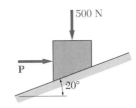

500 N

P

20°

**Fig. P8.1**

45 lb

P

40°

30°

**Fig. P8.3**

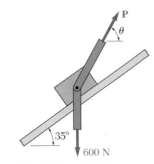

P

θ

35°

600 N

**Fig. P8.6 and P8.7**

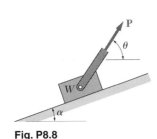

P

θ

W

α

**Fig. P8.8**

# PROBLEMS

**8.1** The coefficients of friction between the block and the incline are $\mu_s = 0.30$ and $\mu_k = 0.25$. Determine whether the block is in equilibrium and find the magnitude and direction of the friction force when $P = 150$ N.

**8.2** Solve Prob. 8.1, assuming that $P = 400$ N.

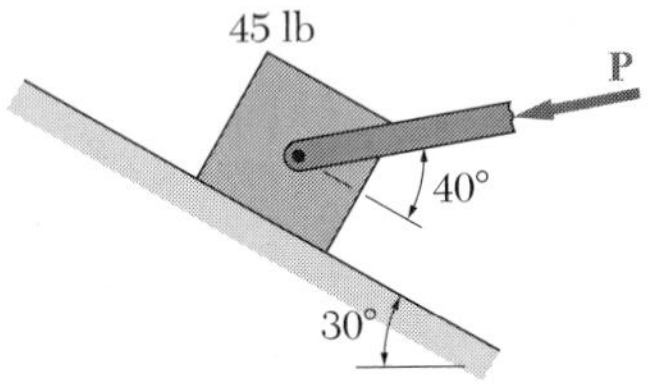

**8.3** The coefficients of friction between the 45-lb block and the incline are $\mu_s = 0.40$ and $\mu_k = 0.30$. Determine whether the block is in equilibrium and find the magnitude and direction of the friction force when $P = 100$ lb.

**8.4** Solve Prob. 8.3, assuming that $P = 60$ lb.

**8.5** For the block of Prob. 8.3, find the smallest value of $P$ required (*a*) to start the block up the incline, (*b*) to keep it moving up, (*c*) to prevent it from moving down.

**8.6** The coefficients of friction between the block and the rail are $\mu_s = 0.25$ and $\mu_k = 0.20$. Knowing that $\theta = 60°$, determine the smallest value of $P$ required (*a*) to start the block up the rail, (*b*) to keep it moving up, (*c*) to prevent it from moving down.

**8.7** The coefficients of friction between the block and the rail are $\mu_s = 0.25$ and $\mu_k = 0.20$. Find the magnitude and direction of the smallest force **P** required (*a*) to start the block up the rail, (*b*) to keep the block from moving down.

**8.8** A block of weight $W = 60$ lb rests on a rough plane as shown. Knowing that $\alpha = 20°$ and $\mu_s = 0.30$, determine the magnitude and direction of the smallest force **P** required (*a*) to start the block up the plane, (*b*) to prevent the block from moving down.

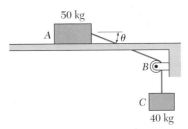

50 kg

A

θ

B

C

40 kg

**Fig. P8.9**

**8.9** Two blocks are connected by a cable as shown. Knowing that the coefficient of static friction between block $A$ and the horizontal surface is 0.50, determine the value of $\theta$ for which motion is impending.

**8.10**  The 40-kg block $A$ hangs from a cable as shown. Pulley $C$ is connected by a short link to block $E$, which rests on a horizontal rail. Knowing that the coefficient of static friction between block $E$ and the rail is 0.30, and neglecting the weight of block $E$ and the friction in the pulleys, determine the maximum allowable value of $\theta$ if the system is to remain in equilibrium.

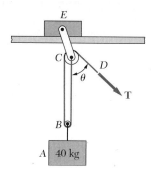

**Fig. P8.10**

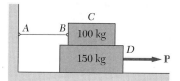

**Fig. P8.11 and P8.12**

**8.11**  The coefficients of friction are $\mu_s = 0.25$ and $\mu_k = 0.20$ between all surfaces of contact. Determine the smallest force **P** required to start block $D$ moving if ($a$) block $C$ is restrained by cable $AB$ as shown, ($b$) cable $AB$ is removed.

**8.12**  The coefficients of friction are $\mu_s = 0.25$ and $\mu_k = 0.20$ between all surfaces of contact. Knowing that $P = 800$ N, determine ($a$) the resultant of the friction forces exerted on block $D$ if block $C$ is restrained as shown, ($b$) the friction force exerted by the ground on block $D$ if cable $AB$ is removed.

**8.13**  Three 4-kg packages $A$, $B$, and $C$ are placed on a conveyor belt which is at rest. Between the belt and both packages $A$ and $C$ the coefficients of friction are $\mu_s = 0.30$ and $\mu_k = 0.20$; between package $B$ and the belt the coefficients are $\mu_s = 0.10$ and $\mu_k = 0.08$. The packages are placed on the belt so that they are in contact with each other and at rest. Determine which, if any, of the packages will move and the friction force acting on each package.

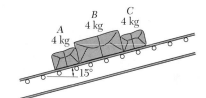

**Fig. P8.13**

**8.14**  Solve Prob. 8.13, assuming that package $B$ is placed to the right of both packages $A$ and $C$.

**8.15**  Block $A$ weighs 50 lb and block $B$ weighs 25 lb. Knowing that the coefficient of static friction is 0.15 between all surfaces of contact, determine the value of $\theta$ for which motion is impending.

**8.16**  Solve Prob. 8.15, assuming that the coefficient of static friction is 0.15 between the two blocks, and zero between block $B$ and the incline.

**8.17**  For the system of Prob. 8.15, and assuming that the coefficients of friction are $\mu_s = 0.15$ and $\mu_k = 0.12$ between the two blocks, and zero between block $B$ and the incline, determine the magnitude of the friction force between the two blocks when ($a$) $\theta = 25°$, ($b$) $\theta = 35°$.

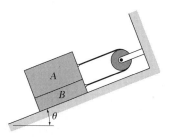

**Fig. P8.15**

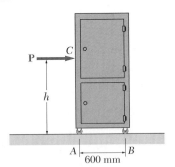

**Fig. P8.18 and P8.19**

**8.18**  A 60-kg cabinet is mounted on casters which can be locked to prevent their rotation. The coefficient of friction is 0.30. If $h = 800$ mm, determine the magnitude of the force **P** required to move the cabinet to the right (*a*) if all casters are locked, (*b*) if the casters at *B* are locked and the casters at *A* are free to rotate, (*c*) if the casters at *A* are locked and the casters at *B* are free to rotate.

**8.19**  A 60-kg cabinet is mounted on casters which can be locked to prevent their rotation. The coefficient of friction between the floor and each caster is 0.30. Assuming that the casters at both *A* and *B* are locked, determine (*a*) the force **P** required to move the cabinet to the right, (*b*) the largest allowable value of *h* if the cabinet is not to tip over.

**8.20**  A half section of pipe, weighing 200 lb, is to be moved to the right along the floor without tipping. Knowing that the coefficient of friction between the pipe and the floor is 0.40, determine (*a*) the largest allowable value of $\alpha$, (*b*) the corresponding tension *T*.

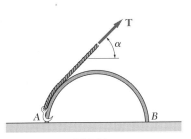

**Fig. P8.20 and P8.21**

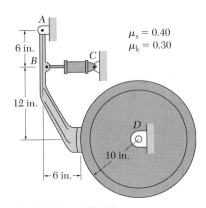

**Fig. P8.22 and P8.23**

**8.21**  A half section of pipe, weighing 200 lb, is pulled by a cable as shown. The coefficient of friction between the pipe and the floor is 0.40. If $\alpha = 30°$, determine (*a*) the tension *T* required to move the pipe, (*b*) whether the pipe will slide or tip.

**8.22**  A couple of moment 60 lb · ft is applied to the drum. Determine the smallest force which must be exerted by the hydraulic cylinder if the drum is not to rotate, when the applied couple is directed (*a*) clockwise, (*b*) counterclockwise.

**8.23**  The hydraulic cylinder exerts on point *B* a force of 500 lb directed to the right. Determine the moment of the friction force about the axle of the drum when the drum is rotating (*a*) clockwise, (*b*) counterclockwise.

**8.24**  Find the moment of the largest couple **M** which may be applied to the cylinder if it is not to spin. The cylinder has a weight *W* and a radius *r*, and the coefficient of static friction $\mu_s$ is the same at *A* and *B*.

**8.25**  The cylinder has a weight *W* and a radius *r*. Express in terms of *W* and *r* the moment of the largest couple **M** which may be applied to the cylinder if it is not to spin, assuming the coefficient of static friction to be (*a*) zero at *A* and 0.40 at *B*, (*b*) 0.30 at *A* and 0.40 at *B*.

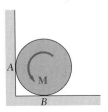

**Fig. P8.24 and P8.25**

**\*8.26** A cord is attached to and partially wound around a cylinder of weight $W$ and radius $r$ which rests on an incline as shown. Knowing that the coefficient of static friction between the cylinder and the incline is 0.30, find (a) the smallest allowable value of $\theta$ if the cylinder is to remain in equilibrium, (b) the corresponding value of the tension in the cord.

**\*8.27** A cord is attached to and partially wound around a cylinder of weight $W$ and radius $r$ which rests on an incline as shown. Knowing that $\theta = 40°$, find (a) the tension in the cord, (b) the smallest value of the coefficient of static friction between the cylinder and the incline for which equilibrium is maintained.

**8.28** A car is stopped with its front wheels resting against a curb when its driver starts the engine and tries to drive over the curb. Knowing that the radius of the wheels is 280 mm, that the coefficient of static friction between the tires and the pavement is 0.90, and that 60 percent of the weight of the car is distributed over its front wheels and 40 percent over its rear wheels, determine the largest curb height $h$ that the car can negotiate, assuming (a) front-wheel drive, (b) rear-wheel drive.

**8.29** Solve Prob. 8.28, assuming that the weight of the car is equally distributed over its front and rear wheels.

**8.30** The end $A$ of a slender, uniform rod of length $L$ and weight $W$ bears on a horizontal surface, while its end $B$ is supported by a cord $BC$. Knowing that the coefficients of friction are $\mu_s = 0.35$ and $\mu_k = 0.25$, determine (a) the maximum value of $\theta$ for which equilibrium is maintained, (b) the corresponding value of the tension in the cord.

**8.31** Determine whether the rod of Prob. 8.30 is in equilibrium when $\theta = 30°$ and find the magnitude and direction of the friction force exerted on the rod.

**8.32** A slender rod of length $L$ is lodged between peg $C$ and the vertical wall and supports a load $\mathbf{P}$ at end $A$. Knowing that $\theta = 30°$ and that the coefficient of static friction is 0.20 at both $B$ and $C$, find the range of values of the ratio $L/a$ for which equilibrium is maintained.

**8.33** Solve Prob. 8.32, assuming that the coefficient of static friction is 0.20 at $B$ and zero at $C$.

**8.34** Solve Prob. 8.32, assuming that the coefficient of static friction is 0.20 at $C$ and zero at $B$.

**8.35** A slender rod of length $L$ is lodged between peg $C$ and the vertical wall and supports a load $\mathbf{P}$ at end $A$. Knowing that $L = 10a$ and $\theta = 30°$ and that the coefficients of friction are $\mu_s = 0.20$ and $\mu_k = 0.15$ at $C$, and zero at $B$, determine whether the rod is in equilibrium.

**8.36** Solve Prob. 8.35, assuming that $L = 5a$ and $\theta = 30°$ and that the coefficients of friction are $\mu_s = 0.20$ and $\mu_k = 0.15$ at $B$, and zero at $C$.

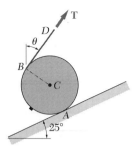

**Fig. P8.26 and P8.27**

**Fig. P8.28**

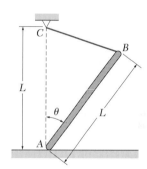

**Fig. P8.30**

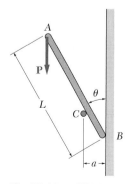

**Fig. P8.32 and P8.35**

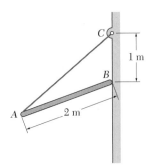

**Fig. P8.37**

**8.37** A 10-kg rod $AB$, of length 2 m and of uniform cross section, is held in equilibrium as shown, with one end against a vertical wall and the other attached to a cord. Knowing that the coefficients of friction between the rod and the wall are $\mu_s = 0.35$ and $\mu_k = 0.25$, determine the range of values of the length $L$ of the cord for which equilibrium is maintained.

**8.38** Assuming that the length of cord $AC$ in Prob. 8.37 is $L = 2.50$ m, determine whether the rod is in equilibrium and find the magnitude and direction of the friction force exerted on the rod.

**8.39** Solve Prob. 8.37, assuming that the coefficients of friction between the rod and the wall are $\mu_s = 0.60$ and $\mu_k = 0.45$.

**8.40** The shear shown is used to cut and trim electronic-circuit-board laminates. Knowing that the coefficient of kinetic friction between the blade and the vertical guide is 0.25, determine the force exerted by the edge $E$ of the blade on the laminate.

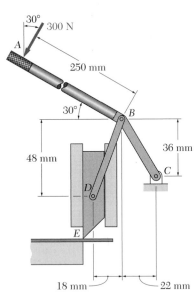

**Fig. P8.40**

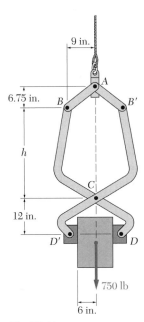

**Fig. P8.41**

**8.41** The friction tongs shown are used to lift a 750-lb casting. Knowing that $h = 30$ in., determine the smallest allowable value of the coefficient of static friction between the casting and blocks $D$ and $D'$.

**8.42** For the friction tongs of Prob. 8.41, determine the smallest allowable value of $h$ if the coefficient of static friction between the casting and blocks $D$ and $D'$ is 0.30.

**8.43** Two large cylinders, each of radius $r = 600$ mm, rotate in opposite directions and form the main elements of a crusher for stone aggregate. The distance $d$ is set equal to the maximum desired size of the crushed aggregate. If $d = 25$ mm and $\mu_s = 0.30$, determine the size $s$ of the largest stones which will be pulled through the crusher by friction alone.

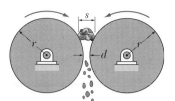

**Fig. P8.43**

**8.44** A safety device used by workers climbing ladders fixed to high structures consists of a rail attached to the ladder and a sleeve which may slide on the flange of the rail. A chain connects the worker's belt to the end of an eccentric cam which may rotate about an axle attached to the sleeve at $C$. Determine the smallest allowable common value of the coefficient of static friction between the flange of the rail, the pins at $A$ and $B$, and the eccentric cam if the sleeve is not to slide down when the chain is pulled vertically downward.

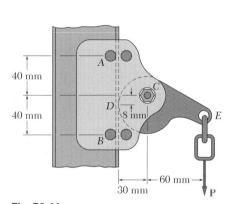

**Fig. P8.44**

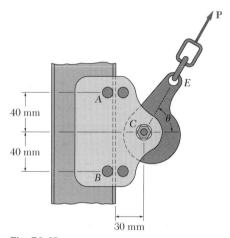

**Fig. P8.45**

**8.45** To be of practical use, the safety sleeve described in the preceding problem must be free to slide along the rail when pulled upward. Determine the largest allowable value of the coefficient of static friction between the flange of the rail and the pins at $A$ and $B$ if the sleeve is to be free to slide when pulled as shown in the figure, assuming $(a)$ $\theta = 70°$, $(b)$ $\theta = 60°$, $(c)$ $\theta = 50°$.

**8.46** The uniform rod $AB$ lies in a vertical plane with its ends resting against the surfaces $AC$ and $BC$. Knowing that the coefficient of static friction is 0.30 between the rod and each of the surfaces, determine the range of values of $\theta$ corresponding to equilibrium when $(a)$ $\alpha = 45°$, $(b)$ $\alpha = 30°$.

**8.47** The uniform rod $AB$ lies in a vertical plane with its ends resting against the surfaces $AC$ and $BC$. Determine the range of values of $\theta$ corresponding to equilibrium in terms of $\alpha$ and the angle of static friction $\phi_s$ between the rod and the two surfaces. Consider the case when $(a)$ $\alpha > 2\phi_s$, $(b)$ $\alpha \leq 2\phi_s$.

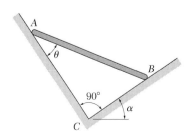

**Fig. P8.46 and P8.47**

**8.48** The slender rod $AB$ of length $l$ = 300 mm is attached to a collar at $A$ and rests on a wheel located at a vertical distance $a$ = 50 mm from the horizontal rod on which the collar slides. Neglecting the friction at $C$ and knowing that the coefficient of static friction between the collar and the horizontal rod is 0.25, determine the range of values of $Q$ for which equilibrium is maintained when $P$ = 60 N and $\theta$ = 20°.

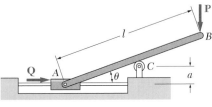

**Fig. P8.48**

**8.49** Solve Prob. 8.48, assuming that the wheel at $C$ is frozen in its bearings and that the coefficient of static friction between the wheel and rod $AB$ is also 0.25.

**8.50** The mechanism shown is acted upon by the force **P** of magnitude $P$ = 20 lb. Knowing that $\theta$ = 25° and that the coefficient of static friction between collar $C$ and the horizontal rod is 0.30, determine the range of values of $Q$ for which equilibrium is maintained.

**Fig. P8.50**

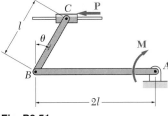

**Fig. P8.51**

**8.51** Denoting by $\mu_s$ the coefficient of static friction between the collar $C$ and the horizontal rod, determine the largest moment of the couple **M** for which equilibrium is maintained. Explain what happens if $\mu_s \geq \tan \theta$.

**\*8.52** Two 2.5-m beams are pin-connected to $D$ and support two loads **P** and **Q** as shown. Knowing that the coefficient of static friction is zero at $A$ and 0.25 at $B$ and $C$, determine the smallest value of $P$ for which equilibrium is maintained when $Q$ = 550 N. (*Hint.* Note that $C$ moves up when $B$ moves to the right.)

**\*8.53** Solve Prob. 8.52, assuming that the coefficient of static friction is 0.25 at all surfaces of contact.

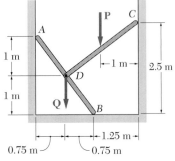

**Fig. P8.52**

**\*8.54** Two slender rods of negligible weight are pin-connected at $A$ and attached to the 20-lb block $B$ and the 60-lb block $C$ as shown. The coefficient of static friction is 0.60 between all surfaces of contact. Determine the range of values of $P$ for which equilibrium is maintained.

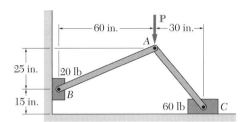

**Fig. P8.54 and P8.55**

**\*8.55** Two slender rods of negligible weight are pin-connected at $A$ and attached to the 20-lb block $B$ and the 60-lb block $C$ as shown. A vertical load of magnitude $P = 300$ lb is applied at $A$. Determine (a) the minimum value of the coefficient of static friction (at both $B$ and $C$) for which equilibrium is maintained, (b) whether sliding is impending at $B$ or at $C$.

**\*8.56** Solve Prob. 8.55, assuming that $P = 200$ lb.

**\*8.57** A 3-m plank lies on two rollers $A$ and $B$, located 1.5 m apart on a 10° incline. Roller $A$ rotates slowly clockwise at a constant speed; roller $B$ is idle. Initially, the plank is placed on the rollers with its center of gravity at $A$ ($x = 0$) and its end $C$ at $B$. Knowing that the coefficients of friction between the board and roller $A$ are $\mu_s = 0.40$ and $\mu_k = 0.30$, (a) determine the value reached by $x$ when the plank starts slipping on roller $A$, (b) describe what happens next.

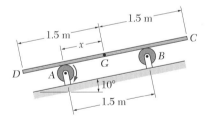

**Fig. P8.57**

**\*8.58** Solve Prob. 8.57, assuming that roller $B$ is frozen in its bearings and that the coefficients of friction are the same at $B$ as at $A$.

**\*8.59** The mathematical model shown, consisting of blocks connected by springs, has been developed for the analysis of a certain structure. The weight of each block is $W = 5$ N, the constant of each spring is $k = 20$ N/m, and the coefficient of friction between the base and each block is 0.40. Knowing that the initial tension in each spring is zero, construct a graph showing the magnitude of the force **P** versus the position of block $A$ as $P$ increases from zero to 5 N and then decreases to zero.

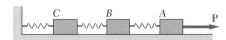

**Fig. P8.59**

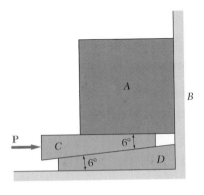

(a)

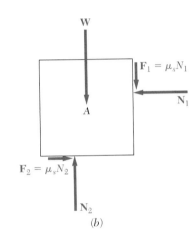

(b)

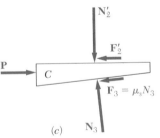

(c)

**Fig. 8.7**

## 8.5. WEDGES

Wedges are simple machines used to raise large stone blocks and other heavy loads. These loads may be raised by applying to the wedge a force usually considerably smaller than the weight of the load. Besides, because of the friction existing between the surfaces in contact, a wedge, if properly shaped, will remain in place after being forced under the load. Wedges may thus be used advantageously to make small adjustments in the position of heavy pieces of machinery.

Consider the block $A$ shown in Fig. 8.7a. This block rests against a vertical wall $B$ and is to be raised slightly by forcing a wedge $C$ between block $A$ and a second wedge $D$. We want to find the minimum value of the force **P** which must be applied to the wedge $C$ to move the block. We shall assume that the weight **W** of the block is known. It may have been given in pounds or determined in newtons from the mass of the block expressed in kilograms.

The free-body diagrams of block $A$ and of wedge $C$ have been drawn in Fig. 8.7b and c. The forces acting on the block include its weight and the normal and friction forces at the surfaces of contact with wall $B$ and wedge $C$. The magnitudes of the friction forces $\mathbf{F}_1$ and $\mathbf{F}_2$ are equal, respectively, to $\mu_s N_1$ and $\mu_s N_2$ since the motion of the block must be started. It is important to show the friction forces with their correct sense. Since the block will move upward, the force $\mathbf{F}_1$ exerted by the wall on the block must be directed downward. On the other hand, since the wedge $C$ moves to the right, the relative motion of $A$ with respect to $C$ is to the left and the force $\mathbf{F}_2$ exerted by $C$ on $A$ must be directed to the right.

Considering now the free body $C$ in Fig. 8.7c, we note that the forces acting on $C$ include the applied force **P** and the normal and friction forces at the surfaces of contact with $A$ and $D$. The weight of the wedge is small compared with the other forces involved and may be neglected. The forces $\mathbf{N}_2'$ and $\mathbf{F}_2'$ acting on $C$ are equal and opposite to the forces $\mathbf{N}_2$ and $\mathbf{F}_2$ acting on $A$; the friction force $\mathbf{F}_2'$ must therefore be directed to the left. We check that the force $\mathbf{F}_3$ is also directed to the left.

The total number of unknowns involved in the two free-body diagrams may be reduced to four if the friction forces are expressed in terms of the normal forces. Expressing that block $A$ and wedge $C$ are in equilibrium will provide four equations which may be solved to obtain the magnitude of **P**. It should be noted that in the example considered here, it will be more convenient to replace each pair of normal and friction forces by their resultant. Each free body is then subjected to only three forces, and the problem may be solved by drawing the corresponding force triangles (see Sample Prob. 8.4).

## 8.6. SQUARE-THREADED SCREWS

Square-threaded screws are frequently used in jacks, presses, and other mechanisms. Their analysis is similar to that of a block sliding along an inclined plane.

Consider the jack shown in Fig. 8.8. The screw carries a load **W** and is supported by the base of the jack. Contact between screw and base takes place along a portion of their threads. By applying a force **P** on the handle, the screw may be made to turn and to raise the load **W**.

The thread of the base has been unwrapped and shown as a straight line in Fig. 8.9a. The correct slope was obtained by plotting horizontally the product $2\pi r$, where $r$ is the mean radius of the thread, and vertically the *lead L* of the screw, i.e., the distance through which the screw advances in one turn. The angle $\theta$ this line forms with the horizontal is the *lead angle*. Since the force of friction between two surfaces in contact does not depend upon the area of contact, the two threads may be assumed to be in contact over a much smaller area than they actually are and the screw may be represented by the block shown in Fig. 8.9a. It should be noted, however, that in this analysis of the jack, the friction between cap and screw is neglected.

The free-body diagram of the block should include the load **W**, the reaction **R** of the base thread, and a horizontal force **Q** having the same effect as the force **P** exerted on the handle. The force **Q** should have the same moment as **P** about the axis of the screw and its magnitude should thus be $Q = Pa/r$. The force **Q**, and thus the force **P** required to raise the load **W**, may be obtained from the free-body diagram shown in Fig. 8.9a. The friction angle is taken equal to $\phi_s$ since the load will presumably be raised through a succession of short strokes. In mechanisms providing for the continuous rotation of a screw, it may be desirable to distinguish between the force required to start motion (using $\phi_s$) and that required to maintain motion (using $\phi_k$).

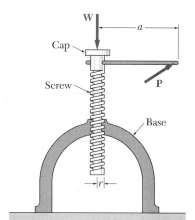

**Fig. 8.8**

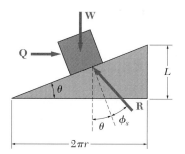

(a) Impending motion upward

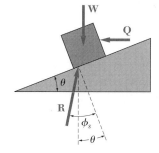

(b) Impending motion downward with $\phi_s > \theta$

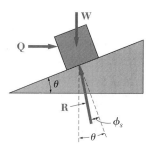

(c) Impending motion downward with $\phi_s < \theta$

**Fig. 8.9** Block-and-incline analysis of a screw.

If the friction angle $\phi_s$ is larger than the lead angle $\theta$, the screw is said to be *self-locking;* it will remain in place under the load. To lower the load, we must then apply the force shown in Fig. 8.9b. If $\phi_s$ is smaller than $\theta$, the screw will unwind under the load; it is then necessary to apply the force shown in Fig. 8.9c to maintain equilibrium.

The lead of a screw should not be confused with its *pitch.* The lead was defined as the distance through which the screw advances in one turn; the pitch is the distance measured between two consecutive threads. While lead and pitch are equal in the case of *single-threaded* screws, they are different in the case of *multiple-threaded* screws, i.e., screws having several independent threads. It is easily verified that for double-threaded screws, the lead is twice as large as the pitch; for triple-threaded screws, it is three times as large as the pitch, etc.

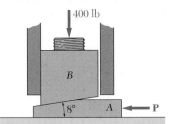

400 lb

B

8°   A ← **P**

# SAMPLE PROBLEM 8.4

The position of the machine block $B$ is adjusted by moving the wedge $A$. Knowing that the coefficient of static friction is 0.35 between all surfaces of contact, determine the force **P** required (*a*) to raise block $B$, (*b*) to lower block $B$.

**Solution.**   For each part, the free-body diagrams of block $B$ and wedge $A$ are drawn, together with the corresponding force triangles, and the law of sines is used to find the desired forces. We note that since $\mu_s = 0.35$, the angle of friction is

$$\phi_s = \tan^{-1} 0.35 = 19.3°$$

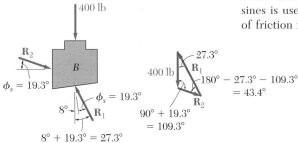

### *a.* Force P to Raise Block
*Free Body: Block B*

$$\frac{R_1}{\sin 109.3°} = \frac{400\ \text{lb}}{\sin 43.4°}$$

$$R_1 = 549\ \text{lb}$$

*Free Body: Wedge A*

$$\frac{P}{\sin 46.6°} = \frac{549\ \text{lb}}{\sin 70.7°}$$

$$P = 423\ \text{lb} \qquad\qquad \mathbf{P} = 423\ \text{lb} \leftarrow \blacktriangleleft$$

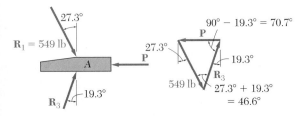

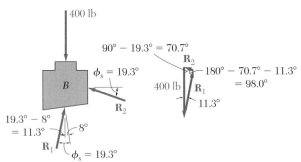

### *b.* Force P to Lower Block
*Free Body: Block B*

$$\frac{R_1}{\sin 70.7°} = \frac{400\ \text{lb}}{\sin 98.0°}$$

$$R_1 = 381\ \text{lb}$$

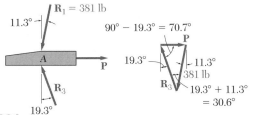

*Free Body: Wedge A*

$$\frac{P}{\sin 30.6°} = \frac{381\ \text{lb}}{\sin 70.7°}$$

$$P = 206\ \text{lb} \qquad\qquad \mathbf{P} = 206\ \text{lb} \rightarrow \blacktriangleleft$$

# SAMPLE PROBLEM 8.5

A clamp is used to hold two pieces of wood together as shown. The clamp has a double square thread of mean diameter equal to 10 mm and with a pitch of 2 mm. The coefficient of friction between threads is $\mu_s = 0.30$. If a maximum torque of 40 N · m is applied in tightening the clamp, determine (a) the force exerted on the pieces of wood, (b) the torque required to loosen the clamp.

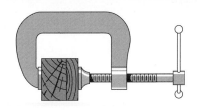

### a. Force Exerted by Clamp.

The mean radius of the screw is $r = 5$ mm. Since the screw is double-threaded, the lead $L$ is equal to twice the pitch: $L = 2(2 \text{ mm}) = 4$ mm. The lead angle $\theta$ and the friction angle $\phi_s$ are obtained by writing

$$\tan \theta = \frac{L}{2\pi r} = \frac{4 \text{ mm}}{10\pi \text{ mm}} = 0.1273 \qquad \theta = 7.3°$$

$$\tan \phi_s = \mu_s = 0.30 \qquad \phi_s = 16.7°$$

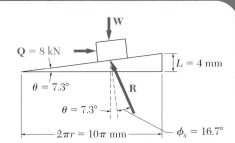

The force **Q** which should be applied to the block representing the screw is obtained by expressing that its moment $Qr$ about the axis of the screw is equal to the applied torque.

$$Q(5 \text{ mm}) = 40 \text{ N} \cdot \text{m}$$

$$Q = \frac{40 \text{ N} \cdot \text{m}}{5 \text{ mm}} = \frac{40 \text{ N} \cdot \text{m}}{5 \times 10^{-3} \text{ m}} = 8000 \text{ N} = 8 \text{ kN}$$

The free-body diagram and the corresponding force triangle may now be drawn for the block; the magnitude of the force **W** exerted on the pieces of wood is obtained by solving the triangle.

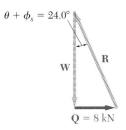

$$W = \frac{Q}{\tan (\theta + \phi_s)} = \frac{8 \text{ kN}}{\tan 24.0°}$$

$$W = 17.97 \text{ kN} \quad \blacktriangleleft$$

### b. Torque Required to Loosen Clamp.

The force **Q** required to loosen the clamp and the corresponding torque are obtained from the free-body diagram and force triangle shown.

$$Q = W \tan (\phi_s - \theta) = (17.97 \text{ kN}) \tan 9.4°$$
$$= 2.975 \text{ kN}$$

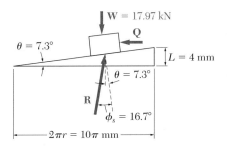

$$\text{Torque} = Qr = (2.975 \text{ kN})(5 \text{ mm})$$
$$= (2.975 \times 10^3 \text{ N})(5 \times 10^{-3} \text{ m}) = 14.87 \text{ N} \cdot \text{m}$$

$$\text{Torque} = 14.87 \text{ N} \cdot \text{m} \quad \blacktriangleleft$$

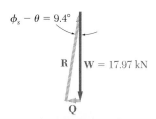

# PROBLEMS

**8.60** The machine part *ABC* is supported by a frictionless hinge at *B* and a 10° wedge at *C*. Knowing that the coefficient of static friction is 0.20 at both surfaces of the wedge, determine (*a*) the force **P** required to move the wedge to the left, (*b*) the components of the corresponding reaction at *B*.

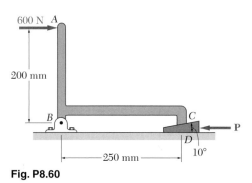

**Fig. P8.60**

**8.61** Solve Prob. 8.60, assuming that the wedge is to be moved to the right.

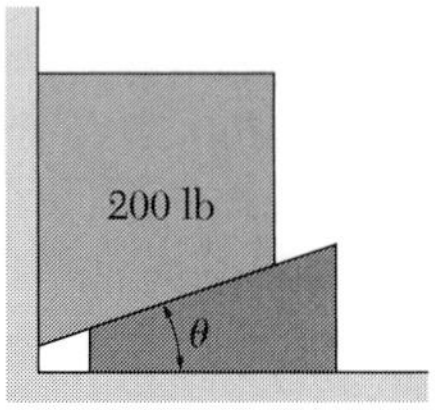

**8.62** A 200-lb block rests as shown on a wedge of negligible weight. Knowing that the coefficient of static friction is 0.30 at all surfaces of contact, determine the angle *θ* for which sliding is impending and compute the corresponding value of the normal force exerted on the block by the vertical wall.

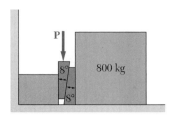

**Fig. P8.62**

**8.63** Solve Prob. 8.62, assuming (*a*) that the coefficient of static friction is zero between the block and the vertical wall and 0.30 at the other two surfaces of contact, (*b*) that the coefficient of static friction is zero between the block and the wedge and 0.30 at the other two surfaces of contact.

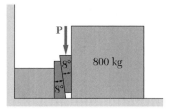

**Fig. P8.64**

**Fig. P8.65**

**8.64 and 8.65** Two 8° wedges of negligible weight are used to move and position the 800-kg block. Knowing that the coefficient of static friction is 0.25 at all surfaces of contact, determine the smallest force **P** which should be applied as shown to one of the wedges.

**8.66** The elevation of the end of the steel floor beam is adjusted by means of the steel wedges $E$ and $F$. The steel base plate $CD$ has been welded to the lower flange of the beam; the end reaction of the beam is known to be 15,000 lb. The coefficient of static friction is 0.30 between two steel surfaces and 0.60 between steel and concrete. If horizontal motion of the base plate is prevented by the force $Q$, determine (a) the force $\mathbf{P}$ required to raise the beam, (b) the corresponding force $\mathbf{Q}$.

**8.67** Solve Prob. 8.66, assuming that the end of the beam is to be lowered.

**8.68** A wedge $A$ of negligible weight is to be driven between two 40-kg plates $B$ and $C$. The coefficient of static friction between all surfaces of contact is 0.35. Determine the magnitude of the force $\mathbf{P}$ required to start moving the wedge (a) if the plates are equally free to move, (b) if plate $C$ is securely bolted to the surface.

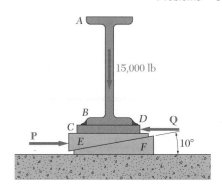

Fig. P8.66

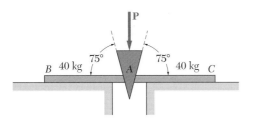

Fig. P8.68

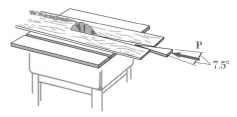

Fig. P8.69

**8.69** A 15° wedge is forced into a saw cut to prevent binding of the circular saw. The coefficient of static friction between the wedge and the wood is 0.25. Knowing that a horizontal force $\mathbf{P}$ of magnitude 140 N was required to insert the wedge, determine the magnitude of the forces exerted on the board by the wedge after it has been inserted.

**8.70** The spring of the door latch has a constant of 300 N/m and in the position shown exerts a 2.4-N force on the bolt. The coefficient of static friction between the bolt and the guide plate is 0.40; all other surfaces are well lubricated and may be assumed frictionless. Determine the magnitude of the force $\mathbf{P}$ required to start closing the door.

**8.71** In Prob. 8.70, determine the angle which the face of the bolt should form with the line $BC$ if the force $\mathbf{P}$ required to close the door is to be the same for both the position shown and the position when $B$ is almost at the guide plate.

**8.72** A 5° wedge is to be forced under a 1400-lb machine base at $A$. Knowing that the coefficient of static friction is 0.20 at all surfaces, (a) determine the force $\mathbf{P}$ required to move the wedge, (b) indicate whether the machine will move.

**8.73** Solve Prob. 8.72, assuming that the wedge is to be forced under the machine base at $B$ instead of $A$.

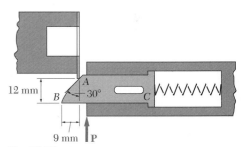

Fig. P8.70

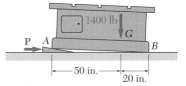

Fig. P8.72

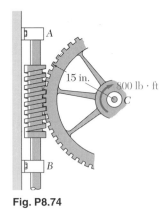

**Fig. P8.74**

**8.74**  The square-threaded worm gear shown has a mean radius of 2 in. and a pitch of $\frac{1}{2}$ in. The large gear is subjected to a constant clockwise torque of 800 lb · ft. Knowing that the coefficient of static friction between gear teeth is 0.10, determine the torque which must be applied to shaft $AB$ in order to rotate the large gear counterclockwise. Neglect friction in the bearings at $A$, $B$, and $C$.

**8.75**  In Prob. 8.74, determine the torque which must be applied to shaft $AB$ in order to rotate the large gear clockwise.

**8.76**  The main features of a screw-luffing crane are shown. Distances $AD$ and $CD$ are each 9 m. The position of the 6-Mg boom $CDE$ is controlled by the screw $ABC$, which is double-threaded at each end (left-handed thread at $A$, right-handed thread at $C$). Each thread has a pitch of 16 mm and a mean diameter of 200 mm. If the coefficient of static friction is 0.08, determine the moment of the couple which must be applied to the screw $(a)$ to raise the boom, $(b)$ to lower the boom.

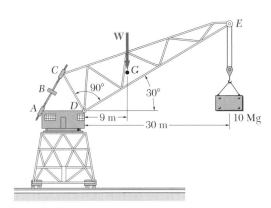

**Fig. P8.76**

**8.77**  Derive the following formulas relating the load **W** and the force **P** exerted on the handle of the jack discussed in Sec. 8.6: $(a)$ $P = (Wr/a) \tan(\theta + \phi_s)$, to raise the load; $(b)$ $P = (Wr/a) \tan(\phi_s - \theta)$, to lower the load if the screw is self-locking; $(c)$ $P = (Wr/a) \tan(\theta - \phi_s)$, to hold the load if the screw is not self-locking.

**8.78**  In the vise shown, the screw is single-threaded in the upper member; it passes through the lower member and is held by a frictionless washer. The pitch of the screw is 3 mm, its mean radius is 12 mm, and the coefficient of static friction is 0.15. Determine the magnitude $P$ of the forces exerted by the jaws when a 60-N · m torque is applied to the screw.

**8.79**  Solve Prob. 8.78, assuming that the screw is single-threaded at both $A$ and $B$ (right-handed thread at $A$ and left-handed thread at $B$).

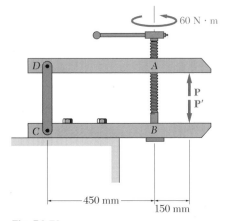

**Fig. P8.78**

**8.80** The vise shown consists of two members connected by two double-threaded screws of mean radius 0.20 in. and pitch 0.06 in. The lower member is threaded at $A$ and $B$ ($\mu_s = 0.35$), but the upper member is not threaded. It is desired to apply two equal and opposite forces of 100 lb on the blocks held between the jaws. (*a*) What screw should be adjusted first? (*b*) What is the maximum torque applied in tightening the second screw?

**8.81** Solve part *b* of Prob. 8.80, assuming that the wrong screw is adjusted first.

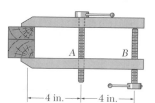

**Fig. P8.80**

**8.82** In the machinist's vise shown, the movable jaw $D$ is rigidly attached to the tongue $AB$ which fits loosely into the fixed body of the vise. The screw is single-threaded into the fixed base and has a mean diameter of 15 mm and a pitch of 6 mm. The coefficient of static friction is 0.25 between the threads and also between the tongue and the body. Neglecting bearing friction between the screw and the movable head, determine the torque which must be applied to the handle in order to produce a clamping force of 4 kN.

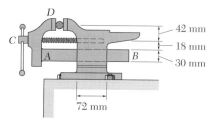

**Fig. P8.82**

**8.83** In Prob. 8.82, a clamping force of 4 kN was obtained by tightening the vise. Determine the torque which must be applied to the screw to loosen the vise.

## *8.7. JOURNAL BEARINGS. AXLE FRICTION

Journal bearings are used to provide lateral support to rotating shafts and axles. Thrust bearings, which will be studied in the next section, are used to provide axial support to shafts and axles. If the journal bearing is fully lubricated, the frictional resistance depends upon the speed of rotation, the clearance between axle and bearing, and the viscosity of the lubricant. As indicated in Sec. 8.1, such problems are studied in fluid mechanics. The methods of this chapter, however, may be applied to the study of axle friction when the bearing is not lubricated or only partially lubricated. We may then assume that the axle and the bearing are in direct contact along a single straight line.

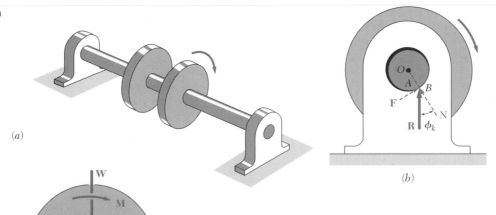

(a)

(b)

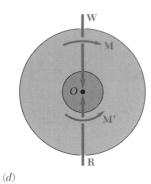

(c)

(d)

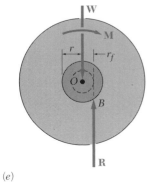

(e)

**Fig. 8.10**

Consider two wheels, each of weight **W**, rigidly mounted on an axle supported symmetrically by two journal bearings (Fig. 8.10*a*). If the wheels rotate, we find that to keep them rotating at constant speed, it is necessary to apply to each of them a couple **M**. A free-body diagram has been drawn in Fig. 8.10*c*, which represents one of the wheels and the corresponding half axle in projection on a plane perpendicular to the axle. The forces acting on the free body include the weight **W** of the wheel, the couple **M** required to maintain its motion, and a force **R** representing the reaction of the bearing. This force is vertical, equal, and opposite to **W** but does not pass through the center *O* of the axle; **R** is located to the right of *O* at a distance such that its moment about *O* balances the moment **M** of the couple. Contact between the axle and bearing, therefore, does not take place at the lowest point *A* when the axle rotates. It takes place at point *B* (Fig. 8.10*b*) or, rather, along a straight line intersecting the plane of the figure at *B*. Physically, this is explained by the fact that when the wheels are set in motion, the axle "climbs" in the bearings until slippage occurs. After sliding back slightly, the axle settles more or less in the position shown. This position is such that the angle between the reaction **R** and the normal to the surface of the bearing is equal to the angle of kinetic friction $\phi_k$. The distance from *O* to the line of action of **R** is thus $r \sin \phi_k$, where *r* is the radius of the axle. Writing that $\Sigma M_O = 0$ for the forces acting on the free body considered, we obtain the moment *M* of the couple required to overcome the frictional resistance of one of the bearings:

$$M = Rr \sin \phi_k \tag{8.5}$$

Observing that, for small values of the angle of friction, $\sin \phi_k$ may be replaced by $\tan \phi_k$, that is, by $\mu_k$, we write the approximate formula

$$M \approx Rr\mu_k \tag{8.6}$$

In the solution of certain problems, it may be more convenient to let the line of action of **R** pass through *O*, as it does when the axle does not rotate. A couple M' of the same moment *M* as the couple **M** but of opposite sense must then be added to the reaction **R** (Fig. 8.10*d*). This couple represents the frictional resistance of the bearing.

In case a graphical solution is preferred, the line of action of **R** may be readily drawn (Fig. 8.10*e*) if we note that it must be tangent to a circle centered at *O* and of radius

$$r_f = r\sin \phi_k \approx r\mu_k \tag{8.7}$$

This circle is called the *circle of friction* of the axle and bearing and is independent of the loading conditions of the axle.

## *8.8. THRUST BEARINGS. DISK FRICTION

Thrust bearings are used to provide axial support to rotating shafts and axles. They are of two types: (1) end bearings and (2) collar bearings (Fig. 8.11). In the case of collar bearings, friction forces develop between the two ring-shaped areas which are in contact. In the case of end bearings, friction takes place over full circular areas, or over ring-shaped areas when the end of the shaft is hollow. Friction between circular areas, called *disk friction*, also occurs in other mechanisms, such as *disk clutches*.

To obtain a formula which is valid in the most general case of disk friction, we shall consider a rotating hollow shaft. A couple **M** keeps the shaft rotating at constant speed while a force **P** maintains it in contact with a fixed bearing (Fig. 8.12).Contact between the shaft and the bearing takes place over a ring-shaped area of inner radius $R_1$ and outer radius $R_2$. Assuming that the pressure between the two surfaces in contact is uniform, we find that the magnitude of the normal force $\Delta N$ exerted on an element of area $\Delta A$ is $\Delta N = P \Delta A/A$, where $A = \pi(R_2^2 - R_1^2)$, and that the magnitude of the friction force $\Delta F$ acting on $\Delta A$ is $\Delta F = \mu_k \Delta N$. Denoting by $r$ the distance from the axis of the shaft to the element of area $\Delta A$, we express as follows the moment $\Delta M$ of $\Delta F$ about the axis of the shaft:

$$\Delta M = r \Delta F = \frac{r\mu_k P \Delta A}{\pi(R_2^2 - R_1^2)}$$

The equilibrium of the shaft requires that the moment *M* of the couple applied to the shaft be equal in magnitude to the sum of the moments of the friction forces $\Delta F$. Replacing $\Delta A$ by the infinitesimal element $dA = r\, d\theta\, dr$ used with polar coordinates, and integrating over the area of

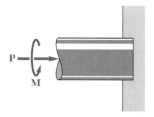

(*a*) End bearing

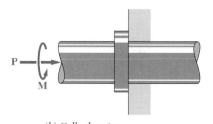

(*b*) Collar bearing

**Fig. 8.11** Thrust bearings.

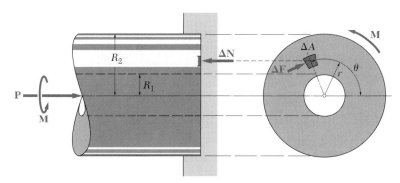

**Fig. 8.12**

contact, we thus obtain the following expression for the moment $M$ of the couple required to overcome the frictional resistance of the bearing:

$$M = \frac{\mu_k P}{\pi (R_2^2 - R_1^2)} \int_0^{2\pi} \int_{R_1}^{R_2} r^2 \, dr \, d\theta$$

$$= \frac{\mu_k P}{\pi (R_2^2 - R_1^2)} \int_0^{2\pi} \tfrac{1}{3} (R_2^3 - R_1^3) \, d\theta$$

$$M = \tfrac{2}{3} \mu_k P \frac{R_2^3 - R_1^3}{R_2^2 - R_1^2} \tag{8.8}$$

When contact takes place over a full circle of radius $R$, formula (8.8) reduces to

$$M = \tfrac{2}{3} \mu_k PR \tag{8.9}$$

The value of $M$ is then the same as would be obtained if contact between shaft and bearing took place at a single point located at a distance $2R/3$ from the axis of the shaft.

The largest torque which may by transmitted by a disk clutch without causing slippage is given by a formula similar to (8.9), where $\mu_k$ has been replaced by the coefficient of static friction $\mu_s$.

### *8.9. WHEEL FRICTION. ROLLING RESISTANCE

The wheel is one of the most important inventions of our civilization. Its use makes it possible to move heavy loads with relatively little effort. Because the point of the wheel in contact with the ground at any given instant has no relative motion with respect to the ground, the wheel eliminates the large friction forces which would arise if the load were in direct contact with the ground. In practice, however, the wheel is not perfect, and some resistance to its motion exists. This resistance has two distinct causes. It is due (1) to a combined effect of axle friction and friction at the rim and (2) to the fact that the wheel and the ground deform, with the result that contact between wheel and ground takes place, not at a single point, but over a certain area.

To understand better the first cause of resistance to the motion of a wheel, we shall consider a railroad car supported by eight wheels mounted on axles and bearings. The car is assumed to be moving to the right at constant speed along a straight horizontal track. The free-body diagram of one of the wheels is shown in Fig. 8.13a. The forces acting on the free body include the load $\mathbf{W}$ supported by the wheel and the normal reaction $\mathbf{N}$ of the track. Since $\mathbf{W}$ is drawn through the center $O$ of the axle, the frictional resistance of the bearing should be represented by a counterclockwise couple $\mathbf{M}$ (see Sec. 8.7). To keep the free body in equilibrium, we must add two equal and opposite forces $\mathbf{P}$ and $\mathbf{F}$, forming a clockwise couple of moment $M$. The force $\mathbf{F}$ is the friction force exerted by the track on the wheel, and $\mathbf{P}$ represents the force which should be applied to the wheel to keep it rolling at constant speed. Note that the forces $\mathbf{P}$ and $\mathbf{F}$ would not exist if there were no friction between wheel and track. The couple $\mathbf{M}$

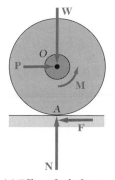

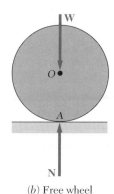

(a) Effect of axle friction           (b) Free wheel           (c) Rolling resistance

**Fig. 8.13**

representing the axle friction would then be zero; the wheel would slide on the track without turning in its bearing.

The couple **M** and the forces **P** and **F** also reduce to zero when there is no axle friction. For example, a wheel which is not held in bearings and rolls freely and at constant speed on horizontal ground (Fig. 8.13b) will be subjected to only two forces: its own weight **W** and the normal reaction **N** of the ground. No friction force will act on the wheel, regardless of the value of the coefficient of friction between wheel and ground. A wheel rolling freely on horizontal ground should thus keep rolling indefinitely.

Experience, however, indicates that the wheel will slow down and eventually come to rest. This is due to the second type of resistance mentioned at the beginning of this section, known as the *rolling resistance*. Under the load **W**, both the wheel and the ground deform slightly, causing the contact between wheel and ground to take place over a certain area. Experimental evidence shows that the resultant of the forces exerted by the ground on the wheel over this area is a force **R** applied at a point $B$, which is not located directly under the center $O$ of the wheel, but slightly in front of it (Fig. 8.13c). To balance the moment of **W** about $B$ and to keep the wheel rolling at constant speed, it is necessary to apply a horizontal force **P** at the center of the wheel. Writing $\Sigma M_B = 0$, we obtain

$$Pr = Wb \tag{8.10}$$

where $r$ = radius of wheel
     $b$ = horizontal distance between $O$ and $B$

The distance $b$ is commonly called the *coefficient of rolling resistance*. It should be noted that $b$ is not a dimensionless coefficient since it represents a length; $b$ is usually expressed in inches or in millimeters. The value of $b$ depends upon several parameters in a manner which has not yet been clearly established. Values of the coefficient of rolling resistance vary from about 0.01 in. or 0.25 mm for a steel wheel on a steel rail to 5.0 in. or 125 mm for the same wheel on soft ground.

## SAMPLE PROBLEM 8.6

A pulley of diameter 4 in. can rotate about a fixed shaft of diameter 2 in. The coefficients of static and kinetic friction between the pulley and shaft are both assumed equal to 0.20. Determine (*a*) the smallest vertical force **P** required to raise a 500-lb load, (*b*) the smallest vertical force **P** required to hold the load, (*c*) the smallest horizontal force **P** required to raise the same load.

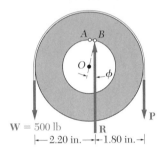

W = 500 lb

|——2.20 in.——|——1.80 in.——|

### *a*. Vertical Force P Required to Raise the Load.

When the forces in both parts of the rope are equal, contact between the pulley and shaft takes place at *A*. When **P** is increased, the pulley rolls around the shaft slightly and contact takes place at *B*. The free-body diagram of the pulley when motion is impending is drawn. The perpendicular distance from the center *O* of the pulley to the line of action of **R** is

$$r_f = r \sin \phi \approx r\mu \qquad r_f \approx (1 \text{ in.})0.20 = 0.20 \text{ in.}$$

Summing moments about *B*, we write

$$+\uparrow \Sigma M_B = 0: \qquad (2.20 \text{ in.})(500 \text{ lb}) - (1.80 \text{ in.})P = 0$$
$$P = 611 \text{ lb} \qquad\qquad \mathbf{P = 611 \text{ lb}\downarrow} \blacktriangleleft$$

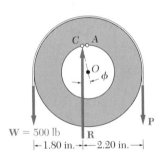

W = 500 lb

|—1.80 in.—|——2.20 in.——|

### *b*. Vertical Force P to Hold the Load.

As the force **P** is decreased, the pulley rolls around the shaft and contact takes place at *C*. Considering the pulley as a free body and summing moments about *C*, we write

$$+\uparrow \Sigma M_C = 0: \qquad (1.80 \text{ in.})(500 \text{ lb}) - (2.20 \text{ in.})P = 0$$
$$P = 409 \text{ lb} \qquad\qquad \mathbf{P = 409 \text{ lb}\downarrow} \blacktriangleleft$$

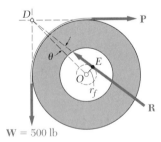

W = 500 lb

### *c*. Horizontal Force P to Raise the Load.

Since the three forces **W**, **P**, and **R** are not parallel, they must be concurrent. The direction of **R** is thus determined from the fact that its line of action must pass through the point of intersection *D* of **W** and **P**, and must be tangent to the circle of friction. Recalling that the radius of the circle of friction is $r_f = 0.20$ in., we write

$$\sin \theta = \frac{OE}{OD} = \frac{0.20 \text{ in.}}{(2 \text{ in.})\sqrt{2}} = 0.0707 \qquad \theta = 4.1°$$

From the force triangle, we obtain

$$P = W \cot (45° - \theta) = (500 \text{ lb}) \cot 40.9°$$
$$= 577 \text{ lb} \qquad\qquad \mathbf{P = 577 \text{ lb} \rightarrow} \blacktriangleleft$$

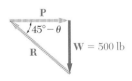

W = 500 lb

# PROBLEMS

**8.84** A hot-metal ladle and its contents have a mass of 60 Mg. Knowing that the coefficient of static friction between the hooks and the pinion is 0.30, determine the tension in cable *AB* required to start tipping the ladle.

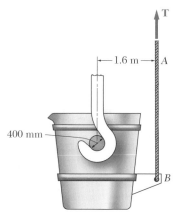

**Fig. P8.84**

**8.85** A lever of negligible weight is loosely fitted onto a 75-mm-diameter fixed shaft. It is observed that the lever will just start rotating if a 2-kg mass is added at *C*. Determine the coefficient of static friction between the shaft and the lever.

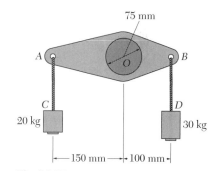

**Fig. P8.85**

**8.86** A windlass of diameter 6 in. is used to raise or lower a 100-lb load. It is supported by two poorly lubricated bearings ($\mu_s = 0.40$) of diameter 2 in. For each of the two positions shown, determine the magnitude of the force **P** required to start raising the load.

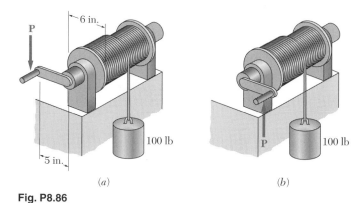

**Fig. P8.86**

**8.87** For the windlass of Prob. 8.86, and for each of the two positions shown, determine the magnitude of the smallest force **P** which will hold the load in place.

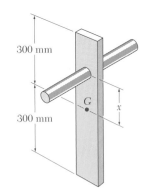

**300 mm**

**300 mm**

$G$

$x$

**Fig. P8.88**

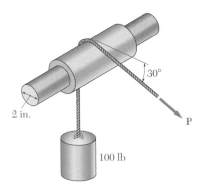

30°

2 in.

100 lb

**P**

**Fig. P8.90**

**8.88** A uniform bar of length 600 mm is supported by a horizontal shaft of diameter 50 mm. The shaft fits loosely in a circular hole in the bar and the coefficient of static friction is 0.40. If $x = 150$ mm and the shaft is slowly rotated, determine the angle at which the bar will slip.

**8.89** For the bar of Prob. 8.88, determine the maximum value of $x$ for which the bar will not slip when the shaft is slowly rotated through a full turn.

**8.90** A bushing of 3-in. outside diameter fits loosely on a horizontal 2-in.-diameter shaft. If a force **P** forming an angle of 30° with the horizontal and having a magnitude $P = 150$ lb is required to start raising the 100-lb load, determine the coefficient of static friction between the shaft and the bushing. Assume that the rope does not slip on the bushing.

**8.91** Knowing that the coefficient of static friction between the shaft and the bushing of Prob. 8.90 is 0.30, determine the magnitude of the smallest force **P** required (a) to start raising the 100-lb load, (b) to hold the 100-lb load in place.

**8.92** A loaded railroad car has a mass of 30 Mg and is supported by eight wheels of 800-mm diameter with 125-mm-diameter axles. Knowing that the coefficients of friction are $\mu_s = 0.020$ and $\mu_k = 0.015$, determine the horizontal force required (a) to get the car in motion, (b) to keep it moving with a constant velocity. Neglect the rolling resistance between the wheels and the track.

**8.93** A scooter is to be designed to roll down a 2 percent slope at a constant speed. Assuming that the coefficient of kinetic friction between the 25-mm-diameter axles and the bearings is 0.10, determine the required diameter of the wheels. Neglect the rolling resistance between the wheels and the ground.

**8.94** A couple of moment $M = 100$ N · m is required to start the vertical shaft rotating. Determine the coefficient of static friction.

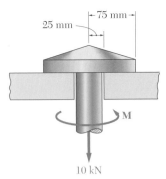

25 mm

75 mm

$M$

10 kN

**Fig. P8.94**

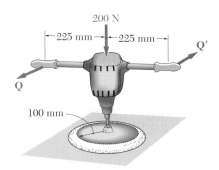

200 N

225 mm — 225 mm

$Q'$

$Q$

100 mm

**Fig. P8.95**

**8.95** A vertical force of 200 N is applied to a 6-kg electric polisher as it is operated on a horizontal surface. Knowing that the coefficient of kinetic friction is 0.25, determine the magnitude $Q$ of the forces required to prevent rotation of the handles. Assume the force between the disk and surface to be uniformly distributed.

**8.96** Four springs, each of constant 30 lb/in., are used to press plate $AB$ against the head on the vertical rod. In the position shown each spring has been compressed 2 in. The plate $AB$ may move vertically, but it is constrained against rotation by bolts which pass through holes cut in the plate. Knowing that $\mathbf{P} = 0$ and that the coefficient of friction is 0.30, determine the moment of the couple $\mathbf{M}$ required to rotate the rod.

**8.97** In Prob. 8.96, determine the required magnitude of the force $\mathbf{P}$ if the rod is to rotate when the moment of the couple $\mathbf{M}$ is 180 lb · in. For what magnitude of the force $\mathbf{P}$ is the rod most easily rotated?

**\*8.98** As the surfaces of shaft and bearing wear out, the frictional resistance of a thrust bearing decreases. It is generally assumed that the wear is directly proportional to the distance traveled by any given point of the shaft, and thus to the distance $r$ from the point to the axis of the shaft. Assuming, then, that the normal force per unit area is inversely proportional to $r$, show that the moment $M$ of the couple required to overcome the frictional resistance of a worn-out end bearing (with contact over the full circular area) is equal to 75 percent of the value given by formula (8.9) for a new bearing.

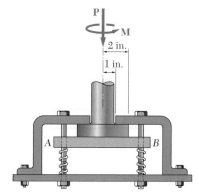

**Fig. P8.96**

**\*8.99** Assuming that bearings wear out as indicated in Prob. 8.98, show that the moment $M$ of the couple required to overcome the frictional resistance of a worn-out collar bearing is

$$M = \tfrac{1}{2}\mu_k P(R_1 + R_2)$$

where $P$ = magnitude of the total axial force
$R_1, R_2$ = inner and outer radii of collar

**\*8.100** Assuming that the pressure between the surfaces of contact is uniform, show that the moment $M$ of the couple required to overcome frictional resistance for the conical pivot shown is

$$M = \frac{2}{3}\frac{\mu_k P}{\sin\theta}\frac{R_2^3 - R_1^3}{R_2^2 - R_1^2}$$

**8.101** Solve Prob. 8.95, assuming that the normal force per unit area between the disk and the surface varies linearly from a maximum at the center to zero at the circumference of the disk.

**8.102** A circular disk of diameter 6 in. rolls at a constant velocity down a 2 percent incline. Determine the coefficient of rolling resistance.

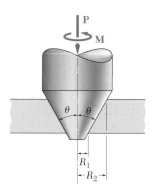

**Fig. P8.100**

**8.103** Determine the horizontal force required to move a 2500-lb automobile along a horizontal road at a constant velocity. Neglect all forms of friction except rolling resistance, and assume the coefficient of rolling resistance to be 0.05 in. The diameter of each tire is 23 in.

**8.104** Solve Prob. 8.92, including the effect of a coefficient of rolling resistance of 0.5 mm.

**8.105** Solve Prob. 8.93, including the effect of a coefficient of rolling resistance of 1.75 mm.

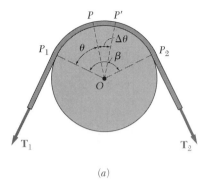

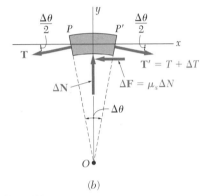

**Fig. 8.14**

## 8.10. BELT FRICTION

Consider a flat belt passing over a fixed cylindrical drum (Fig. 8.14a). We propose to determine the relation existing between the values $T_1$ and $T_2$ of the tension in the two parts of the belt when the belt is just about to slide toward the right.

Let us detach from the belt a small element $PP'$ subtending an angle $\Delta\theta$. Denoting by $T$ the tension at $P$ and by $T + \Delta T$ the tension at $P'$, we draw the free-body diagram of the element of the belt (Fig. 8.14b). Besides the two forces of tension, the forces acting on the free body are the normal component $\Delta \mathbf{N}$ of the reaction of the drum and the friction force $\Delta \mathbf{F}$. Since motion is assumed to be impending, we have $\Delta F = \mu_s \, \Delta N$. It should be noted that if $\Delta\theta$ is made to approach zero, the magnitudes $\Delta N$, $\Delta F$, and the *difference* $\Delta T$ between the tension at $P$ and the tension at $P'$ will also approach zero; the value $T$ of the tension at $P$, however, will remain unchanged. This observation helps in understanding our choice of notations.

Choosing the coordinate axes shown in Fig. 8.14b, we write the equations of equilibrium for the element $PP'$:

$$\Sigma F_x = 0: \qquad (T + \Delta T)\cos\frac{\Delta\theta}{2} - T\cos\frac{\Delta\theta}{2} - \mu_s \, \Delta N = 0 \qquad (8.11)$$

$$\Sigma F_y = 0: \qquad \Delta N - (T + \Delta T)\sin\frac{\Delta\theta}{2} - T\sin\frac{\Delta\theta}{2} = 0 \qquad (8.12)$$

Solving Eq. (8.12) for $\Delta N$ and substituting into (8.11), we obtain after reductions

$$\Delta T \cos\frac{\Delta\theta}{2} - \mu_s(2T + \Delta T)\sin\frac{\Delta\theta}{2} = 0$$

We shall now divide both terms by $\Delta\theta$; as far as the first term is concerned, this will be simply done by dividing $\Delta T$ by $\Delta\theta$. The division of the second term is carried out by dividing the terms in the parentheses by 2 and the sine by $\Delta\theta/2$. We write

$$\frac{\Delta T}{\Delta\theta}\cos\frac{\Delta\theta}{2} - \mu_s\left(T + \frac{\Delta T}{2}\right)\frac{\sin(\Delta\theta/2)}{\Delta\theta/2} = 0$$

If we now let $\Delta\theta$ approach 0, the cosine approaches 1 and $\Delta T/2$ approaches zero as noted above. On the other hand, the quotient of $\sin(\Delta\theta/2)$ over $\Delta\theta/2$ approaches 1, according to a lemma derived in all calculus textbooks. Since the limit of $\Delta T/\Delta\theta$ is by definition equal to the derivative $dT/d\theta$, we write

$$\frac{dT}{d\theta} - \mu_s T = 0 \qquad \frac{dT}{T} = \mu_s \, d\theta$$

We shall now integrate both members of the last equation obtained from $P_1$ to $P_2$ (Fig. 8.14a). At $P_1$, we have $\theta = 0$ and $T = T_1$; at $P_2$, we have $\theta = \beta$ and $T = T_2$. Integrating between these limits, we write

$$\int_{T_1}^{T_2}\frac{dT}{T} = \int_0^{\beta}\mu_s \, d\theta$$

$$\ln T_2 - \ln T_1 = \mu_s \beta$$

or, noting that the left-hand member is equal to the natural logarithm of the quotient of $T_2$ and $T_1$,

$$\ln\frac{T_2}{T_1} = \mu_s\beta \tag{8.13}$$

This relation may also be written in the form

$$\frac{T_2}{T_1} = e^{\mu_s\beta} \tag{8.14}$$

The formulas we have derived apply equally well to problems involving flat belts passing over fixed cylindrical drums and to problems involving ropes wrapped around a post or capstan. They may also be used to solve problems involving band brakes. In such problems, it is the drum which is about to rotate, while the band remains fixed. The formulas may also be applied to problems involving belt drives. In these problems, both the pulley and the belt rotate; our concern is then to find whether the belt will slip, i.e., whether it will move *with respect* to the pulley.

Formulas (8.13) and (8.14) should be used only if the belt, rope, or brake is *about to slip*. Formula (8.14) will be used if $T_1$ or $T_2$ is desired; formula (8.13) will be preferred if either $\mu_s$ or the angle of contact $\beta$ is desired. We should note that $T_2$ is always larger than $T_1$; $T_2$ therefore represents the tension in that part of the belt or rope which *pulls*, while $T_1$ is the tension in the part which *resists*. We should also observe that the angle of contact $\beta$ must be expressed in *radians*. The angle $\beta$ may be larger than $2\pi$; for example, if a rope is wrapped $n$ times around a post, $\beta$ is equal to $2\pi n$.

If the belt, rope, or brake is actually slipping, formulas similar to (8.13) and (8.14), but involving the coefficient of kinetic friction $\mu_k$, should be used. If the belt, rope, or brake does not slip and is not about to slip, none of these formulas may be used.

The belts used in belt drives are often V-shaped. Such a belt, called a *V belt*, is shown in Fig. 8.15a. It is seen that contact between belt and pulley takes place along the sides of the groove. The relation existing between the values $T_1$ and $T_2$ of the tension in the two parts of the belt when the belt is just about to slip may again be obtained by drawing the free-body diagram of an element of belt (Fig. 8.15b and c). Equations similar to (8.11) and (8.12) are derived, but the magnitude of the total friction force acting on the element is now $2\,\Delta F$, and the sum of the $y$ components of the normal forces is $2\,\Delta N \sin(\alpha/2)$. Proceeding as above, we obtain

$$\frac{T_2}{T_1} = e^{\mu_s\beta/\sin(\alpha/2)} \tag{8.15}$$

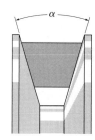

(a)

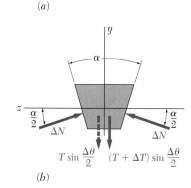

(b)

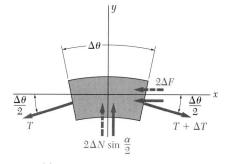

(c)

**Fig. 8.15**

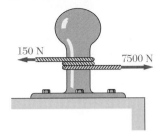

# SAMPLE PROBLEM 8.7

A hawser thrown from a ship to a pier is wrapped two full turns around a capstan. The tension in the hawser is 7500 N; by exerting a force of 150 N on its free end, a dock worker can just keep the hawser from slipping. (*a*) Determine the coefficient of friction between the hawser and the capstan. (*b*) Determine the tension in the hawser that could be resisted by the 150-N force if the hawser were wrapped three full turns around the capstan.

***a*. Coefficient of Friction.** Since slipping of the hawser is impending, we use Eq. (8.13):

$$\ln \frac{T_2}{T_1} = \mu_s \beta$$

Since the hawser is wrapped two full turns around the capstan, we have

$$\beta = 2(2\pi \text{ rad}) = 12.6 \text{ rad}$$
$$T_1 = 150 \text{ N} \qquad T_2 = 7500 \text{ N}$$

Therefore,

$$\mu_s \beta = \ln \frac{T_2}{T_1}$$

$$\mu_s (12.6 \text{ rad}) = \ln \frac{7500 \text{ N}}{150 \text{ N}} = \ln 50 = 3.91$$

$$\mu_s = 0.31 \quad \blacktriangleleft$$

***b*. Hawser Wrapped Three Turns around Capstan.** Using the value of $\mu_s$ obtained in part *a*, we have now

$$\beta = 3(2\pi \text{ rad}) = 18.9 \text{ rad}$$
$$T_1 = 150 \text{ N} \qquad \mu_s = 0.31$$

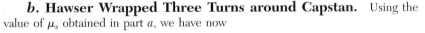

Substituting these values into Eq. (8.14), we obtain

$$\frac{T_2}{T_1} = e^{\mu_s \beta}$$

$$\frac{T_2}{150 \text{ N}} = e^{(0.31)(18.9)} = e^{5.86} = 350$$

$$T_2 = 52\,500 \text{ N} \qquad\qquad T_2 = 52.5 \text{ kN} \quad \blacktriangleleft$$

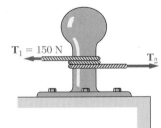

$T_1 = 150 \text{ N}$ $\qquad$ $T_2$

# SAMPLE PROBLEM 8.8

A flat belt connects pulley $A$, which drives a machine tool, to pulley $B$, which is attached to the shaft of an electric motor. The coefficients of friction are $\mu_s = 0.25$ and $\mu_k = 0.20$ between both pulleys and the belt. Knowing that the maximum allowable tension in the belt is 600 lb, determine the largest torque which can be exerted by the belt on pulley $A$.

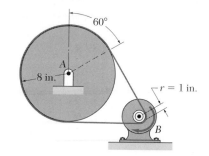

**Solution.** Since the resistance to slippage depends upon the angle of contact $\beta$ between pulley and belt, as well as upon the coefficient of static friction $\mu_s$, and since $\mu_s$ is the same for both pulleys, slippage will occur first on pulley $B$, for which $\beta$ is smaller.

**Pulley B.** Using Eq. (8.14) with $T_2 = 600$ lb, $\mu_s = 0.25$, and $\beta = 120° = 2\pi/3$ rad, we write

$$\frac{T_2}{T_1} = e^{\mu_s \beta} \qquad \frac{600 \text{ lb}}{T_1} = e^{0.25(2\pi/3)} = 1.688$$

$$T_1 = \frac{600 \text{ lb}}{1.688} = 355.4 \text{ lb}$$

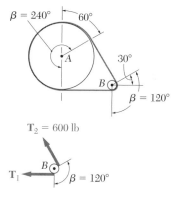

**Pulley A.** We draw the free-body diagram of pulley $A$. The couple $\mathbf{M}_A$ is applied to the pulley by the machine tool to which it is attached and is equal and opposite to the torque exerted by the belt. We write

$$+\circlearrowleft \Sigma M_A = 0: \qquad M_A - (600 \text{ lb})(8 \text{ in.}) + (355.4 \text{ lb})(8 \text{ in.}) = 0$$
$$M_A = 1957 \text{ lb} \cdot \text{in.} \qquad\qquad M_A = 163.1 \text{ lb} \cdot \text{ft} \quad \blacktriangleleft$$

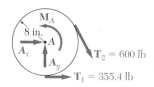

**Note.** We may check that the belt does not slip on pulley $A$ by computing the value of $\mu_s$ required to prevent slipping at $A$ and verifying that it is smaller than the actual value of $\mu_s$. From Eq. (8.13) we have

$$\mu_s \beta = \ln \frac{T_2}{T_1} = \ln \frac{600 \text{ lb}}{355.4 \text{ lb}} = 0.524$$

and, since $\beta = 240° = 4\pi/3$ rad,

$$\frac{4\pi}{3} \mu_s = 0.524 \qquad \mu_s = 0.125 < 0.25$$

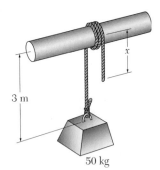

**Fig. P8.107**

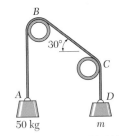

**Fig. P8.110 and P8.111**

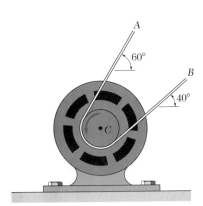

**Fig. P8.112**

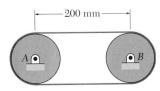

**Fig. P8.113**

# PROBLEMS

**8.106** A hawser is wrapped two full turns around a capstan head. By exerting a force of 800 N on the free end of the hawser, a seaman can resist a force of 60 kN on the other end of the hawser. Determine (*a*) the coefficient of static friction, (*b*) the number of times the hawser should be wrapped around the capstan if a 200-kN force is to be resisted by the same 800-N force.

**8.107** A rope having a mass per unit length of 0.8 kg/m is wound $2\frac{1}{2}$ times around a horizontal rod. What length $x$ of rope should be left hanging if a 50-kg load is to be supported? The coefficient of static friction between the rope and the rod is 0.25.

**8.108** Assume that the bushing of Prob. 8.90 has become frozen to the shaft and cannot rotate. Determine the coefficient of static friction between the bushing and the rope if a force **P** of magnitude 150 lb is required to start raising the 100-lb load.

**8.109** Assume that the bushing of Prob. 8.90 has become frozen to the shaft and cannot rotate. If the coefficient of static friction between the bushing and the rope is 0.30, determine the magnitude of the smallest force **P** required (*a*) to start raising the 100-lb load, (*b*) to hold the 100-lb load in place.

**8.110** A rope *ABCD* is looped over two pipes as shown. Knowing that the coefficient of static friction is 0.25, determine (*a*) the smallest value of the mass $m$ for which equilibrium is possible, (*b*) the corresponding tension in portion *BC* of the rope.

**8.111** A rope *ABCD* is looped over two pipes as shown. Knowing that the coefficient of static friction is 0.25, determine (*a*) the largest value of the mass $m$ for which equilibrium is possible, (*b*) the corresponding tension in portion *BC* of the rope.

**8.112** A flat belt is used to transmit the 25-lb · ft torque developed by an electric motor. The drum in contact with the belt has a diameter of 6 in., and the coefficient of static friction between the belt and the drum is 0.30. Determine the minimum allowable value of the tension in each part of the belt if the belt is not to slip.

**8.113** A flat belt is used to transmit a torque from pulley *A* to pulley *B*. The radius of each pulley is 50 mm and the coefficient of static friction is 0.30. Determine the largest torque which can be transmitted if the allowable belt tension is 3 kN.

**8.114** Solve Prob. 8.113, assuming that the belt is looped around the pulleys in a figure 8.

**8.115** A flat belt passes over two idler pulleys and under a rotating drum of diameter 8 in. The axle of the drum is free to move vertically in a slot, and a spring keeps the drum in contact with the belt. What is the minimum force which should be exerted by the spring if slippage is not to occur when a 30-lb · ft torque is applied to the drum? The coefficient of static friction between belt and drum is 0.25.

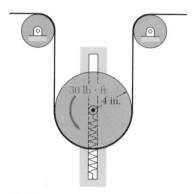

**Fig. P8.115**

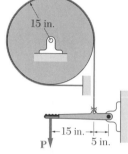

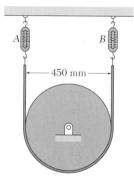

**Fig. P8.116**

**8.116** A band brake is used to control the speed of a flywheel as shown. The coefficients of friction are $\mu_s = 0.30$ and $\mu_k = 0.25$. What torque should be applied to the flywheel to keep it rotating clockwise at a constant speed when $P = 10$ lb?

**8.117** Solve Prob. 8.116, assuming that the flywheel rotates counterclockwise.

**8.118** The setup shown is used to measure the power output of a small turbine. The coefficient of kinetic friction is 0.15. When the flywheel is at rest, the reading of each spring scale is 75 N. What will be the reading of each scale when the flywheel is rotating clockwise? Assume that the belt is of constant length.

**8.119** A brake drum of radius $r = 125$ mm is rotating counterclockwise when a force **P** of magnitude 50 N is applied at $A$. Knowing that the coefficient of kinetic friction is 0.40, determine the moment about $O$ of the friction forces applied to the drum when $a = 200$ mm and $b = 250$ mm.

**8.120** Knowing that $r = 125$ mm and $a = 200$ mm, determine the maximum value of the coefficient of kinetic friction for which the brake is not self-locking when the drum rotates counterclockwise.

**8.121** Knowing that the coefficient of kinetic friction is 0.40, determine the minimum value of the ratio $a/r$ for which the brake is not self-locking. Assume that $a > r$ and that the drum revolves counterclockwise.

**Fig. P8.118**

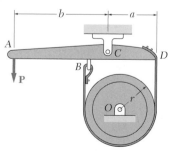

**Fig. P8.119, P8.120, and P8.121**

**8.122** A cable is placed around three pipes, each of 4-in. outside diameter, located in the same horizontal plane. Two of the pipes are fixed and do not rotate; the third pipe is rotated slowly. Knowing that the coefficients of friction are $\mu_s = 0.30$ and $\mu_k = 0.25$ for each pipe, determine the largest weight W which can be raised (a) if only pipe A is rotated, (b) if only pipe B is rotated, (c) if only pipe C is rotated.

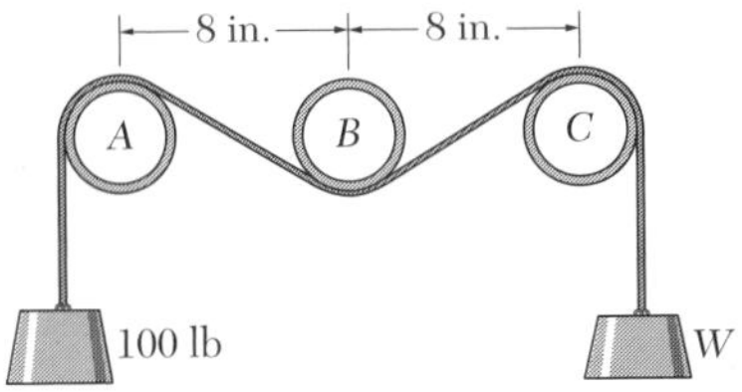

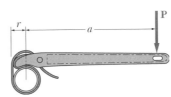

**Fig. P8.122**

**8.123** A cord is placed over two cylinders, each of 3-in. diameter. Knowing that the coefficients of friction are $\mu_s = 0.25$ and $\mu_k = 0.20$, determine the largest weight W which can be raised when cylinder B is rotated slowly and cylinder A is kept fixed.

**8.124** The strap wrench shown is used to grip the pipe firmly and at the same time not mar the external surface of the pipe. Knowing that $a = 250$ mm and $r = 50$ mm, determine the minimum coefficient of static friction for which the wrench will be self-locking.

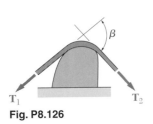

**Fig. P8.123**

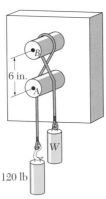

**Fig. P8.124 and P8.125**

**8.125** Denoting by $\mu_s$ the coefficient of static friction between the strap and the pipe, determine the minimum value of the ratio $r/a$ for which the strap wrench will be self-locking.

**8.126** Prove that Eqs. (8.13) and (8.14) are valid for any shape of surface provided that the coefficient of friction is the same at all points of contact.

**8.127** Complete the derivation of Eq. (8.15), which relates the tension in both parts of a V belt.

**Fig. P8.126**

**8.128** Solve Prob. 8.112, assuming that the flat belt and pulley are replaced by a V belt and V pulley with $\alpha = 28°$. (The angle $\alpha$ is as shown in Fig. 8.15a.)

# REVIEW AND SUMMARY
# FOR CHAPTER 8

This chapter was devoted to the study of *dry friction,* i.e., to problems involving rigid bodies which are in contact along *nonlubricated surfaces.*

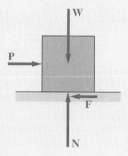

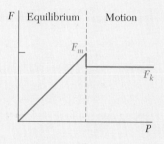

**Fig. 8.16**

Considering a block resting on a horizontal surface [Sec. 8.2] and applying to it a horizontal force **P**, we note that the block at first does not move. This shows that a *friction force* **F** must have developed to balance **P** (Fig. 8.16). As the magnitude of **P** is increased, the magnitude of **F** also increases until it reaches a maximum value $F_m$. If **P** if further increased, the block starts sliding and the magnitude of **F** drops from $F_m$ to a lower value $F_k$. Experimental evidence shows that $F_m$ and $F_k$ are proportional to the normal component $N$ of the reaction of the surface. We have

$$F_m = \mu_s N \qquad F_k = \mu_k N \qquad (8.1, 8.2)$$

where $\mu_s$ and $\mu_k$ are called, respectively, the *coefficient of static friction* and the *coefficient of kinetic friction.* These coefficients depend upon the nature and the condition of the surfaces in contact. Approximate values of the coefficients of static friction were given in Table 8.1.

It is sometimes convenient to replace the normal force **N** and the friction force **F** by their resultant **R** (Fig. 8.17). As the friction force increases and reaches its maximum value $F_m = \mu_s N$, the angle $\phi$ that **R** forms with the normal to the surface increases and reaches a maximum value $\phi_s$, called the *angle of static friction.* If motion actually takes place, the magnitude of **F** drops to $F_k$; similarly the angle $\phi$ drops to a lower value $\phi_k$, called the *angle of kinetic friction.* As shown in Sec. 8.3, we have

$$\tan \phi_s = \mu_s \qquad \tan \phi_k = \mu_k \qquad (8.3, 8.4)$$

Static and kinetic friction

Angles of friction

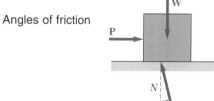

**Fig. 8.17**

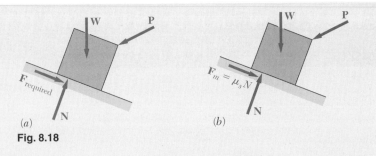

(a)

(b)

**Fig. 8.18**

**Problems involving friction**

When solving equilibrium problems involving friction, we should keep in mind that the magnitude $F$ of the friction force is equal to $F_m = \mu_s N$ *only if the body is about to slide* [Sec. 8.4]. *If motion is not impending, $F$ and $N$ should be considered as independent unknowns* to be determined from the equilibrium equations (Fig. 8.18$a$). We should also check that the value of $F$ required to maintain equilibrium is not larger than $F_m$; if it were, the body would move and the magnitude of the friction force would be $F_k = \mu_k N$ [Sample Prob. 8.1]. On the other hand, *if motion is known to be impending, $F$ has reached its maximum value* $F_m = \mu_s N$ (Fig. 8.18$b$) and this expression may be substituted for $F$ in the equilibrium equations [Sample Prob. 8.3]. When only three forces are involved in a free-body diagram, including the reaction **R** of the surface in contact with the body, it is usually more convenient to solve the problem by drawing a force triangle [Sample Prob. 8.2].

When a problem involves the analysis of the forces exerted on each other by *two bodies A and B*, it is important to show the friction forces with their correct sense. The correct sense for the friction force exerted by $B$ on $A$, for instance, is opposite to that of the *relative motion* (or impending motion) of $A$ with respect to $B$ [Fig. 8.6 of Sec. 8.4].

**Wedges and screws**

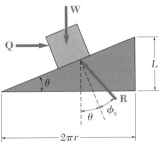

**Fig. 8.19** Block-and-incline analysis of a screw.

**Other applications**

In the second part of the chapter we considered a number of specific engineering applications where dry friction plays an important role. In the case of *wedges*, which are simple machines used to raise heavy loads [Sec. 8.5], two or more free-body diagrams were drawn and care was taken to show each friction force with its correct sense [Sample Prob. 8.4]. The analysis of *square-threaded screws*, which are frequently used in jacks, presses, and other mechanisms, was reduced to the analysis of a block sliding on an incline by unwrapping the thread of the screw and showing it as a straight line [Sec. 8.6]. This is done again in Fig. 8.19, where $r$ denotes the *mean radius* of the thread, $L$ the *lead* of the screw, i.e., the distance through which the screw advances in one turn, $W$ the load, and where $Qr$ is equal to the torque exerted on the screw. It was noted that in the case of multiple-threaded screws the lead $L$ of the screw is *not* equal to its pitch, which is the distance measured between two consecutive threads.

Other engineering applications considered in this chapter were *journal bearings* and *axle friction* [Sec. 8.7], *thrust bearings* and *disk friction*, [Sec. 8.8], *wheel friction* and *rolling resistance* [Sec. 8.9], and *belt friction* [Sec. 8.10].

In solving a problem involving a *flat belt* passing over a fixed cylinder, it is important to first determine the direction in which the belt slips or is about to slip. If the drum is rotating, the motion or impending motion of the belt should be determined *relative* to the rotating drum. For instance, if the belt of Fig. 8.20 is about to slip to the right relative to the drum, the friction forces exerted by the drum on the belt will be directed to the left and the tension will be larger in the right-hand portion of the belt than in the left-hand portion. Denoting by $T_2$ the larger tension, by $T_1$ the smaller tension, by $\mu_s$ the coefficient of static friction, and by $\beta$ the angle (in radians) subtended by the belt, we derived in Sec. 8.10 the formulas

**Belt friction**

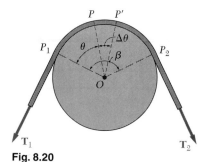

$$\ln \frac{T_2}{T_1} = \mu_s \beta \qquad (8.13)$$

$$\frac{T_2}{T_1} = e^{\mu_s \beta} \qquad (8.14)$$

**Fig. 8.20**

which were used in solving Sample Probs. 8.7 and 8.8. If the belt actually slips on the drum, the coefficient of static friction $\mu_s$ should be replaced by the coefficient of kinetic friction $\mu_k$ in both of these formulas.

# Review Problems

**8.129**   The axle of the pulley is frozen and cannot rotate with respect to the block. Knowing that the coefficient of static friction between cable *ABCD* and the pulley is 0.30, determine (*a*) the maximum allowable value of $\theta$ if the system is to remain in equilibrium, (*b*) the corresponding reactions at *A* and *D*. (Assume that the straight portions of the cable meet at point *E*.)

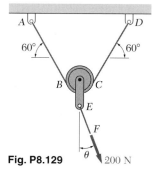

**Fig. P8.129**

**8.130**   A 15-m rope passes over a small horizontal shaft; one end of the rope is attached to a 2.4-kg bucket and the excess rope is coiled inside the bucket. The coefficient of static friction between the rope and the shaft is 0.30, and the rope has a mass per unit length of 800 g/m. If the shaft is slowly rotated, determine how

**Fig. P8.130**

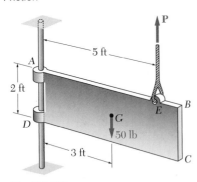

**Fig. P8.131**

far the bucket (*a*) can be raised before the rope slips on the shaft, (*b*) can be lowered before slipping occurs.

**8.131** The 50-lb plate *ABCD* is attached to collars at *A* and *D* which may slide on the vertical rod. Knowing that the coefficient of static friction is 0.40 between both collars and the rod, determine whether the plate is in equilibrium in the position shown when the magnitude of the vertical force applied at *E* is (*a*) *P* = 0, (*b*) *P* = 20 lb.

**8.132** In Prob. 8.131, determine the range of values of the magnitude *P* of the vertical force applied at *E* for which the plate will move downward.

**8.133** Knowing that the coefficient of friction between the joists and the plank *AB* is 0.30, determine the smallest distance *a* for which the plank will slip at *C*.

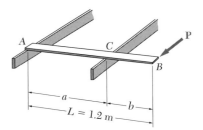

**Fig. P8.133 and P8.134**

**8.134** A 1.2-m plank of mass 3 kg rests on two joists. Knowing that the coefficient of static friction between the plank and the joists is 0.30, determine the magnitude of the horizontal force required to move the plank when (*a*) *a* = 750 mm, (*b*) *a* = 900 mm.

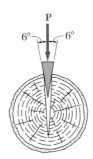

**Fig. P8.135**

**8.135** A 12° wedge is used to split a log. The coefficient of friction between the wedge and the wood is 0.35. Knowing that a force **P** of magnitude 2.5 kN was required to insert the wedge, determine the magnitude of the forces exerted on the log by the wedge after it has been inserted.

**8.136** The uniform rod *AB* of radius *a* and weight *W* is attached to collars at *A* and *B* which may slide on the rods shown. Denoting by $\mu_s$ the coefficient of static friction between each collar and the rod upon which it may slide, determine the smallest value of $\mu_s$ for which the rod will remain in equilibrium in the position shown.

**Fig. P8.136**

**8.137** Two blocks $A$ and $B$ are connected by a cable as shown. Knowing that the coefficient of static friction at all surfaces of contact is 0.30 and neglecting the friction of the pulleys, determine the magnitude $P$ of the smallest force required to move the blocks.

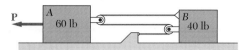

**Fig. P8.137**

**8.138** Solve Prob. 8.137, assuming that the force **P** is applied to block $B$ and is directed horizontally to the right.

**8.139** A slender rod of length $L$ is lodged between peg $C$ and the horizontal ceiling and supports a load **P** at end $B$. Knowing that $\theta = 30°$ and that the coefficient of static friction is 0.35 at both $A$ and $C$, find the range of values of the ratio $L/a$ for which equilibrium is maintained.

**8.140** A slender rod of length $L$ is lodged between peg $C$ and the horizontal ceiling and supports a load **P** at end $B$. Knowing that $L = 8a$, $\theta = 15°$, and that the coefficients of friction are $\mu_s = 0.60$ and $\mu_k = 0.50$ at $A$, and zero at $C$, determine whether the rod is in equilibrium and find the value of the friction force at $A$.

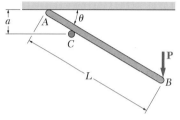

**Fig. P8.139 and P8.140**

# COMPUTER PROBLEMS

**8.C1** The 40-kg block $A$ hangs from a cable as shown. Pulley $C$ is connected by a short link to block $E$, which rests on a horizontal rail. Denoting by $\mu_s$ the coefficient of static friction between block $E$ and the rail, and neglecting the weight of block $E$ and the friction in the pulleys, write a computer program and use it to calculate the value of $\theta$ for which motion impends for values of $\mu_s$ from 0 to 0.55 at 0.05 intervals.

**8.C2** The coefficient of static friction between all surfaces of contact is 0.15. Knowing that block $B$ weighs 25 lb, write a computer program and use it to calculate the value of $\theta$ for which motion is impending for weights of block $A$ from 0 to 60 lb at 5-lb intervals.

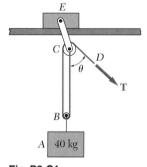

**Fig. P8.C1**

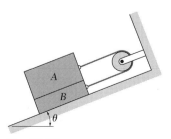

**Fig. P8.C2**

**8.C3** A rope is looped over a fixed pipe at $B$ and is connected to two blocks as shown. The coefficient of static friction is 0.25 between the pipe and the rope and 0.20 between block $C$ and the horizontal surface. Initially block $C$ is directly under the pipe ($x = 0$); the magnitude of the force **P** is slowly increased until $x = 60$ ft. Write a computer program and use it to calculate $P$ for values of $x$ from 0 to 60 ft at 5-ft intervals.

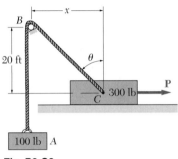

**Fig. P8.C3**

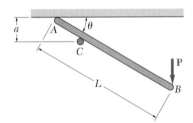

**Fig. P8.C4**

**8.C4** A slender rod of length $L$ is lodged between peg $C$ and the horizontal ceiling and supports a load **P** at end $B$. (a) Knowing that the coefficient of static friction $\mu_s$ is the same at $A$ and $C$, show that the following relation exists between the angle of static friction $\phi_s = \tan^{-1}\mu_s$ and the angle $\theta$ for which sliding is impending to the right:

$$\cot 2\phi_s = \cot\theta - \frac{a}{2L\sin^2\theta\cos\theta}$$

(b) Use this relation to write a computer program and calculate $\mu_s$ for values of $\theta$ from 15 to 60° at 5° intervals, assuming (1) $L = 4a$, (2) $L = 8a$. (c) Determine to three significant figures the value of $\theta$ for which sliding impends to the right, assuming $\mu_s = 0.300$ and (1) $L = 4a$, (2) $L = 8a$.

# Distributed Forces: Moments of Inertia

## 9.1. INTRODUCTION

In Chap. 5, we analyzed various systems of forces distributed over an area or volume. The three main types of forces considered were (1) weights of homogeneous plates of uniform thickness (Secs. 5.3 through 5.6), (2) distributed loads on beams and hydrostatic forces (Secs. 5.8 and 5.9), and (3) weights of homogeneous three-dimensional bodies (Secs. 5.10 and 5.11). In the case of homogeneous plates, the magnitude $\Delta W$ of the weight of an element of plate was proportional to the area $\Delta A$ of the element. In the case of distributed loads on beams, the magnitude $\Delta W$ of each elementary weight was represented by an element of area $\Delta A = \Delta W$ under the load curve; in the case of hydrostatic forces on submerged rectangular surfaces, a similar procedure was followed. In the case of homogeneous three-dimensional bodies, the magnitude $\Delta W$ of the weight of an element of the body was proportional to the volume $\Delta V$ of the element. Thus, in all cases considered in Chap. 5, the distributed forces were proportional to the elementary areas or volumes associated with them. The resultant of these forces, therefore, could be obtained by summing the corresponding areas or volumes and the moment of the resultant about any given axis could be determined by computing the first moments of the areas or volumes about that axis.

In the first part of this chapter, we shall consider distributed forces $\Delta \mathbf{F}$ whose magnitudes depend not only upon the element of area $\Delta A$ on which they act but also upon the distance from $\Delta A$ to some given axis. More precisely, the magnitude of the force per unit area $\Delta F/\Delta A$ will vary linearly with the distance to the axis. As we shall see in the next section, forces of this type are found in the study of the bending of beams and in problems involving submerged nonrectangular surfaces. Assuming that the elementary forces involved are distributed over an area $A$ and vary linearly with the distance $y$ to the $x$ axis, we shall find that while the magnitude of their resultant $\mathbf{R}$ depends upon the first moment $Q_x = \int y \, dA$ of the area $A$, the location of the point where $\mathbf{R}$ should be applied depends upon the *second*

351

*moment,* or *moment of inertia,* $I_x = \int y^2 \, dA$ of the same area with respect to the $x$ axis. We shall learn to compute the moments of inertia of various areas with respect to given $x$ and $y$ axes. We shall also introduce the *polar moment of inertia* $J_o = \int r^2 \, dA$ of an area, where $r$ is the distance from the element of area $dA$ to a point $O$. To facilitate our computations, we shall establish a relation between the moment of inertia $I_x$ of an area $A$ with respect to a given $x$ axis and its moment of inertia $I_{x'}$ with respect to a parallel centroidal $x'$ axis (parallel-axis theorem). We shall also study the transformation of the moments of inertia of a given area under a rotation of axes (Secs. 9.9 and 9.10).

In the second part of the chapter, we shall determine the moments of inertia of various *masses* with respect to a given axis. As we shall see in Sec. 9.11, the moment of inertia of a given mass about an axis $AA'$ is defined as the integral $I = \int r^2 \, dm$, where $r$ is the distance from the axis $AA'$ to the element of mass $dm$. Moments of inertia of masses are encountered in dynamics in problems involving the rotation of a rigid body about an axis. To facilitate the computation of mass moments of inertia, we shall introduce the parallel-axis theorem (Sec. 9.12) and learn to determine the moments of inertia of thin rectangular and circular plates (Sec. 9.13).

## MOMENTS OF INERTIA OF AREAS

### 9.2. SECOND MOMENT, OR MOMENT OF INERTIA, OF AN AREA

In the first part of this chapter, we shall consider distributed forces $\Delta\mathbf{F}$ of magnitude $\Delta F$ proportional to the element of area $\Delta A$ on which they act and varying linearly with the distance from $\Delta A$ to some given axis.

Consider, for example, a beam of uniform cross section, subjected to two equal and opposite couples applied at each end of the beam. Such a beam is said to be in *pure bending,* and it is shown in mechanics of materials that the internal forces in any section of the beam are distributed forces whose magnitudes $\Delta F = ky \, \Delta A$ vary linearly with the distance $y$ from an axis passing through the centroid of the section. This axis, represented by the $x$ axis in Fig. 9.1, is known as the *neutral axis* of the section. The forces on one side of the neutral axis are forces of compression and on the other side, forces of tension, while on the neutral axis itself the forces are zero.

The magnitude of the resultant $\mathbf{R}$ of the elementary forces $\Delta\mathbf{F}$ over the entire section is

$$R = \int ky \, dA = k \int y \, dA$$

The last integral obtained is recognized as the *first moment* $Q_x$ of the section about the $x$ axis; it is equal to $\bar{y}A$ and to zero, since the centroid of the section is located on the $x$ axis. The system of the forces $\Delta\mathbf{F}$ thus reduces to a couple. The magnitude $M$ of this couple (bending moment) must be equal to the sum of the moments $\Delta M_x = y \, \Delta F = ky^2 \Delta A$ of the elementary forces. Integrating over the entire section, we obtain

$$M = \int ky^2 \, dA = k \int y^2 \, dA$$

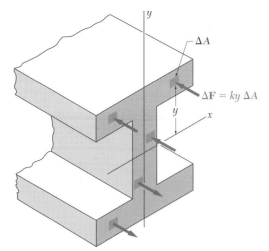

**Fig. 9.1**

The last integral is known as the *second moment*, or *moment of inertia*,[†] of the beam section with respect to the $x$ axis and is denoted by $I_x$. It is obtained by multiplying each element of area $dA$ by the *square of its distance* from the $x$ axis and integrating over the beam section. Since each product $y^2\,dA$ is positive, whether $y$ is itself positive or negative (or zero if $y$ is zero), the integral $I_x$ will always be different from zero and positive.

Another example of second moment, or moment of inertia, of an area is provided by the following problem of hydrostatics: A vertical circular gate used to close the outlet of a large reservoir is submerged under water as shown in Fig. 9.2. What is the resultant of the forces exerted by the water on the gate, and what is the moment of the resultant about the line of intersection of the plane of the gate with the water surface ($x$ axis)?

If the gate were rectangular, the resultant of the forces of pressure could be determined from the pressure curve, as was done is Sec. 5.9. Since the gate is circular, however, a more general method must be used. Denoting by $y$ the depth of an element of area $\Delta A$ and by $\gamma$ the specific weight of water, the pressure at the element is $p = \gamma y$, and the magnitude of the elementary force exerted on $\Delta A$ is $\Delta F = p\,\Delta A = \gamma y\,\Delta A$. The magnitude of the resultant of the elementary forces is thus

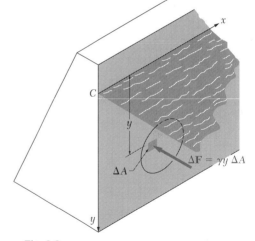

**Fig. 9.2**

$$R = \int \gamma y\,dA = \gamma \int y\,dA$$

and may be obtained by computing the first moment $Q_x = \int y\,dA$ of the area of the gate with respect to the $x$ axis. The moment $M_x$ of the resultant must be equal to the sum of the moments $\Delta M_x = y\,\Delta F = \gamma y^2 \Delta A$ of the

[†] The term second moment is more proper than the term moment of inertia, since, logically, the latter should be used only to denote integrals of mass (see Sec. 9.11). In common engineering practice, however, moment of inertia is used in connection with areas as well as masses.

elementary forces. Integrating over the area of the gate, we have

$$M_x = \int \gamma y^2 \, dA = \gamma \int y^2 \, dA$$

Here again, the integral obtained represents the second moment, or moment of inertia, $I_x$ of the area with respect to the $x$ axis.

## 9.3. DETERMINATION OF THE MOMENT OF INERTIA OF AN AREA BY INTEGRATION

We have defined in the preceding section the second moment, or moment of inertia, of an area $A$ with respect to the $x$ axis.

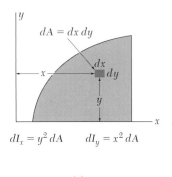

$dI_x = y^2 \, dA$   $dI_y = x^2 \, dA$

(a)

**Fig. 9.3**

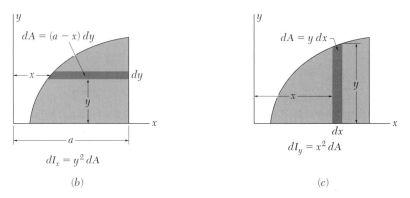

$dI_x = y^2 \, dA$

(b)

$dI_y = x^2 \, dA$

(c)

Defining in a similar way the moment of inertia $I_y$ of the area $A$ with respect to the $y$ axis, we write (Fig. 9.3a)

$$I_x = \int y^2 \, dA \qquad I_y = \int x^2 \, dA \qquad (9.1)$$

These integrals, known as the *rectangular moments of inertia* of the area $A$, may be more easily compute if we choose for $dA$ a thin strip parallel to one of the axes of coordinates. To computer $I_x$, the strip is chosen parallel to the $x$ axis, so that all the points forming the strip are at the same distance $y$ from the $x$ axis (Fig. 9.3b); the moment of inertia $dI_x$ of the strip is then obtained by multiplying the area $dA$ of the strip by $y^2$. To compute $I_y$, the strip is chosen parallel to the $y$ axis so that all the points forming the strip are at the same distance $x$ from the $y$ axis (Fig. 9.3c.); the moment of inertia $dI_y$ of the strip is $x^2 \, dA$.

***Moment, of Inertia of a Rectangular Area.*** As an example, we shall determine the moment of inertia of a rectangle with respect to its base (Fig. 9.4). Dividing the rectangle into strips parallel to the x axis, we obtain

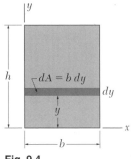

**Fig. 9.4**

$$dA = b \, dy \qquad dI_x = y^2 b \, dy \qquad I_x = \int_0^h b y^2 \, dy = \tfrac{1}{3} b h^3 \qquad (9.2)$$

***Computing $I_x$ and $I_y$ from the Same Elementary Strips.*** The formula just derived may be used to determine the moment of inertia $dI_x$ with respect to the $x$ axis of a rectangular strip parallel to the $y$ axis such as the one shown in Fig. 9.3c. Making $b = dx$ and $h = y$ in formula (9.2), we write

$$dI_x = \tfrac{1}{3} y^3 \, dx$$

On the other hand, we have

$$dI_y = x^2 \, dA = x^2 y \, dx$$

The same element may thus be used to compute the moments of inertia $I_x$ and $I_y$ of a given area (Fig. 9.5).

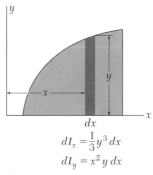

$$dI_x = \frac{1}{3} y^3 \, dx$$
$$dI_y = x^2 y \, dx$$

**Fig. 9.5**

## 9.4. POLAR MOMENT OF INERTIA

An integral of great importance in problems concerning the torsion of cylindrical shafts and in problems dealing with the rotation of slabs is

$$J_O = \int r^2 \, dA \tag{9.3}$$

where $r$ is the distance from the element of area $dA$ to the pole $O$ (Fig. 9.6). This integral is the *polar moment of inertia* of the area $A$ with respect to $O$.

The polar moment of inertia of a given area may be computed from the rectangular moments of inertia $I_x$ and $I_y$ of the area if these integrals are already known. Indeed, noting that $r^2 = x^2 + y^2$, we write

$$J_O = \int r^2 \, dA = \int (x^2 + y^2) dA = \int y^2 \, dA + \int x^2 \, dA$$

that is,

$$J_O = I_x + I_y \tag{9.4}$$

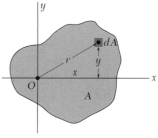

**Fig. 9.6**

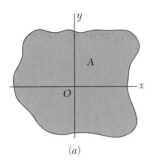

(a)

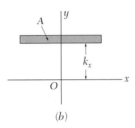

(b)

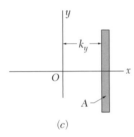

(c)

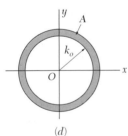

(d)

**Fig. 9.7**

## 9.5. RADIUS OF GYRATION OF AN AREA

Consider an area $A$ which has a moment of inertia $I_x$ with respect to the $x$ axis (Fig. 9.7$a$). Let us imagine that we concentrate this area into a thin strip parallel to the $x$ axis (Fig. 9.7$b$). If the area $A$, thus concentrated, is to have the same moment of inertia with respect to the $x$ axis, the strip should be placed at a distance $k_x$ from the $x$ axis, defined by the relation

$$I_x = k_x^2 A$$

Solving for $k_x$, we write

$$k_x = \sqrt{\frac{I_x}{A}} \qquad (9.5)$$

The distance $k_x$ is referred to as the *radius of gyration* of the area with respect to the $x$ axis. We may define in a similar way the radii of gyration $k_y$ and $k_O$ (Fig. 9.7$c$ and $d$); we write

$$I_y = k_y^2 A \qquad k_y = \sqrt{\frac{I_y}{A}} \qquad (9.6)$$

$$J_O = k_O^2 A \qquad k_O = \sqrt{\frac{J_O}{A}} \qquad (9.7)$$

Substituting for $J_O$, $I_x$, and $I_y$ in terms of the radii of gyration in the relation (9.4), we observe that

$$k_O^2 = k_x^2 + k_y^2 \qquad (9.8)$$

**Example.** As an example, let us compute the radius of gyration $k_x$ of the rectangle shown in Fig. 9.8. Using formulas (9.5) and (9.2), we write

$$k_x^2 = \frac{I_x}{A} = \frac{\frac{1}{3}bh^2}{bh} = \frac{h^2}{3} \qquad k_x = \frac{h}{\sqrt{3}}$$

The radius of gyration $k_x$ of the rectangle is shown in Fig. 9.8. It should not be confused with the ordinate $\bar{y} = h/2$ of the centroid of the area. While $k_x$ depends upon the *second moment*, or moment of inertia, of the area, the ordinate $\bar{y}$ is related to the *first moment* of the area.

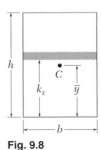

**Fig. 9.8**

# SAMPLE PROBLEM 9.1

Determine the moment of inertia of a triangle with respect to its base.

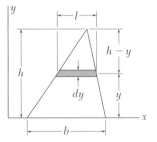

**Solution.** A triangle of base $b$ and height $h$ is drawn; the $x$ axis is chosen to coincide with the base. A differential strip parallel to the $x$ axis is chosen. Since all portions of the strip are at the same distance from the $x$ axis, we write

$$dI_x = y^2\,dA \qquad dA = l\,dy$$

From the similar triangles, we have

$$\frac{l}{b} = \frac{h-y}{h} \qquad l = b\,\frac{h-y}{h} \qquad dA = b\,\frac{h-y}{h}\,dy$$

Integrating $dI_x$ from $y = 0$ to $y = h$, we obtain

$$I_x = \int y^2\,dA = \int_0^h y^2 b\,\frac{h-y}{h}\,dy = \frac{b}{h}\int_0^h (hy^2 - y^3)\,dy$$

$$= \frac{b}{h}\left[ h\frac{y^3}{3} - \frac{y^4}{4} \right]_0^h \qquad\qquad I_x = \frac{bh^3}{12} \quad \blacktriangleleft$$

# SAMPLE PROBLEM 9.2

(*a*) Determine the centroidal polar moment of inertia of a circular area by direct integration. (*b*) Using the result of part *a*, determine the moment of inertia of a circular area with respect to a diameter.

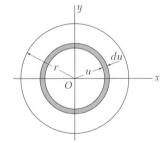

**a. Polar Moment of Inertia.** An annual differential element of area is chosen. Since all portions of this differential area are at the same distance from the origin, we write

$$dJ_O = u^2\,dA \qquad dA = 2\pi u\,du$$

$$J_O = \int dJ_O = \int_0^r u^2 (2\pi u\,du) = 2\pi \int_0^r u^3\,du$$

$$J_O = \frac{\pi}{2}\,r^4 \quad \blacktriangleleft$$

**b. Moment of Inertia with Respect to a Diameter.** Because of the symmetry of the circular area we have $I_x = I_y$. We then write

$$J_O = I_x + I_y = 2I_x \qquad \frac{\pi}{2}\,r^4 = 2I_x \qquad I_{\text{diameter}} = I_x = \frac{\pi}{4}\,r^4 \quad \blacktriangleleft$$

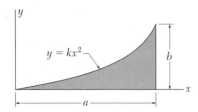

## SAMPLE PROBLEM 9.3

(a) Determine the moment of inertia of the shaded area shown with respect to each of the coordinate axes. This area has also been considered in Sample Prob. 5.4. (b) Using the results of part a, determine the radius of gyration of the shaded area with respect to each of the coordinate axes.

**Solution.** Referring to Sample Prob. 5.4, we obtain the following expressions for the equation of the curve and the total area:

$$y = \frac{b}{a^2} x^2 \qquad A = \tfrac{1}{3}ab$$

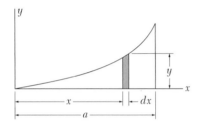

***Moment of Inertia $I_x$.*** A vertical differential element of area is chosen. Since all portions of this element are *not* at the same distance from the $x$ axis, we must treat the element as a thin rectangle. The moment of inertia of the element with respect to the $x$ axis is

$$dI_x = \tfrac{1}{3}y^3\,dx = \frac{1}{3}\left(\frac{b}{a^2}x^2\right)^3 dx = \frac{1}{3}\frac{b^3}{a^6}x^6\,dx$$

$$I_x = \int dI_x = \int_0^a \frac{1}{3}\frac{b^3}{a^6}x^6\,dx = \left[\frac{1}{3}\frac{b^3}{a^6}\frac{x^7}{7}\right]_0^a$$

$$I_x = \frac{ab^3}{21} \qquad \blacktriangleleft$$

***Moment of Inertia $I_y$.*** The same vertical differential element of area is used. Since all portions of the element are at the same distance from the $y$ axis, we write

$$dI_y = x^2\,dA = x^2(y\,dx) = x^2\left(\frac{b}{a^2}x^2\right)dx = \frac{b}{a^2}x^4\,dx$$

$$I_y = \int dI_y = \int_0^a \frac{b}{a^2}x^4\,dx = \left[\frac{b}{a^2}\frac{x^5}{5}\right]_0^a$$

$$I_y = \frac{a^3b}{5} \qquad \blacktriangleleft$$

***Radii of Gyration $k_x$ and $k_y$***

$$k_x^2 = \frac{I_x}{A} = \frac{ab^3/21}{ab/3} = \frac{b^2}{7} \qquad k_x = \sqrt{\tfrac{1}{7}}\,b \qquad \blacktriangleleft$$

$$k_y^2 = \frac{I_y}{A} = \frac{a^3b/5}{ab/3} = \tfrac{3}{5}a^2 \qquad k_y = \sqrt{\tfrac{3}{5}}\,a \qquad \blacktriangleleft$$

# PROBLEMS

**9.1 through 9.4** Determine by direct integration the moment of inertia of the shaded area with respect to the $y$ axis.

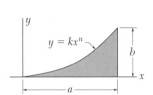

**Fig. P9.1 and P9.5**

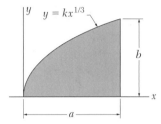

**Fig. P9.2 and P9.6**

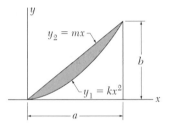

**Fig. P9.3 and P9.7**

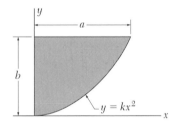

**Fig. P9.4 and P9.8**

**9.5 through 9.8** Determine by direct integration the moment of inertia of the shaded area with respect to the $x$ axis.

**9.9 and 9.10** Determine the moment of inertia and radius of gyration of the shaded area shown with respect to the $x$ axis.

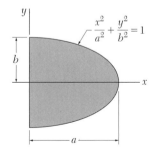

**Fig. P9.9 and P9.11**

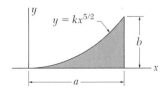

**Fig. P9.10 and P9.12**

**9.11 and 9.12** Determine the moment of inertia and radius of gyration of the shaded area shown with respect to the $y$ axis.

**9.13** Determine the moment of inertia and radius of gyration of the shaded area shown with respect to the $x$ axis.

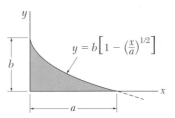

$$y = b\left[1 - \left(\tfrac{x}{a}\right)^{1/2}\right]$$

**Fig. P9.13 and P9.14**

**9.14** Determine the moment of inertia and radius of gyration of the shaded area shown with respect to the $y$ axis.

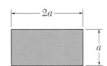

**Fig. P9.15**

**9.15** Determine the polar moment of inertia and the polar radius of gyration of the rectangle shown with respect to the midpoint of one of its (*a*) longer sides, (*b*) shorter sides.

**9.16** Determine the polar moment of inertia and the polar radius of gyration of the trapezoid shown with respect to point $P_1$.

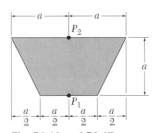

**Fig. P9.16 and P9.17**

**9.17** Determine the polar moment of inertia and the polar radius of gyration of the trapezoid shown with respect to point $P_2$.

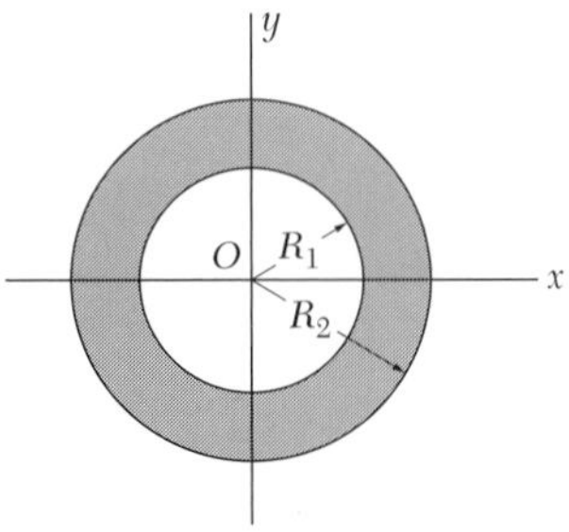

**9.18** (*a*) Determine by direct integration the polar moment of inertia of the annular area shown. (*b*) Using the result of part *a*, determine the moment of inertia of the given area with respect to the $x$ axis.

**9.19** (*a*) Show that the polar radius of gyration $k_O$ of the annular area shown is approximately equal to the mean radius $R_m = (R_1 + R_2)/2$ for small values of the thickness $t = R_2 - R_1$. (*b*) Determine the percentage error introduced by using $R_m$ in place of $k_O$ for values of $t/R_m$ respectively equal to $1$, $\frac{1}{2}$, and $\frac{1}{10}$.

**Fig. P9.18 and P9.19**

**\*9.20** Prove that the centroidal polar moment of inertia of a given area $A$ cannot be smaller than $A^2/2\pi$. (*Hint.* Compare the moment of inertia of the given area with the moment of inertia of a circle of the same area and same centroid.)

## 9.6. PARALLEL-AXIS THEOREM

Consider the moment of inertia $I$ of an area $A$ with respect to an axis $AA'$ (Fig. 9.9). Denoting by $y$ the distance from an element of area $dA$ to $AA'$, we write

$$I = \int y^2 \, dA$$

Let us now draw an axis $BB'$ parallel to $AA'$ through the centroid $C$ of the area; this axis is called a *centroidal axis*. Denoting by $y'$ the distance from

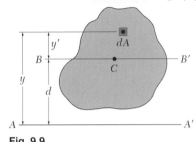

**Fig. 9.9**

the element $dA$ to $BB'$, we write $y = y' + d$, where $d$ is the distance between the axes $AA'$ and $BB'$. Substituting for $y$ in the integral representing $I$, we write

$$I = \int y^2 \, dA = \int (y' + d)^2 \, dA$$
$$= \int y'^2 \, dA + 2d \int y' \, dA + d^2 \int dA$$

The first integral represents the moment of inertia $\bar{I}$ of the area with respect to the centroidal axis $BB'$. The second integral represents the first moment of the area with respect to $BB'$; since the centroid $C$ of the area is located on that axis, the second integral must be zero. Finally, we observe that the last integral is equal to the total area $A$. We write therefore

$$I = \bar{I} + Ad^2 \tag{9.9}$$

This formula expresses that the moment of inertia $I$ of an area with respect to any given axis $AA'$ is equal to the moment of inertia $\bar{I}$ of the area with respect to a centroidal axis $BB'$ parallel to $AA'$ *plus* the product $Ad^2$ of the area $A$ and of the square of the distance $d$ between the two axes. This theorem is known as the *parallel-axis theorem*. Substituting $k^2 A$ for $I$ and $\bar{k}^2 A$ for $\bar{I}$, the theorem may also be expressed in the following way:

$$k^2 = \bar{k}^2 + d^2 \tag{9.10}$$

A similar theorem may be used to relate the polar moment of inertia $J_O$ of an area about a point $O$ and the polar moment of inertia $\bar{J}_C$ of the same area about its centroid $C$. Denoting by $d$ the distance between $O$ and $C$, we write

$$J_O - \bar{J}_C + Ad^2 \quad \text{or} \quad k_O^2 = \bar{k}_C^2 + d^2 \tag{9.11}$$

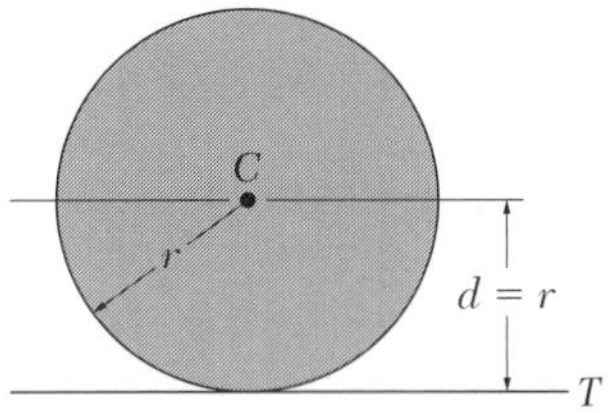

Fig. 9.10

***Example 1.*** As an application of the parallel-axis theorem, we shall determine the moment of inertia $I_T$ of a circular area with respect to a line tangent to the circle (Fig. 9.10). We found in Sample Prob. 9.2 that the moment of inertia of a circular area about a centroidal axis is $\bar{I} = \frac{1}{4}\pi r^4$. We may write, therefore,

$$I_T = \bar{I} + Ad^2 = \tfrac{1}{4}\pi r^4 + \pi r^2 r^2 = \tfrac{5}{4}\pi r^4$$

***Example 2.*** The parallel-axis theorem may also be used to determine the centrodial moment of inertia of an area when the moment of inertia of this area with respect to some parallel axis is known. Consider, for instance, a triangular area (Fig. 9.11). We found in Sample Prob. 9.1 that the moment of inertia of a triangle with respect to its base $AA'$ is equal to $\frac{1}{12}bh^3$. Using the parallel-axis theorem, we write

$$I_{AA'} = \bar{I}_{BB'} + Ad^2$$
$$\bar{I}_{BB'} = I_{AA'} - Ad^2 = \tfrac{1}{12}bh^3 - \tfrac{1}{2}bh(\tfrac{1}{3}h)^2 = \tfrac{1}{13}bh^3$$

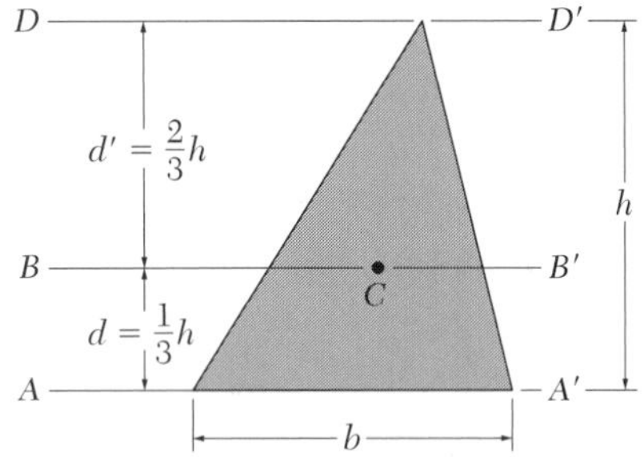

Fig. 9.11

It should be observed that the product $Ad^2$ was *subtracted* from the given moment of inertia in order to obtain the centroidal moment of inertia of the triangle. While this product is *added* in transferring *from* a centrodial axis to a parallel axis, it should be *subtracted* in transferring *to* a centrodial axis. In other words, the moment of inertia of an area is always smaller with respect to a centroidal axis than with respect to any other parallel axis.

Returning to Fig. 9.11, we observe that the moment of inertia of the triangle with respect to a line $DD'$ drawn through a vertex may be obtained by writing

$$I_{DD'} = \bar{I}_{BB'} + Ad'^2 + \tfrac{1}{36}bh^3 + \tfrac{1}{2}bh(\tfrac{2}{3}h)^2 = \tfrac{1}{4}bh^3$$

Note that $I_{DD'}$ *could not* have been obtained directly from $I_{AA'}$. The parallel-axis theorem can be applied only if one of the two parallel axes passes through the centroid of the area.

## 9.7. MOMENTS OF INERTIA OF COMPOSITE AREAS

Consider a composite area $A$ made of several component areas $A_1$, $A_2$, etc. Since the integral representing the moment of inertia of $A$ may be subdivided into integrals computed over $A_1$, $A_2$, etc., the moment of inertia of $A$ with respect to a given axis will be obtained by adding the moments of inertia of the areas $A_1$, $A_2$, etc., with respect to the same axis. The moment of inertia of an area made of several of the common shapes shown in Fig. 9.12 may thus be obtained from the formulas given in that figure. Before adding the moments of inertia of the component areas, however, the parallel-axis theorem should be used to transfer each moment of inertia to the desired axis. This is shown in Sample Probs. 9.4 and 9.5.

The properties of the cross sections of various structural shapes are given in Fig. 9.13. As noted in Sec. 9.2, the moment of inertia of a beam section about its neutral axis is closely related to the value of the internal forces. The determination of moments of inertia is thus a prerequisite to the analysis and design of structural members.

It should be noted that the radius of gyration of a composite area is *not* equal to the sum of the radii of gyration of the component areas. In order to determine the radius of gyration of a composite area, it is necessary first to compute the moment of inertia of the area.

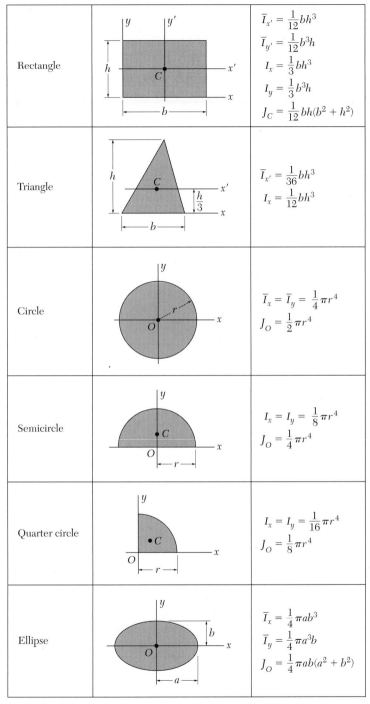

| Rectangle | $\bar{I}_{x'} = \frac{1}{12}bh^3$ |
| | $\bar{I}_{y'} = \frac{1}{12}b^3h$ |
| | $I_x = \frac{1}{3}bh^3$ |
| | $I_y = \frac{1}{3}b^3h$ |
| | $J_C = \frac{1}{12}bh(b^2 + h^2)$ |
| Triangle | $\bar{I}_{x'} = \frac{1}{36}bh^3$ |
| | $I_x = \frac{1}{12}bh^3$ |
| Circle | $\bar{I}_x = \bar{I}_y = \frac{1}{4}\pi r^4$ |
| | $J_O = \frac{1}{2}\pi r^4$ |
| Semicircle | $I_x = I_y = \frac{1}{8}\pi r^4$ |
| | $J_O = \frac{1}{4}\pi r^4$ |
| Quarter circle | $I_x = I_y = \frac{1}{16}\pi r^4$ |
| | $J_O = \frac{1}{8}\pi r^4$ |
| Ellipse | $\bar{I}_x = \frac{1}{4}\pi ab^3$ |
| | $\bar{I}_y = \frac{1}{4}\pi a^3b$ |
| | $J_O = \frac{1}{4}\pi ab(a^2 + b^2)$ |

**Fig. 9.12** Moments of inertia of common geometric shapes.

**Fig. 9.13A**  Properties of Rolled-Steel Shapes (U.S. Customary Units).*

| | Designation | Area in$^2$ | Depth in. | Width in. | Axis X-X | | | Axis Y-Y | | |
|---|---|---|---|---|---|---|---|---|---|---|
| | | | | | $\bar{I}_x$, in$^4$ | $\bar{k}_x$, in. | $\bar{y}$, in. | $\bar{I}_y$, in$^4$ | $\bar{k}_y$, in. | $\bar{x}$, in. |
| W Shapes (Wide-Flange Shapes) 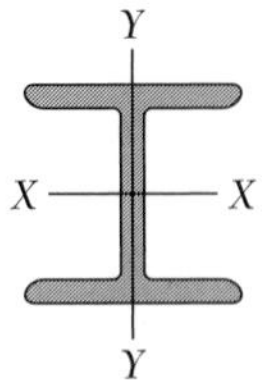 | W18 × 76† | 22.3 | 18.21 | 11.035 | 1330 | 7.73 | | 152 | 2.61 | |
| | W16 × 57 | 16.8 | 16.43 | 7.120 | 758 | 6.72 | | 43.1 | 1.60 | |
| | W14 × 38 | 11.2 | 14.10 | 6.770 | 385 | 5.88 | | 26.7 | 1.55 | |
| | W8 × 31 | 9.13 | 8.00 | 7.995 | 110 | 3.47 | | 37.1 | 2.02 | |
| S Shapes (American Standard Shapes) 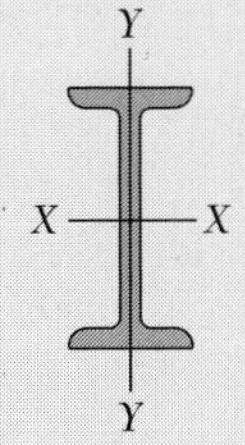 | S18 × 55.7† | 16.1 | 18.00 | 6.001 | 804 | 7.07 | | 20.8 | 1.14 | |
| | S12 × 31.8 | 9.35 | 12.00 | 5.000 | 218 | 4.83 | | 9.36 | 1.00 | |
| | S10 × 25.4 | 7.46 | 10.00 | 4.661 | 124 | 4.07 | | 6.79 | 0.954 | |
| | S6 × 12.5 | 3.67 | 6.00 | 3.332 | 22.1 | 2.45 | | 1.82 | 0.705 | |
| C Shapes (American Standard Channels) 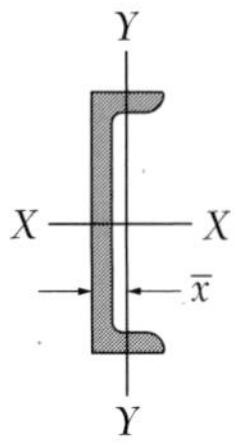 | C12 × 20.7† | 6.09 | 12.00 | 2.942 | 129 | 4.61 | | 3.88 | 0.799 | 0.698 |
| | C10 × 15.3 | 4.49 | 10.00 | 2.600 | 67.4 | 3.87 | | 2.28 | 0.713 | 0.634 |
| | C8 × 11.5 | 3.38 | 8.00 | 2.260 | 32.6 | 3.11 | | 1.32 | 0.625 | 0.571 |
| | C6 × 8.2 | 2.40 | 6.00 | 1.920 | 13.1 | 2.34 | | 0.692 | 0.537 | 0.512 |
| Angles 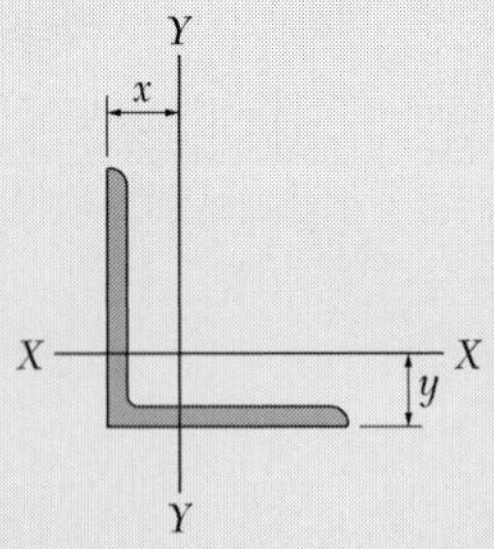 | L6 × 6 × 1‡ | 11.00 | | | 35.5 | 1.80 | 1.86 | 35.5 | 1.80 | 1.86 |
| | L4 × 4 × $\frac{1}{2}$ | 3.75 | | | 5.56 | 1.22 | 1.18 | 5.56 | 1.22 | 1.18 |
| | L3 × 3 × $\frac{1}{4}$ | 1.44 | | | 1.24 | 0.930 | 0.842 | 1.24 | 0.930 | 0.842 |
| | L6 × 4 × $\frac{1}{2}$ | 4.75 | | | 17.4 | 1.91 | 1.99 | 6.27 | 1.15 | 0.987 |
| | L5 × 3 × $\frac{1}{2}$ | 3.75 | | | 9.45 | 1.59 | 1.75 | 2.58 | 0.829 | 0.750 |
| | L3 × 2 × $\frac{1}{4}$ | 1.19 | | | 1.09 | 0.957 | 0.993 | 0.392 | 0.574 | 0.493 |

*Courtesy of the American Institute of Steel Construction, Chicago, Illinois.
†Nominal depth in inches and weight in pounds per foot.
‡Depth, width, and thickness in inches.

**Fig. 9.13B** Properties of Rolled-Steel Shapes (SI Units).

| | Designation | Area mm² | Depth mm | Width mm | Axis X-X | | | Axis Y-Y | | |
|---|---|---|---|---|---|---|---|---|---|---|
| | | | | | $\bar{I}_x$ 10⁶ mm⁴ | $\bar{k}_x$ mm | $\bar{y}$ mm | $\bar{I}_y$ 10⁶ mm⁴ | $\bar{k}_y$ mm | $\bar{x}$ mm |
| W Shapes (Wide-Flange Shapes) 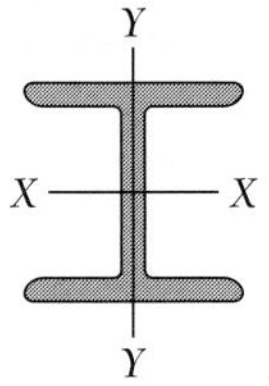 | W460 × 113† | 14400 | 463 | 280 | 554 | 196.3 | | 63.3 | 66.3 | |
| | W410 × 85 | 10800 | 417 | 181 | 316 | 170.7 | | 17.94 | 40.6 | |
| | W360 × 57 | 7230 | 358 | 172 | 160.2 | 149.4 | | 11.11 | 39.4 | |
| | W200 × 46.1 | 5890 | 203 | 203 | 45.8 | 88.1 | | 15.44 | 51.3 | |
| S Shapes (American Standard Shapes) 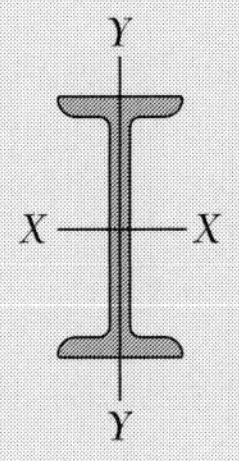 | S460 × 81.4† | 10390 | 457 | 152 | 335 | 179.6 | | 8.66 | 29.0 | |
| | W310 × 47.3 | 6032 | 305 | 127 | 90.7 | 122.7 | | 3.90 | 25.4 | |
| | S250 × 37.8 | 4806 | 254 | 118 | 51.6 | 103.4 | | 2.83 | 24.2 | |
| | S150 × 18.6 | 2362 | 152 | 84 | 9.2 | 62.2 | | 0.758 | 17.91 | |
| C Shapes (American Standard Channels) 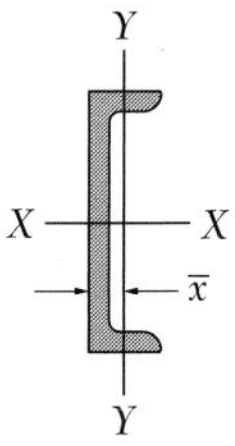 | C310 × 30.8† | 3929 | 305 | 74 | 53.7 | 117.1 | | 1.615 | 20.29 | 17.73 |
| | C250 × 22.8 | 2897 | 254 | 65 | 28.1 | 98.3 | | 0.949 | 18.11 | 16.10 |
| | C200 × 17.1 | 2181 | 203 | 57 | 13.57 | 79.0 | | 0.549 | 15.88 | 14.50 |
| | C150 × 12.2 | 1548 | 152 | 48 | 5.45 | 59.4 | | 0.288 | 13.64 | 13.00 |
| Angles 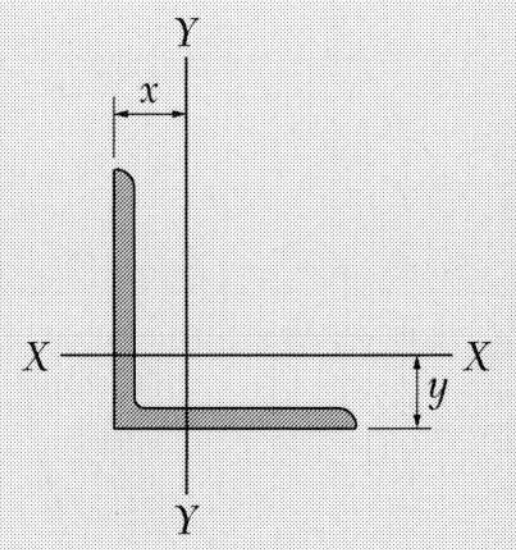 | L152 × 152 × 25.4‡ | 7100 | | | 14.78 | 45.6 | 47.2 | 14.78 | 45.6 | 47.2 |
| | L102 × 102 × 12.7 | 2420 | | | 2.31 | 30.9 | 30.0 | 2.31 | 30.9 | 30.0 |
| | L76 × 76 × 6.4 | 929 | | | 0.516 | 23.6 | 21.4 | 0.516 | 23.6 | 21.4 |
| | L152 × 102 × 12.7 | 3060 | | | 7.24 | 48.6 | 50.5 | 2.61 | 29.2 | 25.1 |
| | L127 × 76 × 12.7 | 2420 | | | 3.93 | 40.3 | 44.5 | 1.074 | 21.1 | 19.05 |
| | L76 × 51 × 6.4 | 768 | | | 0.454 | 24.3 | 25.2 | 0.163 | 14.58 | 12.52 |

†Nominal depth in millimeters and mass in kilograms per meter.
‡Depth, width, and thickness in millimeters.

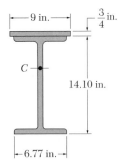

9 in. — $\frac{3}{4}$ in.

$C$

14.10 in.

6.77 in.

# SAMPLE PROBLEM 9.4

The strength of a W14 × 38 rolled-steel beam is increased by attaching a 9 × $\frac{3}{4}$in. plate to its upper flange as shown. Determine the moment of inertia and the radius of gyration of the composite section with respect to an axis through its centroid $C$ and parallel to the plate.

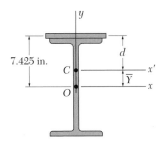

7.425 in.

$y$

$d$

$C$    $x'$

$\overline{Y}$

$O$    $x$

**Solution.** The origin of coordinates $O$ is placed at the centroid of the wide-flange shape, and the distance $\overline{Y}$ to the centroid of the composite section is computed by the methods of Chap. 5. The area of the wide-flange shape is found by referring to Fig. 9.13A. The area and $y$ coordinate of the centroid of the plate are

$$A = (9 \text{ in.})(0.75 \text{ in.}) = 6.75 \text{ in}^2$$
$$\bar{y} = \tfrac{1}{2}(14.10 \text{ in.}) + \tfrac{1}{2}(0.75 \text{ in.}) = 7.425 \text{ in.}$$

| Section | Area, in² | $\bar{y}$, in. | $\bar{y}$A, in³ |
|---|---|---|---|
| Plate | 6.75 | 7.425 | 50.12 |
| Wide-flange shape | 11.20 | 0 | 0 |
| | 17.95 | . . . | 50.12 |

$$\overline{Y}\Sigma A = \Sigma\bar{y}A \qquad \overline{Y}(17.95) = 50.12 \qquad \overline{Y} = 2.792 \text{ in.}$$

*Moment of Inertia.* The parallel-axis theorem is used to determine the moments of inertia of the wide-flange shape and of the plate with respect to the $x'$ axis. This axis is a centroidal axis for the composite section, but *not* for either of the elements considered separately. The value of $\bar{I}_x$ for the wide-flange shape is obtained from Fig. 9.13A.

For the wide-flange shape,

$$I_{x'} = \bar{I}_x + A\overline{Y}^2 = 385 + (11.20)(2.792)^2 = 472.3 \text{ in}^4$$

For the plate,

$$I_{x'} = \bar{I}_x + Ad^2 = (\tfrac{1}{12})(9)(\tfrac{3}{4})^3 + (6.75)(7.425 - 2.792)^2 = 145.2 \text{ in}^4$$

For the composite area,

$$I_{x'} = 472.3 + 145.2 = 617.5 \text{ in}^4$$

$$I_{x'} = 618 \text{ in}^4 \quad \blacktriangleleft$$

### Radius of Gyration

$$k_{x'}^2 = \frac{I_{x'}}{A} = \frac{617.5 \text{ in}^4}{17.95 \text{ in}^2}$$

$$k_{x'} = 5.87 \text{ in.} \quad \blacktriangleleft$$

# SAMPLE PROBLEM 9.5

Determine the moment of inertia of the shaded area with respect to the $x$ axis.

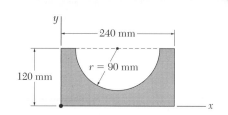

**Solution.** The given area may be obtained by subtracting a half circle from a rectangle. The moments of inertia of the rectangle and of the half circle will be computed separately.

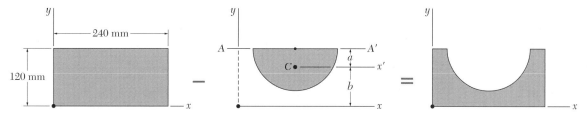

$I_x$ **for Rectangle.** Referring to Fig. 9.12, we obtain

$$I_x = \tfrac{1}{3}bh^3 = \tfrac{1}{3}(240 \text{ mm})(120 \text{ mm})^3 = 138.2 \times 10^6 \text{mm}^4$$

$I_x$ **for the Half Circle.** Referring to Fig. 5.8, we locate the centroid $C$ of the half circle with respect to diameter $AA'$.

$$a = \frac{4r}{3\pi} = \frac{(4)(90 \text{ mm})}{3\pi} = 38.2 \text{ mm}$$

The distance $b$ from the centroid $C$ to the $x$ axis is

$$b = 120 \text{ mm} - a = 120 \text{ mm} - 38.2 \text{ mm} = 81.8 \text{ mm}$$

Referring now to Fig. 9.12, we compute the moment of inertia of the half circle with respect to diameter $AA'$; we also compute the area of the half circle.

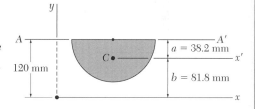

$$I_{AA'} = \tfrac{1}{8}\pi r^4 = \tfrac{1}{8}\pi(90 \text{ mm})^4 = 25.76 \times 10^6 \text{mm}^4$$
$$A = \tfrac{1}{2}\pi r^2 = \tfrac{1}{2}\pi(90 \text{ mm})^2 = 12.72 \times 10^3 \text{mm}^3$$

Using the parallel-axis theorem, we obtain the value of $\bar{I}_{x'}$:

$$I_{AA'} = \bar{I}_{x'} + Aa^2$$
$$25.76 \times 10^6 \text{mm}^4 = \bar{I}_{x'} + (12.72 \times 10^3)(38.2 \text{ mm})^2$$
$$\bar{I}_{x'} = 7.20 \times 10^6 \text{mm}^4$$

Again using the parallel-axis theorem, we obtain the value of $I_x$:

$$I_x = \bar{I}_{x'} + Ab^2 = 7.20 \times 10^6 \text{mm}^4 + (12.72 \times 10^3 \text{mm}^2)(81.8 \text{ mm})^2$$
$$= 92.3 \times 10^6 \text{mm}^4$$

$I_x$ **for Given Area.** Subtracting the moment of inertia of the half circle from that of the rectangle, we obtain

$$I_x = 138.2 \times 10^6 \text{ mm}^4 - 92.3 \times 10^6 \text{ mm}^4$$
$$I_x = 45.9 \times 10^6 \text{ mm}^4 \quad \blacktriangleleft$$

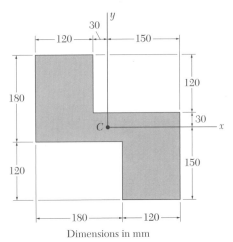

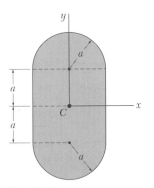

Dimensions in mm

**Fig. P9.21 and P9.22**

**Fig. P9.25**

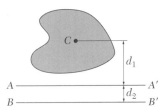

**Fig. P9.27 and P9.28**

**9.21 and 9.23** Determine the moment of inertia and the radius of gyration of the shaded area with respect to the $x$ axis.

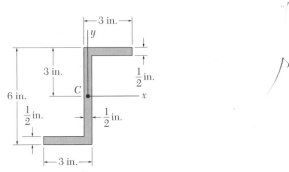

**Fig. P9.23 and P9.24**

**9.22 and 9.24** Determine the moment of inertia and the radius of gyration of the shaded area with respect to the $y$ axis.

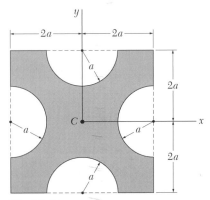

**Fig. P9.26**

**9.25 and 9.26** Determine the moments of inertia of the shaded area shown with respect to the $x$ and $y$ axes when $a = 20$ mm.

**9.27** Determine the shaded area and its moment of inertia with respect to a centroidal axis parallel to $AA'$, knowing that its moments of inertia with respect to $AA'$ and $BB'$ are respectively $4.1 \times 10^6$ mm$^4$ and $6.9 \times 10^6$ mm$^4$, and that $d_1 = 30$ mm and $d_2 = 10$ mm.

**9.28** Knowing that the shaded area is equal to 7500 mm$^2$ and that its moment of inertia with respect to $AA'$ is $31 \times 10^6$ mm$^4$, determine its moment of inertia with respect to $BB'$ for $d_1 = 60$ mm and $d_2 = 15$ mm.

**9.29**  The shaded area shown is equal to $20\,\text{in}^2$, and for $d_1 = 6$ in. the polar moment of inertia of the area with respect to point $B$ is $J_B = 800\,\text{in}^4$. Knowing that $\bar{I}_x = 4\bar{I}_y$, determine $(a)$ $\bar{I}_x$, $(b)$ $J_A$ for $d_2 = 5$ in.

**9.30**  For the shaded area shown, determine the polar moment of inertia with respect to point $D$, knowing that the polar moments of inertia with respect to points $A$ and $B$ are, respectively, $J_A = 4000\,\text{in}^4$ and $J_B = 6240\,\text{in}^4$, and that $d_1 = 8$ in. and $d_2 = 6$ in.

**9.31 and 9.32**  Determine the moments of inertia $\bar{I}_x$ and $\bar{I}_y$ of the area shown with respect to centroidal axes respectively parallel and perpendicular to the side $AB$.

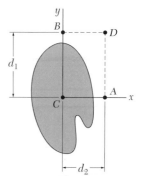

**Fig. P9.29 and P9.30**

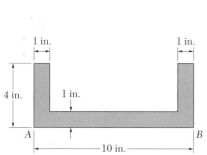

**Fig. P9.31**

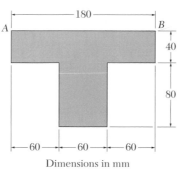

Dimensions in mm

**Fig. P9.32**

**9.33 and 9.34**  Determine the polar moment of inertia of the area shown with respect to $(a)$ point, $O$, $(b)$ the centroid of the area.

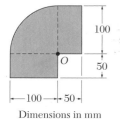

Dimensions in mm

**Fig. P9.33**

**Fig. P9.34**

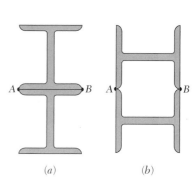

**Fig. P9.35**

**9.35**  Two W8 × 31 rolled sections may be welded at $A$ and $B$ in either of the two ways shown. For each arrangement, determine the moment of inertia of the section with respect to the horizontal centroidal axis.

**9.36** Two $6 \times 4 \times \frac{1}{2}$ in. angles are welded together to form the section shown. Determine the moments of inertia and the radii of gyration of the section with respect to the centroidal axes shown.

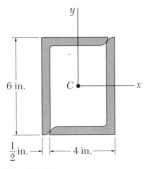

6 in.

$\frac{1}{2}$ in. ⟶ ⟵ 4 in. ⟶

**Fig. P9.36**

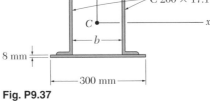

C 200 × 17.1

b

8 mm

300 mm

**Fig. P9.37**

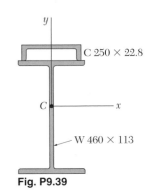

C 250 × 22.8

W 460 × 113

**Fig. P9.39**

**9.37** Two channels and two plates are used to form the column section shown. For $b = 160$ mm, determine the moments of inertia and the radii of gyration of the combined section with respect to the centroidal axes.

**9.38** In Prob. 9.37, determine the distance $b$ for which the centroidal moments of inertia $\bar{I}_x$ and $\bar{I}_y$ of the column section are equal.

**9.39** The strength of the wide-flange rolled section shown is increased by welding a channel to its upper flange. Determine the moments of inertia of the combined section with respect to its centroidal $x$ and $y$ axes.

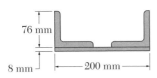

76 mm

8 mm — 200 mm ⟶

**Fig. P9.40**

**9.40** Two $76 \times 76 \times 6.4$ mm angles are welded to an $8 \times 200$ mm steel plate as shown. Determine the moments of inertia of the combined section with respect to centroidal axes respectively parallel and perpendicular to the plate.

**9.41** Two $4 \times 4 \times \frac{1}{2}$ in. angles are welded to a steel plate as shown. For $b = 8$ in., determine the moments of inertia of the combined section with respect to centroidal axes respectively parallel and perpendicular to the plate.

**9.42** Solve Prob. 9.41, assuming that $b = 10$ in.

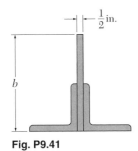

$\frac{1}{2}$ in.

b

**Fig. P9.41**

**9.43 through 9.45** The panel shown forms the end of a trough which is filled with water to the line $AA'$. Referring to Sec. 9.2, determine the depth of the point of application of the resultant of the hydrostatic forces acting on the panel (center of pressure).

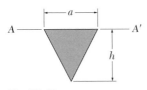

A ———— A'

a

h

**Fig. P9.43**

A ———— A'

r

**Fig. P9.44**

a

A ———————— A'

h

b

**Fig. P9.45**

**9.46**  The center of a vertical circular gate, 1.6 m in diameter, is located 2.4 m below the water surface. The gate is held by three bolts equally spaced as shown. Determine the force on each bolt.

**9.47**  Solve Prob. 9.46, assuming that the center of the gate is located 2 m below the water surface.

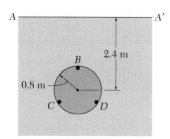

**Fig. P9.46**

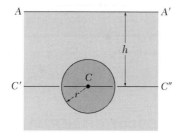

**Fig. P9.48**

**9.48**  A vertical circular gate of radius $r$ is hinged about its diameter $C'C''$. Denoting the specific weight of water by $\gamma$, determine (a) the reaction at each of the two hinges, (b) the moment of the couple required to keep the gate closed.

**\* 9.49**  Determine the $x$ coordinate of the centroid of the volume shown; this volume was obtained by intersecting a circular cylinder by an oblique plane. (*Hint.* The height of the volume is proportional to the $x$ coordinate; consider an analogy between this height and the water pressure on a submerged surface.)

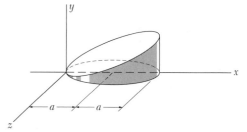

**Fig. P9.49**

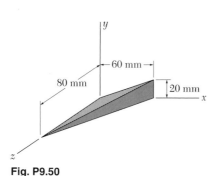

**Fig. P9.50**

**\* 9.50**  Determine the $x$ coordinate of the centroid of the volume shown. (See hint of Prob. 9.49.)

**\* 9.51**  Show that the system of hydrostatic forces acting on a submerged plane area $A$ may be reduced to a force $\mathbf{P}$ at the centroid $C$ of the area and two couples. The force $\mathbf{P}$ is perpendicular to the area and of magnitude $P = \gamma A \bar{y} \sin \theta$, where $\gamma$ is the specific weight of the liquid, and the couples are represented by vectors directed as shown and of magnitude $M_{x'} = \gamma \bar{I}_{x'} \sin \theta$ and $M_{y'} = \gamma \bar{I}_{x'y'} \sin \theta$, where $\bar{I}_{x'y'} = \int x'y'\, dA$ (see Sec. 9.8). Note that the couples are independent of the depth at which the area is submerged.

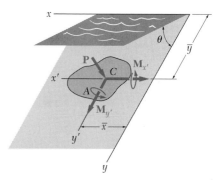

**Fig. P9.51**

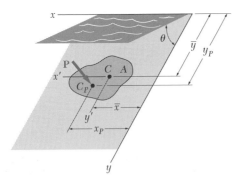

**Fig. P9.52**

Show that the resultant of the hydrostatic forces acting on a submerged plane area $A$ is a force $\mathbf{P}$ perpendicular to the area and of magnitude $P = \gamma A \bar{y} \sin \theta = \bar{p} A$, where $\gamma$ is the specific weight of the liquid and $\bar{p}$ the pressure at the centroid $C$ of the area. Show that $\mathbf{P}$ is applied at a point $C_P$, called the center of pressure, of coordinates $x_P = I_{xy}/A\bar{y}$ and $y_P = I_x/A\bar{y}$, where $I_{xy} = \int xy \, dA$ (see Sec. 9.8). Show also that the difference of ordinates $y_P - \bar{y}$ is equal to $k_{x'}^2/\bar{y}$ and thus depends upon the depth at which the area is submerged.

## *9.8. PRODUCT OF INERTIA
The integral

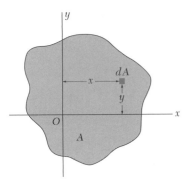

**Fig. 9.14**

$$I_{xy} = \int xy \, dA \tag{9.12}$$

obtained by multiplying each element $dA$ of an area $A$ by its coordinates $x$ and $y$ and integrating over the area (Fig. 9.14), is known as the *product of inertia* of the area $A$ with respect to the $x$ and $y$ axes. Unlike the moments of inertia $I_x$ and $I_y$, the product of inertia $I_{xy}$ may be either positive or negative.

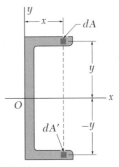

**Fig. 9.15**

When one or both of the $x$ and $y$ axes are axes of symmetry for the area $A$, the product of inertia $I_{xy}$ is zero. Consider, for example, the channel section shown in Fig. 9.15. Since this section is symmetrical with respect to the $x$ axis, we can associate to each element $dA$ of coordinates $x$ and $y$ an element $dA'$ of coordinates $x$ and $-y$. Clearly, the contributions of any pair of elements chosen in this way cancel out, and the integral (9.12) reduces to zero.

A parallel-axis theorem similar to the one established in Sec. 9.6 for moments of inertia may be derived for products of inertia. Consider an area $A$ and a system of rectangular coordinates $x$ and $y$ (Fig. 9.16). Through the centroid $C$ of the area, of coordinates $\bar{x}$ and $\bar{y}$, we draw two *centroidal axes* $x'$ and $y'$ parallel, respectively, to the $x$ and $y$ axes. Denoting by $x$ and $y$ the coordinates of an element of area $dA$ with respect to the original axes, and by $x'$ and $y'$ the coordinates of the same element with respect to the centroidal axes, we write $x = x' + \bar{x}$ and $y = y' + \bar{y}$. Substituting into (9.12), we obtain the following expression for the product of inertia $I_{xy}$:

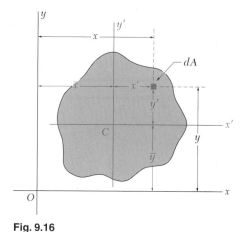

**Fig. 9.16**

$$I_{xy} = \int xy\, dA = \int (x' + \bar{x})(y' + \bar{y})\, dA$$
$$= \int x'y'\, dA + \bar{y}\int x'\, dA + \bar{x}\int y'\, dA + \bar{x}\bar{y}\int dA$$

The first integral represents the product of inertia $\bar{I}_{x'y'}$ of the area $A$ with respect to the centroidal axes $x'$ and $y'$. The next two integrals represent first moments of the area with respect to the centroidal axes; they reduce to zero, since the centroid $C$ is located on these axes. Finally, we observe that the last integral is equal to the total area $A$. We write therefore

$$I_{xy} = \bar{I}_{x'y'} + \bar{x}\bar{y}A \qquad (9.13)$$

## *9.9. PRINCIPAL AXES AND PRINCIPAL MOMENTS OF INERTIA

Consider the area $A$ and the coordinate axes $x$ and $y$ (Fig. 9.17). We assume that the moments and product of inertia

$$I_x = \int y^2\, dA \qquad I_y = \int x^2\, dA \qquad I_{xy} = \int xy\, dA \qquad (9.14)$$

of the area $A$ are known, and we propose to determine the moments and product of inertia $I_{x'}$, $I_{y'}$, and $I_{x'y'}$ of $A$ with respect to new axes $x'$ and $y'$ obtained by rotating the original axes about the origin through an angle $\theta$.

We first note the following relations between the coordinates $x'$, $y'$ and $x$, $y$ of an element of area $dA$:

$$x' = x \cos \theta + y \sin \theta \qquad y' = y \cos \theta - x \sin \theta$$

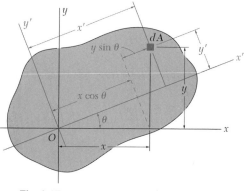

**Fig. 9.17**

Substituting for $y'$ in the expression for $I_{x'}$, we write

$$I_{x'} = \int (y')^2\, dA = \int (y \cos \theta - x \sin \theta)^2\, dA$$
$$= \cos^2 \theta \int y^2\, dA - 2 \sin \theta \cos \theta \int xy\, dA + \sin^2 \theta \int x^2\, dA$$

Taking the relations (9.14) into account, we write

$$I_{x'} = I_x \cos^2 \theta - 2I_{xy} \sin \theta \cos \theta + I_y \sin^2 \theta \qquad (9.15)$$

Similarly, we obtain for $I_{y'}$ and $I_{x'y'}$ the expressions

$$I_{y'} = I_x \sin^2 \theta + 2I_{xy} \sin \theta \cos \theta + I_y \cos^2 \theta \qquad (9.16)$$
$$I_{x'y'} = (I_x - I_y) \sin \theta \cos \theta + I_{xy}(\cos^2 \theta - \sin^2 \theta) \qquad (9.17)$$

Recalling the trigonometric relations

$$\sin 2\theta = 2 \sin \theta \cos \theta \qquad \cos 2\theta = \cos^2 \theta - \sin^2 \theta$$

and

$$\cos^2 \theta = \frac{1 + \cos 2\theta}{2} \qquad \sin^2 \theta = \frac{1 - \cos 2\theta}{2}$$

we may write (9.15), (9.16), and (9.17) as follows:

$$I_{x'} = \frac{I_x + I_y}{2} + \frac{I_x - I_y}{2} \cos 2\theta - I_{xy} \sin 2\theta \qquad (9.18)$$

$$I_{y'} = \frac{I_x + I_y}{2} - \frac{I_x - I_y}{2} \cos 2\theta + I_{xy} \sin 2\theta \qquad (9.19)$$

$$I_{x'y'} = \frac{I_x - I_y}{2} \sin 2\theta + I_{xy} \cos 2\theta \qquad (9.20)$$

We observe, by adding (9.18) and (9.19) member by member, that

$$I_{x'} + I_{y'} = I_x + I_y \qquad (9.21)$$

This result could have been anticipated, since both members of (9.21) are equal to the polar moment of inertia $J_O$.

Equations (9.18) and (9.20) are the parametric equations of a circle. This means that if we choose a set of rectangular axes and plot a point $M$ of abscissa $I_{x'}$ and ordinate $I_{x'y'}$ for any given value of the parameter $\theta$, all the points thus obtained will lie on a circle. To establish this property we shall eliminate $\theta$ from Eqs. (9.18) and (9.20); this is done by transposing $(I_x + I_y)/2$ in Eq. (9.18), squaring both members of Eqs. (9.18) and (9.20), and adding. We write

$$\left(I_{x'} - \frac{I_x + I_y}{2}\right)^2 + I_{x'y'}^2 = \left(\frac{I_x - I_y}{2}\right)^2 + I_{xy}^2 \qquad (9.22)$$

Setting

$$I_{\text{ave}} = \frac{I_x + I_y}{2} \qquad \text{and} \qquad R = \sqrt{\left(\frac{I_x - I_y}{2}\right)^2 + I_{xy}^2} \qquad (9.23)$$

we write the identity (9.22) in the form

$$(I_{x'} - I_{\text{ave}})^2 + I_{x'y'}^2 = R^2 \qquad (9.24)$$

which is the equation of a circle of radius $R$ centered at the point $C$ of abscissa $I_{\text{ave}}$ and ordinate 0 (Fig. 9.18$a$). It may be observed that Eqs. (9.19) and (9.20) are the parametric equations of the same circle. Furthermore, due to the symmetry of the circle about the horizontal axis, the same result would have been obtained if instead of plotting $M$, we had plotted a point $N$ of abscissa $I_{y'}$ and ordinate $-I_{x'y'}$ (Fig. 9.18$b$). This property will be used in Sec. 9.10.

The two points $A$ and $B$ where the circle obtained intersects the axis of abscissas are of special interest: Point $A$ corresponds to the maximum value of the moment of inertia $I_{x'}$, while point $B$ corresponds to its minimum

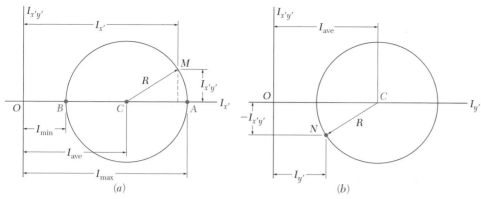

**Fig. 9.18**

value. Besides, both points correspond to a zero value of the product of inertia $I_{x'y'}$. Thus, the values $\theta_m$ of the parameter $\theta$ which correspond to the points $A$ and $B$ may be obtained by setting $I_{x'y'} = 0$ in Eq. (9.20). We obtain†

$$\tan 2\theta_m = -\frac{2I_{xy}}{I_x - I_y} \tag{9.25}$$

This equation defines two values $2\theta_m$ which are 180° apart and thus two values $\theta_m$ which are 90° apart. One of them corresponds to point $A$ in Fig. 9.18$a$ and to an axis through $O$ in Fig. 9.17 with respect to which the moment of inertia of the given area is maximum; the other value corresponds to point $B$ and an axis through $O$ with respect to which the moment of inertia of the area is minimum. The two axes thus defined, which are perpendicular to each other, are called the *principal axes of the area about O*, and the corresponding values $I_{\max}$ and $I_{\min}$ of the moment of inertia are called the *principal moments of inertia of the area about O*. Since the two values $\theta_m$ defined by Eq. (9.25) were obtained by setting $I_{x'y'} = 0$ in Eq. (9.20), it is clear that the product of inertia of the given area with respect to its principal axes is zero.

We observe from Fig. 9.18$a$ that

$$I_{\max} = I_{\text{ave}} + R \quad \text{and} \quad I_{\min} = I_{\text{ave}} - R \tag{9.26}$$

Substituting for $I_{\text{ave}}$ and $R$ from formulas (9.23), we write

$$I_{\max,\min} \frac{I_x + I_y}{2} \pm \sqrt{\left(\frac{I_x - I_y}{2}\right)^2 + I_{xy}^2} \tag{9.27}$$

Unless it is possible to tell by inspection which of the two principal axes corresponds to $I_{\max}$ and which corresponds to $I_{\min}$, it is necessary to substitute one of the values of $\theta_m$ into Eq. (9.18) in order to determine which of

---

†This relation may also be obtained by differentiating $I_{x'}$ in Eq. (9.18) and setting $dI_{x'}/d\theta = 0$.

the two corresponds to the maximum value of the moment of inertia of the area about $O$.

Referring to Sec. 9.8, we note that if an area possesses an axis of symmetry through a point $O$, this axis must be a principal axis of the area about $O$. On the other hand, a principal axis does not need to be an axis of symmetry; whether or not an area possesses properties of symmetry, it will have two principal axes of inertia about any point $O$.

The properties established here hold for any point $O$ located inside or outside the given area. If the point $O$ is chosen to coincide with the centroid of the area, any axis through $O$ is a centroidal axis; the two principal axes of the area about its centroid are referred to as the *principal centroidal axes of the area*.

## *9.10. MOHR'S CIRCLE FOR MOMENTS AND PRODUCTS OF INERTIA

The circle used in the preceding section to illustrate the relations existing between the moments and products of inertia of a given area with respect to axes passing through a fixed point $O$ was first introduced by the German engineer Otto Mohr (1835–1918) and is known as *Mohr's circle*. We shall see that if the moments and product of inertia of an area $A$ are known with respect to two rectangular $x$ and $y$ axes through a point $O$, Mohr's circle may be used to determine graphically (*a*) the principal axes and principal moments of inertia of the area about $O$, or (*b*) the moments and product of inertia of the area with respect to any other pair of rectangular axes $x'$ and $y'$ through $O$.

Consider a given area $A$ and two rectangular coordinate axes $x$ and $y$ (Fig. 9.19*a*). We shall assume that the moments of inertia $I_x$ and $I_y$ and the product of inertia $I_{xy}$ are known, and we shall represent them on a diagram by plotting a point $X$ of coordinates $I_x$ and $I_{xy}$ and a point $Y$ of coordinates $I_y$ and $-I_{xy}$ (Fig. 9.16*b*). If $I_{xy}$ is positive, as assumed in Fig. 9.19*a*, point $X$ is located above the horizontal axis and point $Y$ below, as shown in Fig. 9.19*b*. If $I_{xy}$ is negative, $X$ is located below the horizontal axis and $Y$ above. Joining $X$ and $Y$ by a straight line, we define the point $C$ of intersection of line $XY$ with the horizontal axis and draw the circle of center $C$ and diameter $XY$. Noting that the abscissa of $C$ and the radius of the circle are respectively equal to the quantities $I_{\text{ave}}$ and $R$ defined by the formulas (9.23), we conclude that the circle obtained is Mohr's circle for the given area about point $O$. Thus the abscissas of the points $A$ and $B$ where the circle intersects the horizontal axes represent respectively the principal moments of inertia $I_{\max}$ and $I_{\min}$ of the area.

We also note that, since $\tan(XCA) = 2I_{xy}/(I_x - I_y)$, the angle $XCA$ is equal in magnitude to one of the angles $2\theta_m$ which satisfy Eq. (9.25); thus the angle $\theta_m$ which defines in Fig. 9.19*a* the principal axis $Oa$ corresponding to point $A$ in Fig. 9.19*b* may be obtained by dividing in half the angle $XCA$ measured on Mohr's circle. We further observe that if $I_x > I_y$ and $I_{xy} > 0$, as in the case considered here, the rotation which brings $CX$ into $CA$ is clockwise. But, in that case, the angle $\theta_m$ obtained from Eq. (9.25) and defining the principal axis $Oa$ in Fig. 9.19*a* is negative; thus the rota-

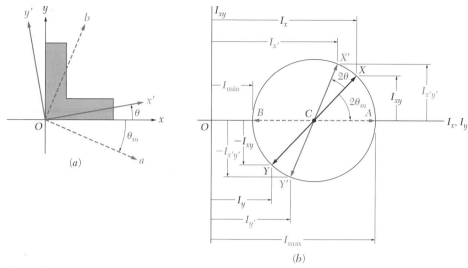

**Fig. 9.19**

tion bringing $Ox$ into $Oa$ is also clockwise. We conclude that the senses of rotation in both parts of Fig. 9.19 are the same; if a clockwise rotation through $2\theta_m$ is required to bring $CX$ into $CA$ on Mohr's circle, a clockwise rotation through $\theta_m$ will bring $Ox$ into the corresponding principal axis $Oa$ in Fig. 9.19*a*.

Since Mohr's circle is uniquely defined, the same circle may be obtained by considering the moments and product of inertia of the area $A$ with respect to the rectangular axes $x'$ and $y'$ (Fig. 9.19*a*). The point $X'$ of coordinates $I_{x'}$ and $I_{x'y'}$ and the point $Y'$ of coordinates $I_{y'}$ and $-I_{x'y'}$ are therefore located on Mohr's circle, and the angle $X'CA$ in Fig. 9.19*b* must be equal to twice the angle $x'Oa$, in Fig. 9.19*a*. Since, as noted above, the angle $XCA$ is twice the angle $xOa$ it follows that the angle $XCX'$ in Fig. 9.19*b* is twice the angle $xOx'$ in Fig. 9.19*a*. The diameter $X'Y'$ defining the moments and product of inertia $I_{x'}$, $I_{y'}$, and $I_{x'y'}$ of the given area with respect to rectangular axes $x'$ and $y'$ forming an angle $\theta$ with the $x$ and $y$ axes may be obtained by rotating through an angle $2\theta$ the diameter $XY$ corresponding to the moments and product of inertia $I_x$, $I_y$, and $I_{xy}$. We note that the rotation which brings the diameter $XY$ into the diameter $X'Y'$ in Fig. 9.19*b* has the same sense as the rotation which brings the $xy$ axes into the $x'y'$ axes in Fig. 9.19*a*.

It should be noted that the use of Mohr's circle is not limited to graphical solutions, i.e., to solutions based on the careful drawing and measuring of the various parameters involved. By merely sketching Mohr's circle and using trigonometry, one may easily derive the various relations required for a numerical solution of a given problem. Actual computations may then be carried out on a calculator (see Sample Prob. 9.8).

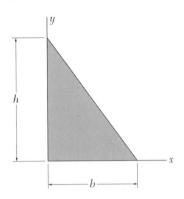

# SAMPLE PROBLEM 9.6

Determine the product of inertia of the right triangle shown (*a*) with respect to the *x* and *y* axes and (*b*) with respect to centroidal axes parallel to the *x* and *y* axes.

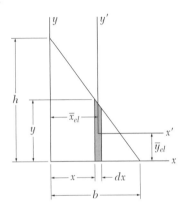

**a. Product of Inertia $I_{xy}$.**  A vertical rectangular strip is chosen as the differential element of area. Using the parallel-axis theorem, we write

$$dI_{xy} = dI_{x'y'} + \bar{x}_{el}\bar{y}_{el}\, dA$$

Since the element is symmetrical with respect to the $x'$ and $y'$ axes, we note that $dI_{x'y'} = 0$. From the geometry of the triangle, we obtain

$$y = h\left(1 - \frac{x}{b}\right) \qquad dA = y\, dx = h\left(1 - \frac{x}{b}\right)dx$$

$$\bar{x}_{el} = x \qquad \bar{y}_{el} = \tfrac{1}{2}y = \tfrac{1}{2}h\left(1 - \frac{x}{b}\right)$$

Integrating $dI_{xy}$ from $x = 0$ to $x = b$, we obtain

$$I_{xy} = \int dI_{xy} = \int \bar{x}_{el}\bar{y}_{el}\, dA = \int_0^b x\left(\tfrac{1}{2}\right)h^2\left(1 - \frac{x}{b}\right)^2 dx$$

$$= h^2\int_0^b \left(\frac{x}{2} - \frac{x^2}{b} + \frac{x^3}{2b^2}\right)dx = h^2\left[\frac{x^2}{4} - \frac{x^2}{3b} + \frac{x^4}{8b^2}\right]_0^b$$

$$I_{xy} = \tfrac{1}{24}b^2h^2 \quad \blacktriangleleft$$

**b. Product of Inertia $I_{x''y''}$.**  The coordinates of the centroid of the triangle are

$$\bar{x} = \tfrac{1}{3}b \qquad \bar{y} = \tfrac{1}{3}h$$

Using the expression for $I_{xy}$ obtained in part *a*, we apply the parallel-axis theorem and write

$$I_{xy} = \bar{I}_{x''y''} + \bar{x}\,\bar{y}A$$
$$\tfrac{1}{24}b^2h^2 = \bar{I}_{x''y''} + \left(\tfrac{1}{3}b\right)\left(\tfrac{1}{3}h\right)\left(\tfrac{1}{2}bh\right)$$
$$\bar{I}_{x''y''} = \tfrac{1}{24}b^2h^2 - \tfrac{1}{18}b^2h^2$$

$$\bar{I}_{x''y''} = -\tfrac{1}{72}b^2h^2 \quad \blacktriangleleft$$

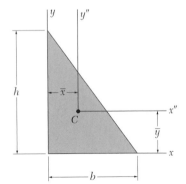

# SAMPLE PROBLEM 9.7

For the section shown, the moments of inertia with respect to the $x$ and $y$ axes have been computed and are known to be

$$I_x = 10.38 \text{ in}^4 \qquad I_y = 6.97 \text{ in}^4$$

Determine (a) the principal axes of the section about $O$, (b) the values of the principal moments of inertia of the section about $O$.

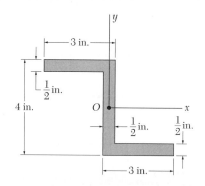

## Solution.
We first compute the product of inertia with respect to the $x$ and $y$ axes. The area is divided into three rectangles as shown. We note that the product of inertia $\bar{I}_{x'y'}$ with respect to centroidal axes parallel to the $x$ and $y$ axes is zero for each rectangle. Using the parallel-axis theorem $I_{xy} = \bar{I}_{x'y'} + \bar{x}\bar{y}A$, we thus find that for each rectangle, $I_{xy}$ reduces to $\bar{x}\bar{y}A$.

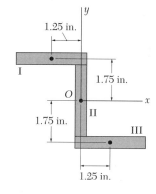

| Rectangle | Area, in² | $\bar{x}$, in. | $\bar{y}$, in. | $\bar{x}\bar{y}A$, in⁴ |
|-----------|-----------|----------------|----------------|------------------------|
| I | 1.5 | −1.25 | +1.75 | −3.28 |
| II | 1.5 | 0 | 0 | 0 |
| III | 1.5 | +1.25 | −1.75 | −3.28 |
| | | | | −6.56 |

$$I_{xy} = \Sigma \bar{x}\bar{y}A = -6.56 \text{ in}^4$$

**a. Principal Axes.** Since the magnitudes of $I_x$, $I_y$, and $I_{xy}$ are known, Eq. (9.25) is used to determine the values of $\theta_m$:

$$\tan 2\theta_m = -\frac{2I_{xy}}{I_x - I_y} = -\frac{2(-6.56)}{10.38 - 6.97} = +3.85$$

$$2\theta_m = 75.4° \text{ and } 255.4°$$

$$\theta_m = 37.7° \qquad \text{and} \qquad \theta_m = 127.7° \blacktriangleleft$$

**b. Principal Moments of Inertia.** Using Eq. (9.27), we write

$$I_{max, min} = \frac{I_x + I_y}{2} \pm \sqrt{\left(\frac{I_x - I_y}{2}\right)^2 + I_{xy}^2}$$

$$= \frac{10.38 + 6.97}{2} \pm \sqrt{\left(\frac{10.38 - 6.97}{2}\right)^2 + (-6.56)^2}$$

$$I_{max} = 15.45 \text{ in}^4 \qquad I_{min} = 1.897 \text{ in}^4 \blacktriangleleft$$

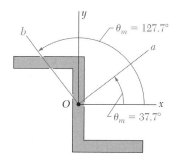

Noting that the area of the section is farther away from the $a$ axis than from the $b$ axis, we conclude that $I_a = I_{max} = 15.45 \text{ in}^4$ and $I_b = I_{min} = 1.897 \text{ in}^4$. This conclusion may be verified by substituting $\theta = 37.7°$ into Eqs. (9.18) and (9.19).

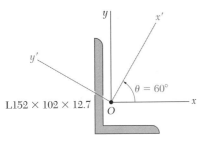

L152 × 102 × 12.7

$\theta = 60°$

## SAMPLE PROBLEM 9.8

For the section shown, the moments and product of inertia with respect to the $x$ and $y$ axes have been computed and are known to be

$$I_x = 7.24 \times 10^6 \text{ mm}^4 \qquad I_y = 2.61 \times 10^6 \text{ mm}^4 \qquad I_{xy} = -2.54 \times 10^6 \text{mm}^4$$

Using Mohr's circle, determine (a) the principal axes of the section about $O$, (b) the values of the principal moments of inertia of the section about $O$, (c) the moments and product of inertia of the section with respect to the $x'$ and $y'$ axes forming an angle of 60° with the $x$ and $y$ axes.

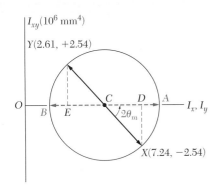

$I_{xy}(10^6 \text{ mm}^4)$

$Y(2.61, +2.54)$

$2\theta_m$

$X(7.24, -2.54)$

**Solution.** We first plot point $X$ of coordinates $I_x = 7.24$, $I_{xy} = -2.54$, and point $Y$ of coordinates $I_y = 2.61$, $-I_{xy} = +2.54$. Joining $X$ and $Y$ by a straight line, we define the center $C$ of Mohr's circle. The abscissa of $C$, which represents $I_{\text{ave}}$, and the radius $R$ of the circle may be measured directly or calculated as follows:

$$I_{\text{ave}} = OC = \tfrac{1}{2}(I_x + I_y) = \tfrac{1}{2}(7.24 \times 10^6 + 2.61 \times 10^6) = 4.925 \times 10^6 \text{ mm}^4$$
$$CD = \tfrac{1}{2}(I_x - I_y) = \tfrac{1}{2}(7.24 \times 10^6 - 2.61 \times 10^6) = 2.315 \times 10^6 \text{ mm}^4$$
$$R = \sqrt{(CD)^2 + (DX)^2} = \sqrt{(2.315 \times 10^6)^2 + (2.54 \times 10^6)^2} = 3.437 \times 10^6 \text{ mm}^4$$

**a. Principal Axes.** The principal axes of the section correspond to points $A$ and $B$ on Mohr's circle and the angle through which we should rotate $CX$ to bring it into $CA$ defines $2\theta_m$. We have

$$\tan 2\theta_m = \frac{DX}{CD} = \frac{2.54}{2.315} = 1.097 \qquad 2\theta_m = 47.6° \nwarrow \qquad \theta_m = 23.8° \nwarrow \quad \blacktriangleleft$$

Thus the principal axis $Oa$ corresponding to the maximum value of the moment of inertia is obtained by rotating the $x$ axis through 23.8° counterclockwise; the principal axis corresponding to the minimum value of the moment of inertia may be obtained by rotating the $y$ axis through the same angle.

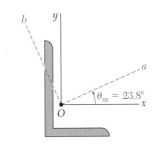

$\theta_m = 23.8°$

**b. Principal Moments of Inertia.** The principal moments of inertia are represented by the abscissas of $A$ and $B$. We have

$$I_{\text{max}} = OA = OC + CA = I_{\text{ave}} + R = (4.925 + 3.437)10^6 \text{ mm}^4$$
$$I_{\text{max}} = 8.36 \times 10^6 \text{ mm}^4 \quad \blacktriangleleft$$
$$I_{\text{min}} = OB = OC - BC = I_{\text{ave}} - R = (4.925 - 3.437)10^6 \text{ mm}^4$$
$$I_{\text{min}} = 1.49 \times 10^6 \text{ mm}^4 \quad \blacktriangleleft$$

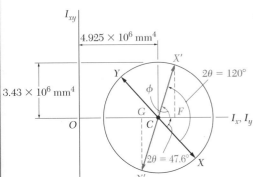

$I_{xy}$

$4.925 \times 10^6 \text{ mm}^4$

$3.43 \times 10^6 \text{ mm}^4$

$2\theta = 120°$

$\phi$

$2\theta = 47.6°$

**c. Moments and Product of Inertia with Respect to $x'y'$ Axes.** The points $X'$ and $Y'$ on Mohr's circle which correspond to the $x'$ and $y'$ axes are obtained by rotating $CX$ and $CY$ through an angle $2\theta = 2(60°) = 120°$ counterclockwise. The coordinates of $X'$ and $Y'$ yield the desired moments and product of inertia. Noting that the angle that $CX'$ forms with the horizontal axis is $\phi = 120° - 47.6° = 72.4°$, we write

$$I_{x'} = OF = OC + CF = 4.925 \times 10^6 \text{ mm}^4 + (3.437 \times 10^6 \text{ mm}^4) \cos 72.4°$$
$$I_{x'} = 5.96 \times 10^6 \text{ mm}^4 \quad \blacktriangleleft$$
$$I_{y'} = OG = OC - GC = 4.925 \times 10^6 \text{ mm}^4 - (3.437 \times 10^6 \text{ mm}^4) \cos 72.4°$$
$$I_{y'} = 3.89 \times 10^6 \text{ mm}^4 \quad \blacktriangleleft$$
$$I_{x'y'} = FX' = (3.437 \times 10^6 \text{ mm}^4) \sin 72.4°$$
$$I_{x'y'} = 3.28 \times 10^6 \text{ mm}^4 \quad \blacktriangleleft$$

# PROBLEMS

**9.53 through 9.56** Determine by direct integration the product of inertia of the given area with respect to the $x$ and $y$ axes.

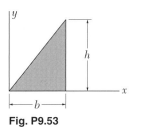

**Fig. P9.53**

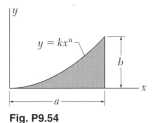

**Fig. P9.54**

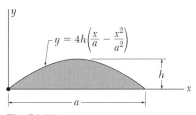

**Fig. P9.55**

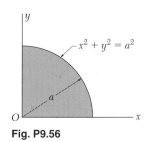

**Fig. P9.56**

**9.57 through 9.62** Using the parallel-axis theorem, determine the product of inertia of the area shown with respect to the centroidal $x$ and $y$ axes.

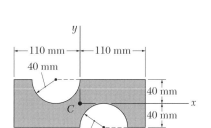

**Fig. P9.57**

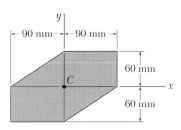

**Fig. P9.58**

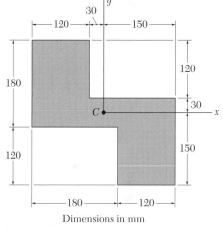

Dimensions in mm

**Fig. P9.59**

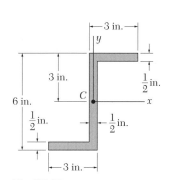

**Fig. P9.60**

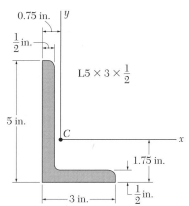

**Fig. P9.61**

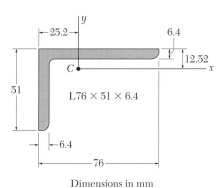

Dimensions in mm

**Fig. P9.62**

**9.63** Determine the moments of inertia and the product of inertia of the quarter circle of Prob. 9.56 with respect to new axes obtained by rotating the $x$ and $y$ axes about $O$ (a) through 30° counterclockwise, (b) through 45° counterclockwise.

**9.64** Determine the moments of inertia and the product of inertia of the area of Prob. 9.60 with respect to new centroidal axes obtained by rotating the $x$ and $y$ axes through 30° counterclockwise.

**9.65** Determine the moments of inertia and the product of inertia of the L 5 × 3 × $\frac{1}{2}$ in. angle cross section of Prob. 9.61 with respect to new centroidal axes obtained by rotating the $x$ and $y$ axes through 30° clockwise. (The moments of inertia $\bar{I}_x$ and $\bar{I}_y$ of the section area given in Fig. 9.13A.)

**9.66** Determine the moments of inertia and the product of inertia of the L 76 × 51 × 6.4 mm angle cross section of Prob. 9.62 with respect to new centroidal axes obtained by rotating the $x$ and $y$ axes through 30° clockwise. (The moments of inertia $\bar{I}_x$ and $\bar{I}_y$ of the section are given in Fig. 9.13B.)

**9.67** Determine the orientation of the principal axes through the centroid and the corresponding values of the moment of inertia for the area of Prob. 9.57. (*Given:* $\bar{I}_x = 6.16 \times 10^6 \, \text{mm}^4$ and $\bar{I}_y = 60.9 \times 10^6 \, \text{mm}^4$.)

**9.68 through 9.70** For the area indicated, determine the orientation of the principal axes through the centroid and the corresponding values of the moment of inertia.

**9.68** Area of Prob. 9.58.
**9.69** Area of Prob. 9.59.
**9.70** Area of Prob. 9.60.

**9.71 and 9.72** For the angle cross section indicated, determine the orientation of the principal axes through the centroid and the corresponding values of the moment of inertia. (The moments of inertia $\bar{I}_x$ and $\bar{I}_y$ of the cross sections are given in Fig. 9.13.)

**9.71** The L 5 × 3 × $\frac{1}{2}$ in. angle of Prob. 9.61.
**9.72** The L 76 × 51 × 6.4 mm angle of Prob. 9.62.

**9.73** Using Mohr's circle, determine the moments of inertia and the product of inertia of the quarter circle of Prob. 9.56 with respect to new axes obtained by rotating the $x$ and $y$ axes about $O$ (a) through 30° counterclockwise, (b) through 45° counterclockwise.

**9.74** Using Mohr's circle, determine the moments of inertia and the product of inertia of the area of Prob. 9.60 with respect to new centroidal axes obtained by rotating the $x$ and $y$ axes through 30° counterclockwise.

**9.75** Using Mohr's circle, determine the moments of inertia and the product of inertia of the L 5 × 3 × $\frac{1}{2}$ in. angle cross section of Prob. 9.61 with respect to new centroidal axes obtained by rotating the $x$ and $y$ axes through 30° clockwise. (The moments of inertia $\bar{I}_x$ and $\bar{I}_y$ of the section are given in Fig. 9.13A.)

**9.76**   Using Mohr's circle, determine the moment of inertia and the product of inertia of the L 76 × 51 × 6.4 mm angle cross section of Prob. 9.62 with respect to new centroidal axes obtained by rotating the $x$ and $y$ axes through 30° clockwise. (The moments of inertia $\bar{I}_x$ and $\bar{I}_y$ of the section are given in Fig. 9.13B.)

**9.77**   Using Mohr's circle, determine the orientation of the principal axes through the centroid and the corresponding values of the moment of inertia for the area of Prob. 9.57. (*Given:* $\bar{I}_x = 6.16 \times 10^6 \text{ mm}^4$ and $\bar{I}_y = 60.9 \times 10^6 \text{ mm}^4$.)

**9.78 through 9.80**   Using Mohr's circle, determine for the area indicated the orientation of the principal axes through the centroid and the corresponding values of the moment of inertia.

**9.78**   Area of Prob. 9.58.
**9.79**   Area of Prob. 9.59.
**9.80**   Area of Prob. 9.60.

**9.81 and 9.82**   Using Mohr's circle, determine for the angle cross section indicated the orientation of the principal axes through the centroid and the corresponding values of the moment of inertia. (The moments of inertia $\bar{I}_x$ and $\bar{I}_y$ of the cross sections are given in Fig. 9.13.)

**9.81**   The L 5 × 3 × $\frac{1}{2}$ in. angle of Prob. 9.61.
**9.82**   The L 76 × 51 × 6.4 mm angle of Prob. 9.62.

**9.83**   Using Mohr's circle show that for any regular polygon (such as a pentagon), (*a*) the moment of inertia with respect to every axis through the centroid is the same, (*b*) the product of inertia with respect to any pair of rectangular axes through the centroid is zero.

**9.84**   The moments and product of inertia of an L 127 × 76 × 12.7 mm angle cross section with respect to two rectangular axes $x$ and $y$ through $C$ are, respectively, $\bar{I}_x = 3.93 \times 10^6 \text{ mm}^4$, $\bar{I}_y = 1.074 \times 10^6 \text{ mm}^4$, and $\bar{I}_{xy} < 0$, with the minimum value of the moment of inertia of the area with respect to any axis through $C$ being $\bar{I}_{min} = 0.655 \times 10^6 \text{ mm}^4$. Using Mohr's circle, determine (*a*) the product of inertia $\bar{I}_{xy}$ of the area, (*b*) the orientation of the principal axes, (*c*) the value of $\bar{I}_{max}$.

**\*9.85**   Prove that the expression $I_{x'}I_{y'} - I_{x'y'}^2$, where $I_{x'}$, $I_{y'}$, and $I_{x'y'}$ represent, respectively, the moments and product of inertia of a given area with respect to two rectangular axes $x'$ and $y'$ through a given point $O$, is independent of the orientation of the $x'$ and $y'$ axes. Considering the particular case when the $x'$ and $y'$ axes correspond to the maximum value of $I_{x'y'}$, show that the given expression represents the square of the tangent drawn from the origin of the coordinates to Mohr's circle.

**\*9.86**   Using the invariance property established in the preceding problem, express the product of inertia $I_{xy}$ of an area $A$ with respect to two rectangular axes through $O$ in terms of the moments of inertia $I_x$ and $I_y$ of $A$ and of the principal moments of inertia $I_{min}$ and $I_{max}$ of $A$ about $O$. Apply the formula obtained to calculate the product of inertia $\bar{I}_{xy}$ of the L 6 × 4 × $\frac{1}{2}$ in. angle cross section shown in Fig. 9.13A, knowing that its minimum moment of inertia is 3.60 in⁴.

# MOMENTS OF INERTIA OF MASSES

## 9.11. MOMENT OF INERTIA OF A MASS

Consider a small mass $\Delta m$ mounted on a rod of negligible mass which may rotate freely about an axis $AA'$ (Fig. 9.20a). If a couple is applied to the system, the rod and mass, assumed initially at rest, will start rotating about $AA'$. The details of this motion will be studied later in dynamics. At present, we wish only to indicate that the time required for the system to reach a given speed of rotation is proportional to the mass $\Delta m$ and to the square of the distance $r$. The product $r^2\,\Delta m$ provides, therefore, a measure of the *inertia* of the system, i.e., of the resistance the system offers when we try to set it in motion. For this reason, the product $r^2\,\Delta m$ is called the *moment of inertia* of the mass $\Delta m$ with respect to the axis $AA'$.

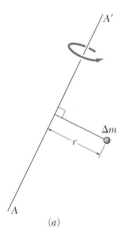

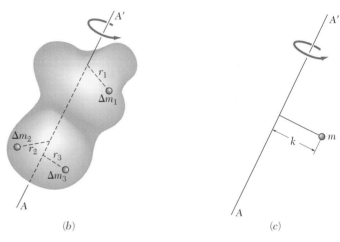

**Fig. 9.20**

Consider now a body of mass $m$ which is to be rotated about an axis $AA'$ (Fig. 9.20b). Dividing the body into elements of mass $\Delta m_1$, $\Delta m_2$, etc., we find that the resistance offered by the body is measured by the sum $r_1^2\,\Delta m_1 + r_2^2\,\Delta m_2 + \cdots$. This sum defines, therefore, the moment of inertia of the body with respect to the axis $AA'$. Increasing the number of elements, we find that the moment of inertia is equal, at the limit, to the integral

$$I = \int r^2\,dm \tag{9.28}$$

The *radius of gyration* $k$ of the body with respect to the axis $AA'$ is defined by the relation

$$I = k^2 m \qquad \text{or} \qquad k = \sqrt{\frac{I}{m}} \tag{9.29}$$

The radius of gyration $k$ represents, therefore, the distance at which the entire mass of the body should be concentrated if its moment of inertia with respect to $AA'$ is to remain unchanged (Fig. 9.20c). Whether it is kept in its original shape (Fig. 9.20b) or whether it is concentrated as shown in Fig. 9.20c, the mass $m$ will react in the same way to a rotation, or *gyration*, about $AA'$.

If SI units are used, the radius of gyration $k$ is expressed in meters and the mass $m$ in kilograms. The moment of inertia of a mass, therefore, will be expressed in $kg \cdot m^2$. If U.S. customary units are used, the radius of gyration is expressed in feet and the mass in slugs, i.e., in $lb \cdot s^2/ft$. The moment of inertia of a mass, then, will be expressed in $lb \cdot ft \cdot s^2$.†

The moment of inertia of a body with respect to a coordinate axis may easily be expressed in terms of the coordinates $x$, $y$, $z$ of the element of mass $dm$ (Fig. 9.21). Noting, for example, that the square of the distance $r$ from

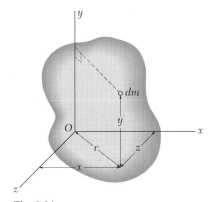

**Fig. 9.21**

the element $dm$ to the $y$ axis is $z^2 + x^2$, we express the moment of inertia of the body with respect to the $y$ axis as

$$I_y = \int r^2 dm = \int \left(z^2 + x^2\right) dm$$

Similar expressions may be obtained for the moments of inertia with respect to the $x$ and $z$ axes. We write

$$I_x = \int \left(y^2 + z^2\right) dm$$

$$I_y = \int \left(z^2 + x^2\right) dm \tag{9.30}$$

$$I_z = \int \left(x^2 + y^2\right) dm$$

---

†It should be kept in mind when converting the moment of inertia of a mass from U.S. customary units to SI units that the base unit pound used in the derived unit $lb \cdot ft \cdot s^2$ is a unit of force (*not* of mass) and should therefore be converted into newtons. We have

$$1 \, lb \cdot ft \cdot s^2 = (4.45 \, N)(0.3048 \, m)(1 \, s)^2 = 1.356 \, N \cdot m \cdot s^2$$

or, since $N = kg \cdot m/s^2$,

$$1 \, lb \cdot ft \cdot s^2 = 1.356 \, kg \cdot m^2$$

## 9.12. PARALLEL-AXIS THEOREM

Consider a body of mass $m$. Let $Oxyz$ be a system of rectangular coordinates with origin at an arbitrary point $O$, and $Gx'y'z'$ a system of parallel *centroidal axes*, i.e., a system with origin at the center of gravity $G$ of the body† and with axes $x'$, $y'$, $z'$, respectively, parallel to $x$, $y$, $z$ (Fig. 9.22). Denoting by $\bar{x}$, $\bar{y}$, $\bar{z}$ the coordinates of $G$ with respect to $Oxyz$, we write the following relations between the coordinates $x$, $y$, $z$ of the element $dm$ with respect to $Oxyz$ and its coordinates $x'$, $y'$, $z'$ with respect to the centroidal axes $Gx'y'z'$:

$$x = x' + \bar{x} \qquad y = y' + \bar{y} \qquad z = z' + \bar{z} \qquad (9.31)$$

Referring to Eqs. (9.30), we may express the moment of inertia of the body with respect to the $x$ axis as follows:

$$I_x = \int (y^2 + z^2)\,dm = \int [(y' + \bar{y})^2 + (z' + \bar{z})^2]\,dm$$

$$= \int (y'^2 + z'^2)\,dm + 2\bar{y}\int y'\,dm + 2\bar{z}\int z'\,dm + (\bar{y}^2 + \bar{z}^2)\int dm$$

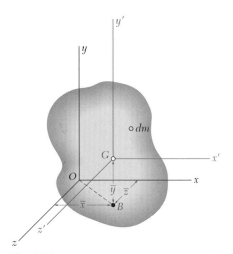

**Fig. 9.22**

The first integral in the expression obtained represents the moment of inertia $\bar{I}_{x'}$ of the body with respect to the centroidal axis $x'$; the second and third integrals represent the first moment of the body with respect to the $z'x'$ and $x'y'$ planes, respectively, and, since both planes contain $G$, the two integrals are zero; the last integral is equal to the total mass $m$ of the body. We write, therefore,

$$I_x = \bar{I}_{x'} + m(\bar{y}^2 + \bar{z}^2) \qquad (9.32)$$

and, similarly,

$$I_y = \bar{I}_{y'} + m(\bar{z}^2 + \bar{x}^2) \qquad I_z = \bar{I}_{z'} + m(\bar{x}^2 + \bar{y}^2) \qquad (9.32')$$

We easily verify from Fig. 9.22 that the sum $\bar{z}^2 + \bar{x}^2$ represents the square of the distance $OB$ between the $y$ and $y'$ axis. Similarly, $\bar{y}^2 + \bar{z}^2$ and $\bar{x}^2 + \bar{y}^2$ represent the squares of the distance between the $x$ and $x'$ axes, and the $z$ and $z'$ axes, respectively. Denoting by $d$ the distance between an arbitrary axis $AA'$ and a parallel centroidal axis $BB'$ (Fig. 9.23), we may, therefore, write the following general relation between the moment of inertia $I$ of the body with respect to $AA'$ and its moment of inertia $\bar{I}$ with respect to $BB'$:

$$I = \bar{I} + md^2 \qquad (9.33)$$

Expressing the moments of inertia in terms of the corresponding radii of gyration, we may also write

$$k^2 = \bar{k}^2 + d^2 \qquad (9.34)$$

where $k$ and $\bar{k}$ represent the radii of gyration about $AA'$ and $BB'$, respectively.

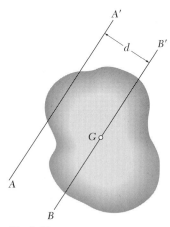

**Fig. 9.23**

† Note that the term centroidal is used to define an axis passing through the center of gravity $G$ of the body, whether or not $G$ coincides with the centroid of the volume of the body.

## 9.13. MOMENTS OF INERTIA OF THIN PLATES

Consider a thin plate of uniform thickness $t$, made of a homogeneous material of density $\rho$ (density = mass per unit volume). The mass moment of inertia of the plate with respect to an axis $AA'$ *contained in the plane* of the plate (Fig. 9.24a) is

$$I_{AA', \text{mass}} = \int r^2 dm$$

Since $dm = \rho t\, dA$, we write

$$I_{AA', \text{mass}} = \rho t \int r^2\, dA$$

But $r$ represents the distance of the element of area $dA$ to the axis $AA'$; the

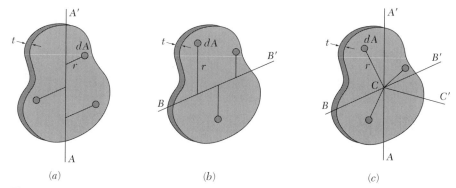

(a)          (b)          (c)

**Fig. 9.24**

integral is therefore equal to the moment of inertia of the area of the plate with respect to $AA'$. We have

$$I_{AA', \text{mass}} = \rho t I_{AA', \text{area}} \tag{9.35}$$

Similarly, we have with respect to an axis $BB'$ perpendicular to $AA'$ (Fig. 9.24b)

$$I_{BB', \text{mass}} = \rho t I_{BB', \text{area}} \tag{9.36}$$

Considering now the axis $CC'$ *perpendicular* to the plate through the point of intersection $C$ of $AA'$ and $BB'$ (Fig. 9.24c), we write

$$I_{CC', \text{mass}} = \rho t J_{C, \text{area}} \tag{9.37}$$

where $J_C$ is the *polar* moment of inertia of the area of the plate with respect to point $C$.

Recalling the relation $J_C = I_{AA'} + I_{BB'}$ existing between polar and rectangular moments of inertia of an area, we write the following relation between the mass moments of inertia of a thin plate:

$$I_{CC'} = I_{AA'} + I_{BB'} \tag{9.38}$$

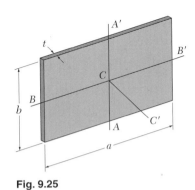

**Fig. 9.25**

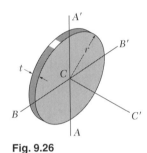

**Fig. 9.26**

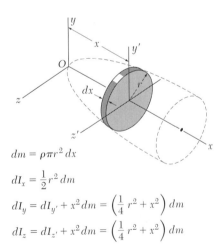

$dm = \rho\pi r^2\, dx$

$dI_x = \frac{1}{2} r^2\, dm$

$dI_y = dI_{y'} + x^2\, dm = \left(\frac{1}{4} r^2 + x^2\right) dm$

$dI_z = dI_{z'} + x^2\, dm = \left(\frac{1}{4} r^2 + x^2\right) dm$

**Fig. 9.27** Determination of the moment of inertia of a body of revolution.

*Rectangular Plate.* In the case of a reactangular plate of sides $a$ and $b$ (Fig. 9.25), we obtain the following mass moments of inertia with respect to axes through the center of gravity of the plate:

$$I_{AA',\,\text{mass}} = \rho t I_{AA',\,\text{area}} = \rho t(\tfrac{1}{12}a^3 b)$$
$$I_{BB',\,\text{mass}} = \rho t I_{BB',\,\text{area}} = \rho t(\tfrac{1}{12}ab^3)$$

Observing that the product $\rho abt$ is equal to the mass $m$ of the plate, we write the mass moments of inertia of a thin rectangular plate as follows:

$$I_{AA'} = \tfrac{1}{12}ma^2 \qquad I_{BB'} = \tfrac{1}{12}mb^2 \tag{9.39}$$
$$I_{CC'} = I_{AA'} + I_{BB'} = \tfrac{1}{12}m(a^2 + b^2) \tag{9.40}$$

*Circular Plate.* In the case of a circular plate, or disk, of radius $r$ (Fig. 9.26), we write

$$I_{AA',\,\text{mass}} = \rho t I_{AA',\,\text{area}} = \rho t(\tfrac{1}{4}\pi r^4)$$

Observing that the product $\rho\pi r^2 t$ is equal to the mass $m$ of the plate and that $I_{AA'} = I_{BB'}$, we write the mass moments of inertia of a circular plate as follows:

$$I_{AA'} = I_{BB'} = \tfrac{1}{4}mr^2 \tag{9.41}$$
$$I_{CC'} = I_{AA'} + I_{BB'} = \tfrac{1}{2}mr^2 \tag{9.42}$$

## 9.14. DETERMINATION OF THE MOMENT OF INERTIA OF A THREE-DIMENSIONAL BODY BY INTEGRATION

The moment of inertia of a three-dimensional body is obtained by computing the integral $I = \int r^2 dm$. If the body is made of a homogeneous material of density $\rho$, we have $dm = \rho\, dV$ and write $I = \rho \int r^2 dV$. This integral depends only upon the shape of the body. Thus, in order to compute the moment of inertia of a three-dimensional body, it will generally be necessary to perform a triple, or at least a double, integration.

However, if the body possesses two planes of symmetry, it is usually possible to determine the body's moment of inertia through a single integration by choosing as an element of mass $dm$ the mass of a thin slab perpendicular to the planes of symmetry. In the case of bodies of revolution, for example, the element of mass should be a thin disk (Fig. 9.27). Using formula (9.42), the moment of inertia of the disk with respect to the axis of revolution may be readily expressed as indicated in Fig. 9.27. Its moment of inertia with respect to each of the other two axes of coordinates will be obtained by using formula (9.41) and the parallel-axis theorem. Integration of the expressions obtained will yield the desired moment of inertia of the body of revolution.

## 9.15. MOMENTS OF INERTIA OF COMPOSITE BODIES

The moments of inertia of a few common shapes are shown in Fig. 9.28. The moment of inertia with respect to a given axis of a body made of several of these simple shapes may be obtained by computing the moments of inertia of its component parts about the desired axis and adding them together. We should note, as we have already noted in the case of areas, that the radius of gyration of a composite body *cannot* be obtained by adding the radii of gyration of its component parts.

| | | |
|---|---|---|
| Slender rod | 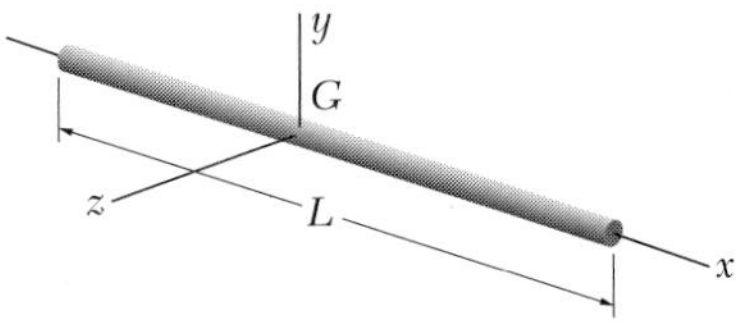 | $I_y = I_z = \frac{1}{12} mL^2$ |
| Thin rectangular plate | 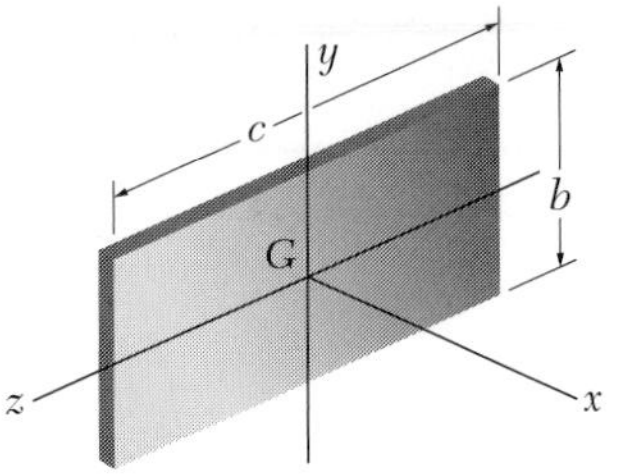 | $I_x = \frac{1}{12} m(b^2 + c^2)$ <br> $I_y = \frac{1}{12} mc^2$ <br> $I_z = \frac{1}{12} mb^2$ |
| Rectangular prism | 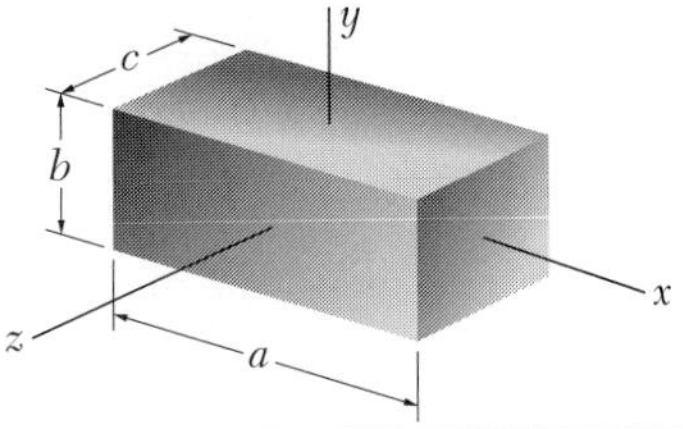 | $I_x = \frac{1}{12} m(b^2 + c^2)$ <br> $I_y = \frac{1}{12} m(c^2 + a^2)$ <br> $I_z = \frac{1}{12} m(a^2 + b^2)$ |
| Thin disk | 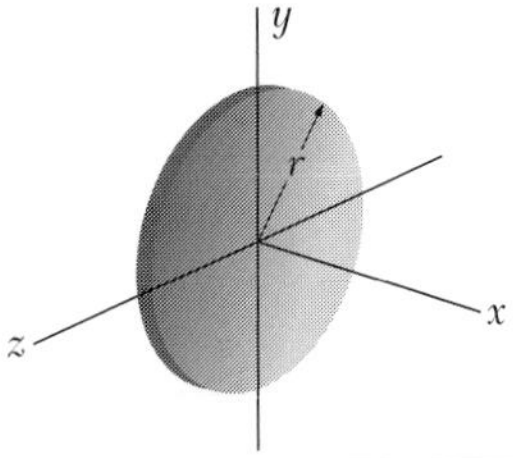 | $I_x = \frac{1}{2} mr^2$ <br> $I_y = I_z = \frac{1}{4} mr^2$ |
| Circular cylinder | 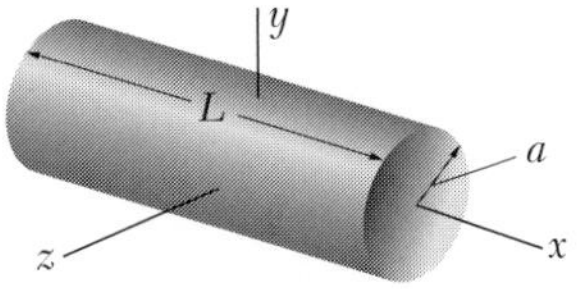 | $I_x = \frac{1}{2} ma^2$ <br> $I_y = I_z = \frac{1}{12} m(3a^2 + L^2)$ |
| Circular cone | 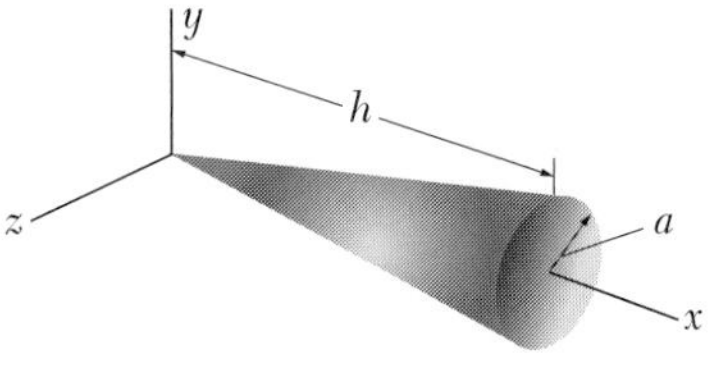 | $I_x = \frac{3}{10} ma^2$ <br> $I_y = I_z = \frac{3}{5} m\left(\frac{1}{4} a^2 + h^2\right)$ |
| Sphere | 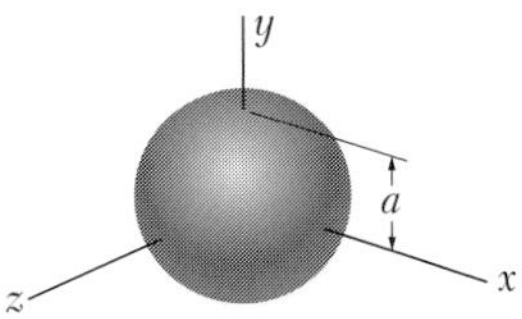 | $I_x = I_y = I_z = \frac{2}{5} ma^2$ |

**Fig. 9.28** Mass moments of inertia of common geometric shapes.

## SAMPLE PROBLEM 9.9

Determine the mass moment of inertia of a slender rod of length $L$ and mass $m$ with respect to an axis perpendicular to the rod and passing through one end of the rod.

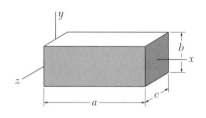

**Solution.** Choosing the differential element of mass shown, we write

$$dm = \frac{m}{L}dx$$

$$I_y = \int x^2\,dm = \int_0^L x^2\frac{m}{L}\,dx = \left[\frac{m}{L}\frac{x^3}{3}\right]_0^L \qquad I_y = \tfrac{1}{3}mL^2 \blacktriangleleft$$

## SAMPLE PROBLEM 9.10

Determine the mass moment of inertia of the homogeneous rectangular prism shown with respect to the $z$ axis.

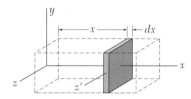

**Solution.** We choose as a differential element of mass the thin slab shown for which

$$dm = \rho bc\,dx$$

Referring to Sec. 9.13, we find that the moment of inertia of the element with respect to the $z'$ axis is

$$dI_{z'} = \tfrac{1}{12}b^2 dm$$

Applying the parallel-axis theorem, we obtain the mass moment of inertia of the slab with respect to the $z$ axis.

$$dI_z = dI_{z'} + x^2\,dm = \tfrac{1}{12}b^2\,dm + x^2\,dm = (\tfrac{1}{12}b^2 + x^2)\rho bc\,dx$$

Integrating from $x = 0$ to $x = a$, we obtain

$$I_z = \int dI_z = \int_0^a (\tfrac{1}{12}b^2 + x^2)\rho bc\,dx = \rho abc\,(\tfrac{1}{12}b^2 + \tfrac{1}{3}a^2)$$

Since the total mass of the prism is $m = \rho abc$, we may write

$$I_z = m(\tfrac{1}{12}b^2 + \tfrac{1}{3}a^2) \qquad\qquad I_z = \tfrac{1}{12}m\left(4a^2 + b^2\right) \blacktriangleleft$$

We note that if the prism is slender, $b$ is small compared to $a$ and the expression for $I_z$ reduces to $\tfrac{1}{3}ma^2$, which is the result obtained in Sample Prob. 9.9 when $L = a$.

## SAMPLE PROBLEM 9.11

Determine the mass moment of inertia of a right circular cone with respect to (a) its longitudinal axis, (b) an axis through the apex of the cone and perpendicular to its longitudinal axis, (c) an axis through the centroid of the cone and perpendicular to its longitudinal axis.

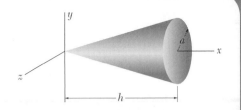

**Solution.** We choose the differential element of mass shown.

$$r = a\frac{x}{h} \qquad dm = \rho\pi r^2\,dx = \rho\pi\frac{a^2}{h^2}x^2\,dx$$

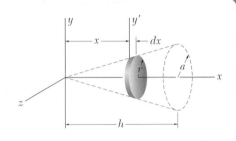

**a. Moment of Inertia $I_x$.** Using the expression derived in Sec. 9.13 for a thin disk, we compute the mass moment of inertia of the differential element with respect to the $x$ axis.

$$dI_x = \tfrac{1}{2}r^2\,dm = \tfrac{1}{2}\left(a\frac{x}{h}\right)^2\left(\rho\pi\frac{a^2}{h^2}x^2\,dx\right) = \tfrac{1}{2}\rho\pi\frac{a^4}{h^4}x^4\,dx$$

Integrating from $x = 0$ to $x = h$, we obtain

$$I_x = \int dI_x = \int_0^h \tfrac{1}{2}\rho\pi\frac{a^4}{h^4}x^4\,dx = \tfrac{1}{2}\rho\pi\frac{a^4}{h^4}\frac{h^5}{5} = \tfrac{1}{10}\rho\pi a^4 h$$

Since the total mass of the cone is $m = \tfrac{1}{3}\rho\pi a^2 h$, we may write

$$I_x = \tfrac{1}{10}\rho\pi a^4 h = \tfrac{3}{10}a^2(\tfrac{1}{3}\rho\pi a^2 h) = \tfrac{3}{10}ma^2 \qquad I_x = \tfrac{3}{10}ma^2 \quad \blacktriangleleft$$

**b. Moment of Inertia $I_y$.** The same differential element will be used. Applying the parallel-axis theorem and using the expression derived in Sec. 9.13 for a thin disk, we write

$$dI_y = dI_{y'} + x^2\,dm = \tfrac{1}{4}r^2\,dm + x^2\,dm = (\tfrac{1}{4}r^2 + x^2)\,dm$$

Substituting the expressions for $r$ and $dm$, we obtain

$$dI_y = \left(\frac{1}{4}\frac{a^2}{h^2}x^2 + x^2\right)\left(\rho\pi\frac{a^2}{h^2}x^2\,dx\right) = \rho\pi\frac{a^2}{h^2}\left(\frac{a^2}{4h^2} + 1\right)x^4\,dx$$

$$I_y = \int dI_y = \int_0^h \rho\pi\frac{a^2}{h^2}\left(\frac{a^2}{4h^2} + 1\right)x^4\,dx = \rho\pi\frac{a^2}{h^2}\left(\frac{a^2}{4h^2} + 1\right)\frac{h^5}{5}$$

Introducing the total mass of the cone $m$, we rewrite $I_y$ as follows:

$$I_y = \tfrac{3}{5}(\tfrac{1}{4}a^2 + h^2)\tfrac{1}{3}\rho\pi a^2 h \qquad I_y = \tfrac{3}{5}m(\tfrac{1}{4}a^2 + h^2) \quad \blacktriangleleft$$

**c. Moment of Inertia $I_{y''}$.** We apply the parallel-axis theorem and write

$$I_y = \bar{I}_{y''} + m\bar{x}^2$$

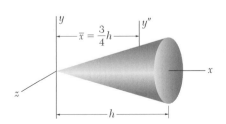

Solving for $\bar{I}_{y''}$ and recalling that $\bar{x} = \tfrac{3}{4}h$, we have

$$\bar{I}_{y''} = I_y - m\bar{x}^2 = \tfrac{3}{5}m(\tfrac{1}{4}a^2 + h^2) - m(\tfrac{3}{4}h)^2$$

$$\bar{I}_{y''} = \tfrac{3}{20}m(a^2 + \tfrac{1}{4}h^2) \quad \blacktriangleleft$$

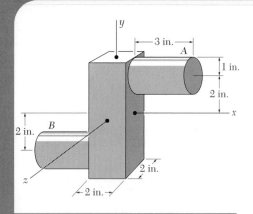

## SAMPLE PROBLEM 9.12

A steel forging consists of a rectangular prism $6 \times 2 \times 2$ in. and of two cylinders of diameter 2 in. and length 3 in., as shown. Determine the mass moments of inertia with respect to the coordinate axes. (Specific weight of steel = 490 lb/ft³.)

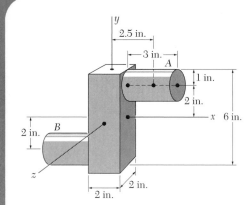

### Computation of Masses
#### Prism

$$V = (2 \text{ in.})(2 \text{ in.})(6 \text{ in.}) = 24 \text{ in}^3$$

$$W = \frac{(24 \text{ in}^3)(490 \text{ lb/ft}^3)}{1728 \text{ in}^3/\text{ft}^3} = 6.81 \text{ lb}$$

$$m = \frac{6.81 \text{ lb}}{32.2 \text{ ft/s}^2} = 0.211 \text{ lb} \cdot \text{s}^2/\text{ft}$$

#### Each Cylinder

$$V = \pi(1 \text{ in.})^2(3 \text{ in.}) = 9.42 \text{ in}^3$$

$$W = \frac{(9.42 \text{ in}^3)(490 \text{ lb/ft}^3)}{1728 \text{ in}^3/\text{ft}^3} = 2.67 \text{ lb}$$

$$m = \frac{2.67 \text{ lb}}{32.2 \text{ ft/s}^2} = 0.0829 \text{ lb} \cdot \text{s}^2/\text{ft}$$

**Mass Moments of Inertia.** The mass moments of inertia of each component are computed from Fig. 9.28, using the parallel-axis theorem when necessary. Note that all lengths should be expressed in feet.

#### Prism

$$I_x = I_z = \tfrac{1}{12}(0.211 \text{ lb} \cdot \text{s}^2/\text{ft})[(\tfrac{6}{12}\text{ft})^2 + (\tfrac{2}{12}\text{ft})^2] = 4.88 \times 10^{-3} \text{lb} \cdot \text{ft} \cdot \text{s}^2$$

$$I_y = \tfrac{1}{12}(0.211 \text{ lb} \cdot \text{s}^2/\text{ft})[(\tfrac{2}{12}\text{ft})^2 + (\tfrac{2}{12}\text{ft})^2] = 0.977 \times 10^{-3} \text{lb} \cdot \text{ft} \cdot \text{s}^2$$

#### Each Cylinder

$$I_x = \tfrac{1}{2}ma^2 + m\bar{y}^2 = \tfrac{1}{2}(0.0829 \text{ lb} \cdot \text{s}^2/\text{ft})(\tfrac{1}{12}\text{ ft})^2$$
$$+ (0.0829 \text{ lb} \cdot \text{s}^2/\text{ft})(\tfrac{2}{12}\text{ ft})^2 = 2.59 \times 10^{-3} \text{ lb} \cdot \text{ft} \cdot \text{s}^2$$

$$I_y = \tfrac{1}{12}m(3a^2 + L^2) + m\bar{x}^2 = \tfrac{1}{12}(0.0829 \text{ lb} \cdot \text{s}^2/\text{ft})[3(\tfrac{1}{12}\text{ ft})^2 + (\tfrac{3}{12}\text{ ft})^2]$$
$$+ (0.0829 \text{ lb} \cdot \text{s}^2/\text{ft})(\tfrac{2.5}{12}\text{ ft})^2 = 4.17 \times 10^{-3} \text{ lb} \cdot \text{ft} \cdot \text{s}^2$$

$$I_z = \tfrac{1}{12}m(3a^2 + L^2) + m(\bar{x}^2 + \bar{y}^2) = \tfrac{1}{12}(0.0829)[3(\tfrac{1}{12})^2 + (\tfrac{3}{12})^2]$$
$$+ (0.0829)[(\tfrac{2.5}{12})^2 + (\tfrac{2}{12})^2] = 6.48 \times 10^{-3} \text{ lb} \cdot \text{ft} \cdot \text{s}^2$$

**Entire Body.** Adding the values obtained,

$$I_x = 4.88 \times 10^{-3} + 2(2.59 \times 10^{-3}) \qquad I_x = 10.06 \times 10^{-3} \text{ lb} \cdot \text{ft} \cdot \text{s}^2 \blacktriangleleft$$
$$I_y = 0.977 \times 10^{-3} + 2(4.17 \times 10^{-3}) \qquad I_y = 9.32 \times 10^{-3} \text{ lb} \cdot \text{ft} \cdot \text{s}^2 \blacktriangleleft$$
$$I_z = 4.88 \times 10^{-3} + 2(6.48 \times 10^{-3}) \qquad I_z = 17.84 \times 10^{-3} \text{ lb} \cdot \text{ft} \cdot \text{s}^2 \blacktriangleleft$$

# SAMPLE PROBLEM 9.13

A thin steel plate 4 mm thick is cut and bent to form the machine part shown. Knowing that the density of steel is 7850 kg/m$^3$, determine the mass moment of inertia of the machine part with respect to the coordinate axes.

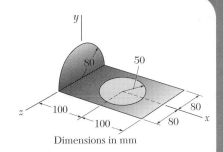

Dimensions in mm

**Solution.** We observe that the machine part consists of a semicircular plate, plus a rectangular plate, minus a circular plate.

### Computation of Masses. *Semicircular Plate*

$$V_1 = \tfrac{1}{2}\pi r^2 t = \tfrac{1}{2}\pi(0.08 \text{ m})^2(0.004 \text{ m}) = 40.21 \times 10^{-6} \text{ m}^3$$
$$m_1 = \rho V_1 = (7.85 \times 10^3 \text{ kg/m}^3)(40.21 \times 10^{-6} \text{ m}^3) = 0.3156 \text{ kg}$$

*Rectangular Plate*

$$V_2 = (0.200 \text{ m})(0.160 \text{ m})(0.004 \text{ m}) = 128 \times 10^{-6} \text{ m}^3$$
$$m_2 = \rho V_2 = (7.85 \times 10^3 \text{ kg/m}^3)(128 \times 10^{-6} \text{ m}^3) = 1.005 \text{ kg}$$

*Circular Plate*

$$V_3 = \pi a^2 t = \pi(0.050 \text{ m})^2(0.004 \text{ m}) = 31.42 \times 10^{-6} \text{ m}^3$$
$$m_3 = \rho V_3 = (7.85 \times 10^3 \text{ kg/m}^3)(31.42 \times 10^{-6} \text{ m}^3) = 0.2466 \text{ kg}$$

### Mass Moments of Inertia.

Using the method presented in Sec. 9.13, we compute the mass moment of inertia of each component.

*Semicircular Plate.* From Fig. 9.28, we observe that for a circular plate of mass $m$ and radius $r$

$$I_x = \tfrac{1}{2}mr^2 \qquad I_y = I_z = \tfrac{1}{4}mr^2$$

Because of symmetry, we note that for a semicircular plate

$$I_x = \tfrac{1}{2}\left(\tfrac{1}{2}mr^2\right) \qquad I_y = I_z = \tfrac{1}{2}\left(\tfrac{1}{4}mr^2\right)$$

Since the mass of the semicircular plate is $m_1 = \tfrac{1}{2}m$, we have

$$I_x = \tfrac{1}{2}m_1 r^2 = \tfrac{1}{2}(0.3156 \text{ kg})(0.08 \text{ m})^2 = 1.010 \times 10^{-3} \text{ kg} \cdot \text{m}^2$$
$$I_y = I_z = \tfrac{1}{4}\left(\tfrac{1}{2}mr^2\right) = \tfrac{1}{4}m_1 r^2 = \tfrac{1}{4}(0.3156 \text{ kg})(0.08 \text{ m})^2 = 0.505 \times 10^{-3} \text{ kg} \cdot \text{m}^2$$

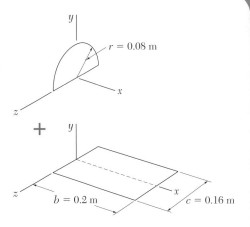

*Rectangular Plate*

$$I_x = \tfrac{1}{12}m_2 c^2 = \tfrac{1}{12}(1.005 \text{ kg})(0.16 \text{ m})^2 = 2.144 \times 10^{-3} \text{ kg} \cdot \text{m}^2$$
$$I_z = \tfrac{1}{3}m_2 b^2 = \tfrac{1}{3}(1.005 \text{ kg})(0.2 \text{ m})^2 = 13.400 \times 10^{-3} \text{ kg} \cdot \text{m}^2$$
$$I_y = I_x + I_z = (2.144 + 13.400)(10^{-3}) = 15.544 \times 10^{-3} \text{ kg} \cdot \text{m}^2$$

*Circular Plate*

$$I_x = \tfrac{1}{4}m_3 a^2 = \tfrac{1}{4}(0.2466 \text{ kg})(0.05 \text{ m})^2 = 0.154 \times 10^{-3} \text{ kg} \cdot \text{m}^2$$
$$I_y = \tfrac{1}{2}m_3 a^2 = m_3 d^2$$
$$= \tfrac{1}{2}(0.2466 \text{ kg})(0.05 \text{ m})^2 + (0.2466 \text{ kg})(0.1 \text{ m})^2 = 2.774 \times 10^{-3} \text{ kg} \cdot \text{m}^2$$
$$I_z = \tfrac{1}{4}m_3 a^2 + m_3 d^2 = \tfrac{1}{4}(0.2466 \text{ kg})(0.05 \text{ m})^2 + (0.2466 \text{ kg})(0.1 \text{ m})^2$$
$$= 2.620 \times 10^{-3} \text{ kg} \cdot \text{m}^2$$

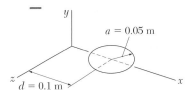

*Entire Machine Part*

$$I_x = (1.010 + 2.144 - 0.154)(10^{-3}) \text{ kg} \cdot \text{m}^2 \qquad I_x = 3.00 \times 10^{-3} \text{ kg} \cdot \text{m}^2 \quad \blacktriangleleft$$
$$I_y = (0.505 + 15.544 - 2.774)(10^{-3}) \text{ kg} \cdot \text{m}^2 \qquad I_y = 13.28 \times 10^{-3} \text{ kg} \cdot \text{m}^2 \quad \blacktriangleleft$$
$$I_z = (0.505 + 13.400 - 2.620)(10^{-3}) \text{ kg} \cdot \text{m}^2 \qquad I_z = 11.29 \times 10^{-3} \text{ kg} \cdot \text{m}^2 \quad \blacktriangleleft$$

# PROBLEMS

**9.87** A thin semicircular plate has a radius $a$ and a mass $m$. Determine the mass moment of inertia of the plate with respect to ($a$) the centroidal axis $BB'$, ($b$) the centroidal axis $CC'$.

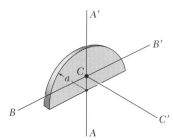

**Fig. P9.87**

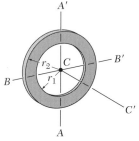

**Fig. P9.88**

**9.88** Determine the mass moment of inertia of a ring of mass $m$, cut from a thin uniform plate, with respect to ($a$) the diameter $AA'$ of the ring, ($b$) the axis $CC'$ perpendicular to the plane of the ring.

**9.89** A thin plate of mass $m$ is cut in the shape of an equilateral triangle of side $a$. Determine the mass moment of inertia of the plate with respect to ($a$) the centroidal axes $AA'$ and $BB'$ in the plane of the plate, ($b$) the centroidal axis $CC'$ perpendicular to the plate.

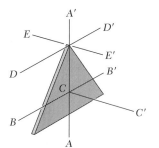

**Fig. P9.89**

**9.90** Determine the mass moments of inertia of the plate of Prob. 9.89 with respect to the axes $DD'$ and $EE'$ parallel to the centroidal axes $BB'$ and $CC'$, respectively.

**9.91** A thin plate of mass $m$ is cut in the shape of a parallelogram as shown. Determine the mass moment of inertia of the plate with respect to ($a$) the $x$ axis, ($b$) the axis $BB'$ prependicular to the plate.

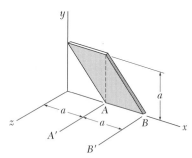

**Fig. P9.91**

**9.92** Determine the mass moment of inertia of the plate of Prob. 9.91 with respect to ($a$) the $y$ axis, ($b$) the axis $AA'$ perpendicular to the plate.

**9.93** The area shown is revolved about the $x$ axis to form a homogeneous solid of revolution of mass $m$. Express the mass moment of inertia of the solid with respect to the $x$ axis in terms of $m$, $a$, and $n$. The expression obtained may be used to verify ($a$) the value given in Fig. 9.28 for a cone (with $n = 1$), ($b$) the answer to Prob. 9.95 (with $n = \frac{1}{2}$).

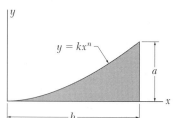

**Fig. P9.93**

**9.94** Determine by direct integration the mass moment of inertia with respect to the $y$ axis of the right circular cylinder shown, assuming a uniform density and a mass $m$.

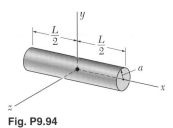

**Fig. P9.94**

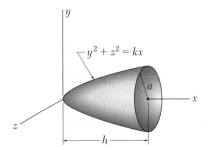

**Fig. P9.95 and P9.96**

**9.95** Determine by direct integration the mass moment of inertia and the radius of gyration with respect to the $x$ axis of the paraboloid shown, assuming a uniform density and a mass $m$.

**9.96** Determine by direct integration the mass moment of inertia and the radius of gyration with respect to the $y$ axis of the paraboloid shown, assuming a uniform density and a mass $m$.

**9.97** Determine by direct integration the mass moment of inertia with respect to the $x$ axis of the pyramid shown, assuming a uniform density and a mass $m$.

**9.98** Determine by direct integration the mass moment of inertia with respect to the $y$ axis of the pyramid shown, assuming a uniform density and a mass $m$.

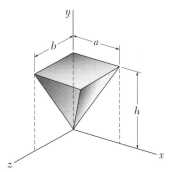

**Fig. P9.97 and P.9.98**

**9.99** A thin steel wire is bent into the shape of a circular arc as shown. Denoting by $m'$ the mass per unit length of the wire, determine the mass moments of inertia of the wire with respect to the coordinate axes.

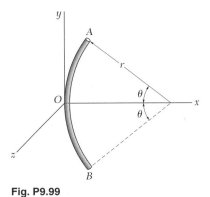

**Fig. P9.99**

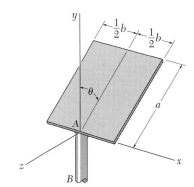

**Fig. P9.100**

**9.100** A thin rectangular plate of mass $m$ is welded to a vertical shaft $AB$ with which it forms an angle $\theta$. Determine by direct integration the mass moment of inertia with respect to $(a)$ the $y$ axis, $(b)$ the $z$ axis.

**9.101** The cross section of a small flywheel is shown. The rim and hub are connected by eight spokes (two of which are shown in the cross section). Each spoke has a cross-sectional area of 160 mm². Determine the mass moment of inertia and radius of gyration of the flywheel with respect to the axis of rotation. (Density of steel = 7850 kg/m³.)

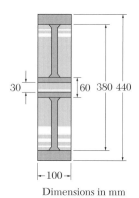

30 | 60  380  440

├─100─┤

Dimensions in mm

**Fig. P9.101**

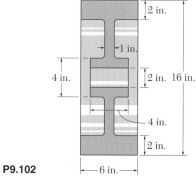

2 in.

←1 in.

4 in.    2 in.  16 in.

4 in.

2 in.

**Fig. P9.102**

├── 6 in. ──┤

**9.102** Determine the mass moment of inertia and the radius of gyration of the steel flywheel shown with respect to the axis of rotation. The web of the flywheel consists of a solid plate 1 in. thick. (Specific weight of steel = 490 lb/ft³.)

**9.103** Knowing that the thin hemispherical shell shown is of mass $m$ and thickness $t$, determine the mass moment of inertia and the radius of gyration of the shell with respect to the $x$ axis. (*Hint.* Consider the shell as formed by removing a hemisphere of radius $r$ from a hemisphere of radius $r + t$; then neglect the terms containing $t^2$ and $t^3$ and keep those terms containing $t$.)

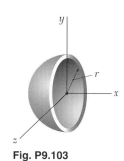

**Fig. P9.103**

**9.104** The machine part shown is formed by machining a conical surface into a circular cylinder. For $b = \frac{1}{2}h$, determine the mass moment of inertia and the radius of gyration of the machine part with respect to the $y$ axis.

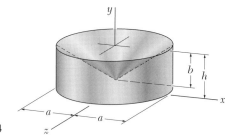

**Fig. P9.104**

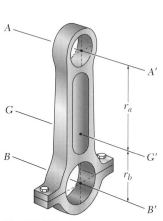

**Fig. P9.105 and P9.106**

**9.105** For the 1.8-kg connecting rod shown, it has been experimentally determined that the mass moments of inertia of the rod with respect to the centerline axes of the bearings $AA'$ and $BB'$ are, respectively, $I_{AA'} = 65$ g·m² and $I_{BB'} = 33$ g·m². Knowing that $r_a + r_b = 275$ mm, determine (*a*) the location of the centroidal axis $GG'$, (*b*) the radius of gyration with respect to axis $GG'$.

**9.106** Knowing that for the 3.1-kg connecting rod shown the mass moment of inertia with respect to axis $AA'$ is 160 g·m², determine the mass moment of inertia with respect to axis $BB'$ for $r_a = 204$ mm and $r_b = 126$ mm.

**9.107**   A circular hole of radius $r$ is to be drilled through the center of a rectangular steel plate to form the machine component shown. Denoting the density of the steel by $\rho$, determine (a) the mass moment of inertia of the component with respect to the axis $BB'$, (b) the value of $r$ for which, given $a$ and $h$, $I_{BB'}$ is maximum, (c) the corresponding value of $I_{BB'}$ and of the radius of gyration $k_{BB'}$.

**9.108**   Determine the mass moment of inertia and the radius of gyration of the steel machine component shown with respect to the axis $BB'$ when $a = 8$ in., $h = 4$ in., and $r = 3$ in. (Specific weight of steel = 490 lb/ft$^3$.)

**9.109 and 9.110**   A section of sheet steel 2 mm thick is cut and bent into the machine component shown. Knowing that the density of steel is 7850 kg/m$^3$, determine the mass moment of inertia of the component with respect to (a) the $x$ axis, (b) the $y$ axis, (c) the $z$ axis.

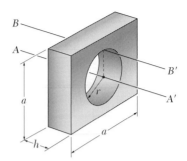

**Fig. P9.107 and P9.108**

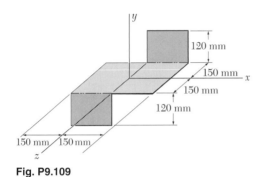

**Fig. P9.109**

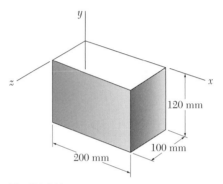

**Fig. P9.110**

**9.111**   A corner reflector for tracking by radar has two sides in the shape of a quarter circle of radius 15 in. and one side in the shape of a triangle. Each part of the reflector is formed from aluminum plate of uniform 0.05-in. thickness. Knowing that the specific weight of the aluminum used is 170 lb/ft$^3$, determine the mass moment of inertia of the reflector with respect to each of the coordinate axes.

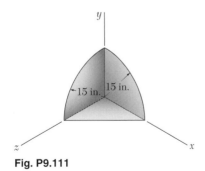

**Fig. P9.111**

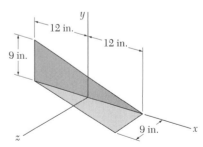

**Fig. P9.112**

**9.112**   A section of sheet steel 0.03 in. thick is cut and bent into the sheet-metal machine component shown. Knowing that the specific weight of steel is 490 lb/ft$^3$, determine the mass moment of inertia of the component with respect to each of the coordinate axes.

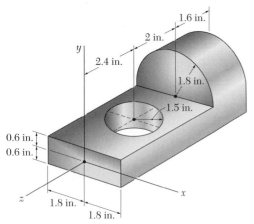

**Fig. P9.114 and P9.115**

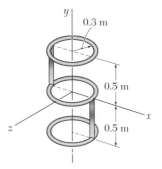

**Fig. P9.117**

**9.113 and 9.114** Determine the mass moment of inertia of the steel machine element shown with respect to the $x$ axis. (Specific weight of steel = 490 lb/ft³; density of steel = 7850 kg/m³.

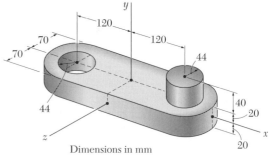

Dimensions in mm

**Fig. P9.113 and P9.116**

**9.115 and 9.116** Determine the mass moment of inertia of the steel machine element shown with respect to the $y$ axis. (Specific weight of steel = 490 lb/ft³; density of steel = 7850 kg/m³.)

**9.117** A homogenous wire with a mass per unit length of 1.5 kg/m is used to form the figure shown. Determine the mass moment of inertia of the wire figure with respect to ($a$) the $x$ axis, ($b$) the $y$ axis.

**9.118** In Prob. 9.117, determine the mass moment of inertia of the wire figure with respect to the $z$ axis.

# REVIEW AND SUMMARY
# FOR CHAPTER 9

In the first part of this chapter we were concerned with the determination of the resultant **R** of forces $\Delta \mathbf{F}$ distributed over a plane area $A$ when their magnitude is proportional to the element of area $\Delta A$ on which they act and, at the same time, varies linearly with the distance $y$ from $\Delta A$ to a given $x$ axis; we thus had $\Delta F = ky\,\Delta A$. We found that the magnitude of the resultant **R** was proportional to the first moment $Q_x = \int y\,dA$ of the area $A$, while the moment of **R** about the $x$ axis was proportional to the *second moment*, or *moment of inertia*, $I_x = \int y^2\,dA$ of $A$ with respect to the same axis [Sec. 9.2].

Rectangular moments of inertia

The *rectangular moments of inertia $I_x$ and $I_y$ of an area $A$* [Sec. 9.3] were obtained by computing the integrals

$$I_x = \int y^2\,dA \qquad I_y = \int x^2\,dA \qquad (9.1)$$

These computations may be reduced to single integrations by choosing for $dA$ a thin strip parallel to one of the axes of coordinates. We also recall that it is possible to compute $I_x$ and $I_y$ from the same elementary strip (Fig. 9.29) by making use of the formula for the moment of inertia of a rectangular area [Sample Prob. 9.3].

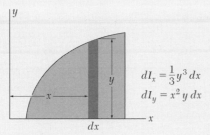

$$dI_x = \frac{1}{3}y^3 \, dx$$
$$dI_y = x^2 y \, dx$$

**Fig. 9.29**

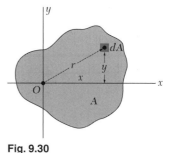

**Fig. 9.30**

**Polar moment of inertia**

The *polar moment of inertia of an area* $A$ with respect to the pole $O$ [Sec. 9.4] was defined as

$$J_O = \int r^2 \, dA \qquad (9.3)$$

where $r$ is the distance from $O$ to the element of area $dA$ (Fig. 9.30). Observing that $r^2 = x^2 + y^2$, we established the relation

$$J_O = I_x + I_y \qquad (9.4)$$

**Radius of gyration**

The *radius of gyration of an area* $A$ with respect to the $x$ axis [Sec. 9.5] was defined as the distance $k_x$ such that $I_x = k_x^2 A$. With similar definitions for the radii of gyration of $A$ with respect to the $y$ axis and with respect to $O$, we had

$$k_x = \sqrt{\frac{I_x}{A}} \qquad k_y = \sqrt{\frac{I_y}{A}} \qquad k_O = \sqrt{\frac{J_O}{A}} \qquad (9.5\text{–}9.7)$$

**Parallel-axis theorem**

The *parallel-axis theorem* was presented in Sec. 9.6. It states that the moment of inertia $I$ of an area with respect to any given axis $AA'$ (Fig. 9.31) is equal to the moment of inertia $\bar{I}$ of the area with respect to a centroidal axis $BB'$ parallel to $AA'$ *plus* the product of the area $A$ and the square of the distance $d$ between the two axes:

$$\bar{I} = \bar{I} + Ad^2 \qquad (9.9)$$

This formula may also be used to determine the moment of inertia $\bar{I}$ of an area with respect to a centroidal axis $BB'$ when its moment of inertia $I$ with respect to a parallel axis $AA'$ is known. In this case, however, the product $Ad^2$ should be *subtracted* from the known moment of inertia $I$.

A similar relation holds between the polar moment of inertia $J_O$ of an area about a point $O$ and the polar moment of inertia $\bar{J}_C$ of the same area about its centroid $C$. Denoting by $d$ the distance between $O$ and $C$, we have

$$J_O = \bar{J}_C + Ad^2 \qquad (9.11)$$

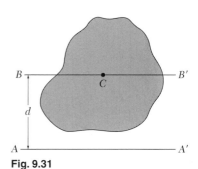

**Fig. 9.31**

**Composite areas**

The parallel-axis theorem may be used very effectively in the computation of the *moment of inertia of a composite area* with respect to a given axis [Sec. 9.7]. Considering each component area separately, we first compute the moment of inertia of each area with respect to a centroidal axis, using whenever possible the data provided in Figs. 9.12 and 9.13; the parallel-axis theorem is then applied to determine the moment of inertia of each component area with respect to the desired axis and the various values obtained are added [Sample Probs. 9.4 and 9.5].

**Product of inertia**

Sections 9.8 to 9.10 were devoted to the transformation of the moments of inertia of an area *under a rotation of the coordinate axes*. First we defined the *product of inertia of an area A* as

$$I_{xy} = \int xy \, dA \tag{9.12}$$

and showed that $I_{xy} = 0$ if the area $A$ is symmetrical with respect to either or both of the coordinate axes. We also derived the *parallel-axis theorem for products of inertia*. We had

$$I_{xy} = \bar{I}_{x'y'} + \bar{x}\bar{y}A \tag{9.13}$$

where $\bar{I}_{x'y'}$ is the product of inertia of the area with respect to centroidal axes $x'y'$ parallel to the axes $xy$, and $\bar{x}$ and $\bar{y}$ the coordinates of the centroid of the area [Sec. 9.8].

**Rotation of axes**

In Sec. 9.9 we computed the moments and product of inertia $I_{x'}$, $I_{y'}$, and $I_{x'y'}$ of an area with respect to axes $x'y'$ obtained by rotating the original coordinate axes $xy$ through an angle $\theta$ counterclockwise (Fig. 9.32), and expressed them in terms of the moments and product of inertia $I_x$, $I_y$, and $I_{xy}$ computed with respect to the original $xy$ axes. We had

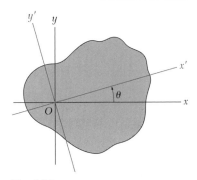

**Fig. 9.32**

$$I_{x'} = \frac{I_x + I_y}{2} + \frac{I_x - I_y}{2} \cos 2\theta - I_{xy} \sin 2\theta \tag{9.18}$$

$$I_{y'} = \frac{I_x + I_y}{2} - \frac{I_x - I_y}{2} \cos 2\theta - I_{xy} \sin 2\theta \tag{9.19}$$

$$I_{x'y'} = \frac{I_x - I_y}{2} \sin 2\theta + I_{xy} \cos 2\theta \tag{9.20}$$

**Principal axes**

The *principal axes of the area about O* were defined as the two axes perpendicular to each other, with respect to which the moment of inertia of the area are, respectively, maximum and minimum. The corresponding values of $\theta$ were denoted by $\theta_m$ and obtained from the formula

$$\tan 2\theta_m = -\frac{2I_{xy}}{I_x - I_y} \tag{9.25}$$

**Principal moments of inertia**

The corresponding maximum and minimum values of $I$ are called the *principal moments of inertia* of the area about $O$; we had

$$I_{\text{max, min}} = \frac{I_x + I_y}{2} \pm \sqrt{\left(\frac{I_x - I_y}{2}\right)^2 + I_{xy}^2} \tag{9.27}$$

We also noted that the corresponding value of the product of inertia is zero.

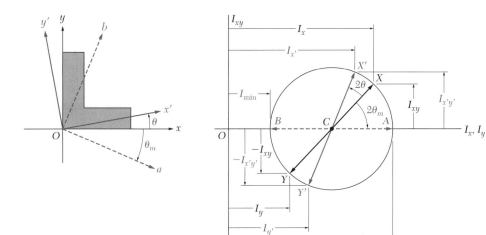

**Fig. 9.33**

The transformation of the moments and product of inertia of an area under a rotation of axes may be represented graphically by drawing *Mohr's circle* [Sec. 9.10]. Given the moments and product of inertia $I_x$, $I_y$, and $I_{xy}$ of the area with respect to the coordinate axes $xy$, we plot points $X$ $(I_x, I_{xy})$ and $Y$ $(I_y, -I_{xy})$ and draw the line joining these two points (Fig. 9.33). This line is a diameter of Mohr's circle and thus defines this circle. As the coordinate axes are rotated through $\theta$, the diameter rotates through *twice that angle* and the coordinates of $X'$ and $Y'$ yield the new values $I_{x'}$, $I_{y'}$, and $I_{x'y'}$ of the moments and product of inertia of the area. Also, the angle $\theta_m$ and the coordinates of points $A$ and $B$ define the principal axes $a$ and $b$ and the principal moments of inertia of the area [Sample Prob. 9.8].

The last part of the chapter was devoted to the determination of *moments of inertia of masses*, which are encountered in dynamics in problems involving the rotation of a rigid body about an axis. The mass moment of inertia of a body with respect to an axis $AA'$ (Fig. 9.34) was defined as

$$I = \int r^2 \, dm \tag{9.28}$$

where $r$ is the distance from $AA'$ to the element of mass [Sec. 9.11], and its *radius of gyration* as

$$k = \sqrt{\frac{I}{m}} \tag{9.29}$$

The moments of inertia of a body with respect to the coordinate axes were expressed as

$$
\begin{aligned}
I_x &= \int (y^2 + z^2) \, dm \\
I_y &= \int (z^2 + x^2) \, dm \\
I_z &= \int (x^2 + y^2) \, dm
\end{aligned}
\tag{9.30}
$$

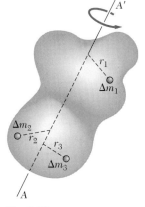

**Fig. 9.34**

We saw that the *parallel-axis theorem* also applies to mass moments of inertia [Sec. 9.12]. Thus, the moment of inertia $I$ of a body with respect to an arbitrary axis $AA'$ (Fig. 9.35), may be expressed as

$$I = \bar{I} + md^2 \tag{9.33}$$

where $\bar{I}$ is the moment of inertia of the body with respect to a centroidal axis $BB'$ parallel to the axis $AA'$, $m$ the mass of the body, and $d$ the distance between the two axes.

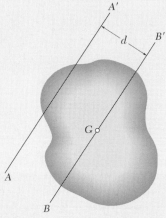

**Fig. 9.35**

The moments of inertia of *thin plates* may be readily obtained from the moments of inertia of their areas [Sec. 9.13]. We found that for a *rectangular plate* the moments of inertia with respect to the axes shown (Fig. 9.36) are

$$I_{AA'} = \tfrac{1}{12}ma^2 \qquad I_{BB'} = \tfrac{1}{12}mb^2 \tag{9.39}$$

$$I_{CC'} = I_{AA'} + I_{BB'} = \tfrac{1}{12}m(a^2 + b^2) \tag{9.40}$$

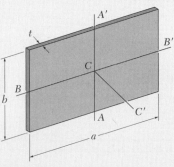

**Fig. 9.36**

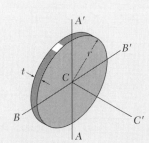

**Fig. 9.37**

while for a *circular plate* (Fig. 9.37) they are

$$I_{AA'} = I_{BB'} = \tfrac{1}{4}mr^2 \tag{9.41}$$

$$I_{CC'} = I_{AA'} + I_{BB'} = \tfrac{1}{2}mr^2 \tag{9.42}$$

When a body possesses *two planes of symmetry*, it is usually possible to determine its moment of inertia with respect to a given axis through a *single integration* by considering an element of mass $dm$ in the shape of a thin plate [Sample Probs. 9.10 and 9.11]. On the other hand, when a body is made of *several common geometric shapes*, its moment of inertia with respect to a given axis may be obtained from the values given in Fig. 9.28 and the application of the parallel-axis theorem [Sample Probs. 9.12 and 9.13].

Composite bodies

# REVIEW PROBLEMS

**9.119** A thin wire has been bent to form a half ring of radius $r$. Denoting the mass of the wire by $m$, determine the mass moment of inertia of the wire with respect to (*a*) the axis $AA'$, (*b*) the axis $BB'$.

**9.120** Determine the moment of inertia of the shaded area with respect to (*a*) the $x$ axis, (*b*) the $y$ axis.

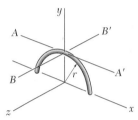

**Fig. P9.119**

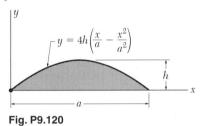

$$y = 4h\left(\frac{x}{a} - \frac{x^2}{a^2}\right)$$

**Fig. P9.120**

**9.121** Determine the moment of inertia of the shaded area with respect to the $y$ axis.

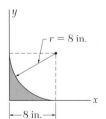

$r = 8$ in.

8 in.

**Fig. P9.121 and P9.122**

**9.122** Determine the product of inertia of the shaded area with respect to the $x$ and $y$ axes. (*Hint.* Use the answer to Prob. 9.56.)

**9.123** Determine the centroidal moments of inertia $\bar{I}_x$ and $\bar{I}_y$ and the centroidal radii of gyration $\bar{k}_x$ and $\bar{k}_y$ for the shaded area shown.

**9.124** Determine the orientation of the principal axes through the centroid $C$ and the corresponding values of the moment of inertia for the shaded area shown.

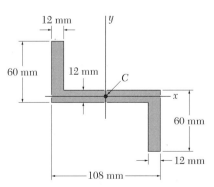

12 mm

60 mm    12 mm    $C$

60 mm

12 mm

108 mm

**Fig. P9.123 and P9.124**

**9.125** A section of sheet steel, 2 mm thick, is cut and bent into the machine component shown. Knowing that the density of steel is 7850 kg/m³, determine the mass moment of inertia and the radius of gyration of the component with respect to the x axis.

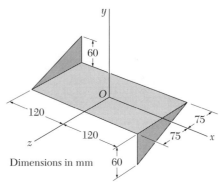

**Fig. P9.125**

**9.126** For the machine component of Prob. 9.125, determine the mass moment of inertia and the radius of gyration with respect to the y axis.

**9.127** Twelve uniform slender rods, each of length l, are welded together to form the cubical figure shown. Denoting by m the total mass of the twelve rods, determine the mass moment of inertia and the radius of gyration of the figure with respect to the x axis.

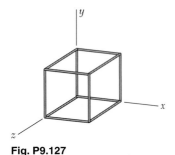

**Fig. P9.127**

**9.128** Determine the polar moment of inertia and the polar radius of gyration of an equilateral triangle of side a with respect to one of its vertices.

**9.129** A thin plate of mass m is cut in the shape of a regular hexagon of side a. Determine the mass moment of inertia and the radius of gyration of the plate with respect to the axis CC′ perpendicular to the plate.

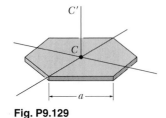

**Fig. P9.129**

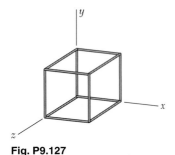

**Fig. P9.130**

**9.130** Determine by approximate means the moment of inertia of the shaded area with respect to (a) the x axis, (b) the y axis.

# COMPUTER PROBLEMS

**9.C1** Write a computer program which, for an area with known moments and product of inertia $I_x$, $I_y$, and $I_{xy}$, can be used to calculate the moments of inertia $I_{x'}$, $I_{y'}$, and $I_{x'y'}$ of the area with respect to axes $x'$ and $y'$ obtained by rotating the original axes counterclockwise through an angle $\theta$. Use this program to compute $I_{x'}$, $I_{y'}$, and $I_{x'y'}$ for the section of Sample Prob. 9.8 for values of $\theta$ from 0 to 90° at 5° intervals.

**9.C2** Write a computer program which, for an area with known moments and product of inertia $I_x$, $I_y$, and $I_{xy}$, can be used to calculate the orientation of the principal axes of the area and the corresponding values of the principal moments of inertia. Use this program to solve (a) Prob. 9.72, (b) parts a and b of Sample Prob. 9.8.

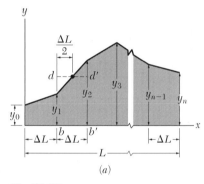

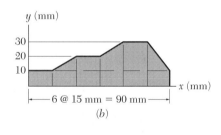

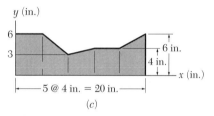

**Fig. P9.C3**

**9.C3** Approximate the plane area shown by a series of rectangles of the form $bb'd'd$ and write a computer program which can be used to calculate the moment of inertia and radius of gyration of the area shown in part a of the figure with respect to each of the coordinate axes. Use this program to compute $I_x$, $k_x$, $I_y$, and $k_y$ for the areas shown in parts b and c of the figure.

**9.C4** The web of the steel flywheel shown consists of a solid plate of thickness $t_2$. Write a computer program which can be used to calculate the mass moment of inertia and the radius of gyration of the flywheel with respect to the axis of rotation. Use this program to solve Prob. 9.102.

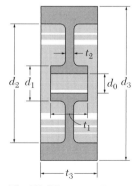

**Fig. P9.C4**

# Method of Virtual Work

## *10.1. INTRODUCTION

In the preceding chapters, problems involving the equilibrium of rigid bodies were solved by expressing that the external forces acting on the bodies were balanced. The equations of equilibrium $\Sigma F_x = 0$, $\Sigma F_y = 0$, $\Sigma M_A = 0$ were written and solved for the desired unknowns. We shall now consider a different method, which will prove more effective for solving certain types of equilibrium problems. This method is based on the *principle of virtual work* and was first formally used by the Swiss mathematician Jean Bernoulli in the eighteenth century.

As we shall see in Sec. 10.3, the principle of virtual work states that if a particle or rigid body, or, more generally, if a system of connected rigid bodies is in equilibrium under various external forces, and if the system is given an arbitrary displacement from that position of equilibrium, the total work done by the external forces during the displacement is zero. This principle may be applied particularly effectively to the solution of problems involving the equilibrium of machines or mechanisms consisting of several connected members.

In the second part of the chapter, the method of virtual work will be applied in an alternate form based on the concept of *potential energy*. As we shall see in Sec. 10.8, if a particle, rigid body, or system of rigid bodies is in equilibrium, then the derivative of its potential energy with respect to a variable defining its position must be zero.

In this chapter, we shall also learn to evaluate the mechanical efficiency of a machine (Sec. 10.5), and to determine whether a given position of equilibrium is stable, unstable, or neutral (Sec. 10.9).

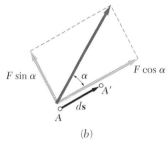

Fig. 10.1

## *10.2. WORK OF A FORCE

We shall first define the terms *displacement* and *work* as they are used in mechanics. Consider a particle which moves from a point $A$ to a neighboring point $A'$ (Fig. 10.1a). The small vector $d\mathbf{s}$ of magnitude $ds$ which joins $A$ to $A'$ is called the *displacement* of the particle. Now, let us assume that a force $\mathbf{F}$, of magnitude $F$ and forming an angle $\alpha$ with the vector $d\mathbf{s}$, is acting on the particle. The *work of the force* $\mathbf{F}$ *corresponding to the displacement* $d\mathbf{s}$ is defined as the scalar quantity

$$dU = F \, ds \cos \alpha \tag{10.1}$$

Being a *scalar quantity,* work has a magnitude and a sign, but no direction. We also note that work should be expressed in units obtained by multiplying units of length by units of force. Thus, if U.S. customary units are used, work should be expressed in ft · lb or in · lb. If SI units are used, work should be expressed in N · m. The unit of work N · m is called a *joule* (**J**).[†]

It appears from (10.1) that the work $dU$ may be considered as the product of the magnitude $ds$ of the displacement and of the component $F \cos \alpha$ of the force $\mathbf{F}$ in the direction of $d\mathbf{s}$ (Fig. 10.1b). If this component and the displacement $d\mathbf{s}$ have the same sense, the work is positive; if they have opposite senses, the work is negative. Three particular cases are of special interest. If the force $\mathbf{F}$ has the same direction as $d\mathbf{s}$, the work $dU$ reduces to $F \, ds$. If $\mathbf{F}$ has a direction opposite to that of $d\mathbf{s}$, the work is $dU = -F \, ds$. Finally, if $\mathbf{F}$ is perpendicular to $d\mathbf{s}$, the work $dU$ is zero.

The work $dU$ of a force $\mathbf{F}$ during a displacement $d\mathbf{s}$ may also be considered as the product of $F$ and of the component $ds \cos \alpha$ of the displacement $d\mathbf{s}$ along $\mathbf{F}$ (Fig. 10.2a). This view is particularly useful in the computation of the work done by the weight $\mathbf{W}$ of a body (Fig. 10.2b). The work of $\mathbf{W}$ is equal to the product of $W$ and of the vertical displacement $dy$ of the center of gravity $G$ of the body. If the displacement is downward, the work is positive; if it is upward, the work is negative.

A number of forces frequently encountered in statics *do not work.* They are forces applied to fixed points ($ds = 0$) or acting in a direction perpendicular to the displacement ($\cos \alpha = 0$). Among the forces which do no work are the following: the reaction at a frictionless pin when the body

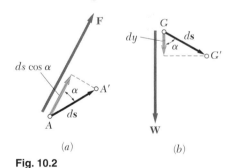

Fig. 10.2

---

[†] The joule is the SI unit of *energy,* whether in mechanical form (work, potential energy, kinetic energy) or in chemical, electrical, or thermal form. We should note that even though N · m = J, the moment of a force must be expressed in N · m, and not in joules, since the moment of a force is not a form of energy.

supported rotates about the pin, the reaction at a frictionless surface when the body in contact moves along the surface, the reaction at a roller moving along its track, the weight of a body when its center of gravity moves horizontally, the friction force acting on a wheel rolling without slipping (since at any instant the point of contact does not move). Examples of forces which *do work* are the weight of a body (except in the case considered above), the friction force acting on a body sliding on a rough surface, and most forces applied on a moving body.

In certain cases, the sum of the work done by several forces is zero. Consider, for example, two rigid bodies *AC* and *BC* connected at *C* by a *frictionless pin* (Fig. 10.3*a*). Among the forces acting on *AC* is the force **F**

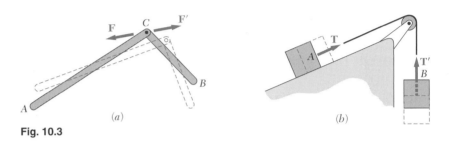

(*a*)                    (*b*)

**Fig. 10.3**

exerted at *C* by *BC*. In general, the work of this force will not be zero, but it will be equal in magnitude and opposite in sign to the work of the force **F**′ exerted by *AC* on *BC*, since these forces are equal and opposite and are applied to the same particle. Thus, when the total work done by all the forces acting on *AB* and *BC* is considered, the work of the two internal forces at *C* cancels out. A similar result is obtained if we consider a system consisting of two blocks connected by an *inextensible cord AB* (Fig. 10.3*b*). The work of the tension force **T** at *A* is equal in magnitude to the work of the tension force **T**′ at *B*, since these forces have the same magnitude and the points *A* and *B* move through the same distance; but in one case the work is positive, and in the other it is negative. Thus, the work of the internal forces again cancels out.

It may be shown that the total work of the internal forces holding together the particles of a rigid body is zero. Consider two particles *A* and *B* of a rigid body and the two equal and opposite forces **F** and **F**′ they exert on each other (Fig. 10.4). While, in general, small displacements *d***s** and *d***s**′ of the two particles are different, the components of these displacements along *AB* must be equal; otherwise, the particles would not remain at the same distance from each other, and the body would not be rigid. Therefore, the work of **F** is equal in magnitude and opposite in sign to the work of **F**′, and their sum is zero.

In computing the work of the external forces acting on a rigid body, it is often convenient to determine the work of a couple without considering separately the work of each of the two forces forming the couple. Consider the two forces **F** and **F**′ forming a couple of moment $M = Fr$ and acting on a rigid body (Fig. 10.5). Any small displacement of the rigid body bringing

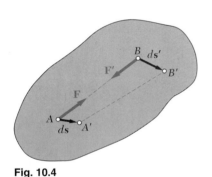

**Fig. 10.4**

A and B, respectively, into $A'$ and $B''$ may be divided into two parts, one in which points A and B undergo equal displacements $ds_1$, the other in which $A'$ remains fixed while $B'$ moves into $B''$ through a displacement $ds_2$ of magnitude $ds_2 = r\,d\theta$. In the first part of the motion, the work of $\mathbf{F}$ is equal in magnitude and opposite in sign to the work of $\mathbf{F}'$, and their sum is zero. In the second part of the motion, only force $\mathbf{F}$ works, and its work is $dU = F\,ds_2 = Fr\,d\theta$. But the product $Fr$ is equal to the moment $M$ of the couple. Thus, the work of a couple of moment $M$ acting on a rigid body is

$$dU = M\,d\theta \tag{10.2}$$

where $d\theta$ is the small angle expressed in radians through which the body rotates. We again note that work should be expressed in units obtained by multiplying units of force by units of length.

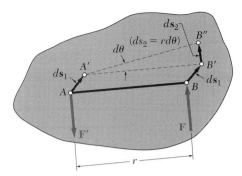

**Fig. 10.5**

## *10.3. PRINCIPLE OF VIRTUAL WORK

Consider a particle acted upon by several forces $\mathbf{F}_1, \mathbf{F}_2, \ldots, \mathbf{F}_n$ (Fig. 10.6). We shall assume that the particle undergoes a small displacement from A to $A'$. This displacement is possible, but it will not necessarily take place. The forces may be balanced and the particle at rest, or the particle may move under the action of the given forces in a direction different from that of $AA'$. The displacement considered is therefore an imaginary displacement; it is called a *virtual displacement* and is denoted by $\delta\mathbf{s}$. The symbol $\delta\mathbf{s}$ represents a differential of the first order; it is used to distinguish the virtual displacement from the displacement $d\mathbf{s}$ which would take place under actual motion. As we shall see, virtual displacements may be used to determine whether the conditions of equilibrium of a particle are satisfied.

The work of each of the forces $\mathbf{F}_1, \mathbf{F}_2, \ldots, \mathbf{F}_n$ during the virtual displacement $\delta\mathbf{s}$ is called *virtual work*. The virtual work of all the forces acting on the particle of Fig. 10.6 is

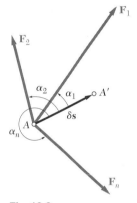

**Fig. 10.6**

$$\delta U = F_1 \cos \alpha_1\, \delta s + F_2 \cos \alpha_2\, \delta s + \cdots + F_n \cos \alpha_n\, \delta s \tag{10.3}$$

or, factoring $\delta s$,

$$\delta U = (F_1 \cos \alpha_1 + F_2 \cos \alpha_2 + \cdots + F_n \cos \alpha_n)\delta s \tag{10.3'}$$

Since the expression in parentheses represents the component of the resultant $\mathbf{R}$ of the forces $\mathbf{F}_1, \mathbf{F}_2, \ldots, \mathbf{F}_n$ along $AA'$, it is seen that the total virtual work of the forces $\mathbf{F}_1, \mathbf{F}_2, \ldots, \mathbf{F}_n$ is equal to the virtual work of their resultant $\mathbf{R}$.

The principle of virtual work for a particle states that, *if a particle is in equilibrium, the total virtual work of the forces acting on the particle is zero for any virtual displacement of the particle*. This condition is necessary: if the particle is in equilibrium, the resultant $\mathbf{R}$ of the forces is zero, and it follows from (10.3') that the total virtual work $\delta U$ is zero. The condition is also sufficient: choosing $AA'$ successively along the $x$ axis and the $y$ axis, and making $\delta U = 0$ in (10.3'), we find that the equilibrium equations $\Sigma F_x = 0$ and $\Sigma F_y = 0$ are satisfied.

In the case of a rigid body, the principle of virtual work states that *if a rigid body is in equilibrium, the total virtual work of the external forces acting on the rigid body is zero for any virtual displacement of the body.* The condition is necessary: if the body is in equilibrium, all the particles forming the body are in equilibrium and the total virtual work of the forces acting on all the particles must be zero; but we have seen in the preceding section that the total work of the internal forces is zero; the total work of the external forces must therefore also be zero. The condition may also be proved to be sufficient.

The principle of virtual work may be extended to the case of a *system of connected rigid bodies.* If the system remains connected during the virtual displacement, *only the work of the forces external to the system need be considered,* since the total work of the internal forces at the various connections is zero.

## *10.4. APPLICATIONS OF THE PRINCIPLE OF VIRTUAL WORK

The principle of virtual work is particularly effective when applied to the solution of problems involving machines or mechanisms consisting of several connected rigid bodies. Consider, for instance, the toggle vise *ACB* of Fig. 10.7*a*, used to compress a wooden block. We wish to determine the

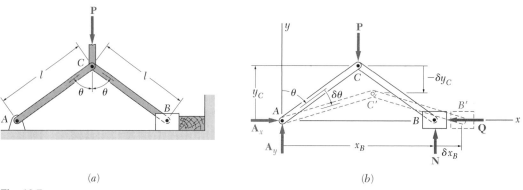

(a)                    (b)

**Fig. 10.7**

force exerted by the vise on the block when a given force **P** is applied at *C*, assuming that there is no friction. Denoting by **Q** the reaction of the block on the vise, we draw the free-body diagram of the vise and consider the virtual displacement obtained by giving to the angle $\theta$ a positive increment $\delta\theta$ (Fig. 10.7*b*). Choosing a system of coordinate axes with origin at *A*, we note that $x_B$ increases while $y_C$ decreases. This is indicated in the figure by means of the positive increment $\delta x_B$ and the negative increment $-\delta y_C$. The reactions $\mathbf{A}_x$, $\mathbf{A}_y$, and **N** will do not work during the virtual displacement considered, and we need only compute the work of **P** and **Q**. Since **Q** and $\delta x_B$ have opposite senses, the virtual work of **Q** is $\delta U_Q = -Q\,\delta x_B$. Since **P** and the increment shown $(-\delta y_C)$ have the same sense, the virtual work of **P** is $\delta U_P = +P(-\delta y_C) = -P\,\delta y_C$. The minus signs obtained could have been predicted by simply noting that the forces **Q** and

**P** are directed opposite to the positive $x$ and $y$ axes, respectively. Expressing the coordinates $x_B$ and $y_C$ in terms of the angle $\theta$ and differentiating, we obtain

$$x_B = 2l \sin \theta \qquad y_C = l \cos \theta$$
$$\delta x_B = 2l \cos \theta \, \delta\theta \qquad \delta y_C = -l \sin \theta \, \delta\theta \qquad (10.4)$$

The total virtual work of the forces **Q** and **P** is thus

$$\delta U = \delta U_Q + \delta U_P = -Q \, \delta x_B - P \, \delta y_C$$
$$= -2Ql \cos \theta \, \delta\theta + Pl \sin \theta \, \delta\theta$$

Making $\delta U = 0$, we obtain

$$2Ql \cos \theta \, \delta\theta = Pl \sin \theta \, \delta\theta \qquad (10.5)$$
$$Q = \tfrac{1}{2}P \tan \theta \qquad (10.6)$$

The superiority of the method of virtual work over the conventional equilibrium equations in the problem considered here is clear: by using the method of virtual work, we were able to eliminate all unknown reactions, while the equation $\Sigma M_A = 0$ would have eliminated only two of the unknown reactions. We may take advantage of this characteristic of the method of virtual work to solve many problems involving machines and mechanisms. *If the virtual displacement considered is consistent with the constraints imposed by the supports and connections, all reactions and internal forces are eliminated and only the work of the loads, applied forces, and friction forces need be considered.*

We shall observe that the method of virtual work may also be used to solve problems involving completely constrained structures, although the virtual displacements considered will never actually take place. Consider, for example, the frame $ACB$ shown in Fig. 10.8$a$. If point $A$ is kept fixed, while $B$ is given a horizontal virtual displacement (Fig. 10.8$b$), we need consider only the work of **P** and $\mathbf{B}_x$. We may thus determine the reaction

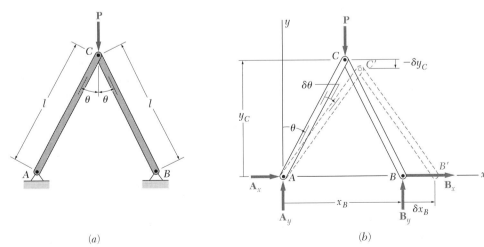

$(a)$ $\qquad\qquad\qquad\qquad\qquad (b)$

**Fig. 10.8**

component $\mathbf{B}_x$ in the same way as the force $\mathbf{Q}$ of the preceding example (Fig. 10.7$b$); we have

$$B_x = -\tfrac{1}{2}P \tan \theta$$

Keeping $B$ fixed and giving to $A$ a horizontal virtual displacement, we may similarly determine the reaction component $\mathbf{A}_x$. The components $\mathbf{A}_y$ and $\mathbf{B}_y$ may be determined by rotating the frame $ACB$ as a rigid body about $B$ and $A$, respectively.

The method of virtual work may also be used to determine the configuration of a system in equilibrium under given forces. For example, the value of the angle $\theta$ for which the linkage of Fig. 10.7 is in equilibrium under two given forces $\mathbf{P}$ and $\mathbf{Q}$ may be obtained by solving Eq. (10.6) for $\tan \theta$.

It should be noted, however, that the attractiveness of the method of virtual work depends to a large extent upon the existence of simple geometric relations between the various virtual displacements involved in the solution of a given problem. When no such simple relations exist, it is usually advisable to revert to the conventional method of Chap. 6.

## *10.5. REAL MACHINES. MECHANICAL EFFICIENCY

In analyzing the toggle vise in the preceding section, we assumed that no friction forces were involved. Thus, the virtual work consisted only of the work of the applied force $\mathbf{P}$ and of the reaction $\mathbf{Q}$. But the work of the reaction $\mathbf{Q}$ is equal in magnitude and opposite in sign to the work of the force exerted by the vise on the block. Equation (10.5), therefore, expresses that the *output work* $2Ql \cos \theta\ \delta\theta$ is equal to the *input work* $Pl \sin \theta\ \delta\theta$. A machine in which input and output work are equal is said to be an "ideal" machine. In a "real" machine, friction forces will always do some work, and the output work will be smaller than the input work.

Consider, for example, the toggle vise of Fig. 10.7$a$, and assume now that a friction force $\mathbf{F}$ develops between the sliding block $B$ and the horizontal plane (Fig. 10.9). Using the conventional methods of statics and

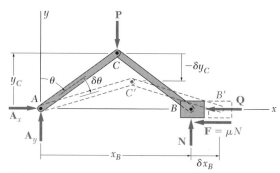

**Fig. 10.9**

summing moments about $A$, we find $N = P/2$. Denoting by $\mu$ the coefficient of friction between block $B$ and the horizontal plane, we have $F = \mu N = \mu P/2$. Recalling formulas (10.4), we find that the total virtual work of the forces $\mathbf{Q}$, $\mathbf{P}$, and $\mathbf{F}$ during the virtual displacement shown in Fig. 10.9 is

$$\delta U = -Q\,\delta x_B - P\,\delta y_C - F\,\delta x_B$$
$$= -2Ql\cos\theta\,\delta\theta + Pl\sin\theta\,\delta\theta - \mu Pl\cos\theta\,\delta\theta$$

Making $\delta U = 0$, we obtain

$$2Ql\cos\theta\,\delta\theta = Pl\sin\theta\,\delta\theta - \mu Pl\cos\theta\,\delta\theta \qquad (10.7)$$

which expresses that the output work is equal to the input work minus the work of the friction force. Solving for $Q$, we have

$$Q = \tfrac{1}{2}P(\tan\theta - \mu) \qquad (10.8)$$

We note that $Q = 0$ when $\tan\theta = \mu$, that is, when $\theta$ is equal to the angle of friction $\phi$, and that $Q < 0$ when $\theta < \phi$. The toggle vise may thus be used only for values $\theta$ larger than the angle of friction.

The *mechanical efficiency* of a machine is defined as the ratio

$$\eta = \frac{\text{output work}}{\text{input work}} \qquad (10.9)$$

Clearly, the mechanical efficiency of an ideal machine is $\eta = 1$, since input and output work are then equal, while the mechanical efficiency of a real machine will always be less than 1.

In the case of the toggle vise we have just analyzed, we write

$$\eta = \frac{\text{output work}}{\text{input work}} = \frac{2Ql\cos\theta\,\delta\theta}{Pl\sin\theta\,\delta\theta}$$

Substituting from (10.8) for $Q$, we obtain

$$\eta = \frac{P(\tan\theta - \mu)l\cos\theta\,\delta\theta}{Pl\sin\theta\,\delta\theta} = 1 - \mu\cot\theta \qquad (10.10)$$

We check that in the absence of friction forces, we would have $\mu = 0$ and $\eta = 1$. In the general case, when $\mu$ is different from zero, the efficiency $\eta$ becomes zero for $\mu\cot\theta = 1$, that is, for $\tan\theta = \mu$, or $\theta = \tan^{-1}\mu = \phi$. We check again that the toggle vise may be used only for values of $\theta$ larger than the angle of friction $\phi$.

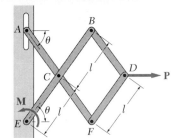

## SAMPLE PROBLEM 10.1

Using the method of virtual work, determine the magnitude of the couple **M** required to maintain the equilibrium of the mechanism shown.

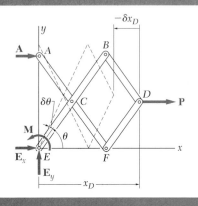

**Solution.**   Choosing a coordinate system with origin at $E$, we write

$$x_D = 3l \cos \theta \qquad \delta x_D = -3l \sin \theta \, \delta\theta$$

***Principle of Virtual Work.***   Since the reactions **A**, $\mathbf{E}_x$, and $\mathbf{E}_y$ will do no work during the virtual displacement, the total virtual work done by **M** and **P** must be zero. Noting that **P** acts in the positive $x$ direction and **M** acts in the positive $\theta$ direction, we write

$$\delta U = 0: \qquad +M \, \delta\theta + P \, \delta x_D = 0$$
$$+M \, \delta\theta + P(-3l \sin \theta \, \delta\theta) = 0$$

$$M = 3Pl \sin \theta \quad \blacktriangleleft$$

## SAMPLE PROBLEM 10.2

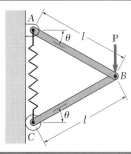

Determine the expressions for $\theta$ and for the tension in the spring which correspond to the equilibrium position of the mechanism. The unstretched length of the spring is $h$, and the constant of the spring is $k$. Neglect the weight of the mechanism.

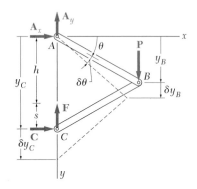

**Solution.**   With the coordinate system shown

$$y_B = l \sin \theta \qquad\qquad y_C = 2l \sin \theta$$
$$\delta y_B = l \cos \theta \, \delta\theta \qquad \delta y_C = 2l \cos \theta \, \delta\theta$$

The elongation of the spring is $\quad s = y_C - h = 2l \sin \theta - h$

The magnitude of the force exerted at $C$ by the spring is

$$F = ks = k(2l \sin \theta - h) \qquad (1)$$

***Principle of Virtual Work.***   Since the reactions $\mathbf{A}_x$, $\mathbf{A}_y$, and **C** do no work, the total virtual work done by **P** and **F** must be zero.

$$\delta U = 0: \qquad P \, \delta y_B - F \, \delta y_C = 0$$
$$P(l \cos \theta \, \delta\theta) - k(2l \sin \theta - h)(2l \cos \theta \, \delta\theta) = 0$$

$$\sin \theta = \frac{P + 2kh}{4kl} \quad \blacktriangleleft$$

Substituting this expression into (1), we obtain

$$F = \tfrac{1}{2}P \quad \blacktriangleleft$$

# SAMPLE PROBLEM 10.3

A hydraulic-lift table is used to raise a 1000-kg crate. It consists of a platform and of two identical linkages on which hydraulic cylinders exert equal forces. (Only one linkage and one cylinder are shown.) Members $EDB$ and $CG$ are each of length $2a$, and member $AD$ is pinned to the midpoint of $EDB$. If the crate is placed on the table, so that half of its weight is supported by the system shown, determine the force exerted by each cylinder in raising the crate for $\theta = 60°$, $a = 0.70$ m, and $L = 3.20$ m. This mechanism has been previously considered in Sample Prob. 6.7.

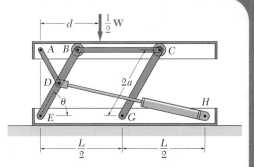

**Solution.** The machine considered consists of the platform and of the linkage, with an input force $\mathbf{F}_{DH}$ exerted by the cylinder and an output force equal and opposite to $\frac{1}{2}\mathbf{W}$.

*Principle of Virtual Work.* We first observe that the reactions at $E$ and $G$ do no work. Denoting by $y$ the elevation of the platform above the base, and by $s$ the length $DH$ of the cylinder-and-piston assembly, we write

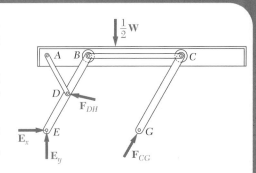

$$\delta U = 0: \qquad -\tfrac{1}{2}W\,\delta y + F_{DH}\,\delta s = 0 \qquad (1)$$

The vertical displacement $\delta y$ of the platform is expressed in terms of the angular displacement $\delta\theta$ of $EDB$ as follows:

$$y = (EB)\sin\theta = 2a\sin\theta$$
$$\delta_y = 2a\cos\theta\,\delta\theta$$

To express $\delta s$ similarly in terms of $\delta\theta$, we first note that by the law of cosines,

$$s^2 = a^2 + L^2 - 2aL\cos\theta$$

Differentiating,

$$2s\,\delta s = -2aL(-\sin\theta)\,\delta\theta$$
$$\delta s = \frac{aL\sin\theta}{s}\,\delta\theta$$

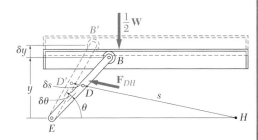

Substituting for $\delta y$ and $\delta s$ into (1), we write

$$(-\tfrac{1}{2}W)2a\cos\theta\,\delta\theta + F_{DH}\frac{aL\sin\theta}{s}\,\delta\theta = 0$$

$$F_{DH} = W\frac{s}{L}\cot\theta$$

With the given numerical data, we have

$$W = mg = (1000\text{ kg})(9.81\text{ m/s}^2) = 9810\text{ N} = 9.81\text{ kN}$$
$$s^2 = a^2 + L^2 - 2aL\cos\theta$$
$$= (0.70)^2 + (3.20)^2 - 2(0.70)(3.20)\cos 60° = 8.49$$
$$s = 2.91\text{ m}$$

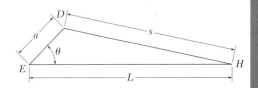

$$F_{DH} = W\frac{s}{L}\cot\theta = (9.81\text{ kN})\frac{2.91\text{ m}}{3.20\text{ m}}\cot 60°$$

$$F_{DH} = 5.15\text{ kN} \blacktriangleleft$$

# PROBLEMS

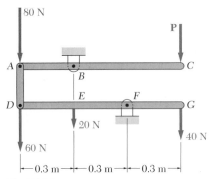

**Fig. P10.2**

**10.1** Determine the mass $m$ which balances the 5-kg block.

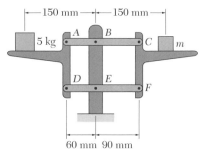

**Fig. P10.1**

**10.2** Determine the magnitude of the force **P** required to maintain the equilibrium of the linkage shown.

**10.3** Determine the horizontal force **P** which must be applied at $A$ to maintain the equilibrium of the linkage.

**10.4** Determine the couple **M** which must be applied to member $ABC$ to maintain the equilibrium of the linkage.

**10.5** Determine the force **P** required to maintain the equilibrium of the linkage shown. All members are of the same length and the wheels at $A$ and $B$ roll freely on the horizontal rod.

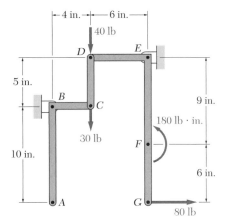

**Fig. P10.3 and P10.4**

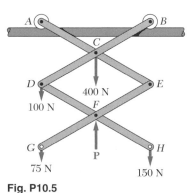

**Fig. P10.5**

**10.6** Solve Prob. 10.5, assuming that the vertical force **P** is applied at point $E$.

**10.7** The mechanism shown is acted upon by the force **P**; derive an expression for the magnitude of the force **Q** required to maintain equilibrium.

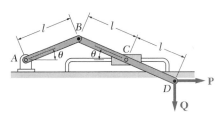

**Fig. P10.7**

**10.8** The mechanism shown is acted upon by the force **P**; derive an expression for the magnitude of the force **Q** required to maintain equilibrium.

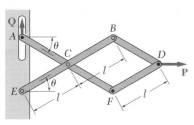

**Fig. P10.8**

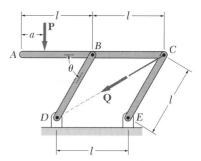

**Fig. P10.9**

**10.9** Knowing that the line of action of the force **Q** passes through point *D*, derive an expression for the magnitude of **Q** required to maintain equilibrium.

**10.10** Rod *AD* is acted upon by a vertical force **P** at end *A* and by two equal and opposite horizontal forces of magnitude *Q* at points *B* and *C*. Derive an expression for the magnitude *Q* of the horizontal forces required for equilibrium.

**10.11** The slender rod *AB* is attached to a collar at *A* and rests on a wheel at *C*. Neglecting the effect of friction, derive an expression for the magnitude of the force **Q** required to maintain equilibrium.

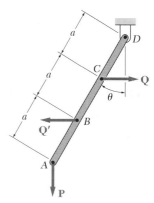

**Fig. P10.10**

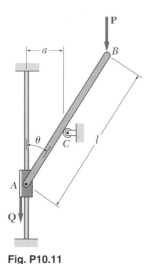

**Fig. P10.11**

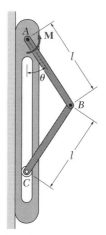

**Fig. P10.13**

**10.12** Solve Prob. 10.11, assuming that the force **P** applied at *B* is horizontal and directed to the right.

**10.13** Each of the uniform rods *AB* and *BC* is of weight *W*. Derive an expression for the moment of the couple **M** required to maintain equilibrium.

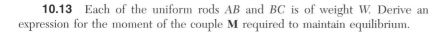

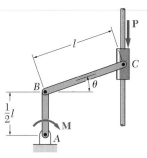

**Fig. P10.14**

**10.14** Derive an expression for the moment of the couple **M** required to maintain the equilibrium of the linkage in the position shown.

**10.15 and 10.16** Knowing that rod $AB$ is of length $2l$, derive an expression for the moment of the couple **M** required to maintain equilibrium.

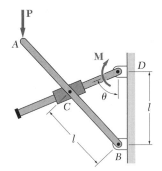

**Fig. P10.16**

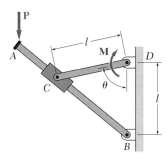

**Fig. P10.15**

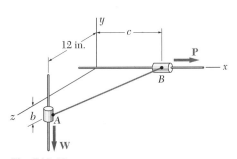

**Fig. P10.19**

**10.17** Solve Prob. 10.15, assuming that the force **P** applied at $A$ is horizontal and directed to the left.

**10.18** Solve Prob. 10.16, assuming that the force **P** applied at $A$ is horizontal and directed to the left.

**10.19** Collar $A$ weighs 12.6 lb and may slide freely on the frictionless vertical rod. Knowing that the length of the wire $AB$ is 22 in., determine the magnitude of **P** required for equilibrium when (a) $c = 4$ in., (b) $c = 12$ in.

**10.20** Solve Prob. 10.19 when (a) $c = 14$ in., (b) $c = 18$ in.

**10.21** In Prob. 10.9, determine the magnitude of the force **Q** required for equilibrium when $l = 500$ mm, $\theta = 70°$, and $P = 350$ N.

**10.22** In Prob. 10.14, determine the moment of the couple **M** required for equilibrium when $l = 500$ mm, $\theta = 35°$, and $P = 500$ N.

**10.23** Determine the value of $\theta$ corresponding to the equilibrium position of the mechanism of Prob. 10.7 when $P = 45$ lb and $Q = 60$ lb.

**10.24** Determine the value of $\theta$ corresponding to the equilibrium position of the mechanism of Prob. 10.8 when $P = 50$ lb and $Q = 160$ lb.

**10.25** Determine the value of $\theta$ corresponding to the equilibrium position of the mechanism of Prob. 10.11 when $P = 200$ N, $Q = 400$ N, $l = 600$ mm, and $a = 100$ mm.

**10.26** Determine the value of $\theta$ corresponding to the equilibrium position of the mechanism of Prob. 10.10 when $P = 500$ N, $Q = 400$ N, and $a = 100$ mm.

**10.27** A vertical load **W** is applied to the linkage at $B$. The constant of the spring is $k$, and the spring is unstretched when $AB$ and $BC$ are horizontal. Neglecting the weight of the linkage, derive an equation in $\theta$, $W$, $l$, and $k$, which must be satisfied when the linkage is in equilibrium.

**10.28** A load **W** of magnitude 900 N is applied to the linkage at $B$. Neglecting the weight of the linkage and knowing that $l = 225$ mm, determine the value of $\theta$ corresponding to equilibrium. The constant of the spring is $k = 2$ kN/m, and the spring is unstretched when $AB$ and $BC$ are horizontal. (*Hint.* Solve the equation obtained for $\theta$ by trial.)

**10.29** Two bars $AD$ and $DG$ are connected by a pin at $D$ and by a spring $AG$. Knowing that the spring is 12 in. long when unstretched and that the constant of the spring is 125 lb/in., determine the value of $x$ corresponding to equilibrium when a 900-lb load is applied at $E$ as shown.

**10.30** Solve Prob. 10.29, assuming that the 900-lb vertical force is applied at $C$ instead of $E$.

**10.31** A load **W** of magnitude 80 lb is applied to the mechanism at $C$. Neglecting the weight of the mechanism, determine the value of $\theta$ corresponding to equilibrium. The constant of the spring is $k = 20$ lb/in., and the spring is unstretched when $\theta = 0$.

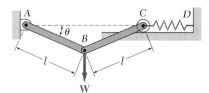

**Fig. P10.27 and P10.28**

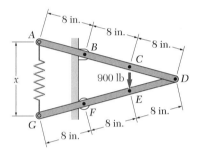

**Fig. P10.29**

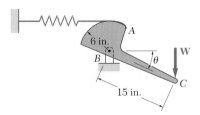

**Fig. P10.31**

**10.32** A force **P** of magnitude 60 lb is applied to end $E$ of cable $CDE$, which passes under pulley $D$ and is attached to the mechanism at $C$. Neglecting the weight of the mechanism and the radius of the pulley, determine the value of $\theta$ corresponding to equilibrium. The constant of the spring is $k = 20$ lb/in., and the spring is unstretched when $\theta = 90°$.

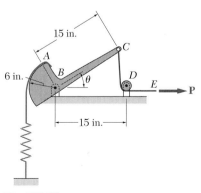

**Fig. P10.32**

**10.33** The lever $AB$ is attached to the horizontal shaft $BC$ which passes through a bearing and is welded to a fixed support at $C$. The torsional spring constant of the shaft $BC$ is $K$; that is, a couple of magnitude $K$ is required to rotate end $B$ through one radian. Knowing that the shaft is untwisted when $AB$ is horizontal, determine the value of $\theta$ corresponding to the position of equilibrium, if $P = 120$ N, $l = 300$ mm, and $K = 16$ N $\cdot$ m/rad.

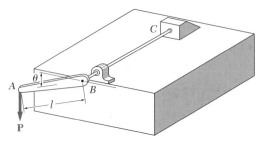

**Fig. P10.33**

**10.34** Solve Prob. 10.33, if $P = 400$ N, $l = 300$ mm, and $K = 16$ N $\cdot$ m/rad. Obtain an answer in each of the following quadrants: $0 < \theta < 90°$, $270° < \theta < 360°$, and $360° < \theta < 450°$. (It is assumed that $K$ remains constant for the angles of twist considered.)

**10.35** The position of boom $ABC$ is controlled by the hydraulic cylinder $BD$. For the loading shown, determine the force exerted by the hydraulic cylinder on pin $B$ when $\theta = 60°$.

**10.36** The position of boom $ABC$ is controlled by the hydraulic cylinder $BD$. For the loading shown, (a) express the force exerted by the hydraulic cylinder on pin $B$ as a function of the length $BD$, (b) determine the largest possible value of the angle $\theta$ if the maximum force that the cylinder can exert on pin $B$ is 12.5 kN.

**10.37** The position of member $ABC$ is controlled by the hydraulic cylinder $CD$. For the loading shown, determine the force exerted by the hydraulic cylinder on pin $C$ when $\theta = 65°$.

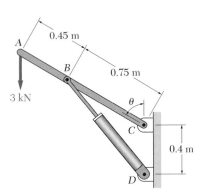

**Fig. P10.35 and P10.36**

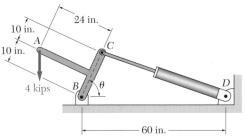

**Fig. P10.37**

**10.38** Solve Prob. 10.37 when $\theta = 115°$.

**10.39** A block of weight $W$ is pulled up a plane forming an angle $\alpha$ with the horizontal by a force $\mathbf{P}$ directed along the plane. If $\mu$ is the coefficient of friction between the block and the plane, derive an expression for the mechanical efficiency of the system. Show that the mechanical efficiency cannot exceed $\frac{1}{2}$ if the block is to remain in place when the force $\mathbf{P}$ is removed.

**10.40**   Derive an expression for the mechanical efficiency of the jack discussed in Sec. 8.6. Show that if the jack is to be self-locking, the mechanical efficiency cannot exceed $\frac{1}{2}$.

**10.41**   Denoting by $\mu$ the coefficient of friction between collar $C$ and the vertical rod, derive an expression for the moment of the largest couple **M** for which equilibrium is maintained in the position shown. Explain what happens if $\mu \geq \tan \theta$.

**10.42**   Knowing that the coefficient of friction between collar $C$ and the vertical rod is 0.30, determine the moment of the largest and smallest couple **M** for which equilibrium is maintained in the position shown, when $\theta = 30°$, $l = 250$ mm, and $P = 400$ N.

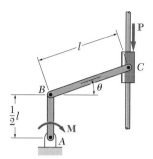

**Fig. P10.41 and P10.42**

**10.43**   Using the method of virtual work, determine the reaction at $E$.

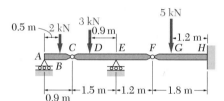

**Fig. P10.43 and P10.44**

**10.44**   Using the method of virtual work, determine separately the force and couple representing the reaction at $H$.

**10.45**   In Prob. 10.5 the force **P** is removed and the linkage is maintained in equilibrium by a cord which is attached to pins $E$ and $H$. Determine the tension in the cord.

**10.46**   A slender rod $AB$, of weight $W$, is attached to blocks $A$ and $B$ which move freely in the guides shown. The blocks are connected by an elastic cord which passes over a pulley at $C$. (*a*) Express the tension in the cord in terms of $W$ and $\theta$. (*b*) Determine the value of $\theta$ for which the tension in the cord is equal to $2W$.

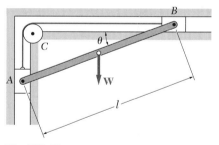

**Fig. P10.46**

**10.47**   Determine the vertical movement of joint $C$ if member $FG$ is lengthened by 1.5 in. (*Hint.* Apply a vertical load at joint $C$ and, using the methods of Chap. 6, compute the force exerted by member $FG$ on joints $F$ and $G$. Then apply the method of virtual work for a virtual displacement making member $FG$ longer. This method should be used only for small changes in the length of members.)

**10.48**   Determine the horizontal movement of joint $C$ if member $FG$ is lengthened by 1.5 in. (Follow hint of Prob. 10.47, applying a *horizontal* force at joint $C$.)

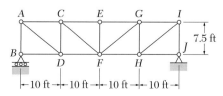

**Fig. P10.47 and P10.48**

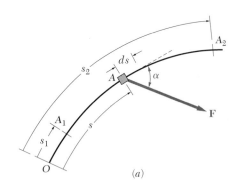

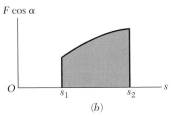

**Fig. 10.10**

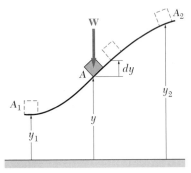

**Fig. 10.11**

## *10.6. WORK OF A FORCE DURING A FINITE DISPLACEMENT

Consider a force $\mathbf{F}$ acting on a particle. The work of $\mathbf{F}$ corresponding to an infinitesimal displacement $d\mathbf{s}$ of the particle was defined in Sec. 10.2 as

$$dU = F\,ds\,\cos\alpha \tag{10.1}$$

The work of $\mathbf{F}$ corresponding to a finite displacement of the particle from $A_1$ to $A_2$ (Fig. 10.10a) is denoted by $U_{1\to2}$ and is obtained by integrating (10.1) along the curve described by the particle:

$$U_{1\to2} = \int_{s_1}^{s_2} (F\cos\alpha)\,ds \tag{10.11}$$

where the variable of integration $s$ measures the distance along the path traveled by the particle. The work $U_{1\to2}$ is represented by the area under the curve obtained by plotting $F\cos\alpha$ against $s$ (Fig. 10.10b). In the case of a force $\mathbf{F}$ of constant magnitude acting in the direction of motion, formula (10.11) yields $U_{1\to2} = F(s_2 - s_1)$.

Recalling from Sec. 10.2 that the work of a couple of moment $M$ during an infinitesimal rotation $d\theta$ of a rigid body is

$$dU = M\,d\theta \tag{10.2}$$

we express as follows the work of the couple during a finite rotation of the body:

$$U_{1\to2} = \int_{\theta_1}^{\theta_2} M\,d\theta \tag{10.12}$$

In the case of a constant couple, formula (10.12) yields

$$U_{1\to2} = M(\theta_2 - \theta_1)$$

***Work of a Weight.*** It was stated in Sec. 10.2 that the work of the weight $\mathbf{W}$ of a body during an infinitesimal displacement of the body is equal to the product of $W$ and of the vertical displacement of the center of gravity of the body. With the $y$ axis pointing upward, the work of $\mathbf{W}$ during a finite displacement of the body (Fig. 10.11) is obtained by writing

$$dU = -W\,dy$$

Integrating from $A_1$ to $A_2$, we have

$$U_{1\to2} = -\int_{y_1}^{y_2} W\,dy = Wy_1 - Wy_2 \qquad (10.13)$$

or

$$U_{1\to2} = -W(y_2 - y_1) = -W\Delta y \qquad (10.13')$$

where $\Delta y$ is the vertical displacement from $A_1$ to $A_2$. The work of the weight **W** is thus equal to *the product of W and of the vertical displacement of the center of gravity of the body.* The work is *positive* when $\Delta y < 0$, that is, *when the body moves down.*

**Work of the Force Exerted by a Spring.** Consider a body $A$ attached to a fixed point $B$ by a spring; it is assumed that the spring is undeformed when the body is at $A_0$ (Fig. 10.12a). Experimental evidence shows that the magnitude of the force **F** exerted by the spring on a body $A$ is proportional to the deflection $x$ of the spring measured from the position $A_0$. We have

$$F = kx \qquad (10.14)$$

where $k$ is the *spring constant,* expressed in N/m if SI units are used, and in lb/ft or lb/in. if U.S. customary units are used. The work of the force **F** exerted by the spring during a finite displacement of the body from $A_1(x = x_1)$ to $A_2(x = x_2)$ is obtained by writing

$$dU = -F\,dx = -kx\,dx$$

$$U_{1\to2} = -\int_{x_1}^{x_2} kx\,dx = \tfrac{1}{2}kx_1^2 - \tfrac{1}{2}kx_2^2 \qquad (10.15)$$

Care should be taken to express $k$ and $x$ in consistent units. For example, if U.S. customary units are used, $k$ should be expressed in lb/ft and $x$ in feet, or $k$ in lb/in. and $x$ in inches; in the first case, the work is obtained in ft · lb; in the second case, in in · lb. We note that the work of the force **F** exerted by the spring on the body is *positive* when $x_2 < x_1$, that is, *when the spring is returning to its undeformed position.*

Since Eq. (10.14) is the equation of a straight line of slope $k$ passing through the origin, the work $U_{1\to2}$ of **F** during the displacement from $A_1$ to $A_2$ may be obtained by evaluating the area of the trapezoid shown in Fig. 10.12b. This is done by computing the values $F_1$ and $F_2$ and multiplying the base $\Delta x$ of the trapezoid by its mean height $\tfrac{1}{2}(F_1 + F_2)$. Since the work of the force **F** exerted by the spring is positive for a negative value of $\Delta x$, we write

$$U_{1\to2} = -\tfrac{1}{2}(F_1 + F_2)\,\Delta x \qquad (10.16)$$

Formula (10.16) is usually more convenient to use than (10.15) and affords fewer chances of confusing the units involved.

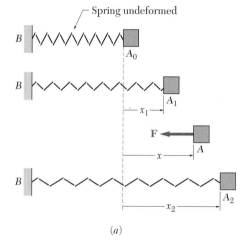

(a)

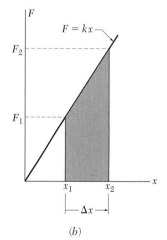

(b)

**Fig. 10.12**

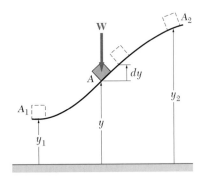

**Fig. 10.11 (repeated)**

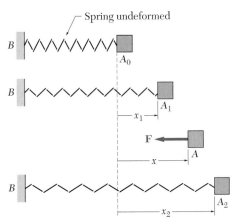

**Fig. 10.12 (a) (repeated)**

## *10.7. POTENTIAL ENERGY

Considering again the body of Fig. 10.11, we note from (10.13) that the work of the weight **W** during a finite displacement is obtained by subtracting the value of the function $Wy$ corresponding to the second position of the body from its value corresponding to the first position. The work of **W** is thus independent of the actual path followed; it depends only upon the initial and final values of the function $Wy$. This function is called the *potential energy* of the body with respect to the *force of gravity* **W** and is denoted by $V_g$. We write

$$U_{1 \to 2} = (V_g)_1 - (V_g)_2 \qquad \text{with } V_g = Wy \qquad (10.17)$$

We note that if $(V_g)_2 > (V_g)_1$, that is, *if the potential energy increases* during the displacement (as in the case considered here), *the work $U_{1 \to 2}$ is negative.* If, on the other hand, the work of **W** is positive, the potential energy decreases. Therefore, the potential energy $V_g$ of the body provides a measure of *the work which may be done* by its weight **W**. Since only the *change* in potential energy, and not the actual value of $V_g$, is involved in formula (10.17), an arbitrary constant may be added to the expression obtained for $V_g$. In other words, the level from which the elevation $y$ is measured may be chosen arbitrarily. Note that potential energy is expressed in the same units as work, i.e., in joules (J) if SI units are used,[†] and in ft · lb or in · lb if U.S. customary units are used.

Considering now the body of Fig. 10.12a, we note from formula (10.15) that the work of the elastic force **F** is obtained by subtracting the value of the function $\frac{1}{2}kx^2$ corresponding to the second position of the body from its value corresponding to the first position. This function is denoted by $V_e$ and is called the *potential energy* of the body with respect to the *elastic force* **F**. We write

$$U_{1 \to 2} = (V_e)_1 - (V_e)_2 \qquad \text{with } V_e = \tfrac{1}{2}kx^2 \qquad (10.18)$$

and observe that during the displacement considered, the work of the force **F** exerted by the spring on the body is negative and the potential energy $V_e$ increases. We should note that the expression obtained for $V_e$ is valid only if the deflection of the spring is measured from its undeformed position.

The concept of potential energy may be used when forces other than gravity forces and elastic forces are involved. It remains valid as long as the elementary work $dU$ of the force considered is an *exact differential*. It is then possible to find a function $V$, called potential energy, such that

$$dU = -dV \qquad (10.19)$$

Integrating (10.19) over a finite displacement, we obtain the general formula

$$U_{1 \to 2} = V_1 - V_2 \qquad (10.20)$$

[†] See footnote, page •••.

which expresses that *the work of the force is independent of the path followed and is equal to minus the change in potential energy.* A force which satisfies Eq. (10.20) is said to be a *conservative force.*

## *10.8. POTENTIAL ENERGY AND EQUILIBRIUM

The application of the principle of virtual work is considerably simplified when the potential energy of a system is known. In the case of a virtual displacement, formula (10.19) becomes $\delta U = -\delta V$. Besides, if the position of the system is defined by a single independent variable $\theta$, we may write $\delta V = (dV/d\theta)\,\delta\theta$. Since $\delta\theta$ must be different from zero, the condition $\delta U = 0$ for the equilibrium of the system becomes

$$\frac{dV}{d\theta} = 0 \qquad (10.21)$$

In terms of potential energy, the principle of virtual work states therefore that *if a system is in equilibrium, the derivative of its total potential energy is zero.* If the position of the system depends upon several independent variables (the system is then said to possess *several degrees of freedom*), the partial derivatives of $V$ with respect to each of the independent variables must be zero.

Consider, for example, a structure made of two members $AC$ and $CB$ and carrying a load $W$ at $C$. The structure is supported by a pin at $A$ and a roller at $B$, and a spring $BD$ connects $B$ to a fixed point $D$ (Fig. 10.13$a$). The constant of the spring is $k$, and it is assumed that the natural length of the spring is equal to $AD$ and thus that the spring is undeformed when $B$ coincides with $A$. Neglecting the friction forces and the weight of the members, we find that the only forces which work during a displacement of the structure are the weight $\mathbf{W}$ and the force $\mathbf{F}$ exerted by the spring at point $B$ (Fig. 10.13$b$). The total potential energy of the system will thus be obtained by adding the potential energy $V_g$ corresponding to the gravity force $\mathbf{W}$ and the potential energy $V_e$ corresponding to the elastic force $\mathbf{F}$.

Choosing a coordinate system with origin at $A$ and noting that the deflection of the spring, measured from its undeformed position, is $AB = x_B$, we write

$$V_e = \tfrac{1}{2}kx_B^2 \qquad V_g = Wy_C$$

Expressing the coordinates $x_B$ and $y_C$ in terms of the angle $\theta$, we have

$$x_B = 2l \sin\theta \qquad y_C = l\cos\theta$$
$$V_e = \tfrac{1}{2}k(2l\sin\theta)^2 \qquad V_g = W(l\cos\theta)$$
$$V = V_e + V_g = 2kl^2\sin^2\theta + Wl\cos\theta \qquad (10.22)$$

The positions of equilibrium of the system are obtained by equating to zero the derivative of the potential energy $V$. We write

$$\frac{dV}{d\theta} = 4kl^2\sin\theta\cos\theta - Wl\sin\theta = 0$$

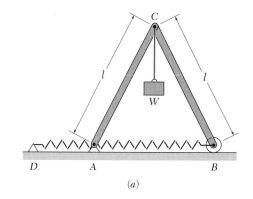

(a)

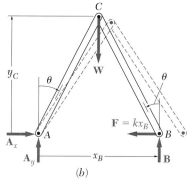

(b)

**Fig. 10.13**

or, factoring $l \sin \theta$,

$$\frac{dV}{d\theta} = l \sin \theta (4kl \cos \theta - W) = 0$$

There are therefore two positions of equilibrium, corresponding respectively to the values $\theta = 0$ and $\theta = \cos^{-1}(W/4kl)$.[†]

## *10.9. STABILITY OF EQUILIBRIUM

Consider the three uniform rods of length $2a$ and weight $\mathbf{W}$ shown in Fig. 10.14. While each rod is in equilibrium, there is an important difference between the three cases considered. Suppose that each rod is slightly disturbed from its position of equilibrium and then released: rod $a$ will move back toward its original position, rod $b$ will keep moving away from its original position, and rod $c$ will remain in its new position. In case $a$, the equilibrium of the rod is said to be *stable;* in case $b$, the equilibrium is said to be *unstable;* and, in case $c$, to be *neutral.*

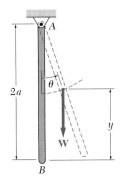

(*a*) Stable equilibrium

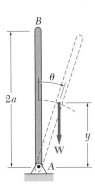

(*b*) Unstable equilibrium

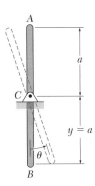

(*c*) Neutral equilibrium    **Fig. 10.14**

Recalling from Sec. 10.7 that the potential energy $V_g$ with respect to gravity is equal to $Wy$, where $y$ is the elevation of the point of application of $\mathbf{W}$ measured from an arbitrary level, we observe that the potential energy of rod $a$ is minimum in the position of equilibrium considered, that the potential energy of rod $b$ is maximum, and that the potential energy of rod $c$ is constant. Equilibrium is thus *stable, unstable,* or *neutral* according to whether the potential energy is *minimum, maximum,* or *constant* (Fig. 10.15).

That the result obtained is quite general may be seen as follows: We first observe that a force always tends to do positive work and thus to decrease the potential energy of the system on which it is applied. Therefore, when a system is disturbed from its position of equilibrium, the forces acting on the system will tend to bring it back to its original position if V is minimum (Fig. 10.15$a$) and to move it farther away if V is maximum (Fig. 10.15$b$). If V is constant (Fig. 10.15$c$), the forces will tend not to move the system either way.

[†] The second position does not exist if $W > 4kl$ (see Prob. 10.81 for a further discussion of the equilibrium of this system).

Recalling from calculus that a function is minimum or maximum according to whether its second derivative is positive or negative, we may summarize as follows the conditions for the equilibrium of a system with one degree of freedom (i.e., a system the position of which is defined by a single independent variable $\theta$):

$$\frac{dV}{d\theta} = 0 \qquad \frac{d^2V}{d\theta^2} > 0 \text{: stable equilibrium}$$

$$\frac{dV}{d\theta} = 0 \qquad \frac{d^2V}{d\theta^2} < 0 \text{: unstable equilibrium}$$

(10.23)

If both the first and the second derivatives of $V$ are zero, it is necessary to examine derivatives of a higher order to determine whether the equilibrium is stable, unstable, or neutral. The equilibrium will be neutral if all

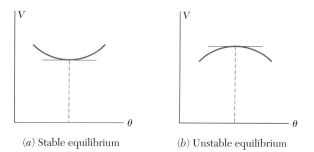

**Fig. 10.15**    (a) Stable equilibrium          (b) Unstable equilibrium          (c) Neutral equilibrium

derivatives are zero, since the potential energy $V$ is then a constant. The equilibrium will be stable if the first derivative found to be different from zero is of even order, and if that derivative is positive. In all other cases the equilibrium will be unstable.

If the system considered possesses *several degrees of freedom,* the potential energy $V$ depends upon several variables and it is thus necessary to apply the theory of functions of several variables to determine whether $V$ is minimum. It may be verified that a system with two degrees of freedom will be stable, and the corresponding potential energy $V(\theta_1, \theta_2)$ will be minimum, if the following relations are satisfied simultaneously:

$$\frac{\partial V}{\partial \theta_1} = \frac{\partial V}{\partial \theta_2} = 0$$

$$\left(\frac{\partial^2 V}{\partial \theta_1 \partial \theta_2}\right)^2 - \frac{\partial^2 V}{\partial \theta_1^2}\frac{\partial^2 V}{\partial \theta_2^2} < 0$$

(10.24)

$$\frac{\partial^2 V}{\partial \theta_1^2} > 0 \qquad \text{or} \qquad \frac{\partial^2 V}{\partial \theta_2^2} > 0$$

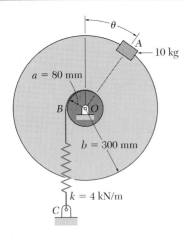

## SAMPLE PROBLEM 10.4

A 10-kg block is attached to the rim of a 300-mm-radius disk as shown. Knowing that spring $BC$ is unstretched when $\theta = 0$, determine the position or positions of equilibrium, and state in each case whether the equilibrium is stable, unstable, or neutral.

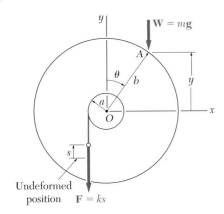

**Potential Energy.** Denoting by $s$ the deflection of the spring from its undeformed position and placing the origin of coordinates at $O$, we obtain

$$V_e = \tfrac{1}{2}ks^2 \qquad V_g = Wy = mgy$$

Measuring $\theta$ in radians, we have

$$s = a\theta \qquad y = b\cos\theta$$

Substituting for $s$ and $y$ in the expressions for $V_e$ and $V_g$, we write

$$V_e = \tfrac{1}{2}ka^2\theta^2 \qquad V_g = mgb\cos\theta$$
$$V = V_e + V_g = \tfrac{1}{2}ka^2\theta^2 + mgb\cos\theta$$

**Positions of Equilibrium.** Setting $dV/d\theta = 0$, we write

$$\frac{dV}{d\theta} = ka^2\theta - mgb\sin\theta = 0$$

$$\sin\theta = \frac{ka^2}{mgb}\theta$$

Substituting $a = 0.08$ m, $b = 0.3$ m, $k = 4$ kN/m, and $m = 10$ kg, we obtain

$$\sin\theta = \frac{(4\text{ kN/m})(0.08\text{ m})^2}{(10\text{ kg})(9.81\text{ m/s}^2)(0.3\text{ m})}\theta$$

$$\sin\theta = 0.8699\,\theta$$

where $\theta$ is expressed in radians. Solving by trial for $\theta$, we find

$$\theta = 0 \qquad \text{and} \qquad \theta = 0.902 \text{ rad}$$
$$\theta = 0 \qquad \text{and} \qquad \theta = 51.7° \quad \blacktriangleleft$$

**Stability of Equilibrium.** The second derivative of the potential energy $V$ with respect to $\theta$ is

$$\frac{d^2V}{d\theta^2} = ka^2 - mgb\cos\theta$$
$$= (4\text{ kN/m})(0.08\text{ m})^2 - (10\text{ kg})(9.81\text{ m/s}^2)(0.3\text{ m})\cos\theta$$
$$= 25.6 - 29.43\cos\theta$$

For $\theta = 0$:   $\dfrac{d^2V}{d\theta^2} = 25.6 - 29.43\cos 0° = -3.83 < 0$

The equilibrium is unstable for $\theta = 0°$ $\quad\blacktriangleleft$

For $\theta = 51.7°$:   $\dfrac{d^2V}{d\theta^2} = 25.6 - 29.43\cos 51.7° = +7.36 > 0$

# PROBLEMS

**10.49**  Using the method of Sec. 10.8, solve Prob. 10.27.

**10.50**  Using the method of Sec. 10.8, solve Prob. 10.28.

**10.51**  Using the method of Sec. 10.8, solve Prob. 10.29.

**10.52**  Using the method of Sec. 10.8, solve Prob. 10.30.

**10.53**  Show that the equilibrium is neutral in Prob. 10.5.

**10.54**  Show that the equilibrium is neutral in Prob. 10.2.

**10.55**  Two uniform rods, each of mass $m$, are attached to gears of equal radii as shown. Determine the positions of equilibrium of the system and state in each case whether the equilibrium is stable, unstable, or neutral.

**10.56**  Two uniform rods, $AB$ and $CD$, are attached to gears of equal radii as shown. Knowing that $m_{AB} = 400$ g and $m_{CD} = 600$ g, determine the positions of equilibrium of the system and state in each case whether the equilibrium is stable, unstable, or neutral.

**10.57**  Two uniform rods, $AB$ and $CD$, of the same length $l$, are attached to gears as shown. Knowing that rod $AB$ weighs 2 lb and that rod $CD$ weighs 3 lb, determine the positions of equilibrium of the system and state in each case whether the equilibrium is stable, unstable, or neutral.

**10.58**  Solve Prob. 10.57, assuming that rods $AB$ and $CD$ each weighs 3 lb.

**10.59**  Using the method of Sec. 10.8, solve Prob. 10.33. Determine whether the equilibrium is stable, unstable, or neutral. (*Hint.* The potential energy corresponding to the couple exerted by a torsion spring is $\frac{1}{2}K\theta^2$, where $K$ is the torsional spring constant and $\theta$ is the angle of twist.)

**10.60**  In Prob. 10.34, determine whether each of the positions of equilibrium is stable, unstable, or neutral. (See hint of Prob. 10.59.)

**10.61**  A load **W** of magnitude 120 lb is applied to the mechanism at $C$. Neglecting the weight of the mechanism, determine the value of $\theta$ corresponding to equilibrium and check that the equilibrium is stable. The constant of the spring is $k = 20$ lb/in., and the spring is unstretched when $\theta = 0$.

**10.62**  Solve Prob. 10.61, assuming that the spring is unstretched when $\theta = 30°$.

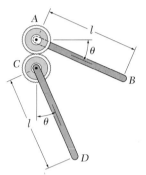

**Fig. P10.55 and P10.56**

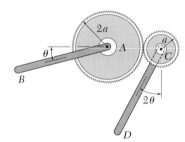

**Fig. P10.57**

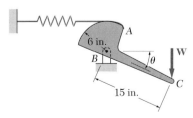

**Fig. P10.61**

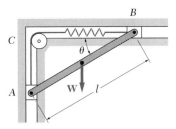

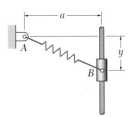

**Fig. P10.63**

**Fig. P10.65**

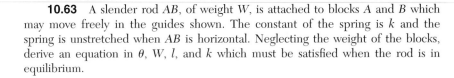

**10.63** A slender rod $AB$, of weight $W$, is attached to blocks $A$ and $B$ which may move freely in the guides shown. The constant of the spring is $k$ and the spring is unstretched when $AB$ is horizontal. Neglecting the weight of the blocks, derive an equation in $\theta$, $W$, $l$, and $k$ which must be satisfied when the rod is in equilibrium.

**10.64** In Prob. 10.63, determine three values of $\theta$ corresponding to equilibrium when $W = 30$ N, $k = 200$ N/m, and $l = 0.5$ m. State in each case whether the equilibrium is stable, unstable, or neutral.

**10.65** A collar $B$, of weight $W$, may move freely along the vertical rod shown. The constant of the spring is $k$ and the spring is unstretched when $y = 0$. (a) Derive an equation in $y$, $W$, $a$, and $k$ which must be satisfied when the collar is in equilibrium. (b) Determine the value of $y$ corresponding to equilibrium when $W = 60$ N, $a = 400$ mm, and $k = 1$ kN/m, and check that the equilibrium is stable.

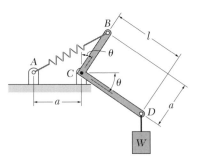

**Fig. P10.66**

**10.66** The constant of spring $AB$ is $k$ and the spring is unstretched when $\theta = 0$. (a) Neglecting the weight of the rigid arm $BCD$, derive an equation in $\theta$, $k$, $a$, $l$, and $W$ which must be satisfied when the arm is in equilibrium. (b) Determine three values of $\theta$ corresponding to equilibrium when $k = 60$ lb/in., $a = 10$ in., $l = 15$ in., and $W = 80$ lb. State in each case whether the equilibrium is stable, unstable, or neutral.

**10.67** Bar $ABC$ is attached to collars $A$ and $B$ which may move freely on the rods shown. The constant of the spring is $k$ and the spring is unstretched when $\theta = 0$. (a) Neglecting the weight of bar $ABC$, derive an equation in $\theta$, $m$, $k$, and $l$ which must be satisfied when bar $ABC$ is in equilibrium. (b) Determine the value of $\theta$ corresponding to equilibrium when $m = 5$ kg, $k = 1.2$ kN/m, and $l = 250$ mm, and check that the equilibrium is stable.

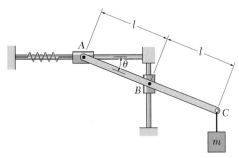

**10.68** Solve Prob. 10.67, assuming that the spring is unstretched when $\theta = 30°$.

**Fig. P10.67**

**10.69**  A vertical load **P** is applied at end $B$ of rod $BC$. The constant of the spring is $k$ and the spring is unstretched when $\theta = 60°$. (*a*) Neglecting the weight of the rod, express the angle $\theta$ corresponding to equilibrium in terms of $P$, $k$, and $l$. (*b*) Determine the values of $\theta$ corresponding to equilibrium when $P = 40$ lb, $k = 20$ lb/in., and $l = 8$ in. State in each case whether the equilibrium is stable, unstable, or neutral.

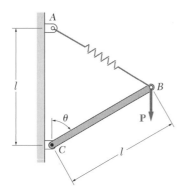

**Fig. P10.69 and P10.70**

**10.70**  A vertical load **P** of magnitude 100 lb is applied at end $B$ of rod $BC$. The constant of the spring is $k = 20$ lb/in. and the spring is unstretched when $\theta = 60°$. Neglecting the weight of the rod and knowing that $l = 8$ in., determine the value of $\theta$ corresponding to equilibrium. State whether the equilibrium is stable, unstable, or neutral.

**10.71**  The internal spring $AC$ is of constant $k$ and is undeformed when $\theta = 45°$. Derive an equation defining the values of $\theta$ corresponding to equilibrium positions.

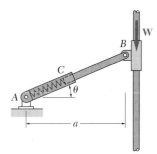

**Fig. P10.71**

**10.72**  In Prob. 10.71, determine the positive values of $\theta$ corresponding to equilibrium positions when $W = 100$ N, $k = 3$ kN/m, and $a = 500$ mm. State in each case whether the equilibrium is stable, unstable, or neutral.

**10.73**  The rod $AB$ is attached to a hinge at $A$ and to two springs, each of constant $k$. If $h = 750$ mm, $d = 400$ mm, and $m = 80$ kg, determine the range of values of $k$ for which the equilibrium of rod $AB$ is stable in the position shown. Each spring can act in either tension or compression.

**10.74**  If $m = 75$ kg, $h = 650$ mm, and the constant of each spring is $k = 2.5$ kN/m, determine the range of values of the distance $d$ for which the equilibrium of rod $AB$ is stable in the position shown. Each spring can act in either tension or compression.

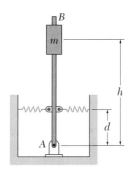

**Fig. P10.73 and P10.74**

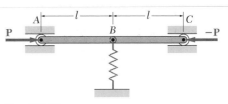

**Fig. P10.75**

**Fig. P10.75**

**10.75** Two bars $AB$ and $BC$ of negligible weight are attached to a single spring of constant $k$ which is undeformed when the bars are horizontal. Determine the range of values of the magnitude $P$ of the two equal and opposite forces **P** and $-$**P** for which the equilibrium of the system is stable in the position shown.

**10.76** Bars $BC$ and $EF$ are connected at $G$ by a pin which is attached to $BC$ and may slide freely in a slot cut in $EF$. Determine the smallest mass $m_1$ for which the equilibrium of the mechanism is stable in the position shown.

**\*10.77 and 10.78** The bars shown, each of length $l$ and of negligible weight, are attached to springs each of constant $k$. The springs are undeformed and the system is in equilibrium when $\theta_1 = \theta_2 = 0$. Determine the range of values of $P$ for which the equilibrium position is stable.

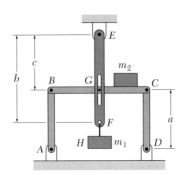

**Fig. P10.76**

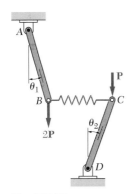

**Fig. P10.78**

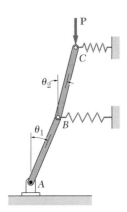

**Fig. P10.77**

**\*10.79** Solve Prob. 10.78, assuming that the vertical force applied at $B$ is increased to 5**P**.

**\*10.80** Solve Prob. 10.77, knowing that $l = 25$ in. and $k = 45$ lb/in.

**10.81** In Sec. 10.8, two positions of equilibrium were obtained for the system shown in Fig. 10.13, namely, $\theta = 0$ and $\theta = \cos^{-1}(W/4kl)$. Show that (a) if $W < 4kl$, the equilibrium is stable in the first position ($\theta = 0$) and unstable in the second; (b) if $W = 4kl$, the two positions coincide and the equilibrium is unstable; (c) if $W > 4kl$, the equilibrium is unstable in the first position ($\theta = 0$) and the second position does not exist. (*Note.* It is assumed that the system must deform as shown and that the system cannot rotate as a single rigid body about $A$ when $A$ and $B$ coincide.)

# REVIEW AND SUMMARY FOR CHAPTER 10

The first part of this chapter was devoted to the *principle of virtual work* and to its direct application to the solution of equilibrium problems. We first defined the *work of a force* **F** *corresponding to the small displacement d***s** [Sec. 10.2] as the scalar quantity

$$dU = F \, ds \cos \alpha \tag{10.1}$$

where $\alpha$ is the angle between **F** and the displacement vector $d\mathbf{s}$ (Fig. 10.16). The work $dU$ is positive if $\alpha < 90°$, zero if $\alpha = 90°$, and negative if $\alpha > 90°$. We also found that the *work of a couple of moment M* acting on a rigid body is

$$dU = M \, d\theta \tag{10.2}$$

where $d\theta$ is the small angle expressed in radians through which the body rotates.

Considering a particle located at $A$ and acted upon by several forces $\mathbf{F}_1$, $\mathbf{F}_2, \ldots, \mathbf{F}_n$ [Sec. 10.3], we imagined that the particle moved to a new position $A'$ (Fig. 10.17). Since this displacement did not actually take place, it was referred to as a *virtual displacement* and denoted by $\delta\mathbf{s}$, while the corresponding work of the forces was called *virtual work* and denoted by $\delta U$. We had

$$\delta U = F_1 \cos \alpha_1 \, \delta s + F_2 \cos \alpha_2 \, \delta s + \cdots + F_n \cos \alpha_n \, \delta s \tag{10.3}$$

The *principle of virtual work* states that, *if a particle is in equilibrium, the total virtual work $\delta U$ of the forces acting on the particle is zero for any virtual displacement of the particle.*

The principle of virtual work may be extended to the case of rigid bodies and systems of rigid bodies. Since it involves *only forces which do work*, its application provides a useful alternative to the use of the equilibrium equations in the solution of many engineering problems. It is particularly effective in the case of machines and mechanisms consisting of connected rigid bodies, since the work of the reactions at the supports is zero and the work of the internal forces at the pin connections cancels out [Sec. 10.4; Sample Probs. 10.1, 10.2, and 10.3].

In the case of *real machines,* however [Sec. 10.5], the work of the friction forces should be taken into account, with the result that the *output work will be less than the input work.* Defining the *mechanical efficiency* of a machine as the ratio

$$\eta = \frac{\text{output work}}{\text{input work}} \tag{10.9}$$

we also noted that for an ideal machine (no friction) $\eta = 1$, while for a real machine $\eta < 1$.

**Work of a force**

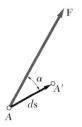

**Fig. 10.16**

**Virtual displacement**

**Principle of virtual work**

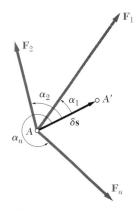

**Fig. 10.17**

**Mechanical efficiency**

Work of a force over a
finite displacement

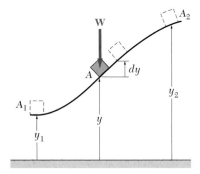

**Fig. 10.18**

In the second part of the chapter we considered the *work of forces corresponding to finite displacements* of their points of application. The work $U_{1\rightarrow2}$ of the force **F** corresponding to a displacement of the particle $A$ from $A_1$ to $A_2$ (Fig. 10.18) was obtained by integrating the right-hand member of Eq. (10.1) along the curve described by the particle [Sec. 10.6]:

$$U_{1\rightarrow2} = \int_{s_1}^{s_2} (F \cos \alpha)\, ds \tag{10.11}$$

Similarly, the work of a couple of moment $M$ corresponding to a finite rotation from $\theta_1$ to $\theta_2$ of a rigid body was expressed as

$$U_{1\rightarrow2} = \int_{\theta_1}^{\theta_2} M\, d\theta \tag{10.12}$$

Work of a weight

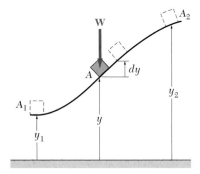

**Fig. 10.19**

The *work of the weight* **W** *of a body* as its center of gravity moves from the elevation $y_1$ to $y_2$ (Fig. 10.19) may be obtained by making $F = W$ and $\alpha = 180°$ in Eq. (10.11):

$$U_{1\rightarrow2} = -\int_{y_1}^{y_2} W\, dy = Wy_1 - Wy_2 \tag{10.13}$$

The work of **W** is therefore positive *when the elevation y decreases.*

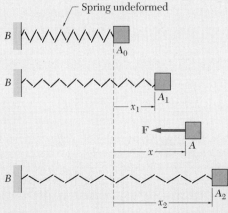

**Fig. 10.20**

Work of the force exerted
by a spring

The *work of the force* **F** *exerted by a spring* on a body $A$ as the spring is stretched from $x_1$ to $x_2$ (Fig. 10.20) may be obtained by making $F = kx$, where $k$ is the constant of the spring, and $\alpha = 180°$ in Eq. (10.11):

$$U_{1\rightarrow2} = -\int_{x_1}^{x_2} kx\, dx = \tfrac{1}{2}kx_1^2 - \tfrac{1}{2}kx_2^2 \tag{10.15}$$

The work of **F** is therefore positive *when the spring is returning to its undeformed position.*

When the work of a force **F** is independent of the path actually followed between $A_1$ and $A_2$, the force is said to be a *conservative force* and its work may be expressed as

$$U_{1 \to 2} = V_1 - V_2 \qquad (10.20)$$

**Potential energy**

where $V$ is the *potential energy* associated with **F**, and $V_1$ and $V_2$ represent the values of $V$ at $A_1$ and $A_2$, respectively [Sec. 10.7]. The potential energies associated, respectively, with the *force of gravity* **W** and the *elastic force* **F** exerted by a spring were found to be

$$V_g = Wy \qquad \text{and} \qquad V_e = \tfrac{1}{2}kx^2 \qquad (10.17, 10.18)$$

When the position of a mechanical system depends upon a single independent variable $\theta$, the potential energy of the system is a function $V(\theta)$ of that variable and it follows from Eq. (10.20) that $\delta U = -\delta V = -(dV/d\theta)\delta\theta$. The condition $\delta U = 0$ required by the principle of virtual work for the equilibrium of the system may thus be replaced by the condition

**Alternative expression for the principle of virtual work**

$$\frac{dV}{d\theta} = 0 \qquad (10.21)$$

The use of Eq. (10.21) may be preferred to the direct application of the principle of virtual work when all the forces involved are conservative [Sec. 10.8; Sample Prob. 10.4].

This approach presents another advantage, since it is possible to determine from the sign of the second derivative of $V$ whether the equilibrium of the system is *stable, unstable,* or *neutral* [Sec. 10.9]. If $d^2V/d\theta^2 > 0$, $V$ is *minimum* and the equilibrium is *stable*; if $d^2V/d\theta^2 < 0$, $V$ is *maximum* and the equilibrium is *unstable*; if $d^2V/d\theta^2 = 0$, it is necessary to examine derivatives of a higher order.

**Stability of equilibrium**

# REVIEW PROBLEMS

**10.82** A spring $AB$ of constant $k$ is attached to two identical gears as shown. A uniform bar $CD$ of mass $m$ is supported by cords wrapped around drums of radius $b$ which are attached to the gears. If the spring is undeformed when $\theta = 0$, obtain an equation defining the angle $\theta$ corresponding to the equilibrium position.

**10.83** For the mechanism of Prob. 10.82 the following numerical values are given: $k = 6$ kN/m, $a = 80$ mm, $b = 60$ mm, $r = 120$ mm, and $m = 50$ kg. Determine the values of $\theta$ corresponding to equilibrium positions and state in each case whether the equilibrium is stable, unstable, or neutral.

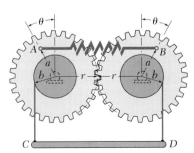

**Fig. P10.82**

**10.84** The pliers shown are used to attach electric connectors to flat cables which carry many separate wires. Knowing that 300-N forces directed along line *a-a* are required to complete the attachment, determine the magnitude *P* of the forces which must be applied to the handles.

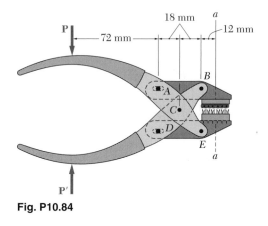

**Fig. P10.84**

**10.85** Collar *B* may slide along rod *AC* and is attached by a pin to a block which may slide in the vertical slot. Derive and expression for the moment of the couple **M** required to maintain equilibrium.

**10.86** Determine the value of $\theta$ corresponding to the equilibrium position of rod *AC* when $R = 10$ in., $P = 50$ lb, and $M = 2000$ lb·in.

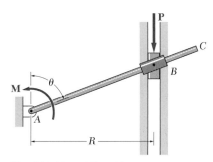

**Fig. P10.85 and P10.86**

**10.87** Collar *A* may slide freely on the balance beam *DCE* which is supported by a frictionless pin at *C*. The position of collar *A* on the beam is controlled by a cord which is attached to a fixed cylindrical support of radius *a*. Knowing that $l_A = l$ when beam *DCE* is horizontal, derive an equation in $\theta$, $m_A$, $m_B$, $l$, and $a$ which must be satisfied when the balance beam is in equilibrium.

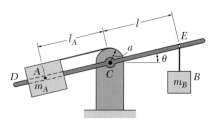

**Fig. P10.87**

**10.88** Determine the value of $\theta$ corresponding to the equilibrium position of the balance beam of Prob. 10.87, when $m_A = 500$ g, $m_B = 450$ g, $l = 200$ mm, and $a = 40$ mm.

**10.89** The horizontal bar $AD$ is attached to two springs of constant $k$ and is in equilibrium in the position shown. Determine the range of values of the magnitude $P$ of the two equal and opposite *horizontal* forces **P** and **P'** for which the equilibrium position is stable (*a*) if $AB = CD$, (*b*) if $AB = 2(CD)$.

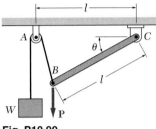

Fig. P10.90

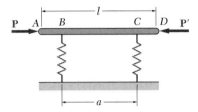

Fig. P10.89

**10.90** (*a*) Derive an equation defining the angle $\theta$ corresponding to the equilibrium position. (*b*) Determine the angle $\theta$ corresponding to equilibrium if $P = 2W$.

**10.91** Two bars $ABC$ and $CD$ are attached to a single spring of constant $k$ which is unstretched when the bars are vertical. Determine the range of values of $P$ for which the equilibrium of the system is stable in the position shown.

**10.92** Blocks $A$ and $B$, of mass $m_A$ and $m_B$, respectively, are connected by a cable $AB$ which passes over a pulley at $C$. Neglecting the effect of friction, derive an equation in $\theta$, $\beta$, $b$, $m_A$, and $m_B$ which must be satisfied when the system is in equilibrium.

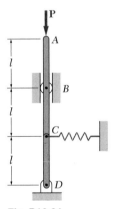

Fig. P10.91

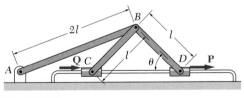

Fig. P10.93

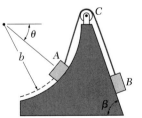

Fig. P10.92

**10.93** The mechanism shown is acted upon by the force **P**; derive an expression for the magnitude of the force **Q** required to maintain equilibrium.

# COMPUTER PROBLEMS

**10.C1** Using the principle of virtual work, write a computer program and use it to calculate the force in member $CD$ for values of $\theta$ from 5 to 120° at 5° intervals.

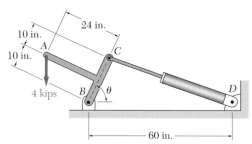

**Fig. P10.C1**

**10.C2** A slender rod $AB$, of weight $W$, is attached to blocks $A$ and $B$ which may move freely in the guides shown. The constant of the spring is $k$ and the spring is unstretched when $AB$ is horizontal. (*a*) Write a computer program which can be used to calculate the potential energy $V$ of this system and its derivative $dV/d\theta$. Neglect the weight of the blocks and place the datum at $B$. (*b*) For $W = 30$ N, $k = 200$ N/m, and $l = 0.5$ m, calculate and plot the potential energy versus $\theta$ for values of $\theta$ from 0 to 90° at 5° intervals. (*c*) By examining the $V$-$\theta$ curve, determine the approximate values of $\theta$ corresponding to equilibrium and state in each case whether the equilibrium is stable, unstable, or neutral. (*d*) Determine, to three significant figures, the values of $\theta$ corresponding to equilibrium.

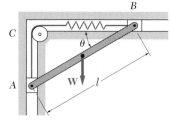

**Fig. P10.C2**

**10.C3** The internal spring $AC$ is of constant $k$ and is undeformed when $\theta = 45°$. (*a*) Write a computer program which can be used to calculate the potential energy $V$ of this system and its derivative $dV/d\theta$. (*b*) For $W = 100$ N, $a = 0.5$ m, $k = 3$ kN/m, calculate and plot the potential energy versus $\theta$ for values of $\theta$ from 0 to 50° at 5° intervals. (*c*) By examining the $V$-$\theta$ curve, determine the approximate values of $\theta$ corresponding to equilibrium and state in each case whether the equilibrium is stable, unstable, or neutral. (*d*) Determine, to three significant figures, the values of $\theta$ corresponding to equilibrium.

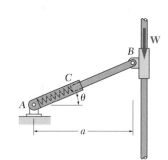

**Fig. P10.C3**

**10.C4** A vertical load $\mathbf{W}$ is applied to the linkage at $B$. The constant of the spring is $k$, and the spring is unstretched when $AB$ and $BC$ are horizontal. (*a*) Neglecting the weight of the linkage, write a computer program which can be used to calculate the load $W$ for which $AB$ forms a given angle $\theta$ with the horizontal. (*b*) Use this program to compute $W$ for values of $\theta$ from 0 to 80° at 5° intervals, assuming $l = 225$ mm and $k = 2$ kN/m. (*c*) Determine, to three significant figures, the value of $\theta$ for which the linkage is in equilibrium when $W = 900$ N, $l = 225$ mm, and $k = 2$ kN/m.

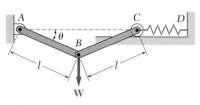

**Fig. P10.C4**

# INDEX

**443**

## CHAPTER 2

**2.2** 659 lb ⦩ 47°.

**2.4** 17.00 kN ⦨ 84.8°.

**2.6** 16.70°.

**2.8** (a) 585 lb. (b) 896 lb.

**2.10** (a) 78.9 N. (b) 169.2 N.

**2.12** $P = 72.1$ N; $\alpha = 44.7°$.

**2.14** 414 lb ⦨ 72.0°.

**2.16** (45 lb) +25.8 lb, +36.9 lb;
(60 lb) +49.1 lb, +34.4 lb;
(75 lb) +48.2 lb, −57.5 lb.

**2.18** (340 N) +300 N, +160 N;
(500 N) −480 N, +140 N.

**2.20** (a) 261 lb. (b) −167.8 lb.

**2.22** −120 lb, +350 lb.

**2.26** 1212 N ⦫ 93.4°.

**2.28** 350 N ⦩ 121°.

**2.30** 25.6 kN.

**2.32** (a) 36.9°. (b) 80 lb.

**2.34** $T_{AC} = 115.5$ lb; $T_{BC} = 231$ lb.

**2.36** $T_{AC} = 265$ N; $T_{BC} = 175.0$ N.

**2.38** $T_{AB} = 236$ N; $F_{BC} = 183.9$ N.

**2.40** $T_A = 231$ lb; $T_B = 577$ lb.

**2.42** $T_{AB} = 600$ N; $T_{AC} = 344$ N.

**2.44** (a) $\alpha = 35°$; $T_{AC} = 410$ N; $T_{BC} = 287$ N.
(b) $\alpha = 55°$; $T_{AC} = T_{BC} = 305$ N.

**2.46** 30 in.

**2.48** $T_{AC} = 200$ lb; $F_{BC} = 229$ lb.

**2.50** 913 N ⦨ 82.5°.

**2.52** (a) 18 lb. (b) 64 lb.

**2.54** 360 m, 321 m, 131.0 m.

**2.56** (a) 24 lb, 40 lb, 15 lb.
(b) 49 lb, $\theta = 32.0°$, $\phi = 35.3°$.
(c) 49 lb, $\theta_x = 60.7°$, $\theta_y = 35.3°$, $\theta_z = 77.2°$.

**2.58** (a) −77.1 N, 69.7 N, −60 N.
(b) 120 N, $\theta = 37.9°$, $\phi = 54.5°$.

**2.60** 1050 N; 51.8°, 107.7°, 43.6°.

**2.62** (a) $F_x = +107.7$ lb, $F_z = -267$ lb; $F = 416$ lb.
(b) 43.9°.

**2.64** −95.7 lb, 16.60 lb, 87.9 lb.

**2.66** −225 lb, +150 lb, +90 lb.

**2.68** $R = 623$ lb; $\theta_x = 37.4°$, $\theta_y = 122.0°$, $\theta_z = 72.6°$.

**2.70** $R = 498$ N; $\theta_x = 68.9°$, $\theta_y = 26.3°$, $\theta_z = 75.1°$.

**2.72** 510 N.

**2.74** 2025 lb.

**2.76** $T_{AB} = 60$ N, $T_{BC} = T_{BD} = 62.5$ N.

**2.78** $T_{AB} = 17$ lb, $T_{BC} = T_{BD} = 8.24$ lb.

**2.80** $T_{AC} = T_{AD} = 6.13$ kN; $T_{AB} = 11.67$ kN.

**2.82** $T_{AB} = T_{AC} = 3.35$ lb; $T_{AD} = 5.80$ lb.

**2.84** $T_{AB} = 31.7$ lb; $T_{AC} = 64.3$ lb.

**2.86** $P = 138$ N; $T_{AB} = 270$ N; $T_{AC} = 196$ N.

**2.88** $P = 131.2$ N; $Q = 29.6$ N.

**2.90** $T_{BAC} = 57.6$ lb; $T_{AD} = 20.1$ lb; $T_{AE} = 36.9$ lb.

**2.92** 605 mm.

**2.94** $x = 134.2$ mm, $z = 67.1$ mm.

**2.96** (a) 991 N. (b) 973 N.

**2.98** 62.8 lb $\leq P \leq$ 258 lb.

**2.100** 141.5°, 124.4°, 74.9°.

**2.102** 179.3 N $< P <$ 669 N.

**2.104** (a) 337 lb ⦨ 10.7°. (b) 296 lb ⦨ 3.1°.

**2.106** $T_{AD} = 40$ lb; $T_{BD} = T_{CD} = 34$ lb.

**2.C2** $\beta = 10°$; $F = 206$ lb, $\alpha = 69.7°$;
$\beta = 20°$: $F = 230$ lb, $\alpha = 46.8°$.

**2.C3** (a) $R = 234$ lb; $\theta_x = 29.7°$, $\theta_y = 61.0°$, $\theta_z = 95.7°$.
(b) $R = 200$ lb; $\theta_x = 57.9°$, $\theta_y = 66.0°$, $\theta_z = 42.0°$.
(c) $R = 196.7$ lb; $\theta_x = 52.7°$, $\theta_y = 55.7°$, $\theta_z = 55.8°$.

**2.C4** $a = 0.8$ m: $h = 2.25$ m, $P = 189.1$ N, $T = 544$ N;
$a = 2$ m: $h = 2.95$ m, $P = 634$ N, $T = 729$ N.

## CHAPTER 3

**3.2** $\alpha = 18.20°$.

**3.4** $116.0 \text{ N} \cdot \text{m} \downarrow$.

**3.6** (a) $27.4 \text{ N} \cdot \text{m} \uparrow$. (b) $228 \text{ N} \nearrow 42.0°$.

**3.8** $20.9 \text{ N} \cdot \text{m} \uparrow$.

**3.10** (a) $230 \text{ lb} \cdot \text{in} \uparrow$. (b) $2.30 \text{ in.}$

**3.12** $68.3 \text{ N} \cdot \text{m} \uparrow$.

**3.14** $156.0 \text{ lb.}$

**3.16** $M_O = -F_x y + F_y x$

**3.18** $M_O = P \dfrac{x_1 y_2 - x_2 y_1}{\sqrt{(x_2 - x_1)^2 + (y_2 - y_1)^2}}$

**3.20** (a) $F_1 = 50 \text{ N}$, (b) $F_2 = 35.4 \text{ N}$.

**3.22** (a) $10 \text{ N} \cdot \text{m} \uparrow$, (b) $10 \text{ N} \cdot \text{m} \uparrow$, (c) $10 \text{ N} \cdot \text{m} \uparrow$.

**3.24** (a) $F_1 = 173.2 \text{ N}$, (b) $F_2 = 141.4 \text{ N}$.

**3.26** $F = 260 \text{ lb} \nearrow 67.4°$, $M = 200 \text{ lb} \cdot \text{in.} \downarrow$.

**3.28** (a) $F = 250 \text{ lb} \searrow 25°$, $M = 57.5 \text{ N} \cdot \text{m} \downarrow$.
(b) $A = 375 \text{ N} \searrow 25°$, $B = 625 \text{ N} \searrow 25°$.

**3.30** $F = 750 \text{ N} \downarrow$; $x = 40 \text{ mm}$.

**3.32** (a) $30°$. (b) $65.7°$

**3.34** $A = 155.3 \text{ lb} \searrow 60°$, $B = 205 \text{ lb} \searrow 60°$.

**3.36** $B = 150 \text{ N} \searrow 45°$, $C = 450 \text{ N} \searrow 45°$.

**3.38** $c$ and $g$.

**3.40** (a) $500 \text{ N} \downarrow$; $2.2 \text{ m}$. (b) $500 \text{ N} \uparrow$; $1 \text{ m}$.
(c) $500 \text{ N} \downarrow$; $0.4 \text{ m}$.

**3.42** (a) $310 \text{ N} \angle 18.8°$, $52.6 \text{ N} \cdot \text{m} \downarrow$.
(b) $d_y = 0.1794 \text{ m}$.

**3.44** (a) $391 \text{ lb} \searrow 39.8°$.
(b) $12 \text{ in. to the right of } C$ and $10 \text{ in. above } C$.

**3.46** (a) $136.0 \text{ N} \searrow 28.0°$. (b) $18.40 \text{ mm to the right of } A$
and $50.0 \text{ mm above } C$.

**3.48** (a) $200 \text{ lb} \cdot \text{in.} \downarrow$. (b) $280 \text{ lb} \cdot \text{in.} \uparrow$.
(c) $440 \text{ lb} \cdot \text{in.} \downarrow$. (d) $40 \text{ lb} \cdot \text{in.} \downarrow$.

**3.50** $R = 329 \text{ kN} \searrow 61.8°$, $6.82 \text{ m to the right of } A$.

**3.52** $P = 60 \text{ N}$; $Q = 74.4 \text{ N}$; $\beta = 53.8°$.

**3.54** (a) $34.3°$ and $82.2°$. (b) $0.564$ and $0.991$.
(c) $22.1°$ and $41.3°$.

**3.56** (a) $23.4 \text{ lb} \uparrow$. (b) $78.3 \text{ lb} \uparrow$.

**3.58** (a) $4.15 \text{ kip}$. (b) $7.35 \text{ kip}$.

**3.60** $62.0°$.

**3.62** $0.5 \text{ kN} \leq Q \leq 11 \text{ kN}$.

**3.64** $1 \text{ kN} \leq Q \leq 3.25 \text{ kN}$.

**3.66** $A = 489 \text{ N} \downarrow$; $C = 669 \text{ N} \searrow 81.4°$.

**3.68** $230 \text{ lb.}$

**370.** (a) $A = 4.27 \text{ kN} \searrow 20.6°$; $B = 4.50 \text{ kN} \uparrow$.
(b) $A = 1.50 \text{ kN} \uparrow$; $B = 6.02 \text{ kN} \searrow 48.4°$.
(c) $A = 2.05 \text{ kN} \searrow 47.0°$; $B = 5.20 \text{ kN} \searrow 60.0°$.

**3.72** (a) $A_x = 82.8 \text{ N} \rightarrow$, $A_y = 228 \text{ N} \uparrow$.
(b) $N_B = 242 \text{ N} \measuredangle 20°$.

**3.74** $T_{BE} = 196.2 \text{ N}$; $A = 73.6 \text{ N} \rightarrow$; $D = 73.6 \text{ N} \leftarrow$.

**3.76** $A = 63.3 \text{ lb} \rightarrow$; $B = 34.6 \text{ lb} \searrow 60°$;
$C = 173.2 \text{ lb} \searrow 60°$.

**3.78** (a) $1842 \text{ N} \rightarrow$. (b) $1250 \text{ N} \nearrow 40°$.

**3.80** (a) $128 \text{ lb}$. (b) $A = 80 \text{ lb} \uparrow$; $B = 64 \text{ lb} \rightarrow$.

**3.82** $A_x = 59.1 \text{ N} \rightarrow$, $A_y = 80.6 \text{ N} \downarrow$, $F_B = 191.4 \text{ N} \nwarrow$

**3.84** $A_x = 1760 \text{ N} \rightarrow$, $A_y = 960 \text{ N} \downarrow$, $F_{BC} = 2490 \text{ N} \searrow 45°$.

**3.86** $\theta = \tan^{-1}(Q/3P)$.

**3.88** (a) $\theta = 2 \sin^{-1}(W/2P)$. (b) $\theta = 29.0°$.

**3.90** (a) $\theta = 2 \sin^{-1}[\frac{1}{2}kl/(kl - P)]$. (b) $83.6°$.

**3.92** (a) $y \left(1 - \dfrac{a}{\sqrt{a^2 + y^2}}\right) = \dfrac{W}{k}$.
(b) $1000 \text{ N/m}$.

**3.94** (1) Completely constrained; determinate;
$A = 12.02 \text{ kips} \angle 56.3°$; $B = 6.67 \text{ kips} \leftarrow$.
(2) Improperly constrained; indeterminate; no equil.
(3) Completely constrained; determinate;
$A = C = 5 \text{ kips} \uparrow$. (4) Completely constrained;
indeterminate; $A_x = 6.67 \text{ kips} \rightarrow$; $B_x = 6.67 \text{ kips} \leftarrow$;
$(A_y + B_y = 10 \text{ kips} \uparrow)$.
(5) Improperly constrained; indeterminate; no equil.
(6) Partially constrained; determinate; equil.;
$A = C = 5 \text{ kips} \uparrow$. (7) Completely constrained;
determinate; $A = 5 \text{ kips} \uparrow$; $B = 8.33 \text{ kips} \searrow 36.9°$;
$C = 6.67 \text{ kips} \rightarrow$. (8) Completely constrained;
indeterminate; $A_y = 5 \text{ kips} \uparrow$.

**3.96** $A = 2230 \text{ N} \searrow 7.73°$; $B = 2210 \text{ N} \rightarrow$.

**3.98** $A_x = 1.58 \text{ kN} \leftarrow$, $A_y = 2.24 \text{ kN} \downarrow$, $F_{BC} = 3.17 \text{ kN} \angle 60°$.

**3.100** $F_{AB} = 3.93 \text{ kip} \searrow 50°$, $C_x = 2.52 \text{ kip} \rightarrow$, $C_y = 4.99 \text{ kip} \uparrow$.

**3.102** $A = 82.5 \text{ lb} \angle 14.04°$; $T = 100 \text{ lb.}$

**3.104** $A_x = 17.14 \text{ lb} \leftarrow$, $A_y = 17.14 \text{ lb} \downarrow$, $F_{BC} = 24.2 \text{ lb} \angle 48°$.

**3.106** $A = 170 \text{ N} \searrow 33.9°$; $C = 160 \text{ N} \angle 28.1°$.

**3.108** (a) $120 \text{ lb}$. (b) $C = 150 \text{ lb} \nearrow 36.9°$.

**3.110** $\theta + \alpha = 90°$.

**3.112** (a) $225 \text{ mm}$. (b) $23.1 \text{ N}$. (c) $12.21 \text{ N} \rightarrow$.

**3.114** $34.3°$.

**3.116** $560 \text{ lb} \cdot \text{in.} \downarrow$.

**3.118** (a) $500 \text{ N} \searrow 60°$, $97.2 \text{ N} \cdot \text{m} \downarrow$.
(b) $A = 1218 \text{ N} \searrow 60°$; $B = 1718 \text{ N} \searrow 60°$.

**3.120** $44.7 \text{ lb} \searrow 26.6°$; $10.61 \text{ in. to the left of } C$,
$5.30 \text{ in. below } C$.

**3.122** (a) $1.530 \text{ kN}$. (b) $A = 2.42 \text{ kN} \nearrow 21.8°$.

**3.124** (a) $T = P$. (b) $B = 2\sqrt{3} P \angle 60°$;
$C = \sqrt{3} P \nearrow 60°$.

**3.126** (a) $A = 0.745 P \searrow 63.4°$; $D = 0.471 P \angle 45°$;
(b) $A = P \rightarrow$; $D = 1.414 P \searrow 45°$.
(c) $A = 0.471 P \angle 45°$. $D = 0.745 P \searrow 63.4°$.
(d) $A = 0.707 P \angle 45°$; $C = 0.707 P \nearrow 45°$; $D = P \uparrow$.

**3.C2** $\beta = 30°$: $\mathbf{R} = 165.2$ N $\measuredangle$ 21.3°,
$\mathbf{M} = 50.9$ N · m $\downarrow$;

$\beta = 150°$: $\mathbf{R} = 80.7$ N $\measuredangle$ 48.1°,
$\mathbf{M} = 22.6$ N · m $\uparrow$.

**3.C3** (a) $\theta = 25°$: $\alpha = 40.0°$ $\measuredangle$; $\theta = 75°$: $\alpha = 60.0°$ $\measuredangle$.
(b) $\theta = 25°$: $\alpha = 22.0°$ $\measuredangle$; $\theta = 75°$: $\alpha = 67.4°$ $\measuredangle$.

**3.C4** $\theta = 25°$: $P = 10.77$ lb;
$\theta = 65°$: $P = 19.96$ lb.

## CHAPTER 4

**4.2** (40 lb) $M_{AB} = 160$ lb · in $\searrow$ $\downarrow M_{GF} = 400$ lb · in. $\nearrow$,
$M_{BC} = 0$, $M_{AC} = 148.6$ lb · in. $\measuredangle$ 21.8°
(50 lb) $M_{AD} = 0$, $M_{GF} = 0$, $M_{BC} = 0$, $M_{AC} = 0$
(80 lb) $M_{AD} = 0$, $M_{GF} = 4.80$ lb · in. $\nearrow$, $M_{BC} = 0$, $M_{AC} = 0$.

**4.4** $M_x = 4.50$ N · m, $M_y = -7.50$ N · m, $M_z = -12.99$ N · m.

**4.6** $M_x = -192.0$ N · m, $M_y = -81.0$ N · m,
$M_z = 135.0$ N · m.

**4.8** 4.88 kN.

**4.10** 4.24 ft.

**4.12** 25.3 N · m.

**4.16** (a) $M = 18.30$ N · m, $\theta_x = 37.9°$, $\theta_y = 52.1°$, $\theta_z = 90°$.
(b) 24.4 N $\measuredangle$ 52.1° at $B$, 24.4 N $\measuredangle$ 52.1° at $C$.

**4.18** $M = 3610$ lb · ft; $\theta_x = 159.4°$, $\theta_y = 90.8°$, $\theta_z = 69.4°$.

**4.20** $F_x = -150$ lb, $F_y = 60$ lb, $F_z = -100$ lb.
$M_x = 0$, $M_y = 1500$ lb · ft, $M_z = 900$ lb · ft.

**4.22** $F_x = 0$, $F_y = -122.9$ N, $F_z = -86.0$ N;
$M_x = 22.6$ N · m, $M_y = 15.49$ N · m,
$M_z = -22.1$ N · m.

**4.24** Force-couple systems at $A$ and $C$.

**4.26** (a) $B_x = B_y = 0$, $B_z = -80$ N; $C_z = -30$ N, $C_y = 0$,
$C_z = 40$ N. (b) $R_y = 0$, $R_z = -40$ N.
(c) Vertical.

**4.28** (a) 60°. (b) $R_x = 15.00$ lb, $R_y = -26.0$ lb,
$R_z = 0$; $M_x = 312$ lb · in., $M_y = M_z = 0$.

**4.30** $R_x = 15.00$ lb, $R_y = -26.0$ lb, $R_z = 0$; $M_z = 312$ lb · in.,
$M_y = 0$, $M_z = 89.7$ lb · in.
(a) Tighten. (b) Loosen.

**4.32** 75 kips at $x = +4.80$ ft and $z = -1.386$ ft.

**4.34** 140 kN; 0.714 m from $A$ on edge $AD$.

**4.36** (a) $R_x = 0$, $R_y = -60$ lb, $R_z = 0$.
(b) $-0.6$ in. (c) $x = 0$, $z = 9$ in.

**4.38** (a) $R_x = 0$, $R_y = -40$ N, $R_z = 0$. (b) 2.5 mm.
(c) $x = -5$ mm, $z = 0$.

**4.40** $A_x = 112$ N, $A_y = 81.3$ N, $A_z = 79.3$ N, $M_{AX} = 32.5$ N · m,
$M_{AY} = -22.4$ N · m, $F_{BC} = 179$ N

**4.42** $A_x = 0$, $A_y = 104$ N, $A_z = 128$ N; $D_y = 52$ N,
$D_z = 160$ N.

**4.44** (a) 750 N. (b) $C_x = 250$ N, $C_y = 750$ N, $C_z = 0$;
$D_x = -1000$ N, $D_y = -300$ N.

**4.46** $A_x = 37.5$ N, $A_y = 23.5$ N, $A_z = 6.5$ N, $M_{AX} = 17.30$
N · m, $M_{AY} = -6.50$ N · m, $T_{BC} = 101.6$ N.

**4.48** $W = 90$ lb at $x = 0.899$ ft, $Z = 2.44$ ft.

**4.50** (a) $T_{BD} = T_{BE} = 16.60$ kN.
(b) $A_x = 17.00$ kN, $A_y = 45.1$ kN, $A_z = 0$.

**4.52** (a) $T_{BD} = 700$ lb; $T_{BC} = 975$ lb.
(b) $A_x = 1500$ lb, $A_y = 425$ lb, $A_z = 0$.

**4.54** $A_x = 22.9$ lb, $A_y = -20$ lb, $A_z = 20$ lb, $M_{AY} = -15.7$ lb · in.,
$M_{AZ} = -91.5$ lb · in., $T_{BC} = 36.4$ lb.

**4.56** $A_x = 31.3$ lb, $A_y = 21.4$ lb, $A_z = 20$ lb, $B_x = -6.27$ lb,
$B_y = 7.71$ lb, $T_{CD} = 33.87$ lb.

**4.58** (a) 3.00 lb. (b) $A_x = -2.94$ lb, $A_y = 6.90$ lb; $B_x = 0$,
$B_y = 7.50$ lb, $B_z = 0$.

**4.60** (a) 689 N. (b) $A_x = 503$ N, $A_y = 424$ N;
$B_x = 150.9$ N, $B_y = 557$ N, $B_z = -229$ N.

**4.62** (a) 1548 N. (b) $A_x = 503$ N, $A_y = 1094$ N;
$B_x = 150.9$ N, $B_y = -113.2$ N, $B_z = -1003$ N.

**4.64** (a) 3.00 lb. (b) $B_x = -2.94$ lb, $B_y = 14.40$ lb,
$B_z = 0$; $(M_B)_x = -20.7$ lb · ft, $(M_B)_y = -8.82$ lb · ft.

**4.66** (a) 689 N. (b) $B_x = 654$ N, $B_y = 981$ N,
$B_z = -229$ N; $(M_B)_x = -883$ N · m,
$(M_B)_y = 1046$ N · m.

**4.68** (a) 12.50 lb; (b) $A_x = 0$, $A_y = 6.00$ lb, $A_z = 4.50$ lb;
$B_x = 10.00$ lb, $B_y = B_z = 0$.

**4.70** (a) $D = 225$ N. (b) $S = 300$ N; $d = 0.75$ m.

**4.72** $B_x = 30.5$ lb, $B_y = 0$, $B_z = -45.0$ lb;
$C_x = 0$, $C_y = 0.5$ lb, $C_z = 45.0$ lb;
$D_x = -30.5$ lb, $D_y = 6.00$ lb, $D_z = 0$.

**4.74** $T_{BD} = 2.90$ kN; $T_{BE} = 5.25$ kN; $T_{CD} = 2.00$ kN.

**4.76** $M_x = 31.1$ kN · m, $M_y = 5.18$ kN · m,
$M_z = -6.90$ kN · m.

**4.78** $A_x = -9600$ lb, $A_z = -2400$ lb; $B_x = 9600$ lb,
$B_z = 2400$ lb; $A_y$ and $B_y$ indeterminate
$(A_y + B_y = 12,000$ lb).

**4.80** $A_x = 0$, $A_y = 375$ N, $A_z = -500$ N;
$B_x = 625$ N, $B_y = 0$, $B_z = 500$ N;
$C_x = -625$ N, $C_y = 375$ N, $C_z = 0$.

**4.82** 60.1 kN $\downarrow$, at $x = 4.99$ m, $z = 3.33$ m.

**4.84** $T_{BD} = T_{BE} = 2200$ lb; $A_x = -720$ lb,
$A_y = 2800$ lb, $A_z = 0$.

**4.86** $T_{EBF} = 35.4$ kN, $T_{CD} = 24.6$ kN.

**4.C1** (b) $F_x = 0$, $F_y = -2.4$ kips, $F_z = -1$ kip;
$M_x = -12$ kip · in., $M_y = 6$ kip · in.,
$M_z = -14.4$ kip · in. (c) $F_x = 420$ lb,
$F_y = -105$ lb, $F_z = 140$ lb; $M_x = 1680$ lb · in.,
$M_y = 3360$ lb · in., $M_z = -2520$ lb · in.

**4.C2** (b) 500 kN at 2.56 m from $AD$ and 2.00 m from $CD$.

**4.C3** $\phi = 12°$: $T_{BD} = 11.63$ kN, $T_{BE} = 20.8$ kN;
$\phi = 24°$: $T_{BD} = 6.16$ kN, $T_{BE} = 24.2$ kN.

**4.C4** $\alpha = 30°$: 250 lb; $A_x = 0$, $A_y = -130.4$ lb,
$A_z = -178.6$ lb; $B_y = 45$ lb, $B_z = 0$.
$\alpha = 75°$: 54.0 lb; $A_x = 0$, $A_y = -7.41$ lb,
$A_z = 12.95$ lb; $B_y = 45$ lb, $B_z = 0$.

# CHAPTER 5

**5.2** $\overline{X} = 1.654$ in., $\overline{Y} = 1.615$ in.
**5.4** $\overline{X} = 2.409$ in., $\overline{Y} = 5$ in.
**5.6** $\overline{X} = 16.21$ mm, $\overline{Y} = 31.9$ mm.
**5.8** $\overline{X} = \overline{Y} = 5.06$ in.
**5.10** $\overline{X} = 2.114$ ft, $\overline{Y} = 2.141$ ft.
**5.12** $\overline{X} = 19.44$ mm, $\overline{Y} = 23.6$ mm.
**5.14** $\overline{X} = 321$ mm, $\overline{Y} = 53.1$ mm.
**5.16** $\overline{X} = \frac{1}{3}a(h_1 + 2h_2)/(h_1 + h_2)$.
**5.18** $\overline{X} = -0.125$ in., $\overline{Y} = 5.54$ in.
**5.20** $\overline{X} = 3.33$ in., $\overline{Y} = 3.33$ in.
**5.22** $42.25 \times 10^3$ mm$^3$ and $-42.25 \times 10^3$ mm$^3$.
**5.24** $\overline{X} = 17.98$ mm, $\overline{Y} = 18.77$ mm.
**5.26** $\overline{X} = 17.78$ mm, $\overline{Y} = 29.76$ mm.
**5.28** $A_x = 0$, $A_y = 11.71$ lb $\uparrow$, $B_y = 13.36$ lb $\uparrow$.
**5.30** 120 mm.
**5.32** (a) $\sqrt{3}\,r$. (b) $\sqrt{5}\,r$.
**5.34** $A_x = 0$, $A_y = 42.4$ lb $\uparrow$, $B_y = 26.5$ lb $\uparrow$.
**5.36** 15.3 mm.
**5.38** $\overline{x} = 0.7424a$; $-1.01$ percent.
**5.40** $\overline{x} = 2.25$ in., $\overline{y} = 1.60$ in.
**5.42** $\overline{x} = 3a/5$, $\overline{y} = 12b/35$.
**5.50** $\overline{y} = 0.48\,h$.
**5.52** $\overline{x} = L/\pi$, $\overline{y} = \pi a/8$.
**5.54** $\overline{x} = h$, $\overline{y} = \frac{1}{4}h$.
**5.56** (a) 584 in$^3$. (b) 679 in$^3$.
**5.58** $A = 4\pi^4 rR$; $V = 2\pi^2 r^2 R$.
**5.60** 494 in$^3$; 11.44 lb.
**5.62** 6.03 m$^2$.
**5.64** $21.3 \times 10^3$ mm$^3$, 153.6 g.
**5.66** 14.52 in$^2$.
**5.68** (a) 158 ft$^3$. (b) 146 ft$^2$.
**5.70** (a) $\pi R^2 h$. (b) $\frac{2}{3}\pi R^2 h$. (c) $\frac{1}{2}\pi R^2 h$. (d) $\frac{1}{3}\pi R^2 h$.
(e) $\frac{1}{6}\pi R^2 h$.
**5.72** $R = 3.6$ kN, 1.28 m to the right of $A$.
$A_x = 0$, $A_y = 1.68$ kN $\uparrow$, $B_y = 1.92$ kN $\uparrow$.
**5.74** $A_x = 0$, $A_y = 350$ lb $\uparrow$, $M_A = 3250$ lb $\cdot$ in. $\uparrow$.
**5.76** $A_x = 0$, $A_y = 4.46$ kN $\uparrow$, $B_y = 13.54$ kN $\uparrow$.
**5.78** $A_x = 0$, $A_y = 120$ lb $\uparrow$, $M_A = 336$ lb $\cdot$ ft $\downarrow$.
**5.80** $\mathbf{B} = 3770$ lb $\uparrow$; $\mathbf{C} = 429$ lb $\uparrow$.
**5.82** $w_A = 10$ kN/m; $w_B = 50$ kN/m.
**5.84** $\mathbf{R}_W = [-44.2, -14.7]$ kN
$\mathbf{R}_G = [44.2, 144.2]$ kN

**5.86** (a) 36.4 kN, 0.864 m below $A$. (b) 21.0 kN $\leftarrow$.
**5.88** $d = 20$ in.
**5.90** $\mathbf{F}_R = [3.74, -3.18]$ kip
**5.92** $\mathbf{F}_R = [4.12, 5.49]$ kip
**5.94** (a) $\mathbf{F}_v = \mathbf{W} = 38.5$ MN down (b) $\mathbf{F}_v = \mathbf{F}_R = 38.5$ MN
down (c) Both methods produce identical results on a
horizontal surface.
**5.96** 4.87 ft.
**5.98** $\mathbf{A} = 153.0$ N $\uparrow$; $\mathbf{B}_x = 176.6$ N $\leftarrow$, $\mathbf{B}_y = 789$ N $\uparrow$.
**5.100** $\frac{1}{8}b$ to the right of base of cone.
**5.102** 27.8 mm above base.
**5.104** $\overline{X} = \overline{Z} = 0$, $\overline{Y} = -0.608h$.
**5.106** $\overline{Y} = -19.02$ mm.
**5.108** $\overline{Z} = 3.47$ in.
**5.110** $\overline{Z} = 51.5$ mm.
**5.112** $\overline{X} = 125$ mm, $\overline{Y} = 167.0$ mm, $\overline{Z} = 33.5$ mm.
**5.114** $\overline{X} = \overline{Z} = 3.62$ in., $\overline{Y} = 4.83$ in.
**5.116** $\overline{X} = 0$, $\overline{Y} = 5.02$ in., $\overline{Z} = 2.57$ in.
**5.118** 33.5 mm above base.
**5.124** $\overline{x} = 5h/8$.
**5.126** $\overline{x} = \frac{1}{2}a(1 - 4/\pi^2)$.
**5.128** $\overline{x} = 3a/8$, $\overline{y} = h/4$, $\overline{z} = 3b/8$.
**5.130** 42.5 mm.
**5.132** $\overline{x} = 0$, $\overline{y} = 5h/16$, $\overline{z} = -a/4$.
**5.134** $\overline{X} = 58.3$ mm, $\overline{Y} = 83.5$ mm.
**5.136** (a) $\overline{X} = 12$ in., $\overline{Y} = -4.67$ in., $\overline{Z} = 0$. (b) 1.807.
**5.138** $\overline{x} = 90$ mm, $\overline{y} = 15$ mm, $\overline{z} = 60$ mm.
**5.140** (a) $+155.5$ m$^3$. (b) $+83.3$ m$^3$.
**5.142** $\overline{X} = 2.53$ in., $\overline{Y} = 0.615$ in., $\overline{Z} = 2$ in.
**5.144** $\cos \alpha = \frac{1}{2}\csc \theta$.
**5.C1** (b) $\overline{X} = 51.25$ mm, $\overline{Y} = 11.04$ mm.
(c) $\overline{X} = 9.57$ in., $\overline{Y} = 2.38$ in.
**5.C2** (a) 341 in$^2$. (b) 615 in$^2$. (c) 1354 in$^2$.
**5.C3** $d = 0.3$ m: ins. $\mathbf{A} = 10.94$ kN $\nearrow$ 71.2°,
$\mathbf{B} = 10.36$ kN $\uparrow$; $d = 2.1$ m: $\mathbf{A} = 105.5$ kN $\nearrow$ 65.8°,
$\mathbf{B} = 103.2$ kN $\uparrow$.
**5.C4** (a) $\overline{X} = \overline{Z} = 186.7$ mm, $\overline{Y} = 76.9$ mm.
(b) $\overline{X} = \overline{Z} = 147.7$ mm, $\overline{Y} = 217$ mm.
(c) $\overline{X} = 135.4$ mm, $\overline{Y} = 141.0$ mm, $\overline{Z} = 163.4$ mm.

# CHAPTER 6

**6.2** $F_{AB} = 261.8$ lb $T$
$F_{BC} = 78.9$ lb $C$
$F_{AC} = 39.4$ lb $T$
**6.4** $F_{AB} = 1000$ lb $T$
$F_{AC} = 800$ lb $C$
$F_{BC} = 600$ lb $C$
$F_{BE} = 800$ lb $T$
$F_{CE} = 1500$ lb $T$

**6.6** $F_{AE} = F_{CD} = 2.31$ kN $C$
$F_{AB} = F_{BC} = 1.155$ kN $T$
$F_{BE} = F_{BD} = 2.31$ kN $T$
$F_{ED} = 2.31$ kN $C$

**6.8** $F_{BC} = 1.6$ kN $T$
$F_{CD} = 1.2$ kN $T$
$F_{AB} = 1.27$ kN $T$
$F_{AD} = 0.40$ kN $C$
$F_{BD} = 1.70$ kN $C$

**6.10** $F_{AB} = 420$ lb $C$; $F_{AC} = F_{CE} = 400$ lb $T$;
$F_{AD} = 260$ lb $C$; $F_{BC} = F_{CD} = 125$ lb $T$;
$F_{BE} = 832$ lb $C$.

**6.12** $F_{AB} = 10$ kN $C$;
$F_{AC} = F_{BC} = F_{CD} = F_{CE} = 11.55$ kN $T$;
$F_{AD} = F_{BE} = 5.77$ kN $C$.

**6.14** Non-zero-force: UV, VB, UB, BC, UC, UL, LC, CH, LH
$F_{BV} = 0.894$ kip $C$
$F_{UV} = 0.447$ kip $T$
$F_{BU} = 0.378$ kip $C$
$F_{BC} = 0.133$ kip $C$
$F_{CU} = 0.067$ kip $T$
$F_{UL} = 0.133$ kip $T$

**6.16** Non-zero-force: BK, BC, CD, JK, DK, DJ, DI, DE, IJ,
EF, FI
$F_{BK} = 8.49$ kN $C$
$F_{JK} = 2.98$ kN $C$
$F_{DK} = 4.71$ kN $C$

**6.18** (a) $A_x = -112.5$ lb, $A_z = -60$ lb; $B_x = 0$,
$B_z = 60$ lb; $C_x = 112.5$ lb, $C_y = 170$ lb.
(b) $F_{AB} = F_{AD} = F_{BC} = 0$; $F_{AC} = 170$ lb $C$;
$F_{AE} = 212.5$ lb $T$; $F_{BD} = 127.5$ lb $T$;
$F_{BE} = F_{CD} = 112.5$ lb $C$; $F_{DE} = 60$ lb $C$.

**6.20** (a) $B_y = 0$, $B_z = -2700$ N; $C_x = -1800$ N,
$C_y = -3375$ N; $D_x = 1800$ N, $D_y = 3375$ N.
(b) $F_{AB} = F_{CD} = 0$; $F_{AC} = 4275$ N $T$;
$F_{AD} = 4275$ N $C$; $F_{BC} = 4270$ N $C$;
$F_{BD} = 4270$ N $T$.

**6.22** $F_{FG} = 5$ kN $T$; $F_{FH} = 20$ kN $T$.

**6.24** $F_{CD} = 4.58$ kip $T$
$F_{CG} = 1.08$ kip $T$

**6.26** (a) neutral axix. (b) 2.00.

**6.28** $F_{FG} = 2.98$ kN $C$
$F_{CF} = 0.943$ kN $T$

**6.30** 1.672 in.

**6.32** $F_{DE} = 6.0$ kN $C$
$F_{IJ} = 6.67$ kN $T$
$F_{DI} = 2.68$ kN $C$

**6.34** $F_{JK} = 7.5$ kN $C$
$F_{IG} = 7.5$ kN $T$
$F_{IJ} = 3$ kN $C$

**6.36** Zero-force: FH, FJ, EJ, EK, BM
$F_{LM} = 13$ kN $C$
$F_{BC} = 11.2$ kN $T$
$F_{BL} = 5.39$ kN $C$

**6.38** $F_{AC} = 11.25$ kN $T$; $F_{BE} = 11.25$ kN $C$.

**6.40** $F_{AE} = 142.5$ kN $C$; $F_{DE} = 170$ kN $T$.

**6.42** $F_{DE} = 0$; $F_{BE} = 10$ kips $T$; $F_{EF} = 5$ kips $T$.

**6.44** $F_{DE} = 1.5$ kips $C$; $F_{BE} = 2.5$ kips $T$;
$F_{DG} = 2.5$ kips $T$.

**6.46** (a) Completely constrained; determinate.
(b) Completely constrained; indeterminate.
(c) Improperly constrained.

**6.48** (a) Completely constrained; determinate.
(b) Improperly constrained.
(c) Completely constrained; indeterminate.

**6.50** $F_{BD} = 300$ N $C$; $C_x = 164$ N $\leftarrow$, $C_y = 288$ N $\downarrow$.

**6.52** $A_x = 900$ lb $\leftarrow$, $A_y = 75$ lb $\uparrow$; $B = 825$ lb $\downarrow$;
$D_x = 900$ lb $\rightarrow$, $D_y = 750$ lb $\uparrow$.

**6.54** (a) 2.08. (b) 2.10

**6.56** $N_B = 120$ ↗ on EBD
$M_A = 2.1$ kN-m ↗
$A_x = 304$ N $\leftarrow$
$A_y = 72$ N $\downarrow$

**6.58** $B = 82$ lb $\downarrow$; $C_x = 75$ lb $\leftarrow$, $C_y = 115$ lb $\uparrow$;
$D_x = 75$ lb $\rightarrow$, $D_y = 12$ lb $\uparrow$.

**6.60** $F_x = 0$; $F_y = 430$ N
$A_y = 170$ N
$B_y = 300$ N
$E_y = 300$ N
$E_x = 560$ N
$D_x = 40$ N
$D_y = 130$ N $\downarrow$

**6.62** (a) $A = 65$ lb ↗ 22.6°; $C = 120$ lb $\rightarrow$;
$G = 60$ lb $\leftarrow$; $I = 25$ lb $\uparrow$.
(b) $A = 65$ lb ↗ 22.6°; $C = 100$ lb $\rightarrow$;
$G = 80$ lb $\leftarrow$; $I = 25$ lb $\uparrow$.

**6.64** (a) $D_x = 750$ N $\rightarrow$, $D_y = 250$ N $\downarrow$;
$E_x = 750$ N $\leftarrow$, $E_y = 250$ N $\uparrow$.
(b) $D_x = 375$ N $\rightarrow$, $D_y = 250$ N $\downarrow$;
$E_x = 375$ N $\leftarrow$, $E_y = 250$ N $\uparrow$.

**6.66** (a) $A'$ and $B'$. (b) 268 N $T$.

**6.68** (a) 176.9 lb $T$.
(b) $C = 262$ lb ∡ 85.0°.

**6.70** $D_x = 13.60$ kN $\rightarrow$, $D_y = 7.50$ kN $\uparrow$;
$E_x = 13.60$ kN $\leftarrow$,
$E_y = 2.70$ kN $\downarrow$.

**6.72** (a) $E_x = 3$ kips $\leftarrow$, $E_y = 1.5$ kips $\uparrow$.
(b) $C_x = 3$ kips $\leftarrow$, $C_y = 6.5$ kips $\uparrow$.

**6.74** (a) At each wheel; $A = 2.5$ kips $\uparrow$;
$B = 37.5$ kips $\uparrow$. (b) $C = 6$ kips $\leftarrow$;
$D_x = 6$ kips $\rightarrow$, $D_y = 15$ kips $\downarrow$.

**6.76**   (*a*) 2125 N.   (*b*) *At each wheel:* $\mathbf{A}$ = 4740 N ↑;
      $\mathbf{B}$ = 2730 N ↑; $\mathbf{C}$ = 3570 N ↑.

**6.78**   $\mathbf{A}_x$ = 1110 lb ←, $\mathbf{A}_y$ = 600 lb ↑;
      $\mathbf{B}_x$ = 1110 lb ←, $\mathbf{B}_y$ = 800 lb ↓;
      $\mathbf{D}_x$ = 2220 lb →, $\mathbf{D}_y$ = 200 lb ↑,

**6.80**   $\mathbf{A}_x$ = 4 kN ←, $\mathbf{A}_y$ = 2 kN ↑; $\mathbf{B}_x$ = 6 kN ←,
      $\mathbf{B}_y$ = 1 kN ↑; $\mathbf{D}_x$ = 10 kN →, $\mathbf{D}_y$ = 3 kN ↓.

**6.82**   $F_{AF} = \frac{1}{2}P$ comp.; $F_{BG} = P/\sqrt{2}$ ten.; $F_{GD} = \sqrt{2}\,P$ ten.;
      $F_{EH} = \frac{1}{2}P$ comp.

**6.84**   $F_{CF}$ = 7.88 kN C; $F_{DG}$ = 5.25 kN T.

**6.86**   $F_{BG}$ = 5.25 kN T; $F_{CH}$ = 2.63 kN C.

**6.88**   $\mathbf{A}$ = 40 lb ↑; $\mathbf{C}_x$ = 360 lb →, $\mathbf{C}_y$ = 40 lb ↓;
      $\mathbf{D}$ = 360 lb ←.

**6.90**   (*a*) $\mathbf{C}_x$ = 90 kips ←, $\mathbf{C}_y$ = 84 kips ↑.
      (*b*) $\mathbf{B}_x$ = 90 kips ←, $\mathbf{B}_y$ = 6 kips ↓.

**6.92**   $\mathbf{A}$ = $P/15$ ↑; $\mathbf{D}$ = $2P/15$ ↑; $\mathbf{E}$ = $8P/15$ ↑; $\mathbf{H}$ = $4P/15$ ↑.

**6.94**   (*a*) Rigid; $\mathbf{A}$ = 2.24 $P$ ⦦ 26.6°; $\mathbf{B}$ = $2P$ →.
      (*b*) and (*c*) Not rigid.

**6.96**   $a \geq 0.6$ m.

**6.98**   (*a*) 96 lb ↓.   (*b*) 100 lb ⦣ 73.7°.

**6.100**   (*a*) 2020 N ↓.   (*b*) 1875 N ⬈ 66.2°.

**6.102**   $\mathbf{D}$ = 9360 N ↓. $\mathbf{F}$ = 6770 N ⬊ 15.1°.

**6.104**   $\mathbf{C}$ = 17.36 kips →; $\mathbf{E}$ = 10.18 kips ⬈ 23.1°.

**6.106**   $\mathbf{A}_x$ = 180 N ←, $\mathbf{A}_y$ = 2250 N ↓;
       $\mathbf{B}$ = 2550 N ⦤ 61.9°; $\mathbf{C}$ = 1020 N ←.

**6.108**   (*a*) 2.14 kN ↓.   (*b*) 5.00 kN ↓.

**6.110**   (*a*) $\mathbf{M}_C$ = 920 lb · in. ↑.   (*b*) $\mathbf{C}_x$ = 30 lb ←,
       $\mathbf{C}_y$ = 40 lb ↑.

**6.112**   (*a*) 3 kN.   (*b*) 3.35 kN ⦤ 26.6°.

**6.114**   60 N.

**6.116**   29.4 lb.

**6.118**   200 N.

**6.120**   35.7 lb ↓.

**6.122**   (*a*) Zero.
       (*b*) $\mathbf{C}_x$ = 5000 lb ←, $\mathbf{C}_y$ = 1200 lb ↓.

**6.124**   $F_{AB}$ = 1.484 kN C; $F_{DE}$ = 31.3 kN T;
       $F_{FI}$ = 4.55 kN C.

**6.126**   (*a*) 420 lb · in.   (*b*) $A$ = 0, $M_A$ = −288 lb · in.;
       $B$ = 0, $M_B$ = −432 lb · in.; axes of couples are parallel
       to the *x* axes.

**6.128**   (*a*) 0.8.   (*b*) $2.6M_A$ ↙.

**6.130**   (*a*) $M_G$ = 26.6 N · m.   (*b*) $B_x$ = 113.7 N,
       $B_y = B_x$ = 0; $C_x$ = −113.7 N, $C_y = C_z$ = 0; $E$ = 0.

**6.132**   4660 lb.

**6.134**   (*a*) 126.0 N · m ↑.   (*b*) 116.9 N · m ↑.

**6.136**   (*a*) 3750 lb T.   (*b*) $\mathbf{H}_x$ = 3000 lb ←; $\mathbf{H}_y$ = 6000 lb ↓.

**6.138**   8.45 kN.

**6.140**   (*a*) $\mathbf{A}$ = 1.8 kips ↑, $\mathbf{M}_A$ = 14.4 kip · ft ↑;
       $\mathbf{D}$ = 1.2 kips ↑.
       (*b*) $\mathbf{A}$ = 0, $\mathbf{M}_A$ = 18 kip · ft ↙; $\mathbf{D}$ = 3 kips ↑.

**6.142**   $\mathbf{A}$ = 666 N ⬐ 63.2°; $\mathbf{F}$ = 325 N ⦤ 22.6°.

**6.144**   $F_{CE}$ = 6 kips T; $F_{DE}$ = 20 kips T; $F_{DF}$ = 18 kips C.

**6.C1**   $\theta = 15°$: $F_{AB}$ = 966 lb T, $F_{AC}$ = 905 lb C,
       $F_{BC}$ = 835 lb T. $\theta = 65°$: $F_{AB}$ = 720 lb T,
       $F_{AC}$ = 860 lb C, $F_{BC}$ = 794 lb T.

**6.C2**   $h$ = 6 m: $F_{EH}$ = 17.50 kN C, $F_{GH}$ = 1.601 kN C,
       $F_{GI}$ = 17.50 kN T.

**6.C3**   (*a*) $a$ = 0.1 m: $F_{CF}$ = 19.48 kN T, $F_{DG}$ = 10.50 kN T.
       $a$ = 0.2 m: $F_{CF}$ = 21.13 kN C, $F_{DG}$ = 13.12 kN T.
       (*b*) Improper constraints.

**6.C4**   (*a*) $\theta = 60°$: $M$ = 149.1 N · m ↑;
       $\theta = 140°$: $M$ = 74.9 N · m ↑.
       (*b*) 75.1°, $M_{max}$ = 156.0 N · m ↑.

## CHAPTER 7

**7.2**   (On *JD*) $\mathbf{F}$ = 1106 lb ⬈ 20.6°; $\mathbf{V}$ = 386 lb ⬊ 69.4°;
     $\mathbf{M}$ = 1650 lb · ft ↓.

**7.4**   (On *JD*) $\mathbf{F}$ = 0; $\mathbf{V}$ = 100 N ↑; $\mathbf{M}$ = 15 N · m ↑.

**7.6**   (On *KCD*) $\mathbf{F}$ = 750 N ↓; $\mathbf{V}$ = 400 N →;
     $\mathbf{M}$ = 170 N · m ↙.

**7.8**   (On *CD*) $\mathbf{F}$ = 270 N →; $\mathbf{V}$ = 90 N ↑;
     $\mathbf{M}$ = 43.2 N · m ↙.

**7.10**   (On *JCD*) $\mathbf{F}$ = 750 lb ←, $\mathbf{V}$ = 0; $\mathbf{M}$ = 6 kip · in. ↙.

**7.12**   (*a*) (On *AC*) $\mathbf{F}$ = 520 N ←;
      $\mathbf{V}$ = 540 N ↑; $\mathbf{M}$ = 0.
      (*b*) (On *AJ*) $\mathbf{F}$ = 520 N ←;
      $\mathbf{V}$ = 540 N ↑; $\mathbf{M}$ = 54 N · m ↑.

**7.14**   (On *AJ*) $\mathbf{F}$ = 1040 lb ⬐ 21.8°; $\mathbf{V}$ = 92.8 lb ⬈ 68.2°;
      $\mathbf{M}$ = 2000 lb · in. ↙.

**7.16**   (On *JC*) $\mathbf{M}$ = 4.28 N · m ↙.

**7.18**   (On *JC*) $\mathbf{F}$ = $(\frac{1}{2}W - W\theta/\pi)\cos\theta$ ⬊;
      $\mathbf{V}$ = $(\frac{1}{2}W - W\theta/\pi)\sin\theta$ ⬋;
      $\mathbf{M}$ = $\frac{1}{2}Wr(1 - \cos\theta) - (Wr/\pi)(\sin\theta - \theta\cos\theta)$ ↙.

**7.20**   0.1786 $W$; $\theta$ = 49.3°.

**7.22**   $M$ = $+\frac{1}{8}wL^2$ at center of the beam.

**7.24**   $V$ = 0, everywhere; $M$ = $+T$ between $B$ and $C$;
      $M$ = 0, eleswhere.

**7.26**   $V_C$ = $-2P$; $M_C$ = $-3Pa$.

**7.28**   $M_D$ = $+41.6$ kip · ft.

**7.30**   $M_C$ = $+180$ N · m.

**7.32**   0.4 m.

**7.34**   $M$ = $-20$ kN · m at center.

**7.36**   $M_C$ = $-6$ kip · ft.

**7.38**   *Just to the right of C:* $V$ = $+120$ N; $M$ = $+24$ N · m.

**7.40**   $M$ = $+400$ lb · in. to the left of $F$.

**7.42**   $M$ = $+4.8$ N · m in portion *CD*.

**7.44**   (*a*) $\frac{1}{3}P$.   (*b*) $\frac{1}{4}P$.

**7.58**   $M_D$ = $+1230$ lb · in.

**7.60**   $M_C$ = $+14$ kN · m.

**7.62**   $M$ = $+5.76$ kN · m. 2.4 m from $A$.

**7.64**   $M$ = $+3125$ lb · ft, 5 ft from $A$.

**7.66** $M = +44.7$ kN $\cdot$ m, 3.28 in from $A$.

**7.68** $V = w_0(L/\pi) \cos(\pi x/L)$; $M = w_0(L/\pi)^2 \sin(\pi x/L)$;
$M_{max} = w_0(L/\pi)^2$, at $x = \frac{1}{2}L$.

**7.70** $V = \frac{1}{2}w_0 L[\frac{1}{3} - (x/L)^2]$; $M = \frac{1}{6}w_0 L^2[(x/L) - (x/L)^3]$;
$M_{max} = +0.0642w_0 L^2$, at $x = 0.577L$.

**7.72** $\mathbf{P} = 23$ kips $\downarrow$, $\mathbf{Q} = 79$ kips $\downarrow$; $V_D = +24$ kips,
$M_F = +252$ kip $\cdot$ ft.

**7.74** $\mathbf{P} = 250$ N $\downarrow$, $\mathbf{Q} = 350$ N $\downarrow$; $V_D = +1.8$ kN,
$M_C = -2.88$ kN $\cdot$ m.

**7.76** $L/6$.

**7.78** 2 m.

**7.80** (a) $\mathbf{E}_x = 25$ kips $\rightarrow$, $\mathbf{E}_y = 12$ kips $\uparrow$.
(b) 27.7 kips.

**7.82** (a) 9 m.   (b) $\mathbf{E}_x = 5$ kN $\rightarrow$, $\mathbf{E}_y = 3$ kN $\uparrow$.

**7.84** $a = 2.55$ m; $b = 5.95$ m.

**7.86** (a) $\mathbf{D}_x = 280$ lb $\rightarrow$, $\mathbf{D}_y = 210$ lb $\uparrow$.
(b) 1170 lb.
(c) $T_{AB} = 1200$ lb, $T_{BC} = 750$ lb, $T_{CD} = 350$ lb.

**7.88** (a) 66,900 kips.   (b) 4353 ft.

**7.90** (a) 4.74 m.   (b) 1760 N.

**7.92** (a) 125 mm.   (b) $-2.9°$.

**7.94** (a) 750 mm from $A$.   (b) 8.57 N.

**7.96** 15.07 kips; 17.79 kips.

**7.102** 49.86 m.

**7.104** (a) 9.89 ft.   (b) 15.38 lb.

**7.106** 15.02 m.   (b) 4.42 kg/m.

**7.108** 80.4 ft; 92.5 lb.

**7.110** 1.822 in.

**7.112** 24.3 lb.

**7.114** (a) $L = 1.326 \, T_m/w$.   (b) 8.04 mi.

**7.116** 0.1408.

**7.118** (a) 138.1 m.   (b) 649 N.

**7.120** $M_C = +1800$ lb $\cdot$ ft; $M_E = -2400$ lb $\cdot$ ft.

**7.122** (a) 2.6 m.   (b) 6.71 kN.

**7.124** (On $FB$) (a) $\mathbf{V} = 0$; $\mathbf{M} = 0.8$ kN $\cdot$ m $\downarrow$.
(b) $\mathbf{V} = 0$; $\mathbf{M} = 0.4$ kN $\cdot$ m $\downarrow$.
(c) $\mathbf{V} = 2$ kN $\downarrow$; $\mathbf{M} = 0.6$ kN $\cdot$ m $\downarrow$.
(d) $\mathbf{V} = 2$ kN $\uparrow$; $\mathbf{M} = 1.4$ kN $\cdot$ m $\downarrow$.

**7.126** (c) $M_D = M_E = +0.8$ kN $\cdot$ m.
(d) $M_D = +0.8$ kN $\cdot$ m; $M_E = +1.2$ kN $\cdot$ m.

**7.128** $M_C = +41$ lb $\cdot$ in.

**7.C1** (a) $V_C = -40$ kN; $M_C = +120$ kN $\cdot$ m.
(b) $V_C = +10$ kN; $M_C = +60$ kN $\cdot$ m.
(c) $V_C = +18$ kN; $M_C = +48$ kN $\cdot$ m.

**7.C2** (c) $V_D = -2.2$ kN; $M_D = +2.2$ kN $\cdot$ m.

**7.C3** (a) $M_B = +3.4$ kN $\cdot$ m; $M_C = 4$ kN $\cdot$ m.

**7.C4** *Prob. 7.77:* $\mathbf{E}_x = 8$ kN $\rightarrow$, $\mathbf{E}_y = 5$ kN $\uparrow$.
*Prob. 7.79:* $\mathbf{E}_x = 20$ kips $\rightarrow$. $\mathbf{E}_y = 11$ kips $\uparrow$.
*Prob. 7.81:* $\mathbf{E}_x = 7.5$ kN $\rightarrow$, $\mathbf{E}_y = 2.25$ kN $\uparrow$.

## CHAPTER 8

**8.2** Block moves up; $\mathbf{F} = 151.7$ N $\nearrow$ 20°.

**8.4** Equilibrium; $\mathbf{F} = 23.5$ lb $\searrow$ 30°.

**8.6** (a) 462 N.   (b) 447 N.   (c) 276 N.

**8.8** (a) 35.9 lb $\measuredangle$ 36.7°.   (b) 3.45 lb $\measuredangle$ 3.30°.

**8.10** 51.8°.

**8.12** (a) Does not move; 800 N $\leftarrow$.
(b) Moves; 491 N $\leftarrow$.

**8.14** All packages move; $\mathbf{F}_A = 7.58$ N $\nearrow$;
$\mathbf{F}_B = 3.03$ N $\nearrow$; $\mathbf{F}_C = 7.58$ N $\nearrow$.

**8.16** 31.0°.

**8.18** (a) 176.7 N.   (b) 147.3 N.   (c) 63.1 N.

**8.20** (a) 66.4°.   (b) 104.3 lb.

**8.22** (a) 468 lb $\rightarrow$.   (b) 612 lb $\rightarrow$.

**8.24** $M = Wr\mu_s(1 + \mu_s)/(1 + \mu_s^2)$.

**8.26** (a) 30.2°.   (b) 0.232 W.

**8.28** (a) 71.9 mm.   (b) 39.9 mm.

**8.30** (a) 38.6°.   (b) 0.330 W.

**8.32** $4.84a \leq L \leq 29.9a$.

**8.34** $5.94a \leq L \leq 12.24a$.

**8.36** Rod not in equilibrium.

**8.38** Rod in equilibrium; $\mathbf{F}_B = 18.39$ N $\uparrow$.

**8.40** 2200 N $\downarrow$.

**8.42** 18.75 in.

**8.44** 0.0533.

**8.46** (a) $11.6° \leq \theta \leq 78.4°$.   (b) $0 \leq \theta \leq 63.4°$.

**8.48** $27.4$ N $\leq Q \leq 51.8$ N.

**8.50** $20.6$ lb $\leq Q \leq 46.0$ lb.

**8.52** 200 N.

**8.54** $190.9$ lb $\leq P \leq 315$ lb.

**8.56** (a) 0.592.   (b) At $B$.

**8.58** (a) 0.479 m.   (b) Plank does not move.

**8.60** (a) 283 N $\leftarrow$.
(b) $\mathbf{B}_x = 413$ N $\leftarrow$, $\mathbf{B}_y = 480$ N $\downarrow$.

**8.62** 33.4°; 55.0 lb.

**8.64** 1430 N $\downarrow$.

**8.66** (a) 12,040 lb $\rightarrow$.   (b) 4500 lb $\leftarrow$.

**8.68** (a) 246 N.   (b) 246 N.

**8.70** 3.05 N.

**8.72** (a) 197.0 lb $\rightarrow$.   (b) Machine will not move.

**8.74** 14.97 lb $\cdot$ ft.

**8.76** (a) 11.71 kN $\cdot$ m.   (b) 2.58 kN $\cdot$ m.

**8.78** 19.64 kN.

**8.80** (a) Screw $A$.   (b) 9.22 lb $\cdot$ in.

**8.82** 14.47 N $\cdot$ m.

**8.84** 21.3 kN.

**8.86** (a) 72.8 lb.   (b) 62.8 lb.

**8.88** 3.55°.

**8.90** 0.367.

**8.92** (a) 920 N. (b) 690 N.

**8.94** 0.185.

**8.96** 208 lb · in.

**8.102** 0.060 in.

**8.104** (a) 1.288 kN. (b) 1.058 kN.

**8.106** (a) 0.344. (b) 2.55 turns.

**8.108** 0.1936.

**8.110** (a) 22.8 kg. (b) 291 N.

**8.112** $T_A = 176.3$ lb; $T_B = 76.3$ lb.

**8.114** 107.4 N · M.

**8.116** 112.4 lb · ft.

**8.118** $T_A = 57.6$ N; $T_B = 92.4$ N.

**8.120** 0.441.

**8.122** (a) 85.5 lb. (b) 48.0 lb. (c) 85.5 lb.

**8.124** 0.253.

**8.128** $T_A = 103.24$ lb; $T_B = 3.24$ lb.

**8.130** (a) 952 mm. (b) 3.95 m.

**8.132** 10 lb < $P$ < 36.7 lb.

**8.134** (a) 2.94 N. (b) 4.41 N.

**8.136** 0.304.

**8.138** 39 lb.

**8.140** Rod in equilibrium; $\mathbf{F} = 0.518P \leftarrow$.

**8.C1** $\mu_s = 0.15$: $\theta = 25.8°$;
$\mu_s = 0.50$: $\theta = 90.0°$.

**8.C2** $W_A = 10$ lb: $\theta = 28.8°$;
$W_A = 40$ lb: $\theta = 55.4°$.

**8.C3** $x = 25$ ft: $P = 175.0$ lb;
$x = 50$ ft: $P = 199.1$ lb.

**8.C4** (b) (1) $\theta = 20°$: $\mu_s = 0.285$;
(2) $\theta = 20°$: $\mu_s = 0.219$.
(c) (1) $\theta = 22.0°$; (2) $\theta = 28.7°$.

# CHAPTER 9

**9.2** $3a^3b/10$.

**9.4** $2a^3b/15$.

**9.6** $ab^3/6$.

**9.8** $2ab^3/7$.

**9.10** $2ab^3/51$; $b\sqrt{7/51}$.

**9.12** $2a^3b/11$; $a\sqrt{7/11}$.

**9.14** $a^3b/21$; $a\sqrt{1/7}$.

**9.16** $43a^4/48$; $a\sqrt{43/72}$.

**9.18** (a) $\frac{1}{2}\pi(R_2^4 - R_1^4)$. (b) $\frac{1}{4}\pi(R_2^4 - R_1^4)$.

**9.22** $403 \times 10^6$ mm$^4$; 92.8 mm.

**9.24** 6.99 in$^4$; 1.127 in.

**9.26** $\bar{I}_x = \bar{I}_y = 2.005 \times 10^6$ mm$^4$.

**9.28** $46.2 \times 10^6$ mm$^4$.

**9.30** 9120 in$^4$.

**9.32** $\bar{I}_x = 13.89 \times 10^6$ mm$^4$; $\bar{I}_y = 20.88 \times 10^6$ mm$^4$.

**9.34** (a) 408 in$^4$. (b) 81.6 in$^4$.

**9.36** $\bar{I}_x = 44.5$ in$^4$, $\bar{I}_y = 27.7$ in$^4$; $\bar{k}_x = 2.16$ in.,
$\bar{k}_y = 1.707$ in.

**9.38** 170.7 mm.

**9.40** $1.595 \times 10^6$ mm$^4$; $17.84 \times 10^6$ mm$^4$.

**9.42** 96.6 in$^4$; 26.6 in$^4$.

**9.44** $3\pi r/16$.

**9.46** $B = 13.15$ kN; $C = D = 17.09$ kN.

**9.48** (a) $\frac{1}{2}\gamma\pi r^2 h$. (b) $\frac{1}{4}\gamma\pi r^4$.

**9.50** 30 mm.

**9.54** $\frac{1}{4}a^2b^2/(n + 1)$.

**9.56** $\frac{1}{8}a^4$.

**9.58** $+12.15 \times 10^6$ mm$^4$.

**9.60** $+10.31$ in$^4$.

**9.62** $+159.6 \times 10^3$ mm$^4$.

**9.64** 13.79 in$^4$, 21.16 in$^4$, $+14.24$ in$^4$.

**9.66** $374 \times 10^3$ mm$^4$, $243 \times 10^3$ mm$^4$, $+206 \times 10^3$ mm$^4$.

**9.68** $+25.1°$; $42.1 \times 10^6$ mm$^4$, $10.51 \times 10^6$ mm$^4$.

**9.70** $-22.3°$; 32.2 in$^4$, 2.77 in$^4$.

**9.72** $+23.8°$; $524 \times 10^3$ mm$^4$, $92.5 \times 10^3$ mm$^4$.

**9.74** 13.79 in$^4$, 21.16 in$^4$, $+14.24$ in$^4$.

**9.76** $374 \times 10^3$ mm$^4$, $243 \times 10^3$ mm$^4$,
$+206 \times 10^3$ mm$^4$.

**9.78** $+25.1°$; $42.1 \times 10^6$ mm$^4$, $10.51 \times 10^6$ mm$^4$.

**9.80** $-22.3°$; 32.2 in$^4$, 2.77 in$^4$.

**9.82** $+23.8°$; $524 \times 10^3$ mm$^4$, $92.5 \times 10^3$ mm$^4$.

**9.84** (a) $-1.171 \times 10^6$ mm$^4$. (b) $+19.7°$.
(c) $4.35 \times 10^6$ mm$^4$.

**9.86** $I_{xy} = \sqrt{I_x I_y - I_{max} I_{min}}$; 6.07 in$^4$.

**9.88** (a) $\frac{1}{4}m(r_1^2 + r_2^2)$. (b) $\frac{1}{2}m(r_1^2 + r_2^2)$.

**9.90** $I_{DD'} = 3ma^2/8$; $I_{EE'} = 5ma^2/12$.

**9.92** (a) $7ma^2/6$. (b) $\frac{1}{2}ma^2$.

**9.94** $m(3a^2 + L^2)/12$.

**9.96** $m(a^2 + 3h^2)/6$; $\sqrt{(a^2 + 3h^2)/6}$.

**9.98** $m(a^2 + b^2)/5$.

**9.100** (a) $m(b^2 + 4a^2 \sin^2\theta)/12$.
(b) $m(b^2 + 4a^2 \cos^2\theta)/12$.

**9.102** 1.743 lb · ft · s$^2$, 6.54 in.

**9.104** $27ma^2/50$; $0.735a$.

**9.106** 80.2 g · m$^2$.

**9.108** 0.215 lb · ft · s$^2$; 4.95 in.

**9.110** (a) 7.11 g · m$^2$. (b) 16.96 g · m$^2$.
(c) 15.27 g · m$^2$.

**9.112** $I_x = 5.35 \times 10^{-3}$ lb · ft · s$^2$;
$I_y = I_z = 21.7 \times 10^{-3}$ lb · ft · s$^2$.

**9.114** $30.5 \times 10^{-3}$ lb · ft · s$^2$.

**9.116** 183.8 g · m$^2$.

**9.118** 2.06 kg · m$^2$.

**9.120** (a) $0.1524 ah^3$. (b) $a^3h/5$.

**9.122** $+25.7$ in$^4$.

**9.124** $-25.7°$; $4.73 \times 10^6$ mm$^4$, $0.477 \times 10^6$ mm$^4$.

**9.126** $6.07 \text{ g} \cdot \text{m}^2$; 92.7 mm.

**9.128** $5\sqrt{3}\, a^4/48$; $a\sqrt{5/12}$.

**9.130** $(a)\ 813 \times 10^3 \text{ mm}^4.$   $(b)\ 4.40 \times 10^6 \text{ mm}^4.$

**9.C3** $(b)\ I_x = 315 \times 10^3 \text{ mm}^4,\ k_x = 13.23 \text{ mm};$
$I_y = 5.70 \times 10^6 \text{ mm}^4,\ k_y = 56.3 \text{ mm}.$
$(c)\ I_x = 719 \text{ in}^4,\ k_x = 2.79 \text{ in,;}$
$I_y = 11{,}880 \text{ in}^4,\ k_y = 11.36 \text{ in}.$

## CHAPTER 10

**10.2** 65 N.

**10.4** $1200 \text{ lb} \cdot \text{in.} \nearrow.$

**10.6** $750 \text{ N} \uparrow.$

**10.8** $Q = \frac{3}{2}P \tan \theta.$

**10.10** $Q = 3P \tan \theta.$

**10.12** $Q = (Pl/a) \cos \theta \sin^2 \theta.$

**10.14** $M = \frac{1}{2}Pl \cot \theta.$

**10.16** $M = 4Pl \sin 2\theta.$

**10.18** $M = -4Pl \sin 2\theta.$

**10.20** $(a)\ 14.70 \text{ lb}.$   $(b)\ 56.7 \text{ lb}.$

**10.22** $178.5 \text{ N} \cdot \text{m}.$

**10.24** $64.9°.$

**10.26** $14.9°.$

**10.28** $52.2°.$

**10.30** $13.2 \text{ in}.$

**10.32** $19.4°.$

**10.34** $79.4°;\ 317.7°;\ 386.1°.$

**10.36** $(a)\ 12(BD).$   $(b)\ 52.8°.$

**10.38** $6.77 \text{ kips} \searrow.$

**10.42** $180.3 \text{ N} \cdot \text{m};\ 57.0 \text{ N} \cdot \text{m}.$

**10.44** $1.361 \text{ kN} \uparrow;\ 550 \text{ N} \cdot \text{m} \nearrow.$

**10.46** $(a)\ T = \frac{1}{2}W/(1 - \tan \theta).$   $(b)\ 36.9°.$

**10.48** $0.469 \text{ in.} \rightarrow.$

**10.56** $33.7°,$ stable; $-146.3°,$ unstable.

**10.58** $14.5°$ and $165.5°,$ stable; $90°$ and $-90°,$ unstable.

**10.60** $79.4°$ and $386.1°,$ stable; $317.7°,$ unstable.

**10.62** $72.7°.$

**10.64** $12.5°$ and $90°,$ stable; $30.8°,$ unstable.

**10.66** $(a)\ \cos \frac{1}{2}\theta - \sin \frac{1}{2}\theta = (1 - Wl/ka^2)\cos \theta.$
$(b)\ 34.2°$ and $145.8°,$ stable; $90°,$ unstable.

**10.68** $52.2°.$

**10.70** $180°,$ stable.

**10.72** $9.6°,$ unstable; $40.3°,$ stable.

**10.74** $d > 309 \text{ mm}.$

**10.76** $m_1 = m_2(c^2/ab).$

**10.78** $P < \frac{1}{2}kl.$

**10.80** $P < 430 \text{ lb}.$

**10.82** $\sin 2\theta = \dfrac{mgb}{2ka^2}.$

**10.84** 60 N.

**10.86** $60°.$

**10.88** $14.2°.$

**10.90** $(a)\ P \cos \theta = W \cos \frac{1}{2}\theta.$   $(b)\ 65.1°.$

**10.92** $\dfrac{\cos \theta}{\cos \frac{1}{2}\theta} = \dfrac{m_B}{m_A}\sin \beta.$

**10.C1** $\theta = 40°\!:\ F_{CD} = 1.872 \text{ kips } T;$
$\theta = 80°\!:\ F_{CD} = 4.44 \text{ kips } T.$

**10.C2** $(b)\ \theta = 15°\!:\ V = -0.678 \text{ J};$
$\theta = 45°\!:\ V = -1.014 \text{ J}.$
$(c)$ and $(d)\ 12.5°$ and $90.0°,$ stable; $30.8°,$ unstable.

**10.C3** $(b)\ \theta = 20°\!:\ V = 64.1 \text{ J};$
$\theta = 40°\!:\ V = 46.4 \text{ J}.$
$(c)$ and $(d)\ 9.59°,$ unstable; $40.3°,$ stable.

**10.C4** $(b)\ \theta = 40°\!:\ W = 353 \text{ N}.$
$\theta = 60°\!:\ W = 1559 \text{ N}.$
$(c)\ 52.2°.$

## SI Prefixes

| Multiplication Factor | Prefix† | Symbol |
|---|---|---|
| $1\ 000\ 000\ 000\ 000 = 10^{12}$ | tera | T |
| $1\ 000\ 000\ 000 = 10^9$ | giga | G |
| $1\ 000\ 000 = 10^6$ | mega | M |
| $1\ 000 = 10^3$ | kilo | k |
| $100 = 10^2$ | hecto‡ | h |
| $10 = 10^1$ | deka‡ | da |
| $0.1 = 10^{-1}$ | deci‡ | d |
| $0.01 = 10^{-2}$ | centi‡ | c |
| $0.001 = 10^{-3}$ | milli | m |
| $0.000\ 001 = 10^{-6}$ | micro | $\mu$ |
| $0.000\ 000\ 001 = 10^{-9}$ | nano | n |
| $0.000\ 000\ 000\ 001 = 10^{-12}$ | pico | p |
| $0.000\ 000\ 000\ 000\ 001 = 10^{-15}$ | femto | f |
| $0.000\ 000\ 000\ 000\ 000\ 001 = 10^{-18}$ | atto | a |

† The first syllable of every prefix is accented so that the prefix will retain its identity. Thus, the preferred pronunciation of kilometer places the accent on the first syllable, not the second.

‡ The use of these prefixes should be avoided, except for the measurement of areas and volumes and for the nontechnical use of centimeter, as for body and clothing measurements.


## Principal SI Units Used in Mechanics

| Quantity | Unit | Symbol | Formula |
|---|---|---|---|
| Acceleration | Meter per second squared | . . . | $m/s^2$ |
| Angle | Radian | rad | † |
| Angular acceleration | Radian per second squared | . . . | $rad/s^2$ |
| Angular velocity | Radian per second | . . . | $rad/s$ |
| Area | Square meter | . . . | $m^2$ |
| Density | Kilogram per cubic meter | . . . | $kg/m^3$ |
| Energy | Joule | J | $N \cdot m$ |
| Force | Newton | N | $kg \cdot m/s^2$ |
| Frequency | Hertz | Hz | $s^{-1}$ |
| Impulse | Newton-second | . . . | $kg \cdot m/s$ |
| Length | Meter | m | ‡ |
| Mass | Kilogram | kg | ‡ |
| Moment of a force | Newton-meter | . . . | $N \cdot m$ |
| Power | Watt | W | $J/s$ |
| Pressure | Pascal | Pa | $N/m^2$ |
| Stress | Pascal | Pa | $N/m^2$ |
| Time | Second | s | ‡ |
| Velocity | Meter per second | . . . | $m/s$ |
| Volume, solids | Cubic meter | . . . | $m^3$ |
|    Liquids | Liter | L | $10^{-3}\ m^3$ |
| Work | Joule | J | $N \cdot m$ |

† Supplementary unit (1 revolution $= 2\pi$ rad $= 360°$).

‡ Base unit.